GEOLOGICALLY STORING CARBON

For our children and grandchildren

GEOLOGICALLY STORING CARBON

LEARNING FROM THE
OTWAY PROJECT EXPERIENCE

EDITOR: PETER J COOK

CSIRO
PUBLISHING

WILEY

Published exclusively in Australia and New Zealand by

CSIRO Publishing
150 Oxford Street (PO Box 1139)
Collingwood VIC 3066
Australia

Email: publishing.sales@csiro.au
Web site: www.publish.csiro.au

**Published in the Americas, Europe and Rest of the World
(excluding Australia and New Zealand) by**

John Wiley & Sons Ltd
The Atrium, Southern Gate
Chichester, West Sussex
PO19 8SQ United Kingdom

Email: customer@wiley.com
Web site: www.wiley.com

For distribution in the rest of the world other than Australia and New Zealand.

Library of Congress Cataloging-in-Publications Data
A catalogue record for this book is available from the British Library and the Library of Congress.
ISBN: 9781118986189

Front cover image: Rebecca Jones, CO2CRC
Back cover image: CO2CRC

Set in Adobe Garamond Pro
Cover design by James Kelly
Text design by Roslyn Paonin
Typeset by Roslyn Paonin, Jo Ashley, Matthew Harris
Index by Indexicana
Printed in China by 1010 Printing International Ltd

Original print edition:
The paper this book is printed on is in accordance with the rules of the Forest Stewardship Council®. The FSC® promotes environmentally responsible, socially beneficial and economically viable management of the world's forests.

CONTENTS

FOREWORD 1

Carbon dioxide capture and storage (CCS) is one of the most important approaches for achieving large and rapid CO_2 emission reductions. Pioneering projects initiated in the 1990s demonstrated that CCS could be safe and effective at an industrial scale. With growing concerns about global climate change, and encouraged by these early successes, governments, industry and academia in Australia, Europe, Japan and North America initiated research programs to accelerate bringing this technology to the global scale needed to make substantial reductions in CO_2 emissions.

The Otway Project, Australia's premier CO_2 storage demonstration project, has identified and tackled the most pressing issues needed to bring this technology to scale. Some of these challenges are purely technical, such as the need for improving site characterisation and monitoring methods. To address these issues, the Otway Project assembled a world-class team of geologists, geochemists, geophysicists, hydrologists and reservoir engineers. Together they designed and implemented a large-scale demonstration project that would provide a replicable model for how to decide which sites are suitable for CO_2 storage and to develop methods for monitoring the fate and transport of injected CO_2. They worked closely with regulators to ensure that the data would satisfy their requirements and leveraged industry capabilities to bring state-of-the-art tools and know-how to the Project. Having successfully demonstrated that this suite of measurement and modelling tools are up to the job, they then went one step farther. They developed an entirely new approach for quantifying the extent of residual gas trapping—an important secondary trapping mechanism that increases storage security over time. On

the strength of these contributions alone, the Otway Project is second to none.

Importantly, however, many new and unforeseen issues emerged as the ambitious project ploughed new ground—cutting across the traditional boundaries of business, law and the social sciences. What is the appropriate business model for public–private partnerships for one-of-a-kind demonstration projects? Who is responsible if something goes wrong? How do you communicate and work cooperatively with the local community? Here, more than any other project, the Otway Project shines light on the full spectrum of challenges and then offers solutions for moving this technology forward.

There are so many unique contributions from this project, from business models to regulatory frameworks, from public engagement to communications research, and from site characterisation to fundamental research on trapping processes. This extraordinary book documents these contributions, with over 40 authors from four continents who provide a rich and detailed account of all aspects of the Project. They generously share more than a decade of learning. This book provides invaluable contributions to advancing the state-of-the-art and practical know-how about CO_2 storage. I thank the authors for this wonderful book and the countless hours of hard work, scientific inspiration, and the fortitude that brought this project to fruition. In the future, our children and grandchildren will also thank them for pushing this important technology for reducing CO_2 emissions forward.

Professor Sally Benson
Director, Global Climate and Energy Project
Stanford University, USA

FOREWORD 2

This book documents the amazing journey of applied research undertaken by the CO2CRC Otway Project over the past decade, involving industry, researcher and government collaboration, local community engagement and political and international interaction. All of this is against a backdrop of rapidly increasing global fossil fuel emissions and growing scientific insights into global warming. CCS promises to be a vital part of any serious global response to climate change and the Otway Project has made and will, no doubt, continue to make a significant contribution to this unfolding endeavour.

The Otway Project is important to Australia as the nation's large fossil fuel reserves are potentially affected in a future requiring substantial emission reductions. Victoria is particularly vulnerable with its substantial but high emission brown coal resources and related energy production. In a world that will ultimately see CCS widely deployed, the Otway Project will be seen to have made a significant contribution to the emerging new discipline of subsurface carbon storage technology; all the more so because of this book.

For Australia, the safe storage of more than 60,000 tonnes of CO_2 serves as a clear example of the practicability of this new technology, while giving added confidence to communities and regulators alike. The storage of CO_2 in a depleted gas field in the Otway Basin combined with a strict encompassing scientific and monitoring regime, has established an outstanding international reputation for the Project. The lessons and insights are set out comprehensively in this volume, from the project design and engineering to the community and regulatory processes. Most importantly it brings together the scientific insights and technologies for carbon storage, building a new discipline on the well-established knowledge and principles of subsurface engineering and geoscience from the oil and gas industry.

The Project is unique because of its scale, its significant in-situ CO_2 reserves, a supportive local community and the established constructive relationship with local regulators. Importantly, the more work that has been done on the site, the better characterised it has become. The rich data sets of subsurface geology, geochemistry, geomechanics and geophysics make it increasingly valuable as a test bed for the study of new techniques and technologies.

Ultimately it has been the people and organisations with vision and drive that have made this Project and its outstanding science a reality. The planning and execution of the experiments, approvals processes, funding, governance, community engagement and analysis and interpretation of results, has required a dedicated team of people with a wide array of experience and skills. The Project has seen the direct and indirect involvement of many experts from industry and some of the world's leading scientists in the field. One person that deserves particular mention is Peter Cook for his leadership, vision and unfailing enthusiasm to make a difference. He has led an outstanding team setting up and running the Project. It has been an inspired personal quest for him, from the time he started the Project, to the effort put into the delivery of this book. Both will be seen as a monumental legacy in the field of CCS.

Dr Richard Aldous
Chief Executive, CO2CRC

PREFACE

A great many scientists, policy makers, environmentalists and the community at large are concerned about the potential climatic impact of increasing concentrations of greenhouse gases in the atmosphere, particularly CO_2. And even those not convinced of a CO_2–climate link, that argue there is too much uncertainty to warrant action, must surely recognise that if we are not certain of the impact, we should stop emitting CO_2 to the atmosphere in ever-increasing quantities. In other words, should not the Precautionary Principle apply? Many point to the importance of renewable energy in mitigating CO_2 emissions and it is certainly true that renewables, along with greater energy efficiency and switching to low intensity fuels, all have a part to play.

But it is unrealistic to believe that the world can completely switch to renewable energy almost overnight. Such a transition will take decades and will require much improved technologies to be developed, critically including far better systems for energy storage. Also, the cost of mitigation and of the move to clean energy cannot be ignored; we cannot spend all of the world's wealth on addressing greenhouse concerns when there are so many other problems needing resolution. Wealthy countries may choose to spend a high proportion of their GDP on clean energy, but some countries, particularly developing countries, may have no alternative but to continue to use fossil fuels for energy because they cannot afford the high cost of alternative energy, or because fossil fuels are their only practical option. In other words there is little prospect of moving totally away from fossil fuels anytime soon. Some may see this as a defeatist view but I see it as a realistic view that should be the starting point for any sensible mitigation strategy.

The International Energy Agency, the IPCC, NGOs, government organisations and many leading scientists, have pointed to the critical importance of deploying CCS (carbon capture and geological storage of carbon dioxide) in a carbon constrained world. Despite the increasing application of renewable energy, the use of fossil fuels continues to grow worldwide. Coal is the fuel of choice of many countries because it is cheap and abundant, but it does inevitably result in increasing emissions of CO_2. In many developed countries gas is seen as the future fuel of choice with the benefit of producing lower carbon emissions, but obviously it too produces significant CO_2 emissions, albeit less than coal produces. Therefore whether the energy fuel of choice is coal or gas, for as long as we continue to use fossil fuels, CCS is the only technology available for making deep cuts in the emissions arising from that use. But despite this, CCS has yet to be widely deployed.

So why is CCS not being widely deployed? The argument is often advanced that CCS is too expensive at an industrial scale. The reality is that CCS is more expensive than some renewable technologies that are in use, but less expensive than others. It is therefore vital to keep CCS in the mix of clean energy technologies, while at the same time working to bring down the cost of CCS and particularly the cost of capture. The claim is also frequently made that CCS is "unproven". The reality is that separation and capture of CO_2 is used extensively in large scale gas processing and tens of millions of tonnes of CO_2 are injected into the subsurface every year as part of enhanced oil recovery operations. There are also thousands of kilometres of CO_2 pipelines already in existence. In other words many parts of the CCS chain are already in operation. What is not yet in operation is integrated CCS applied to large scale coal- or gas-fired electricity generation, although power plants with CCS that are under construction in Canada and the USA will go some way to addressing this criticism.

However much of the challenge to the widespread deployment of CCS lies in a lack of knowledge of storage technology which in turn breeds "fear of the unknown", that in turn can translate into an unwillingness on the part of individuals, communities and governments to countenance CCS as a mitigation option. The town of Barendrecht in Holland is an example of a community living safely and confidently above a large accumulation of natural gas for many years, but unwilling to consider using that same geological structure to store CO_2. Some governments, for example Germany, have banned underground onshore storage of CO_2 because of perceived risks and uncertainty and because of community opposition. There is really only one way to address these concerns and that is by actually undertaking CCS and clearly and convincingly demonstrating in an open and transparent manner, that CO_2 can be safely, securely and effectively stored, and that it is a credible option for addressing greenhouse concerns. Ideally this would be done at a scale needed to make significant cuts in emissions now, and there are examples of this, for example the Sleipner Project in the North Sea, the Weyburn Project and soon the Boundary Dam Project in Canada, but such projects are often constrained by commercial concerns or logistics that inhibit them from providing open access. Similarly they cannot be expected to be used as sites for a wide range of fundamental research into CO_2 storage when they have commercial imperatives that have to be addressed.

For these and other reasons, smaller scale demonstration or pilot projects are often chosen as the way forward. Their cost is modest compared to a large-scale project; they can be undertaken within a reasonable time frame; and they provide an excellent low cost opportunity to undertake research that will in turn address uncertainties that may confront large scale projects. Crucially they also provide the opportunity not only for scientists and engineers to see and test CCS technology under field conditions, but also the chance for politicians, NGOs, regulators and members of the community to see CCS for themselves. The opportunity to "kick the tyres"; the chance to hear firsthand from the scientists how they know what is happening in the subsurface and why they are confident that the CO_2 will not leak to the surface or contaminate aquifers.

It was with all of these issues in mind, that in 2003 I first set about trying to persuade people that we should undertake a CO_2 storage project in Australia at a commercially significant scale. Given Australia's reliance on fossil fuels there was a compelling case for getting a project underway as soon as possible. But as is so often the case with "first of a kind" projects, it was to take several years before the first molecule of CO_2 could be injected. That project was of course the CO2CRC Otway Project.

This book outlines the progress and achievements of the CO2CRC Otway Project, Australia's first CO_2 storage project, over the period from 2003 to 2013. From a personal perspective, the genesis of the Project and this book can be traced back more than 20 years. I was Director of the British Geological Survey at that time and first became interested in geological storage of CO_2 through some of the early work of BGS on North Sea storage opportunities. This in turn led me to initiate the GEODISC Project in Australia in 1998, and then in 2003 as Chief Executive of CO2CRC, to develop the idea of undertaking an actual storage project. But it is one thing to develop the "vision" and it is entirely another thing to turn that vision into the reality of a successful project; that is the part that requires the hard work! The 45 authors of this book were of course a critical component of the Otway team that has achieved so much over the past 10 years.

Geologically Storing Carbon provides a detailed account of the CO2CRC Otway Project, one of the most comprehensive demonstrations of the deep geological storage of carbon dioxide undertaken anywhere. This book of 18 comprehensive chapters written by leading experts in the field is concerned with outstanding science. But it is not just a collection of scientific papers and indeed much of the science has already been published in peer-reviewed journals, which is as it should be. What this book is about in particular is "learning by doing". It describes the organisational, governance and decision-making processes that were used, and outlines how the storage site was chosen and what was done to prove its suitability. It also describes how and why the Project went about securing a source of CO_2 and the various options for processing the gas.

The book also provides insights into the operational details of the Project, including risk assessment and ensuring that the site was geologically suitable for taking the Project forward. A major program of scientific investigation was undertaken, which involved activities as diverse as obtaining fluids from 2 km down a well and 500 m in a water well; sampling a few centimetres in the soil and sampling in the atmosphere. How the samples were obtained and how they were analysed is described in some detail. Monitoring and verification were critical to the Project and a wide range of methods were employed; the book describes how some methods worked well while others did not, and how results were processed and interpreted.

In addition to the technical details, *Geologically Storing Carbon* outlines how the Project was regulated and what it cost, as well as the experience of successfully communicating with the local community and the community at large. While it does not pretend to provide all the answers, hopefully it will assist others who want to develop a demonstration of geological storage, up to the size of perhaps a 100,000 tonne injection. It is a book for geologists, engineers, regulators, project developers, industry, communities, indeed anyone who wants to better understand how a carbon storage project really "works". It is also for people concerned with obtaining an in-depth appreciation of one of the key technology options for decreasing greenhouse emissions to the atmosphere.

Each storage project will have its own particular challenges depending not only on the geology, but also the expertise of the team, the availability and composition of the CO_2 stream and of course the availability of funding. In addition and crucially, the objectives of each project will vary. For some the prime objective will be to just demonstrate storage and certainly that was initially the main objective of the Otway Project and the one discussed in the greatest detail in this book. However as our knowledge at Otway increased, so did our recognition of what we did not know or understand. Therefore as the work progressed, so too did our aspirations to undertake more research and provide more answers. I suspect that other projects will show a similar aspirational progression. This detailed knowledge in turn will hopefully serve to provide the community, industry, government, and stakeholders, with the confidence to progress the deployment of CCS and make deep cuts in the emissions of CO_2 from large energy and industrial sources.

Many people have played a key role in the Otway Project and in the publishing of this book. The acknowledgements section of the book lists many of the key contributors—individuals, researchers, support and technical staff, organisations, companies, governments and others, without whom the Otway Project and this book would have never happened. Because they all played an essential role, one way or the other, and because there are literally hundreds of people who have contributed to the Otway Project and this book in one way or another, it is difficult to pick out particular people to acknowledge, but nonetheless let me do just that from a personal perspective.

During my time as Chief Executive of the APCRC and then CO2CRC, I have been most fortunate in the Chairmen with whom I have worked—the late Dr Alan Reid OA was enthusiastic about pursuing the far-out (for that time) idea of storing Australia's emissions underground. Mr Tim Besley AC, the founding Chair of CO2CRC, was extraordinarily supportive as we took the Otway Project from an idea to reality; his successor, Mr David Borthwick, PSM, subsequently continued that support and Dr Mal Lees, Chair of CPPL, was outstanding in dealing with the practical complexities of undertaking a first-of-a-kind project. The various CO2CRC Boards provided me with the leeway, the freedom of action and the backing that is so essential for a project such as Otway. The CO2CRC Executive was extraordinarily supportive throughout the decade; in the early days of the Project, the efforts of the late David Collins and Andy Rigg and later Sandeep Sharma, were particularly notable. Most recently, my successor as Chief Executive, Dr Richard Aldous, has been extremely supportive of my aim to publish a comprehensive volume on the Otway Project. Finally, from a personal perspective, this book was only possible because of the willingness of my wife Norma to put up with my absences in the field, my need to rush off to yet another meeting, my preoccupation with getting the Project successfully underway and my many evenings and weekends spent writing (and most recently editing) this volume.

Professor Peter J Cook CBE FTSE
University of Melbourne, Australia
Chief Executive, CO2CRC 2003–2011

AUTHORS

Dr Colin Allison
CSIRO Marine and Atmospheric Research
Aspendale, VIC 3195, Australia
colin.allison@csiro.au

Ian S Black, AGR
Address now: ExxonMobil, Melbourne, Australia
ian.s.black@exxonmobil.com

Dr Chris Boreham
Basin Resources Group, Geoscience Australia
GPO Box 378, Canberra, ACT 2601, Australia
chris.boreham@ga.gov.au

Dr Mark Bunch
Australian School of Petroleum
The University of Adelaide, SA 5005, Australia
mark.bunch@adelaide.edu.au

Dr Eva Caspari
Department of Exploration Geophysics
Curtin University, GPO Box U1987
Perth, WA 6845, Australia
eva.caspari@curtin.edu.au

Professor Peter J Cook
Peter Cook Centre for CCS Research
University of Melbourne
Melbourne, VIC 3010, Australia
pjcook@co2crc.com.au

Ms Tess Dance
CSIRO Earth Science and Resource Engineering
26 Dick Perry Ave, Kensington, WA 6151, Australia
tess.dance@csiro.au

Dr Patrice de Caritat
Geoscience Australia
GPO Box 378,Canberra, ACT 2601, Australia
patrice.decaritat@ga.gov.au

Mr Thomas M Daley
Geophysics Department, Earth Science Division
Lawrence Berkeley National Laboratory
1 Cyclotron Rd, Berkeley, CA 94720, USA
tmdaley@lbl.gov

Dr Richard (Ric) Daniel
Australian School of Petroleum
The University of Adelaide
Adelaide, SA 5005, Australia
richard.daniel@adelaide.edu.au

Mr Craig Dugan, Process Group
Address now: Optimal Group Australia Pty Ltd
175 Wellington Road, Clayton, VIC 3168, Australia
craig.dugan@optimalgroup.com.au

Dr Jonathan Ennis-King
CSIRO Earth Science and Resource Engineering
Clayton, VIC 3168, Australia
jonathan.ennis-king@csiro.au

Dr David Etheridge
CSIRO Marine and Atmospheric Research
Aspendale, VIC 3195, Australia
david.etheridge@csiro.au

Dr Paul Fraser
CSIRO Marine and Atmospheric Research
Aspendale, VIC 3195, Australia
paul.fraser@csiro.au

Dr Barry M Freifeld
Hydrogeology Department, Earth Science Division
Lawrence Berkeley National Laboratory
1 Cyclotron Rd, Berkeley, CA 94720, USA
bmfreifeld@lbl.gov

Professor Boris Gurevich
Department of Exploration Geophysics
Curtin University and CSIRO, GPO Box U1987
Bentley, Perth, WA 6845, Australia
b.gurevich@curtin.edu.au

Professor Ralf Haese
Peter Cook Centre for CCS Research
University of Melbourne
Melbourne, VIC 3010, Australia
ralf.haese@unimelb.edu.au

Ms Allison Hortle
CSIRO Earth Science and Resource Engineering
26 Dick Perry Ave, Kensington, WA 6151, Australia
allison.hortle@csiro.au

Dr Charles Jenkins
CSIRO Earth Science and Resource Engineering
Black Mountain, Canberra, ACT 2601, Australia
charles.jenkins@csiro.au

Professor John Kaldi
Australian School of Petroleum
The University of Adelaide, SA 5005, Australia
jkaldi@co2crc.com.au

Dr Anton Kepic
Department of Exploration Geophysics
Curtin University, GPO Box U1987
Perth, WA 6845, Australia
a.kepic@curtin.edu.au

Dr Dirk Kirste
Department of Earth Sciences
Simon Fraser University
8888 University Dr, Burnaby, BC V5A 1S6, Canada
dkirste@sfu.ca

Mr Paul Krummel
CSIRO Marine and Atmospheric Research
Aspendale, VIC 3195, Australia
paul.krummel@csiro.au

Dr Maxim Lebedev
Department of Exploration Geophysics
Curtin University, GPO Box U1987
Perth, WA 6845, Australia
m.lebedev@curtin.edu.au

Dr Malcolm J Lees (retired)
Chief Advisor Technical
RioTinto Coal Australia
Brisbane, Queensland

Dr Ray Leuning
CSIRO Marine and Atmospheric Research
Black Mountain, Canberra, ACT 2601, Australia
ray.leuning@csiro.au

Dr Zoe Loh
CSIRO Marine and Atmospheric Research
Aspendale, VIC 3195, Australia
zoe.loh@csiro.au

Dr Ashok Luhar
CSIRO Marine and Atmospheric Research
Aspendale, VIC 3195, Australia
ashok.luhar@csiro.au

Dr Lincoln Paterson
CSIRO Earth Science and Resource Engineering
Clayton, VIC 3168, Australia
lincoln.paterson@csiro.au

Dr Roman Pevzner
Department of Exploration Geophysics
Curtin University, GPO Box U1987
Perth, WA 6845, Australia
r.pevzner@curtin.edu.au

Dr Matthias Raab
CO2CRC, School of Earth Sciences
University of Melbourne, VIC 3010, Australia
mraab@co2crc.com.au

Ms Namiko Ranasinghe
Department of State Development, Business
and Innovation
121 Exhibition Street
Melbourne, VIC 3000, Australia
namiko.ranasinghe@dsdbi.vic.gov.au

Dr Ulrike Schacht
Australian School of Petroleum
The University of Adelaide, SA 5005, Australia
uschacht@co2crc.com.au

Mr Sandeep Sharma
Schlumberger Carbon Services
256 St Georges Terrace, Perth, WA 6000, Australia
sharmass@bigpond.com

Dr Valeriya Shulakova
CSIRO Earth Science and Resource Engineering
Bentley, WA 6102, Australia
valeriya.shulakova@csiro.au

Dr Anthony Siggins
CSIRO Earth Science and Resource Engineering
Clayton, VIC 3168, Australia
tony.siggins@csiro.au

Mr Rajindar Singh
CO2CRC, School of Earth Sciences
University of Melbourne, VIC 3010, Australia
rssingh@co2crc.com.au

Mr Darren Spencer
CSIRO Marine and Atmospheric Research
Aspendale, VIC 3195, Australia
darren.spencer@csiro.au

Dr Linda Stalker
CSIRO Australian Resources Research Centre
26 Dick Perry Avenue
Kensington, WA 6151, Australia
linda.stalker@csiro.au

Dr Paul Steele
CSIRO Marine and Atmospheric Research
Aspendale, VIC 3195, Australia
paul.steele@csiro.au

Mr Tony Steeper
CO2CRC, 14–16 Brisbane Avenue
Barton, Canberra, ACT 2600, Australia
tsteeper@co2crc.com.au

Dr Eric Tenthorey
Geoscience Australia
GPO Box 378, Canberra, ACT 2601, Australia
eric.tenthorey@ga.gov.au

Dr Milovan Urosevic
Department of Exploration Geophysics
Curtin University, GPO Box U1987
Perth, WA 6845, Australia
m.urosevic@curtin.edu.au

Dr Maxwell N Watson
CO2CRC, School of Earth Sciences
University of Melbourne, VIC 3010, Australia
mwatson@co2crc.com.au

Mr Steve Zegelin
CSIRO Marine and Atmospheric Research
Black Mountain, Canberra, ACT 2601, Australia
steve.zegelin@csiro.au

ACKNOWLEDGEMENTS

The CO2CRC Otway Project was made possible through a range of contributions from many Australian and international researchers, research organisations, universities, companies, government bodies, officials and government ministers, regulators, stakeholders, landholders, contractors, administrative staff and management. The contributions were as diverse as the provision of expertise, the sharing of information and knowledge, the provision of access to farmland, and of course funding. In so many ways the Otway Project was a team effort (with every member and associate of CO2CRC a part of that team) and therefore it is the team rather than individuals that ensured success. Nonetheless it is appropriate to highlight a number of contributors and contributions.

Member companies of the CO2CRC Joint Venture contributed not only funding and in-kind support but also technical, legal and organisational advice. They included Anglo Coal, Australian Coal Research (now Australian National Low Emissions Coal Research and Development—ANLEC R&D), BHP Billiton Petroleum Pty Ltd, BP Developments Australia Pty Ltd, Chevron Australia Pty Ltd, ConocoPhillips, Inpex Browse Ltd, Origin Energy Ltd, QER Pty Ltd, Rio Tinto (through Technological Resources Pty Ltd), Sasol Petroleum International Pty Ltd, Schlumberger Oilfield Australia Pty Ltd, Shell Development (Australia) Pty Ltd, Solid Energy New Zealand Ltd, Stanwell Corporation Ltd, Total S.A., Woodside Energy Pty Ltd and Xstrata Coal Pty Ltd (now Glencore).

Over and above their responsibilities as Members of the CO2CRC Joint Venture, a number of companies also took on the added role of being Members of CO2CRC Pilot Project Ltd (CPPL), with responsibilities not only for operational and related issues, but also for holding some of the potential liabilities associated with the Project. The companies who were Members of CPPL were Anglo Coal, BHP Billiton Petroleum Pty Ltd, BP Developments Australia Pty Ltd, Chevron Australia Pty Ltd, Rio Tinto, Schlumberger Oilfield Australia Pty Ltd, Shell Development (Australia) Pty Ltd, Solid Energy New Zealand Ltd, Woodside Energy Pty Ltd and Xstrata Coal Pty Ltd.

Governmental Members of CO2CRC included the New Zealand Foundation for Research Science and Technology, NSW Department of Primary Industries, Queensland Department of Employment, Economic Development and Innovation, Victorian Department of Primary Industries and the WA Department of Mines and Petroleum.

Research Members of CO2CRC included the Commonwealth Scientific and Industrial Research Organisation (CSIRO), Geoscience Australia, Curtin University, Institute of Geological and Nuclear Sciences, Korean Institute of Geoscience and Mineral Resources (KIGAM), Monash University and the Universities of Adelaide, Melbourne, New South Wales and Western Australia.

There were several Associate Members of CO2CRC; the Lawrence Berkeley National Laboratory in particular was a very major research contributor to the Otway Project. A number of small to medium enterprises contributed their expertise, including the Process Group, Cansyd and URS Australia. EnergyAustralia (previously TRUenergy) provided access to geomechanical data from the Iona Gas Storage Facility. AGR Ltd provided many operational services at the Otway site. Santos Ltd assisted with establishing the suitability of the Otway site.

Funding of site activities and research was obviously critical to being able to undertake the Otway Project. The single largest financial contributor to the Project was the Australian Government through the CRC Program, AusIndustry and the Australian Greenhouse Office. Major funding from the Victorian Government was also critical to the financial viability of the Project. Financial assistance provided through the Australian National Low Emissions Coal Research and Development (ANLEC R&D) which is supported by Australian Coal Association Low Emissions Technology Limited and the Australian Government through the Clean Energy Initiative, was also a vital part of the financial support base. Industry and government Members of CO2CRC listed above were very important financial contributors to the Project; Members of CPPL listed above also provided additional funding over and above that of the other Members. The Otway Project was also financially supported by the GEO-SEQ Project for the Assistant Secretary for Fossil Energy, Office of Coal and Power Systems through the National Energy Technology Laboratory of the U.S. Department of Energy, under LBNL contract DE-AC02-05CH11231. Financial support was also provided by the Carbon Capture Program.

While direct financial support was obviously crucial to undertaking an expensive research project, in-kind support was also very important indeed to the success of the Otway Project. All of the Research Members and the Associates listed above contributed very significantly through their in-kind contribution, particularly through the provision of researcher time, but also including equipment, analyses and other services. The Australian National University, the University of Calgary, Simon Fraser University and the University of Canberra also provided analytical services. Also a number of companies made software available to the Project, including Halliburton (Landmark), Schlumberger (Petrel, Eclipse), CGG (Hampson-Russell), dGB (OpenDTect), Ikon Science (RokDoc), and DECO Geophysical SK (RadExPro).

In addition to the support from organisations, the support of individuals was extraordinarily important to the success of the Otway Project and amongst the most important of these were the farmers who provided access to their land, other local stakeholders and participants in the Community Reference Group. These included John McInerney, Delcie Dumesny, Peter Dumesny, Corey Couch, Karin Couch, Gavin Couch, Ron Brumby, Sue Blake, Andrew Straker, Gillian Blair, Jenny Porteous, Ken Gale, Terry O'Connor, Andrew Gosden and Oliver Moles. Especially important to ensuring effective and productive relations with the local community was the untiring work of Josie McInerney, the Community Liaison Officer for the Project.

A number of officials at the State level played a vital role in regulating the Project in a practical and consultative manner, including John Frame (Environment Protection Authority), Terry McKinley, Geoff Collins and Namiko Ranasinghe (Department of Primary Industries), Bala Balendran (Energy Safe Victoria), Martin Kent, Mick Fennessy, Lynda Hardy (Southern Rural Water) and Peter Greenham (Department of Infrastructure). Other officials who contributed greatly to the Otway Project, by facilitating State support, included Dale Seymour, Richard Aldous and Cliff Kavonic (Department of Primary Industries). The enthusiasm of the then Victorian Minister for Primary Industries, Mr Peter Batchelor, was an important element in enabling the Project to go ahead.

Many officials at the Federal level similarly played a vital part in helping the Project to go forward including Tania Constable, John Hartwell, John Ryan, members of the CRC Program (Department of Industry) and Gerry Morvell (Australian Greenhouse Office). The Federal Minister for Industry, the Honourable Ian Macfarlane MP, has been a strong and much valued supporter of the Project, as has the Honourable Martin Ferguson MP.

The external review of the Otway Project coordinated by John Gale of the IEA Greenhouse Gas Programme and including Sally Benson, Susan Hovorka and Malcolm Wilson, was much appreciated by the members of the Project.

The contribution and dedication of Board members, too many to name individually, and the untiring efforts of the Board Chairs—the late Alan Reid (APCRC, 1991–2003), Tim Besley (2003–2010), David Borthwick (2010 to present) and Mal Lees (CPPL, 2005–2010) have been especially valued by the Centre. Members of the CO2CRC Executive who have played a major role in the Otway Project include Sandeep Sharma, Matthias Raab, Andy Rigg, John Kaldi, Barry Hooper, the late David Collins, Carole Peacock, and since August 2011 Richard Aldous (CEO).

Perhaps the most difficult part of this Acknowledgement is to attempt to do justice to the efforts of so many dedicated scientists, technicians and many other specialists involved in the Otway Project. The 45 authors of this volume of course deserve special mention and are listed elsewhere in the book and in the individual chapters. A great many other CO2CRC people contributed to xxii

the Project in a variety of ways and their names can be found at www.co2crc.com.au. There are others who contributed to the Otway Project and deserve mention, who were not officially part of the Project or who have now moved to other organisations. These include in geology and geochemistry—Lynton Spencer, Frank La Pedalina, Catherine Gibson-Poole, Ernie Perkins, Claire Rogers, Sandrine Vidal-Gilbert, Peter van Ruth, David Dewhurst, Thomas Berly, Dennis van Puyvelde, Jürgen Streit, Jacques Sayers, Myles Regan, Simon Mockler, Alex Moisis, Mark Pavloudis, Dominic Pepicelli, Piotr Sapa, Peter Tingate, Adrian Tuitt, Mario Werner, Dae Gee Huh and his colleagues at KIGAM. In reservoir engineering—Qiang (Josh) Xu, Geoff Weir, Martin Leahy and Jim Underschultz. In geophysics—Kevin Dodds, Jonathan Ajo-Franklin,

Brian Evans, Andrej Bóna, Aleksander Dzunic, Christian Dupuis, Ruiping Li, David Lumley, Donald Sherlock, Aline Gendrin, Shoichi Nakanishi, Allan Campbell, Les Nutt, Shujaat Ali, Vladimir Tcheverda, Evgeny Landa, Yusouf Al-Jabri, Putri Wisman and Faisal Abdulkader Alonaizi. In communications—Carmel Anderson. In atmospheric monitoring—Scott Coram, Ray Langenfelds, Rebecca Gregory, David Thornton, Mick Meyer and Dale Hughes. In risk assessment—Adrian Bowden and Donna Pershke.

Finally, thanks to Jo Ashley, Roslyn Paonin and Matthew Harris for their graphic design skills and Julia Stuthe and Tracey Millen of CSIRO Publishing and their colleagues at Wiley for helping to make this book a reality.

Peter Cook, Mal Lees, Sandeep Sharma

1. DEVELOPING THE PROJECT

1.1 Introduction

The foundation for the CO2CRC Otway Project was established as long ago as March 1998, when it was first proposed to the Board of the then Australian Petroleum Cooperative Research Centre (APCRC) that a programme be established to look at the opportunities in Australia for "geologically disposing" of carbon dioxide, with an initial focus on high-CO_2 natural gas, but with the intention to also look at the opportunities for applying the technology more broadly, to address what was perceived as Australia's looming greenhouse gas issue. In 1998, this was not an issue of broad community or political import and therefore it was not possible to get funding from the CRC Programme, despite attempts to do so. Nonetheless the Board continued to support the concept. A workshop was held in Perth, Western Australia to discuss geosequestration in late 1998, under the aegis of Chevron (who was at that stage increasingly interested in the technology for the proposed Gorgon Project) and subsequently, a number of oil and gas companies (BHPB, BP, Chevron, Shell,

Woodside), together with the Australian Greenhouse Office, agreed to provide some funding to get work underway.

In 1999, the GEODISC (Geological Disposal of CO_2) Project was initiated by the APCRC, with the specific objective of assessing on a continent-wide basis, what the opportunities were likely to be for the geological storage of carbon dioxide in Australia (Cook et al. 2000). In order to make that assessment, a team of earth scientists was assembled by the Centre, drawing on the original participants in the APCRC (CSIRO, Curtin University, University of NSW, University of Adelaide), together with new members of the team from Geoscience Australia. The outcome of that work, which extended over 4 years, was to convincingly show that there were indeed opportunities to apply the technology in Australia and, as part of the GEODISC Project, a very preliminary analysis of the storage potential of Australia was undertaken, the first such exercise attempted for an entire continent. The results, which were summarised in a series of publications and APCRC reports, clearly suggested that Australia did indeed have the potential opportunity to apply what was then known as geosequestration (carbon capture and storage, or CCS) on a large scale. By 2001–02, greenhouse gas concerns in government and the community at large were increasingly evident and the GEODISC findings had a

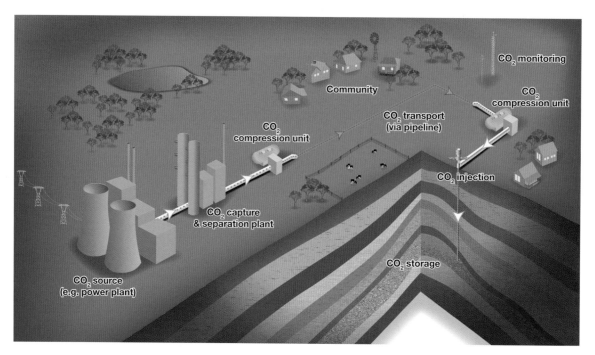

Figure 1.1: A simplified overview of carbon capture and storage.

major impact on government thinking on the possible options for decreasing emissions.

In 2003, a new Centre, the Cooperative Research Centre for Greenhouse Gas Technologies (CO2CRC) was formed out of the APCRC, to undertake applied research into the capture and geological storage of carbon dioxide. The support base was broadened to now include the oil, gas, coal, power and service industries initially through Australian Coal Research, BHPBilliton, BP, Chevron, Rio Tinto, Shell, Stanwell Corporation, Schlumberger, Woodside and Xstrata Coal, with subsequent support from Anglo American, Conoco Phillips, Sasol, Solid Energy and Total. Unlike GEODISC, which was only concerned with the geosciences, a much wider range of CCS science and engineering was brought into the CRC via expertise within CSIRO, Curtin University, Geosciences Australia, GNS New Zealand, Monash University, University of Adelaide, University of New South Wales and the University of Melbourne, as well as the Australian Greenhouse Office, together with several small enterprises (Cansyd, Process Group, URS). Close and much valued links were also developed with Lawrence Berkeley National Laboratory.

Right from the start it was recognised that a key element of the new CRC had to be to not just talk about geological storage of CO_2, but to actually demonstrate it, if possible with the complete CCS chain of capture, transport and storage as shown conceptually in Figure 1.1. Geological storage of CO_2 was already underway in 2003 through the Sleipner Project in offshore Norway and also at Weyburn in Canada as part of an enhanced oil recovery (EOR) project. Plans were well advanced for the Frio Brine Project in Texas and a number of other demonstration projects were also at the planning stage, though none in Australia at that time. Therefore in 2003, CO2CRC started to look at the feasibility of a small-scale CCS project in Australia.

1.2 Developing an Australian project

1.2.1 The first practical steps

The concept of an Australian demonstration/pilot project was first developed by CO2CRC around the technical

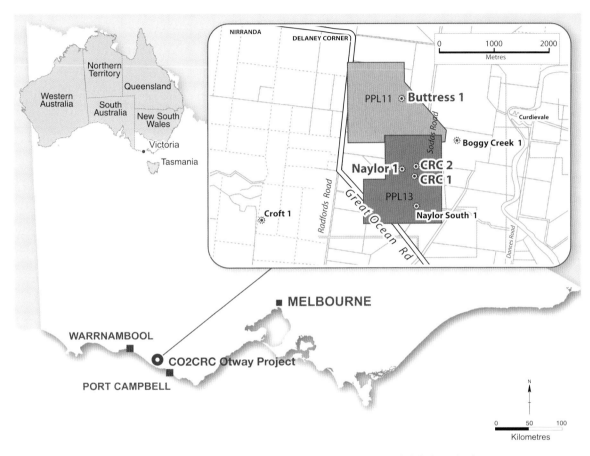

Figure 1.2: Location of the petroleum tenements in the Otway Basin that ultimately provided the basis for the Otway Project.

components embedded in Figure 1.1, with all aspects of the CCS chain to be demonstrated at a commercially significant scale. It soon became apparent that this was an impractical aspiration for a small-scale project in the short to medium term. One of the key constraints on a number of the proposed overseas small-scale injection projects was the lack of CO_2 and as a result of this it was necessary for some projects to buy food grade CO_2 at high cost, which in turn limited the scale of the injections to just a few hundred to a few thousand tonnes in many cases. If an Australian pilot project was to be undertaken at the commercially significant scale the CO2CRC wanted, it was important to ensure that there was ready access to an affordable source of CO_2 at sufficient scale (around 100,000 t of CO_2) and that it was available now. While CO2CRC had plans for also undertaking a pilot capture project, there was no prospect of using a power station

or an industrial plant as the source of CO_2 for at least 5 years and possibly longer. It was therefore necessary to have a "surrogate" for the major CO_2 source shown in Figure 1.1.

It was decided to focus on securing a natural gas-related source of CO_2, whether a high CO_2 natural gas such as is found in the Cooper Basin or relatively pure geological CO_2 such as is found in the Otway Basin. After considering a number of possible options, attention soon turned to the Otway Basin, a natural gas-producing basin in western Victoria, where a number of natural CO_2 accumulations were known to exist in tenements licensed to the Santos-Beach JV (Figure 1.2). In 2004 CO2CRC made a visit to the Santos data room visit and examined PPL-11 (known to have a high CO_2 content in the Buttress well) as a potential source and with prospects for an injection site.

Options short-listed for injection and storage included the Croft and Naylor depleted gas fields (Figure 1.3). The Buttress Field had been logged and cased by Santos, but not perforated and tested, before it was suspended as a potential CO_2 producer (Frederickson 2002). Based on the available data, there was a reasonable level of confidence that it could provide the 100,000 t of CO_2 required for a small-scale storage project.

In 2004, Santos decided to sell its entire portfolio of onshore Otway gas fields, including the above options, to Origin Energy (OE). In early 2005 OE acquired 90% of the Santos share in all these tenements; with the Santos/Origin sale there was a related purchase agreement outlining that PPL-11 would be sold to the CRC and funds from the sale would flow back to Santos. If that were not to happen, Santos indicated it would find an alternate buyer for PPL-11. Through subsequent discussions, it became evident that OE were willing to sell only the Naylor depleted gas field to CO2CRC (up until that time, CO2CRC was also interested in the Croft depleted gas field as a CO_2 storage site). The Naylor Field was a small depleted gas field with original gas in place estimated at 170 million standard cubic metres (6 billion standard cubic feet). From May 2002 to February 2004, Naylor-1 produced a total of 112 million standard cubic metres (3.965 billion standard cubic feet) of natural gas from the Waarre "C" and "A" units. The well was suspended in 2004 after the well started to produce water and the field was considered depleted (Bowden and Rigg 2004). As a potential sink site based on the volumes of natural gas produced, Naylor was deemed to be adequate for a storage project. However because of the wish of OE to retain the Croft Field, PPL-10 needed to be partitioned in order to separate the holdings of the two fields, Croft and Naylor. Accordingly, a new PPL-13 was defined (Figure 1.3).

Therefore CO2CRC now had the essential components for a pilot or demonstration storage project. Through the purchase of the Petroleum Production Licence PPL-11, it had an unproduced natural gas well (Buttress-1) known to be high in carbon dioxide (95,000 t of CO_2 at the P90 level, 250,000 t at the P50 level and 950,000 t at the P10 level).

What was not then known (because the well had not been produced up to that time) was the actual composition of the natural gas. However as the well was close to the Boggy Creek gas field, which produced CO_2 for the food industry, it was considered likely that the composition of the gas (by weight) would be of the order of 90% CO_2 and 10% methane. In the event, the composition was closer to 80%:20%. Nonetheless this provided the basis for using the Buttress Field as the source of CO_2.

In addition, CO2CRC (initially through CMPL, subsequently through CPPL—see Section 1.3) was in a position to purchase PPL-13, which included the depleted gas field, Naylor-1, which was seen as a suitable site for testing the geological storage of CO_2. A significant amount of residual methane remained in the field but a preliminary assessment by CO2CRC suggested that the abandoned Naylor structure would provide a suitable storage site for up to 100,000 t of CO_2 and also that it might be possible to use the existing Naylor well as a monitoring well.

What was not clear at that stage was how the PPLs would be held by CO2CRC or how a project of the scope envisaged, would be undertaken by CO2CRC. It meant that CO2CRC, a research consortium, was potentially taking on responsibilities and risks akin to those of a small oil and gas company. This clearly needed careful consideration particularly as each member of the CO2CRC had a different attitude to risk. Nonetheless, despite these uncertainties, the opportunity offered in the Otway Basin was seen to be so important to the future of CO2CRC that it was decided to go ahead with the purchase of the two properties, PPL-11 and PPL-13, from Origin Energy. It was also decided at that time to develop a more appropriate corporate structure for CO2CRC that would meet the needs of the proposed project and address any related concerns of the members of the joint venture.

1.2.2   Naming the project

At first glance, the naming of a project would seem to be a relatively unimportant issue, but this is not necessarily correct, for the name becomes a unique identifier through which the project becomes widely known and with which the scientists, the engineers and other staff identify. Therefore getting the right name was important.

Initially, without giving the matter a great deal of thought, the project was called the Otway Basin Pilot Project. This was

Figure 1.3: Partitioning of PPL-10 into a new PPL-13 and a reduced PPL-10.

soon abbreviated to OBPP, which the scientists were quite comfortable using, but which meant absolutely nothing to anybody other than those closely acquainted with the Project. It was therefore decided to call it the CO2CRC Otway Project, sometimes used in full, particularly when used for the first time, but then abbreviated to the Otway Project, which for many people immediately identified the area where it was being undertaken. This then was the name that entered into general use.

A nomenclatural complication arose in 2006–07: by this time it was evident that the site offered the opportunity to undertake other field experiments and activities beyond the initial activity of storing CO_2 in the Waarre Formation. It was therefore decided to retain the name Otway or Otway Project, but to add Stage 1 to the initial project

and Stage 2 to the proposed second project, with the further option of a Stage 3 and a Stage 4, etc., at the Otway site. However, this became further complicated by the need to divide Stage 2 into 3 phases leading to the need to have Stages 2A, 2B and 2C. Table 1.1 summarises the nomenclature and the activities undertaken. For the most part, this book is concerned with Otway Stage 1 although a number of chapters make reference to Stage 2A or Stage 2B. Stage 2C is only mentioned briefly. There are no firm plans for a Stage 3, although options are under consideration.

A further nomenclatural issue which arose was whether to refer to the Project as a demonstration project or a pilot project? There is in fact no definition of these terms and they are often used interchangeably. Notionally the

Table 1.1: The various phases of the Otway Project 2003–13.

CO2CRC Otway Project	
Stage 1	Injection of 66,000 t of CO_2-rich gas into the Waarre Formation at the CRC-1 well and related monitoring (completed)
Stage 2A	Drilling of new injection well CRC-2 into the Paaratte Formation and site characterisation (completed)
Stage 2B	Injection of 140 t of pure CO_2 into the Waarre Formation and determination of residual saturation (completed)
Stage 2C	Injection of up to 30,000 t of CO_2-rich gas into the Paaratte Formation with related seismic monitoring (planned)
Stage 3	Under consideration

term "pilot" could perhaps be reserved for projects of less than 10,000 t of CO_2 and "demonstration" for projects of 10,000–100,000 t, but there is no agreed convention. Because of the wish to inject up to 100,000 t (an amount arrived at somewhat arbitrarily) at Otway, the apt but perhaps somewhat vague term "commercially significant" was often used to distinguish it from, say, a small injectivity experiment, where only tens of tonnes of CO_2 were injected. The IEAGHG programme in its recent review of storage projects (Cook et al. 2013) used the term "small-scale" for projects of less than 100,000 t. More recently the 100,000 t quantity has acquired greater significance in that it is the upper limit set within EU legislation for the size of projects that could be dealt with under research-related legislation rather than under the much more onerous regulations relating to large (commercial) scale injection of CO_2. There obviously continues to be a degree of arbitrariness about these terms. The issue was avoided in the case of the Otway project by not including "Pilot" or "Demonstration" in the name of the Project, not least because this made it possible to have a continuum of projects in the future, with a range of scales and objectives all referred to under the name of the Otway Project.

One final nomenclature issue was whether to refer to the gas injected from the Buttress Field as CO_2 or CO_2-rich gas or Buttress gas. As already mentioned and as discussed

in greater detail in Chapter 4 and subsequently, the gas injected was approximately 80% CO_2. Therefore while 66,000 t of "gas" was injected, the amount of pure CO_2 actually injected was approximately 60,000 t. Nevertheless throughout the Project (and in this book), for the sake of brevity, the course followed was to refer to the total amount of gas injected, as "CO_2". This was a reasonable course to follow, given that the methane was totally miscible in the CO_2 and within the pressure and temperature conditions encountered in the Project, it had no significant impact on the behaviour of the CO_2.

1.2.3 Developing the science programme

In first putting forward an Otway Project to funding agencies in the Federal and State systems in 2004, the objectives were spelt out in a largely non-scientific way, namely "to demonstrate that CO_2 capture and storage is a viable, safe and secure greenhouse gas abatement option in Australia". This objective was then underpinned by several sub-objectives needed to achieve this overriding objective, namely:

> "effectively separating and capturing CO_2

> safely transporting CO_2 from source to sink

> safely storing CO_2 in the subsurface

> safely abandoning the facilities and restoring the site

> communicating with all stakeholders

> conducting the project within approved time and budget

> capturing all research outcomes".

It was of course the last point which highlighted that research was to be done at the site, and the fact it was put last did not in any way reflect a lack of importance. Rather, its position reflected the reality of a funding climate focused on practical outcomes rather than the research needed to provide those outcomes. Nonetheless it was recognised by all involved that innovative high quality research was really what the proposal was all about.

Throughout the negotiations to obtain PPL-11 and PPL-13, the underpinning driver was the science that CO2CRC proposed to undertake. By 2001–02, Sleipner and other projects in Europe and North America had demonstrated the viability of underground geological storage of CO_2, but they were limited in number and none were in Australia. By 2003, with significant funding available to CO2CRC, there was the potential opportunity to demonstrate that CO_2 could be geologically stored under Australian conditions. The scientists were confident that it could be, based on overseas experience, but it really was not sufficient to tell stakeholders, or the community at large, that "it has worked in Texas and therefore it will work in Australia". There was a need to demonstrate that it could work under Australian conditions and to provide people with the opportunity to "kick the tyres"! In other words, an important initial driver related to communications and demonstrating that the technology could and would work. However, this should not then be misconstrued as primarily a "public relations exercise" for CCS, in that right from the start it was recognised that there had to be world-class science underpinning all aspects of the project.

CO2CRC already had access to leading scientists through its predecessor organisation, the APCRC, and the related

GEODISC programme. Researchers, together with the CO2CRC Executive, started to develop the detailed research proposal for what was to become the CO2CRC Otway Project Stage 1. The key scientific drivers for the Project were to demonstrate under Australian conditions:

> safe geological storage of CO_2 at a "commercially significant" scale (defined as up to 100,000 t)

> the successful application of a range of monitoring techniques to confirm effective geological storage of CO_2

> to undertake all of the above with no adverse environmental consequences.

To meet these objectives required the application of a range of scientific, engineering and organisational skills. A number of these were already available within the broader range of activities undertaken by CO2CRC; some of the expertise was only deployed in the Otway Project for a limited time; some was deployed virtually full time (Figure 1.4).

With increasing confidence in the imminent purchase of the Otway tenements PPL-11 and PPL-13, the scientific planning was able to be more clearly defined. One of the major benefits of purchasing the tenements was that a great deal of information was acquired on prior gas production, subsurface geology, seismic reflection profiles and so on from the previous operators. It was now known what sort of geology would be encountered, the presence of a potentially suitable reservoirs and their depths (Figure 1.5), the potential rates of injection, the likelihood of an effective seal, the capacity of the structure and the presence of faults. In other words, it was possible to start the high level characterisation of the site (Dance et al. 2009) and on this basis, start to develop the full science programme.

In particular it was now possible to gain a better idea of the feasibility of the original concept of injecting up to 100,000 t of CO_2 and of monitoring the behaviour of the CO_2 plume. The operational concept was quite straightforward in that it involved producing CO_2 from the Buttress well, treating that CO_2 as necessary, transporting by pipeline to an injection well (CRC-1) and then injecting

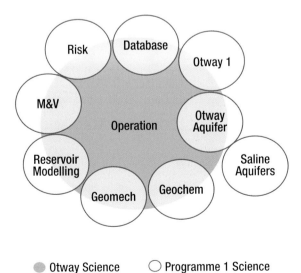

● Otway Science   ○ Programme 1 Science

Figure 1.4: Relationship between Otway science and activities within the broader CO2CRC programme.

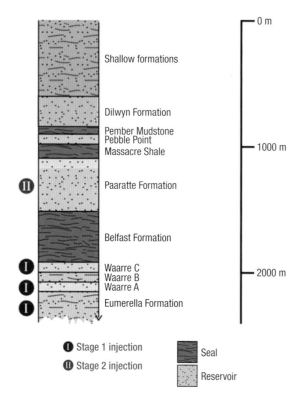

Figure 1.5: Preliminary stratigraphic column available for the Otway site. More definitive stratigraphy was obtained by drilling the CRC-1 well (see Chapter 5).

and storing the CO_2 in a suitable geological formation, within the depleted Naylor gas field (Figure 1.6). The CO_2 would also be monitored, using suitable technologies. Although superficially quite straightforward, there were of course many scientific unknowns, which needed to be addressed if the objectives were to be satisfactorily addressed by CO2CRC research.

There were two key objectives (discussed in detail in Chapter 9) in monitoring the CO_2, namely to monitor whether or not stored CO_2 was remaining within the reservoir interval (integrity monitoring) and to monitor with a view to ensuring that the community and the regulators could be confident that CO_2 was not leaking into aquifers or into the atmosphere (assurance monitoring). This required surface or subsurface systems that would monitor the deep storage formation (the reservoir), the overlying seal, the aquifers, the soils and the atmosphere. These are discussed in detail in Chapters 10–14 and also

by Jenkins et al. (2011), Dodds et al. (2009) and Hennig et al. (2008). It was also necessary to devise programmes that would ensure that the site was well characterised (Chapter 5), that the seals would be suitable (Chapter 6), the geomechanical properties were known and understood (Chapter 7) and that any risk were clearly defined (see Chapter 8 and later in this chapter). Integral to all of these was the need to develop a programme of reservoir modelling and monitoring (Chapter 16). Finally it was also necessary to ensure that these programmes could be integrated not only in terms of the science but also in terms of the practicality of undertaking then in a timely and effective manner, which meant of course that the operations team and the science team had to work very closely together (discussed later in this chapter). By 2008, all of these objectives were achieved.

Given the nature of research and researchers, the Otway Stage 1 Project was not completed before consideration started to be given to using the undoubted science potential of the Otway site to address other questions of importance to CCS. It was of course essential from a research management perspective to ensure that enthusiastic researchers did not start running off to undertake the next exciting research project before the current one was successfully completed! At the same time, it was not realistic to expect a research programme to run until everything was finished and only then start to plan for a new phase and obtain the new funding. One of the key reasons why this would have been quite unrealistic for the Otway site (and probably for most other long-term CCS research projects) is that, whether any research was done at the site or not, it cost of the order of $1 million a year to maintain the site, ensure it was secure, undertake maintenance of equipment, and so on. Therefore from this perspective alone there was a compelling reason to have a rolling programme of research at the site if at all possible. The other problem that would have arisen, had there been a significant hiatus between Stage 1 and Stage 2, was that much of the expertise assembled to undertake Stage 1 would have been lost, with a major adverse impact on Stage 2.

Therefore in 2006, proposals for a new phase of Otway science, to be known as Otway Stage 2, were developed initially as a funding proposal to the Federal government

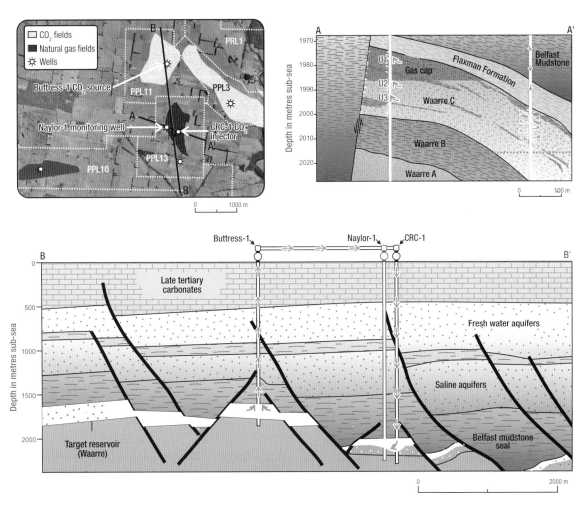

Figure 1.6: North-south cross-section through the fault-bound CO_2 source and sink intervals overlaying seals and aquifers (after Jenkins et al. 2011).

CRC programme for supplementary funding. Stage 1 involved injection and storage of CO_2 in a depleted gas field. This is a significant storage option (Underschultz et al. 2010; Cook et al. 2011) but volumetrically is regarded as much less important as a storage option than saline aquifers (IPCC 2005). However, while trapping of CO_2 and related gases in a closed geological structure (such as Naylor) is relatively well understood, storage in a saline aquifer is not. In particular the trapping of the CO_2 is dependent on residual capillary trapping and dissolution trapping (IPCC 2005).

With this in mind, the main objective of the Stage 2 proposal was to investigate residual capillary trapping

(Zhang et al. 2011). Because Otway Stage 1 was within a depleted gas field with abundant residual methane, it was not suitable for investigating residual trapping of CO_2. Therefore Stage 2 had to include the drilling of a new well to provide access to a saline aquifer with no residual hydrocarbons. The interval chosen was the Paaratte Formation, which was somewhat shallower than the Waarre Formation (Figure 1.5) and relatively poorly investigated (largely because of its lack of hydrocarbons). Therefore an important aspect of the study had to be to first geologically characterise the poorly known Paaratte Formation, before the investigation of residual trapping could be carried out. The need to inject CO_2 into a saline aquifer was also seen as an opportunity to investigate the use of seismic imaging to establish the distribution of CO_2 within a saline aquifer.

Closer investigation of the research objectives showed them to be sound but the research to achieve them needed significant modification for several reasons. First it was concluded that it would be difficult to do the necessary Stage 2 field experiments using Buttress gas because of the presence of methane; it was therefore decided to use food-grade CO_2, which added significantly to the costs. Because of this, it was decided to only inject a small quantity of CO_2 (around 140 t). This work (Stage 2B) is discussed in detail in Chapter 17. This decision meant that it would not be possible to detect such a small quantity of CO_2 using seismic methods. Therefore what had by this stage become known as Stage 2C required a new injection of Buttress gas (up to 30,000 t) into the Paaratte Formation via CRC-2. All of these considerations led to a reassessment of the scientific methods to be deployed and also a sharper definition of the scientific objectives, which could be summarised as:

> geologically characterise a saline aquifer (in this case the Paaratte Formation) in detail

> establish residual gas saturation of CO_2 in a saline aquifer via a single well huff and puff-type test using various measurement options

> verify numerical simulations and predictions by field measurements

> establish the distribution of CO_2 within a saline aquifer by remote measurements

> use time-lapse seismic anisotropy to verify pore-pressure change, fluid migration and saturation

> develop a multi-level monitoring system.

Chapter 17 clearly indicates that the first three objectives were met through Stages 2A and 2B. Stage 2C has yet to be undertaken, but the research team has confidence that the remaining objectives can be met.

Is there to be a Stage 3? Possibly, with the uncertainty inevitably arising because of funding uncertainties! However

what is not in doubt is the potential to use the Otway site and the Project facilities to answer a range of other important CCS-related questions, such as developing a better understanding of the impact of CO_2 on pre-existing faults or the potential for induced seismicity, or the opportunities for remediation if leakage were to occur. The likelihood of such events taking place is very small, but nonetheless it is prudent to investigate them to be able answer community concerns if they were to occur. The Otway site offers that opportunity.

1.3   Developing a suitable corporate structure

The initial structure of CO2CRC developed in 2002 (Figure 1.7), was in the form of a conventional unincorporated joint venture (JV). In common with many other joint ventures, two other incorporated entities were formed to provide services to the JV. The first entity, CO2CRC Management Pty Ltd (CMPL), was a management company set up to handle finance and accounting, employ staff, and generally deal with financial governance issues. The other entity, Innovative Carbon Technologies Pty Ltd (ICTPL), subsequently renamed CO2CRC Technologies Pty Ltd (CO2TECH), was established primarily to hold intellectual property (IP) rights on behalf of the members and to provide a vehicle to commercialise IP arising from the work of CO2CRC. Both companies were proprietary non-reporting entities. By 2003, membership of CO2CRC stood at 11 industry and government participants, each with a seat on the JV

Figure 1.7: Initial structure of CO2CRC and related entities in 2003.

Board and eight Research Participants who were represented as a group by three elected JV Board members. The Chief Executive was also a Board member and there was an independent Chair. The Board of CO2TECH was similar but not identical in its composition (not all members of the JV wished to be involved in commercialisation) but the CEO, the Chair and several Directors were common to both Boards. The Board of CMPL, the management company, was significantly smaller. Again there was some commonality of Directorship with other Centre Boards.

This structure remained largely in place throughout the first term of the CO2CRC with only minor changes. The principal change to the JV was an amendment to the Corporations Regulations to allow unincorporated CRCs to allow up to 50 partners (previously limited to 20). This was important to CO2CRC as its membership grew significantly after 2003.

However, as pointed out previously, a joint venture did not provide the best structure for managing the new responsibilities of CO2CRC. One of the objectives of the CO2CRC was not only to carry out research but to take that research to an applied level by practical demonstration of CCS. The medium-scale field demonstration envisaged was at semi-industrial scale which would:

> demonstrate the engineering technology to safely store carbon dioxide but also research and measure many aspects of the sequestration process

> involve governments in the formation of the statutory and regulatory framework necessary for the industrial application of the technology

> provide a practical example to the community of a successful application of the technology and thereby enhance community acceptance.

These requirements were seen as likely to have a significant impact on the structure of CO2CRC and require the establishment of an operating entity.

In May 2004 a draft proposal was submitted to the CO2CRC Joint Venture participants. At this stage the proposed structure for carrying out the project consisted of a Project Steering Committee coordinated and managed by CO2CRC. It envisaged that an operator would need to be appointed to deal with day-to-day operations, while the CO2CRC would undertake the research activities. It envisaged that operational liability would be managed by insurance and long-term liability would be borne by the State Government. A funding requirement of some $12–13 million was envisaged at that time (see Section 1.4). As a holding operation CO2CRC's incorporated commercial entity, CMPL, was used to sign some of the preliminary agreements including service agreements to undertake a well test on the CO_2 well, but this was not seen by the Board as the best option for the long-term.

Initially it was hoped that an experienced operating company would agree to manage the envisaged Otway Project on behalf of CO2CRC, for a fee. However, the fee that the companies required to undertake the project was far greater than the amount of funding that CO2CRC believed it had available. In retrospect, it is clear the costs envisaged by the operating companies were more realistic than the early estimates of CO2CRC. Even more significantly, in all the options considered, the operating company was required to take on all risks and liabilities associated with the Project and no single company was willing to do this. In hindsight, such an arrangement, even if it had been possible, would probably have not met the needs of the JV or the Project and most likely would have unduly inhibited the research aspirations of CO2CRC.

It was apparent that the undertaking of the Project was going to expose the CO2CRC joint venture to operational and financial risks, potential exposure to long-term liability, and the usual obligations associated with ownership and operation of petroleum tenements. An unincorporated JV was legally not able to unilaterally accept these risks on behalf of its members and in subsequent joint venture discussions it became apparent that the nature of the Project was such that not all joint venturers would be prepared or be allowed to participate in such a project. Core participant research organisations and industry bodies such as ACARP were unable to assume any exposure to potential liabilities because of their structure or charter.

In the period mid-2004 to mid-2005 there was slow progress in acquisition of the interest in tenements PPL-13 and PPL-11 with ownership, pre-emptive rights issues and possible operatorship of the project all intertwined issues.

By March 2005 the prospect of forming a new entity to take on the Project (referred to at that stage as the Otway Basin Pilot Project or OBPP) was floated and industry members were polled to see which ones were prepared to accept such a role. The responses were mixed and reflected traditional legal advice from company lawyers. Discussions continued in an attempt to resolve various issues, legal advice was sought on the nature of the appropriate entity, and this culminated in a formal proposal at the subsequent June JV Board meeting (Figure 1.8) based on legal advice, detailing options for the formation of an entity to manage the operation of the Project. This included incorporated and unincorporated options but the recommendation was to proceed with the formation of an incorporated entity to carry out the Project.

A new special purpose company was recommended for the following reasons:

> administratively simple

> structure, ownership and responsibilities clear

> works well with existing CO2CRC entities

> no presumptions on operator

> provides maximum flexibility

> limited liability from having a separate legal entity

> shares risk equitably (but only by the members of this entity).

There appeared to be no adverse tax implications from such an option. The risk sharing was subsequently the subject of much discussion and much later became an issue relevant to the transition to CO2CRC Ltd. The Board of CO2CRC was asked to consider immediate formation of the company so that negotiations with governments could be successfully concluded and the Project could get underway.

Following this CO2CRC Board meeting, it became apparent that most of the industry participants supported the formation of an incorporated entity to carry out the project. These members became known as "The Coalition of the Willing"—or COW for short! However, there were varying views on several issues including the mechanisms to handle project liability, especially the long-term liability (LTL) associated with sequestered CO_2. The process of establishing a formal company proceeded slowly and therefore in the interim it was necessary for the COW to informally take on advisory role in moving the project forward. It was apparent by this time that the members of the "COW" would essentially be comprised of the coal

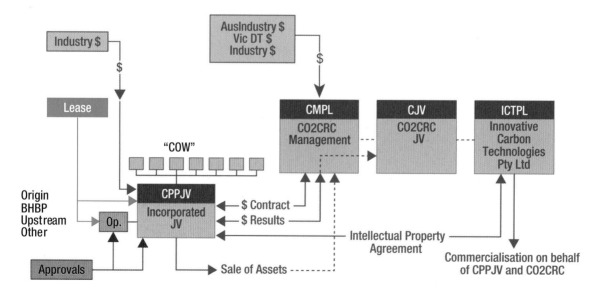

Figure 1.8: 2005 incorporated option proposed for the undertaking of the Project.

and oil/gas industry participants in the CO2CRC JV. Other participants would not or could not participate and insisted that when the new incorporated OBPP vehicle was formed it should be independent to the extent that no liability for operational activities or long-term storage could flow to the CO2CRC JV. This meant that the Project company could not be an agent for the JV. That role had to continue to be fulfilled by CMPL.

There were many meetings of the COW led initially by an oil industry representative and later by a coal industry person. Some of the issues that were prominent in the COW discussions were driven by the criteria these members needed to resolve in order to sell the proposition to their respective companies. These issues included:

> dealing with operational liability issues including cost overruns and defining a mechanism to handle this

> dealing with long-term liability (LTL) issues surrounding storage of sequestered CO_2

> determining what criteria might be required prior to a decision to inject gas for sequestration

> establishing operational controls appropriate to the Project. It was clear that the company needed to control its own destiny and required a complete mandate to operationally manage the Project and all site activities. Other issues included

  – details and specification of the scientific research was to remain the domain of the CO2CRC JV but site activities were to be under the control of the Project company

  – defining the statutory and regulatory regime under which the sequestration would be carried out

  – determining who would be the operator of the project and under what terms and conditions

  – establishing the company Constitution and Member's Agreement

  – defining the relationship between the Project company and the CO2CRC JV.

Legal advisors were commissioned to prepare draft criteria for the incorporation of a not-for-profit company, limited by guarantee, to be known as CO2CRC Pilot Project Ltd (CPPL). There were many meetings of the COW in which "Term Sheets" covering the various issues were debated and for the most part were resolved. The process was tedious because the various members of the COW were companies whose legal advisors and head office processes were widely scattered and held differing criteria for approvals for entry to such a project. By December 2005 there was resolution of issues to the point where the company could be incorporated and the first Members formally join. The expectation was that there would be 10 Members—broadly five from each of the coal and oil/gas sectors.

1.4 Formation of CO2CRC Pilot Project LTD

On 28 November 2005, CPPL was formally incorporated with the broad purpose outlined in the Constitution to:

> undertake the CPPL Project

> operate, or engage others to operate, facilities to undertake the CPPL Project.

Although a Pty Ltd option was considered, the structure of a public company limited by guarantee was seen to have several advantages:

> the structure was typical for non-profit entities

> facilitates entry and exit—no need to buy or sell shares

> perhaps more amenable to tax exemption.

The Member's Agreement and Constitution defined the detail but the COW had resolved the following issues:

> Each Member had equal accountability to share operating liability and any potential cost overruns. It was recognised that, despite the best endeavours, a real operating situation (such as an emergency) may require utilisation of resources that exceeded the capacity of the CO2CRC JV to support

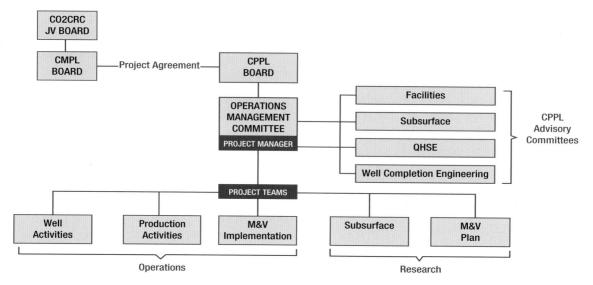

Figure 1.9: The CPPL organisational chart.

technically and/or financially. However, the expectation was that these issues could be managed with traditional risk assessment and careful planning. Reputational risk issues were discussed at length and were a key consideration.

> The issue of how to deal with long-term liability of sequestered CO_2 could not be resolved at the time and the COW's expectation was that this liability should be accepted by the State or Federal Government. In order to move the project forward the COW resolved to defer this decision by putting a special clause defining that it must be a unanimous decision of the Board to proceed to injection. In practical effect, this meant that each Director would need to satisfy the Member Company of this decision and effectively made it a Member's decision at a future point in time. The Company (CPPL) also would own the tenements and have the right to manage all activity on those tenements including appointment of operator and operational staff and be able to develop and install any necessary systems required for the Project.

> A Project Agreement would be put in place with CO2CRC Management Pty Ltd (CMPL as agent for the CO2CRC JV) which detailed

the contractual arrangement to carry out the programme of works. Implicit in this arrangement was an understanding that all funds required for the project activity had to be identified and available (from CMPL) prior to commitment of those funds by CO2CRC Pilot Project Ltd. Injection of CO_2 required consent of both CMPL and CPPL before any injection would take place.

These concepts were embodied in the various legal documents such as the Constitution, Member's Agreement, and the Project Agreement between CPPL and CMPL. The relationship between CPPL and the CO2CRC entities is shown in Figure 1.9.

The first meeting of the Company (CPPL) was held in early December 2005 and at that time there were five Members. Other participants were seeking approval to join but due to geographic diversity and different approval systems it was to take another 12 months before all 10 Members were in place.

At the initial meeting the industry representative that had been leading the meetings of the COW was elected Chairman. At that time it seemed that the path forward would be for the Chairman role to move to a petroleum industry representative because of the relevance of petroleum skills. This proved to be a wrong assumption because in practice the skills required to get the project established were more related to statutory and regulatory approval, land use and landholder issues, local government and community arrangements and cultural heritage. These skills were just as relevant in a mining background as they were for the petroleum industry.

Early tasks for CPPL included the appointment of a Project Manager. In conjunction with the Project Manager the Board then established the operating structure for CPPL as shown in Figure 1.9.

This structure also defined the interface with the research activity to be undertaken by the CO2CRC JV. Figure 1.9 shows a line management structure within CPPL which included an Operations Management Committee. The members of this committee included the Project Manager and various people with relevant skills seconded from the Members. The committee typically had three or four members and people who had skills appropriate for the work activity at any particular time were seconded. This committee had delegated authorities in excess of the Project Manager and could meet at short notice to provide support and advice and if necessary commit funds and resources to an agreed level. Beyond that level referral was to the Chairman and Board. However, it is important to note that CPPL required the CO2CRC JV to provide those funds via CMPL before they were committed by CPPL. In other words, for the most part, CO2CRC JV continued to be the vehicle for raising funds to undertake the research programme, to develop the research programme (and provide the researchers to undertake that research), and obtain the funds to enable CPPL to undertake the Otway operations.

The Board of CPPL had 10 Directors including the Chairman. This Board was a good size and contained a good spread of technical, operations, project management and financial skills and also provided an avenue to access skills and resources to support the various aspects of the Project through the Members. While there was some turnover in representation there was a stable nucleus of Directors throughout the period of the Stage 1 Project. The CEO and the Chair of CO2CRC JV attended the meetings of the CPPL Board to ensure close coordination between CO2CRC science and CPPL operations.

Four Advisory Committees were established by CPPL to provide advice and support to the Project Manager but these committees did not have any line management function or authority to commit funds or resources. Again membership of these committees came from people in both the Members and researchers who had appropriate skills. One of the principal early tasks was to establish a HSE (health, safety and environment) system for the project in conjunction with the appointment of an operator.

In the early period there were many activities running in parallel. CPPL in its formative stages and after incorporation had a significant role in all of the following activities:

> purchase, ownership and management of the two petroleum tenements

> establishing the regulatory regime in conjunction with the various levels of government and government authorities, including establishing the appropriate legal path to accomplish a regime satisfactory to CPPL and its Members

> establishing the company systems including safety, financial, project budgeting, project management and project controls and financial audit. In this area it shared resources with the CRC utilising the CRC Business Manager as Company Secretary

> establishing lease agreements and compensation agreements with landholders

> establishing contracts for the provision of services including drilling and construction activities

> managing statutory requirements such as compliance with the requirements of the Federal EPBC, State Environmental and Rural Water authorities and Cultural Heritage legislation

> managing access to site by contractors, researchers and visitors and the HSE requirements of the site.

In parallel with these responsibilities, the CO2CRC JV had the responsibility for:

> identifying the research priorities for the Otway Project

> seeking the necessary funding for the scientific and operational aspects of the Project

> liaising with funders and other stake holders to enable the Otway Project to progress

> staffing and undertaking the research activities

> developing the monitoring and verification regime that would both meet the scientific objectives of the Project and enable CPPL to meet its statutory obligations

> communicating outcomes from the Otway Project via the scientific literature, the media and other outlets.

During this period it was clear that neither the Federal Government nor the Victorian Government were prepared to accept outright liability associated with injected CO_2. However under the arrangements with the Victorian Government, it became evident that should the Project meet all of the Agreed Statutory requirements, then normal processes for ultimate relinquishment of the tenements would apply.

At the CPPL Board level, extensive Risk Assessment was undertaken prior to the injection phase to assure the Members that the activity would be carried out safely and without liability or reputational damage. This was summarised on one page that effectively became the final check list. On the basis of this assessment the decision was made to proceed with injection under the terms of the relevant agreements with the Members and CMPL.

Obviously it was essential that CO2CRC JV and CPPL worked together as seamlessly as possible to achieve the objectives of the Project. This was achieved in part by having joint membership/attendees of the various Boards. On occasions there were inevitable tensions arising from the competing priorities of operational needs and research aspirations, but these were minor and the collaborative arrangement worked well.

CO2CRC operated under this structure for approximately 5 years during which time the Otway Stage 1 Project was successfully carried out. The large amount of effort establishing the structure and systems was well justified and the Project was completed within the financial constraints and without any lost time injury to contractors or research staff.

Following the completion of Stage 1 of the Otway Project in 2008, CO2CRC took early steps to seek an extension for a further funding term beyond 2010 under the CRC Programme. The Government advisors had earlier stated a preference for any extension to be carried out under an incorporated entity rather than an unincorporated joint venture. Also there was recognition of the inevitable complexities of having four entities—CO2CRC JV, CMPL, CO2TECH and CPPL—all with a range of responsibilities. In the case of CPPL those responsibilities were solely to the Otway Project, but in the case of the other three entities the responsibilities covered not only Otway but also a wider range of storage activities, capture activities, CCS economics, education, training and commercialisation. Nonetheless the question needed to be asked—was there a more cost effective way to handle the affairs of Otway and the other CCS activities, using a simpler corporate structure?

CPPL had been set up as a special purpose vehicle to demonstrate CO_2 storage at Otway and clearly it had achieved that purpose. Under the original arrangements, the intention was to close activities at Otway once statutory obligations had been completed and then disband CPPL. Up to 2008–09, the members of CPPL had carried project risk on behalf of all participants of the CO2CRC Joint Venture (which now numbered around 30 participants) and they considered that any future burden should be more equally shared amongst all participants. By this time, CO2CRC had already been successful in obtaining supplementary funds to undertake Otway Stage 2 (see later) and therefore there was recognition of the need by CPPL and CO2CRC more broadly, to consider longer term arrangements. In addition, in 2009, CO2CR was successful in obtaining new funding for a further 5 years of research for the period 2010–15.

Various options for the incorporation of the new phase of CO2CRC were proposed, but the only option that addressed all requirements, including the concerns of the Members of CPPL regarding liability, was either the currently unincorporated JV (CO2CRC) became incorporated, or CPPL becoming the incorporated entity for the period beyond 2015. It was decided to take the latter course (Figure 1.10).

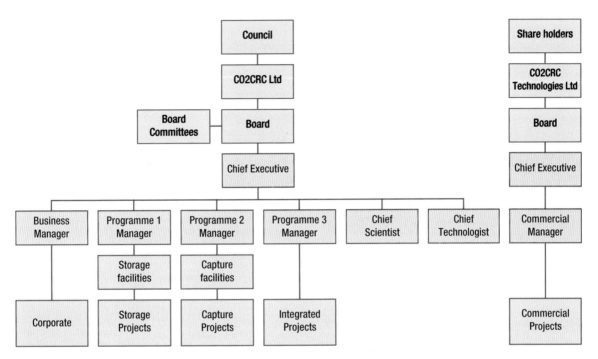

Figure 1.10: The CO2CRC corporate structure used from 2010 onwards.

There were many practical reasons for this decision, including substantial savings in the cost of transfer of assets and agreements. Also, CPPL was the formal holder of the petroleum tenements, held leases and agreements with Landholders and Governments, and had established systems for the financial, operational, and safety requirements that would be necessary for the proposed future activities of CO2CRC, both in storage and capture. This change also facilitated the establishment of a new and more effective Board structure for the CRC, with only 9 to 11 Directors (a number that had proved very effective in CPPL), rather than the 20 or more directors under the JV constitution. At the same time all of the CRC participants would be assimilated as shareholder members.

From the point of view of CPPL Members it also provided the basis for other CRC participants to join the company and to equally participate in risk sharing for future project activity. It also provided the opportunity to establish a much simpler structure for the new CRC since the combined roles of CPPL, CMPL and the original JV could all be achieved through a single entity, by making some changes to the Constitution and Member's Agreement

and then renaming CPPL as CO2CRC Ltd. However, there was an ongoing need at that time for CO2TECH in terms of management of IP and therefore this entity was retained (Figure 1.10).

These changes were put into effect following the success of the rebid for the CO2CRC 2010–15 extension and the first meeting of CO2CRC Ltd was held in early 2010.

1.5 Funding the project

When the original CO2CRC bid for funding was lodged with the Federal Government's CRC Programme in 2001 (with funding proposed to start in 2003), there was every intention to undertake a pilot or demonstration storage project, and this was flagged in the proposal. The original funding did allow for researcher costs, but no request was made for funding an actual field project at that time. There were two reasons for this: first, a pilot storage project was an aspiration and in 2001–02 there was no clear idea of how and where such a project could be undertaken (such a project had never been undertaken in Australia before). Second, there was a ceiling on the

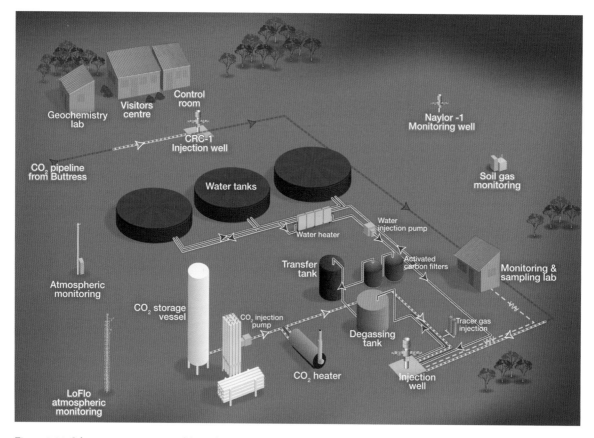

Figure 1.11: Schematic representation of the surface installations at the Otway site (locations labelled in blue indicate surface installations that existed in Otway Stage 1; all other facilities were established for Stage 2).

funding available for a Cooperative Research Centre bid and any bid for major capital costs would have taken the proposal far over the limit, thereby jeopardising the entire bid. Therefore a very conscious decision was taken to seek new funding for any pilot or demonstration bid only when the proposal for a CRC was successful. This successful bid was announced in 2002 and the Centre commenced in mid-2003.

There was now the opportunity for CO2CRC to develop a project proposal of substance. The project in the Otway Basin as initially proposed was fairly simple with the major components consisting of the CO_2 source (Buttress) with some separation of CO_2 from the gas, a pipeline, an injection well (CRC-1) and a single monitoring well (the existing Naylor-1 well). The facilities were greatly expanded for Stage 2 (Figure 1.11).

It was to be accompanied by basic baseline surveys before the commencement of injection of CO_2 and extensive monitoring during and post-injection. While still lacking detail, it did allow CO2CRC for the first time to develop some approximate costings. At this stage (2004), the project cost/budget was estimated at $12.8 million (Table 1.2)

The funding request did not address Otway-related research costs (mostly researcher time) as all such costs had been covered in the original bid for the CRC in 2002. If the budget proposal in 2004–05 had included this, then the original cost would have been of the order of around $20 million. The cost of purchasing the Buttress property ($1 million) was met from existing CO2CRC funds at an early stage in the negotiations. This was a risky move to the extent that there was no certainty that the remainder of the funds would be found, or that the project would go ahead. Nonetheless, access to the Buttress CO_2 was seen as so critical to the success of a future pilot project that it was seen as a risk worth taking. This proved to be the right move.

With $13 million promised by the funders (Federal Government, Victorian State Government and industry) plus the funding already held by CO2CRC, it was now possible to start on developing more detailed plans and costings. Two things happened at this time: first, and somewhat inevitably, real costs proved to be much higher than the original estimates. Second, researchers began to realise that Otway presented some exceptional research opportunities, with the inevitable rise in costs that an expanded science programme entailed. Fortunately the Australian Greenhouse Office agreed to provide some additional funding ($8.8 million) to undertake enhanced monitoring as part of the Project. However, the Project was still some way from having assured budgets and funds that would allow a final investment decision, despite by this time being at an advanced stage of research planning.

Table 1.2: Initial costings for the CO2CRC Otway Project Stage 1 as submitted to potential funders in late 2004–early 2005.

Budget	$ millon
Production and separation of CO_2	6.8
Pipeline	0.7
Injection well	1.5
Monitoring equipment	1.0
Baseline surveys	0.2
Operation of maintenance	0.7
Monitoring	0.6
Planning	0.15
Project management	0.8
Abandonment	0.35
Total	**12.8**
Contributions	
Federal Government	5
Victorian Government	4
Industry	4

This level of financial uncertainty, while not uncommon in a research environment, was new to most of the industry-based CPPL Board, as was the need to operate without a "banker". Research managers are used to "living dangerously" and existing from one research grant to the next. In industry, project managers have access to project funding through their parent organisations and have considerable flexibility to modify specifications, change schedules, and allocate resources to achieve best outcomes for the project. In the case of the Otway Project, much of this flexibility did not exist. It was seen by CPPL as a "real project in an unreal funding environment"! Funds were often tied to specific milestones or goals or available only from grants in specific fiscal periods, or in some instances funds were tied to particular project activities. This restricted CPPL's ability to make changes to get the best operational outcomes, as it was up to CO2CRC to seek out the extra funds required, or make cuts elsewhere in the research programme in order to act (through CMPL) as the project "banker".

One of the major potential costs proved to be for the CO_2 separation plant. Buttress production well tests on the composition and characteristics of the source gas revealed a higher methane content than expected. While the difference was modest, it posed some design, engineering and cost issues to remove the excess methane. In order to execute the operations in a timely and cost effective manner, it was decided to build a simplified surface plant (see Chapter 4). Given that the CO_2-rich gas did not contain any hydrogen sulphide or mercury, it was concluded that the injection of the mixed gas would be possible and would not in any way compromise the research objectives of the Project related to monitoring and verification. This decision resulted in major cost savings.

Early estimates of drilling costs proved overly optimistic. There were several reasons for this: first, the period 2004–07 was a time of massive inflation in costs related to all resource projects. Second, there was a shortage of rigs due to a major increase in drilling activity particularly in Queensland. Third, the repositioning costs proved to be far in excess of what was estimated. Could these budgetary uncertainties have been better handled? Not really, to the extent that the escalation in costs was outside the control of CO2CRC. It was also the period prior to the 2008 Global Financial Crisis, when it was difficult to retain staff (in a booming economy) when they were being offered ever higher salaries by other employers. The programme of research could have been cut, to save funds, but then that would have defeated the whole purpose of the Otway Project. It was therefore decided to seek additional funding.

The industry participants in CPPL agreed to provide an additional $4 million, the US Department of Energy (via NETL and LBNL) was able to fund some of the Otway activities, as was the Korean Geological Survey (KIGAM). The Federal Government, through the CRC Programme,

provided some supplementary funds in 2007 for Otway Stage 2 but this also helped with some Stage 1 costs. Finally, CO2CRC undertook a major reassessment of its programme priorities in all areas, in order to reallocate additional funds to Otway. Together, these measures provided sufficient funding to take Stage 1 forward and lay the foundations for Stage 2.

By early 2007, the revised operational budget was on a firm basis, with accurate capex (capital expenditure) and opex (operating expenses) costings of approximately $22 million (Table 1.3). This represented an escalation in

Table 1.3: Stage 1 budget estimates for 2007 and final costings at completion.

Description	Revised Total Budget (Oct 07)	Forecast Final Cost At Completion
Naylor-1 well	800,000	799,919
Monitoring and verification	891,000	892,176
CRC-1 well	4,829,000	4,822,183
Buttress-1 well	571,000	565,133
Pipeline	1,527,000	1,526,871
Process plant	2,793,000	2,928,693
Permits/licenses	252,000	252,043
Process group	1,796,000	1,810,185
Project management	1,905,000	2,005,077
Abandonment	900,000	900,000
Opex total	1,450,000	1,440,000
Ops contingency	201,277	−9,704
Total up	**17,915,277**	**17,942,280**
Scope change	75,000	325,433
Management (legal/bank fees, etc.)	641,000	590,000
Operations (regulatory/landowner permits, etc.)	971,000	729,000
Tenements	2,655,000	2,655,000
Total operations (including scope change)	**22,257,277**	**21,906,576**
CRC Executive OBPP	2,049,000	2,086,000
CRC Geoscience	1,216,000	1,246,000
CRC M&V personnel	1,496,000	1,496,000
CRC M&V research	2,267,000	2,467,000
Monitoring atmosphere	620,000	620,000
Monitoring geochemistry	785,000	877,000
Monitoring geophysics	862,000	970,000
CRC outreach and risk	181,000	327,000
Research and legal contingency	244,000	0
Total science	**7,453,000**	**7,622,000**
Total operations and science	**29,710,277**	**29,528,576**

costs of $9 million or approximately 70%. These figures proved to be close to the final figures (Table 1.3). The research and science management costs were of the order of $7.5 million (Table 1.3) but were more elastic than the operational costs. The in-kind contribution from the research organisations, based on the 50:50 formula applied by the CRC to all its research costs, would have added an additional $7.5 million to these costs. In addition a significant part of the Otway effort was not "booked" by the researchers, either because it was undertaken outside normal hours or because they were too busy! Further, there was significant reallocation of activities such as communications away from more "general" CCS communication to communications specifically targeted at the Otway Project, because the Project provide an unparalleled real-world opportunity for communication. So, rather than the figure of $327,000 shown in Table 1.3 for communication and outreach, that figure is likely to be a million dollars or more over the life of the Project. Finally a great deal of effort was not booked against the Otway Project in the first couple of years when the Project was still at a very preliminary stage. Taking all this into account, a more realistic figure for the cost of Otway Stage 2 science was of the order of $20 million, bringing the total cost of the complete Otway Project Stage 1 to in excess of $40 million.

Was the budgeting for Otway Stage 2 better, i.e. more accurate, than for Stage 1? One of the consequences of a change of government in 2007 was that it introduced a significant element of uncertainty into the CRC Programme, which was not finally resolved until late 2009, when it was agreed to fund a new phase of CO2CRC for the period of 2010–15. This then allowed the Centre to more confidently plan for the next stage of the Otway Project. By this time, the Project team was of course far more aware of the actual costs of drilling a well, moving a rig, or of installing surface plant. In addition better systems were in place for cost control. Further, a realistic amount (20% or more, depending on the item or activity) was set aside for contingencies.

In the case of Stage 2A, even though CRC-2 was shallower than CRC-1, the cost of drilling CRC-2 was significantly higher ($8 million) largely because far more coring was undertaken and the well completion was more complex.

Operations expenditure and facilities management cost approximately $1.2 million and the real cost of the science (including approximately $500K of in-kind support) was of the order of $1 million, bringing the total cost of Stage 2A to approximately $10 million (Table 1.4).

Stage 2B did not include any drilling, but did include some complex downhole operations with extensive new surface plant (Figure 1.11) and innovative field experiments (see Chapter 17), which meant moving into relatively unknown territory in terms of budgeting and costs. Consequently once again the final costs were rather different to the preliminary budget. Total operations and related capital expenditure were approximately $15 million. The scientific activities related specifically to 2B cost approximately $3.2 million in cash, but if in-kind contributions are included total science expenditure is approximately $6 million. Therefore the complete cost of Stage 2B operations and science was an estimated $23 million. The estimated costs of all the Otway activities to 2013 is $74 million (Table 1.4). The estimate currently available for the cost of the proposed Stage 2C is $15 million.

There are several observations that should be made about these costs. First, they should be regarded as indicative, with uncertainties arising in the split between categories, particularly costs incurred in the early evolution of Stage 1. Accurately capturing the scientific costs was particularly difficult because some research was relevant not only to Otway but also to non-Otway related CCS research and any split between the two was at times quite arbitrary. There was also at times difficulty in making a precise split between capex and opex, particularly during Stage 2B, when complex surface construction and operations

Table 1.4: Summary project costs for Otway Stage 1, Stage 2A and Stage 2B in A$ millions.

Stage	Capex	Opex	Science	Miscell	Total
Otway 1	13.2	3.4	20	4.2	40.8
Otway 2A	8.0	1.2	1.1	–	10.3
Otway 2B	5.3	9.8	6.4	1.2	22.7
Total	**26.5**	**14.4**	**27.5**	**5.4**	**$73.8**
Otway 2C (early estimate)					$15

tended to be a continuum, rather than discrete activities. A significant part of the science costs were in the form of in-kind contributions from universities and research bodies; CO2CRC was able to capture its own cash costs for that research, but the in-kind costs were not always accurately booked by the contributing organisation. Finally and not surprisingly, there was no hard and fast boundary between the various stages of Otway. Some of the work undertaken during Stage 1 contributed to Stage 2A, which in turn contributed to Stage 2B.

Despite these limitations, the numbers are very useful and, as far as is known, provide some of the first publicly available breakdowns of costs for a successfully completed storage project. They indicate that, in this particular case, capex and science costs were about equal and opex was about half the cost of the capex or the science. Obviously where a project is able to use a pre-existing well, then there are significant savings, although as demonstrated by the problems encountered in using the pre-existing Naylor-1 well, this can also lead to significant compromises. It should of course always be borne in mind that costs will be very site-specific. Drilling costs were relatively high in the Otway Basin compared to many parts of the United States, for example. It was necessary to purchase the petroleum tenements in order to proceed with the Otway Project, at significant cost to the project. Conversely, that provided the Project with access to the large quantity of CO_2 needed for Stage 1; to have purchased that CO_2 from the nearby commercial CO_2 plant would have cost approximately \$13 million. Assuming Stage 2C goes ahead, then there will be further savings of \$6 million on CO_2 costs. Therefore Otway acquired a very significant asset of long-term research significance, through the original purchase of the Buttress Field.

One of the costs for which provision were made by CO2CRC, but which do not show in the figures provided here, was the cost of abandonment and remediation of the site. When Stage 1A was first mooted, provision was made for this, and was initially set at \$900,000. In 2009, at the start of Stage 2A this was increased to \$1.2 million. However, subsequently the scope of the project expanded, more wells were drilled and costs escalated. At the present time, based on more accurate costings, \$3.5 million has been set aside for abandonment and remediation, assuming

an abandonment date of 2020. It is possible the site will be abandoned before that date. Equally well, there may be activities beyond that date which could further delay the need to expend any money on abandonment or remediation. The other uncertainty is the extent to which these costs might ultimately be offset in part, or wholly, by the sale of assets, such as the compressor, but also including the petroleum tenements. Conversely the Otway site offers many opportunities for future geosciences research and society could perhaps be better served by keeping the site for that purpose.

It is also important to point out that the Otway Project probably did not make sufficient provision, in terms of cost and time, to fully write up all activities at the end of each stage. Nor was there adequate provision to curate all the data and ensure ongoing accessibility. Most if not all other projects suffer from the same deficiency, which, it has to be said, is a shortcoming of research funding systems in many instances. Finally, as pointed out earlier the funding system that applies to most research projects often leads to a start-stop-start approach to activities, while the next tranche of funding is sought. The Otway Project was better than most in this respect in that the CRC system under which it was funded guaranteed funding for 7 years. A shortcoming was that there was no provision for inflation in that funding, which posed significant difficulties to the Otway Project during the period 2003–10. However, perhaps the greatest difficulties during this time were totally outside the control of CO2CRC, namely the Global Financial Crisis in 2008, and changes in policy associated with the change in government in 2008–09. Indeed the final observation that can be made is that, despite these changes, CO2CRC was able to successfully deliver Australia's first CO_2 storage project, a testament to the robustness of the systems that were in place to manage the operations and the science required to meet the objectives of the CO2CRC Otway Project.

1.6   Designing the Otway Project

Developing a CCS project needs to follow established project development practices, with defined decision points to enable the project to proceed in a staged fashion. An important premise, assuming that an assured source of

CO$_2$ has been identified, is that, in order to get sufficient investment confidence along the whole chain, it is necessary to have an up front investment in finding and de-risking the storage element. This de-risking requires early screening via desk-top studies and numerical modelling, but these alone are unlikely to give the information required, or to sufficiently de-risk the project to allow a final investment decision. Therefore, acquisition of new field data to address key residual risks and uncertainties is required. This is a critical path activity in seeking a licence to store CO$_2$.

Development of a storage (and transport) scheme can be divided into two investment periods. The first of these is a preliminary "speculative" exploration or "identify and assess" phase aimed at finding and assessing a site sufficiently to justify investment in later phases (i.e. a stage-gated capital investment project). The second stage focuses on transport and needs to be concurrent with the development of the companion capture project. In the case of the Otway Project a very deliberate decision was made to first obtain the source of CO$_2$ and examine the feasibility of transport, before examining the storage options. However, it has to be also pointed out that enough information was gleaned at an early (data room) stage to know that there were several feasible storage options in the vicinity.

A generic project development process is shown in Figure 1.12. Each phase is associated with an increasing level of capital exposure and ends with a decision gate which, depending on technical, commercial, economic and regulatory confidence, will yield a STOP, REVIEW or GO decision.

1.6.1 Identifying and assessing the Otway opportunity (2004–05)

The early part of the first phase typically comprised desk-top compilation studies aimed at creating a number of options for further in-depth investigation. For a site to "qualify" for a project, numerous sub-elements of the site (geotechnical, economic, commercial, environmental political) must all work. At Otway, these were all subject to consideration of risk and uncertainty. Prior to 2004, an Australia-wide study of sedimentary basins, conducted by the CO2CRC predecessor APCRC (see Section 1.1), assessed regions (Rigg et al. 2001) for their suitability for the safe, long-term storage of CO$_2$.

Subsequently, a number of potential sites for a pilot project were considered and the Otway site ranked highly due to the proximity of depleted natural gas fields and high CO$_2$ fields. The latter could serve as a source and allow for demonstration of storage in the depleted fields (Section 1.1).

In the absence of legislation governing CCS and the long-term liability associated with storage, cooperation from the Victorian State Government was sought regarding defining the legislative framework for the project. To address asset ownership, including ownership of the petroleum leases containing the Buttress-1 and Naylor-1 wells, and on-site operational liabilities and to undertake all operational aspects of the Otway Project, the special purpose company, CO2CRC Pilot Project Limited (CPPL), was established in November 2005 (Section 1.2).

Hand in hand with defining the property rights, a number of other processes were initiated which were critical for developing the project:

> an "asset team" was formed, led by a Programme Manager and staffed by people with the appropriate geotechnical skill set

> a contracting strategy was designed to allow specific project tasks to be contracted

> early public discussions and community consultations started referring to the project

Figure 1.12: The generic project development process used for the Otway Project.

as a "potential" one should all the approvals including the support of the shire and community be obtained

> project governance and approval processes were defined

> a draft budget was prepared and funding gaps defined to allow additional capital raising efforts.

1.6.2 Select concept: feasibility through to FEED (2005)

In contrast to the previous more speculative phase, which could have resulted in null findings (hence the need for more than one option), the "select concept" phase was expected to result in a higher level of confidence and therefore justified higher levels of investment.

The project was also breaking new ground in terms of testing the existing regulatory frameworks and finding solutions to the long-term liability associated with the storage of CO_2. In consultation with the Victorian Department of Primary Industries (DPI) and the Victorian Environment Protection Agency (EPA), a process to permit the Project was proposed using the Research Development and Demonstration (RD&D) approval provision under the Victorian *Environment Protection Act*, while recognising that more comprehensive legislative cover would be necessary in the future for any commercial geosequestration projects. This provision covered projects (such as the Otway Project) that were limited in scale, duration and environmental impact. Therefore it provided a simple mechanism to develop and test new technologies with legal certainty through the issuing of an RD&D approval. This is discussed in some detail in Chapter 3. It was agreed that conditions of approval would be established between the EPA and CO2CRC, underpinned by a set of key performance indicators (KPIs). Once the monitoring results demonstrated that the agreed KPIs had been met, then the organisation would be considered to have met all of its obligations under that approval (Table 1.5).

The Project work scope was divided into four project phases (Table 1.6), which reflected the focus of the Project on storage and related monitoring activities. This matrix provided a legislative pathway and gave certainty to permitting the Project for activities ranging from working-over and drilling wells, to constructing surface facilities and laying pipelines. It also provided access mechanisms to enable the CRC to conduct monitoring activities and negotiate with relevant landowners.

Table 1.5: The key performance indicators (KPIs) required by the Victorian Environmental Protection Agency.

Phase	Key Performance Indicator (KPI)
1	1. Establish injection and migration models and uncertainties
1A	2. Environmental impacts within SEPP bounds 3. Injection/migration within model prediction bounds
2	4. Verified stable plume within model prediction bounds a. Measurements show no evidence of CO_2 beyond secondary containment in Naylor-1 b. Air samples collected over deep water wells show no evidence of injected CO_2 c. Air samples collected over Naylor-1 (over a few days) show no evidence of injected CO_2 5. Appropriate decommissioning certificate from authorities a. Wells decommissioned as per regulation b. Sites restored as per regulation
3	6. No evidence of injected CO_2 over 2 years a. Air samples collected over deep water wells show no evidence of injected CO_2 b. Air samples collected over Naylor-1, (over a few days) show no evidence of injected CO_2
4	7. No evidence of injected CO_2 over 2 years a. Air samples collected over deep water wells show no evidence of injected CO_2

Table 1.6: Otway Stage 1 showing the various phases from pre-injection to post-closure.

	Phase 1A Pre-injection 2005–07	Phase 1B Injection 2007–08 approx.	Phase 2 Post-injection Post-2008 approx.	Phase 3 Post-closure 2009+	Phase 4 Longer term
Surface activities	Plant, gathering line, baseline monitoring	Atmospheric, seismic, geochemical monitoring	Atmospheric, seismic, geochemical monitoring, closure	Surface monitoring	Surface monitoring
Legislation	Petroleum	Petroleum (production) EPA (injection)	Petroleum, EPA	EPA	EPA
Risk management	Insurance	Insurance	Insurance	TBA	TBA
Subsurface activities	Well operations, new well drilling and completions, logging	Injection, well operations, M&V	Logging and sampling, well operations, M&V	None	None
Legislation	Petroleum, EPA	Petroleum (production) EPA (injection)	Petroleum (closure) EPA (M&V KPIs)	EPA	EPA
Risk management	Operational: insurance Reservoir: operational control	Operational: insurance Reservoir: operational control	Operational: insurance Reservoir: operational control	Operational: TBA Reservoir: TBA	Operational: TBA Reservoir: TBA

A conceptual plant design was constructed based on an estimated 90% CO_2 content of Buttress. This involved distillation and refrigeration columns to purify the CO_2 to 97% and use of available methane for co-generation (see Chapter 4).

Multiple reservoir geo-models were constructed in PETREL™ and provided the basis on which reservoir simulations were undertaken (Chapter 5). Formation properties were estimated from the basic wire-line log interpretations of the Waarre C in Naylor-1 and Naylor South-1. History matching was performed using the available production wellhead pressure data from Naylor-1 and a reasonable match was achieved (Sharma et al. 2007). These models allowed further development of the permitting conditions with the EPA, including agreement on the KPIs.

Risk assessment of the Otway Project was carried out at all stages, considering both the engineered and natural systems. The engineered systems consisted of the wells, the processing plant, and the gathering line while the natural system included the geology, the reservoir, the overlying and underlying sealing formations and the groundwater flow regimes (see Chapter 8). Overall, it was considered that risks associated with the capture, transport and injection components of the project were well understood by the oil and gas industry, with robust engineering design methodologies, established procedures

and continuous improvement management practices. Conversely, risks associated with the long-term storage of CO_2 were considered to be an active area of research. A quantitative risk assessment was performed using the RISQUE method and it was concluded that the Otway site was capable of achieving a proposed benchmark of 99% containment of the injected CO_2 for 1000 years (Bowden and Rigg 2004) in the target reservoir. The site was therefore considered acceptable on that basis (see Chapter 8).

A comprehensive monitoring and verification concept plan was developed to support the primary project objectives to safely transport, inject and store CO_2, and in addition to safely decommission facilities and restore all disturbed surface sites. The main drivers for the monitoring and verification plan were based on site characterisation, risk assessment and meeting the requirements of the key performance indicators agreed with regulatory authorities as part of the project approval (see Chapter 9).

The geotechnical and engineering works helped identify the key uncertainties so that a new data acquisition programme could be planned. Key uncertainties in Buttress were around the gas composition and pressures, which needed to be resolved before a target injection well location could be finalised.

Concurrently, a range of project planning tasks continued during this phase including:

> the contracting strategy was approved and a lead contractor selected on the basis of qualification conditions

> public discussions and community consultations continued and always included members of the Victorian Government

> a project risk register was constructed

> budget and schedules were refined, allowing for a new data acquisition and uncertainty reduction plan.

1.6.3 Define: detailed engineering design to FID (2005–06)

The aim of this phase was to reduce uncertainty and finalise the project design before entering into the expensive execution and construction phase.

A well test was performed at Buttress and the fluid composition and reservoir pressures were determined. Given the results of the pressure tests, confidence increased in the field being able to supply the 100,000 t of CO_2 needed for the Project. The fluid composition analysis determined that the CO_2 content was lower than anticipated (79%); this had major implications on the plant design and an iteration of the initial design concept was undertaken. The options are considered in detail in Chapter 4, but can be summarised as:

> base case plant design (distillation/refrigeration to inject 97% purity CO_2)

> alternative plant design: dehydration and compression option (while not much water production was expected during the project operations phase, it was recognised it could be an issue over time)

> common aspects to both of the above, relating to wax drop out and potential blockage of tubulars etc. (considerations from a process perspective were likely to add to the cost, i.e. injecting a pore

point depressant (ppd), setting up a pipeline for pigging, heat tracing and jacketing on pipe)

> injection of gas (as produced from Buttress) straight into Naylor (given the pressure differentials; it was expected that injection would be possible for the first few months, with subsequent support through a compressor or multiphase pump).

Following a detailed evaluation (see Chapter 4) and budgetary considerations, it was decided to inject the Buttress gas directly into Naylor. As the gas was not dehydrated, the 2″ pipeline was constructed from stainless steel. The process envelope for the systems at early time, mid time and late time was constructed to ensure that the compressor specifications were selected to ensure dense phase transport. Detailed engineering commenced around this option and the compressor was sited as close as possible to the Buttress well.

Reservoir saturation and well seismic surveys were recorded in Naylor-1 to determine the reservoir pressures and fluid contacts. The offset seismic data were used to try and map the gas-water contact spatially as well as vertically. During this process it emerged that the original history match was not as predicted, as the reservoir pressure was higher than anticipated. This led to a broader study and a dual aquifer simulation model being developed. A good history match was obtained and this became the basis for selecting the new (CRC-1) injection well target location, at a distance of around 300 m along the line of maximum dip (south-east) and with an anticipated CO_2 breakthrough time of between 4 and 9 months at Naylor.

The M&V plan was updated and baseline data acquisition of water wells and soil gas commenced. The well completion for Naylor-1 (to allow it to function as a monitoring well) was designed and the injection well design was finalised. All technical work was peer reviewed by the industry support advisory group under the CPPL governance system.

Concurrently, a range of project planning tasks continued during this phase:

> landowner lease agreements were finalised; there were difficulties with one particular landowner,

requiring the Victorian Government to carry out a compulsory acquisition (see Chapter 3)

> a health, safety and environmental management system (HSEMS) was developed for the Project and mapped across the system used by the lead contractor to ensure all elements were covered

> project approvals were obtained from the relevant Victorian State Government authorities

> public discussions and community consultations continued and always included members of the Victorian Government (see Chapter 3)

> the project risk register was updated

> budget and schedules were refined and adequate contingency built in for field activities.

The end of this phase was the Final Investment Decision (FID) which was contingent on technical success and economic and regulatory confidence in delivery. This received Board approval in late 2006.

1.6.4 Execute, construct and commission (2006–08)

This phase was broken into individual sub-projects relating to the drilling of the injection well, completing Naylor-1 as a monitoring well and constructing and commissioning the compression station and the injection pipeline.

Each of these sub-projects had their own unique challenges. The injection well CRC-1 had to be pre-terminated in the Eumerella Formation (but well below the target reservoir interval in the Waarre Formation) as there were concerns regarding hole stability. Fortunately this contingency had been anticipated and the project objectives were not adversely impacted. Good quality log and core data were obtained, allowing the picking of the injection target and the refining of the reservoir model. The Naylor-1 completion was extremely challenging as a lot of instrumentation was deployed into a small borehole in a live gas well. In addition to the custom design of the completion itself, an enormous amount of effort was required to define adequate safety procedures for conveyance of the instrumentation into the well (see Chapter 12).

Wet weather impacted on the pipeline construction and boring machines had to be deployed to complete certain sections. There were delays in getting power to the site; this in turn impacted on the plant schedule. Commissioning took longer than initially expected as additional vibration tests and process-related changes, such as providing thermal cladding to exposed pipes, had to be performed. The delays totalled around 3 months, with the site ready for injection at the end of March 2008.

Additional project planning tasks continued during this phase, including:

> refining the monitoring plan based on the updated reservoir model; this included finalising the selection and the quantities of tracers to be injected

> developing a Journey Management Plan to the site to manage the increased traffic

> developing an operations and Emergency Response Plan for the site

> designing a visitors' centre and making provision for future educational needs

> obtaining project approvals from the relevant Victorian State Government authorities

> public discussions and community consultations continued

> updating the project risk register

> refining of budget and schedules at the end of the construction phase in preparation for the operations phase.

1.6.5 Operate (2008–09)

Commencing in March 2008, CO_2-rich gas was injected into the Waaree C Formation through the injection well CRC-1, and a full range of monitoring and verification activities were carried out (see Chapter 10). The operations phase was planned to last for 2 years with a target volume of 100,000 t of CO_2 injected. However, after 18 months of injection, it was considered that the project objectives had been met, with around 66,000 t of gas (approximately 60,000 t of CO_2) injected.

Under the initial approvals the CRC tenure would have expired in July 2010. Consequently the monitoring and verification programme was structured to ensure that the Project objectives were satisfied within that time frame. The site would then have been decommissioned following the cessation of injection. However, funding for CO2CRC was extended to 2015 and a second stage (Stage 2) of the Otway project was launched, retaining the current assets for ongoing use. Decommissioning was thus deferred and its budget was carried forward on the CRC balance sheet (see Section 1.5).

1.6.6   Operations relating to Stage 2

As pointed out in Section 1.5, there were significant delays in the period 2008–09, due to a change in government and related funding uncertainties for CO2CRC. Consequently it was not until very early in 2010 that the new operational stage could be commenced and a new well (CRC-2) drilled. However, it is important to note that monitoring activities related to the Stage 1 injection continued after the Stage 1 injection had been completed. Indeed at the time of writing (January 2014) monitoring activities related to Stage 1 are still underway. In other words the ongoing nature of the research has meant that there was no sharp and well-defined end to Stage 1 activities.

Otway Stage 2 was based on the idea of injection into the Parattee Formation, with the objective of understanding trapping mechanisms in a saline aquifer. The research programme was developed by CO2CRC and its Members and refined through assurance reviews by the sponsoring companies. Stage 2 was structured to ensure that Stage 1 was not in anyway compromised. Managing the Project expenses in-line with the available grant funding required that Stage 2 was divided into separate phases. This was not ideal from an operational perspective but did allow the work to progress in a staged manner that was logical, though not necessarily the most cost-effective way to proceed, Therefore Stage 2 was constructed as a series of building blocks with clearly defined research objectives, with the potential to be enhanced beyond 2010. Programmatically, the Stage 2 effort was broken into discrete stages aimed at fulfilling specific objectives of the overall research programme.

The stages were as follows:

Scope objective

2A Characterisation of Parattee:

> detailed site characterisation with support from experimental data with laboratory determination of relative permeability and geophysical properties from cores obtained from new well (CRC-2).

2B Residual saturation tests:

> investigating residual saturation processes using a Huff-n-Puff (inject/soak/back-produce) type of CO_2 injection testing methodology.

2C Limited volume injection:

> demonstrate that injection in an unconfined aquifer is safe and can be monitored reliably, that the CO_2 is residually trapped and ultimately dissolves.

Initially Stage 2B and 2C were considered as part of a single test design, using a notional 10,000 t of CO_2. However, as the theoretical work advanced it became evident that the investigation of residual processes could be undertaken with smaller quantities of CO_2 than 10,000 t (Zhang et al. 2011). While this simplified the 2B process, the volumes injected for 2B would not be sufficient for 2C from a surface detectability perspective using seismic techniques. As a result, Stage 2C was defined as a separate stage to enable planning to progress, based on a larger scale "seismic-detectable" injection of CO_2.

Stage 2A was a relatively straight-forward activity (Figure 1.13(a)) focused on drilling and completing the new well (CRC-2). Consequently the pathway followed for Stage 1 and the drilling of CRC-1 was followed quite closely and the CRC-2 well was successfully drilled and cored in January–February 2010 under a water licence, with approvals from Southern Rural Water and the EPA (Singh and Steeper 2011). At the same time approval in principle was also given for Stage 2B and Stage 2C.

Stage 2B involved major new surface installations (Figure 1.11), sophisticated downhole engineering (Figure 1.13(b)) and complex science that was unlike anything previously undertaken at the Otway site, or indeed anywhere else (Chapter 17). Throughout 2010 and into 2011, major new

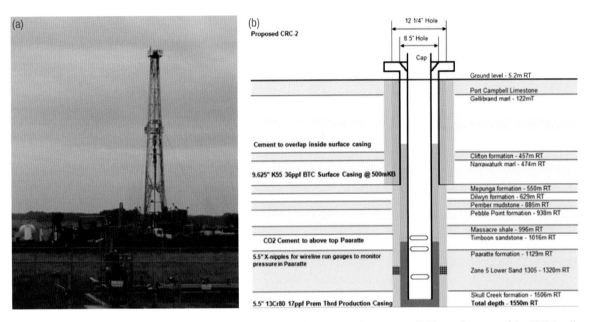

Figure 1.13: (a) Drilling of the CRC-2 well by Hunt Energy using Moduspec Rig, February 2010. (b) The configuration of the CRC-2 well.

Table 1.7: The approvals required for Stage 2 activities.

Activity	Approval/Permits	Regulator	Application Process
Production of CO₂	Production plan	DPI	- *Petroleum Act 2000* (DPI)
Compression and transport of CO₂ 1. Plant (compressor) 2. Gathering line 3. Other facilities (centre, etc.)	Planning approval, gathering line approval	DSE, DPI, Moyne Shire, DOI	- *Petroleum Act 2000* (DPI) - Ministerial Amendment request of the *Planning and Environment Act 1987* (Moyne Shire/DSE) - *Exemption of Pipeline Act 2005* (DPI) - *Cultural Heritage Act* (DPI) - Compensation agreement: consent to land access - Project of Significance and Compulsory Aquisition (DOI) - Exemption of Rural Fire Service (CFA)
Drilling of new well	Drilling licence	SRW	- Submit EMP, SPM and drilling plan. Well being drilled underwater well licence.
Injection of CO₂ (CRC-2)	Disposal approval	SRW, EPA	- *Water Act 1989* Sections 76 and 67: Application for approval to dispose of matter by means of a bore - Compensation agreement: consent to land access
Storage of CO₂	Storage approvals	EPA	- *Environmental Protection Act 1970*: Research development and demonstration (RDD) approval (EPA)
Monitoring and verification 1. Atmospheric 2. Water wells 3. Downhole (Naylor-1) Monitoring	Planning approvals, compensation agreement, DSE access rights	EPA, DSE, Moyne Shire	- Ministerial amendment request of the *Planning and Environment Act 1987* (Moyne Shire/DSE) - Consent to use (SOBN) bores (DSE) - Compensation agreement: consent to land access

surface facilities were established at the site, with the actual field experiments undertaken between June and September 2011. Because of the novelty of the experiments there was a need for particularly close collaboration between the researchers, the operators and all levels of management, as well as with the regulators (Table 1.7). Nonetheless much the same project development and approvals template was followed as in Stage 1 (Figure 1.12).

1.7 Project liability and risk

1.7.1 Categories of risks and liabilities

The issue of risk associated with as storage project is considered in some detail in Chapter 8. There is general agreement that at well characterised CO_2 storage sites, there is a low level of risk. Nonetheless the issue of how to handle (and equitably share) risk and liability was a major concern to the Board; it had a major impact on the final structure of CO2CRC and related entities and also resulted in a significant delay in reaching agreement amongst the stakeholders.

It was important that all aspects of risk and liability were considered in formulating the Otway operations. The nature of the risks associated with the Otway Project was considered to result in potential liability in two broad categories:

> liability associated with long-term storage of sequestered CO_2

> operational risks and liabilities associated with engineering activities including design, construction, and operation of the project and management of the petroleum tenements.

The former category was novel in the case of Otway, because no statutory regulations existed to define the boundaries of compliance and the initial issue confronting the Project was to understand how this category of risk and potential liability should be managed. On the other hand potential for liability arising from operational activity was a risk that was very familiar to petroleum and mining operators; there were well established methods of dealing with this risk.

1.7.2 Liability associated with sequestered CO_2

Companies routinely take on liability arising from projects, but sequestered CO_2 is assumed to be stored indefinitely so that the company carrying out the sequestration process is unlikely to still be in existence over comparable time frames. Indeed, CPPL was formed as a special purpose vehicle only for the duration of the Project. The simple solution to the issue, as far as the Project was concerned,

would have been for government(s) to assume this liability in the public interest. However, difficulties arise in this approach, partly because of the time frames involved. The legal position in terms of sequestered CO_2 was complex as indicated by the comprehensive legal advice obtained by CPPL at the time, but in practical terms a successful project had to be based around credibility and reputation.

The proposition then put forward was that if the company gave the government certain undertakings in respect to the performance of the Project and was able to deliver those outcomes, then the government should treat the tenements and regulatory matters in much the same way as a normal petroleum tenement, allowing relinquishment at project completion. This was accepted as the way forward.

Table 1.8: The Hazardous Operations/Hazards Identification (HAZOP/HAZID) compilation for Otway Stage 1.

No.	HAZOP/HAZID/SIL/Risk Assessment	Date
1	HAZOP of commissioning procedures	1/08/07
2	HAZOP of tracer chemical injection report	8/08/07
3	Naylor-1 workcover and CRC-1 completion HAZID and risk	8/08/07
4	CO2CRC Project—detailed HAZOP follow-up report	9/08/07
5	Buttress well test CO_2 dispersion assessment	30/05/06
6	Buttress-1 well test HAZOP	9/06/06
7	HAZARD risk register	14/11/06
8	Wellhead ESD SIL assessment	2/03/07
9	CO2CRC Project—operational phase HAZID and risk assessment	5/12/06
10	Gathering line construction HAZID worksheets Close out report	4/04/07
11	Compressor station construction—HAZID workshop	3/04/07
12	CO2CRC Project—operational phase HAZOP	5/12/07
13	CO2CRC Project—detailed HAZOP report	5/04/07
14	CO2CRC Project—plant SIL assessment	17/04/07
15	Buttress and Naylor well work—HAZID and risk assessment	19/05/07
16	CO2CRC HAZOP shop drawings	29/11/06

1.7.3   Operational risks and liabilities

Operational liability was a much more familiar issue for companies and governments to deal with compared to the sequestration issue. The risks were those associated with conventional engineering design, construction and operations. A review of the Project showed that none of the activities was novel in the context of petroleum or mining activities. Consequently conventional mitigation measures were utilised in the Project. In summary these included establishment of typical industry based systems and processes including a HSE system (Table 1.9) and reporting protocol, conventional HAZOP/HAZID analysis (Table 1.8), peer review of all major technical and engineering assumptions, and a system of financial budgets, controls and reporting requirements.

From its inception the Otway Project was developed in a stage-gated fashion (Figure 1.12), assessing the risks and developing mitigation plans in accordance with the ALARP (as low as reasonably possible) principle. The process followed was to use the widely accepted qualitative risk matrix approach (Chapter 8). Project risks were considered across the range and included those associated with producing, transporting and storing CO_2 safely. The plant

Table 1.9: Comparison of systems to ensure HSE conformity between CPPL and the main contractor.

Main Contractor \ CCPL	Risk management	Standard for HSE planning	Training and competency	Contractor management	Document management	Communication and consultation	Conduct of work, management and performance standards	Environmental standards	Management of change	Disaster planning	Incident reporting	HSE audits	Management review and evaluation
Management system policy, objectives and planning		■											
Organisation and responsibilty								■					
Information management and document control					■								
Control of service provision				■									
Risk assessment and risk management	■												
Communication and consultation						■							
Resources management			■										
Contractor and support services				■									
Procurement						■							
Design, construction and commissioning							■						
Control of operations							■						
Maintenance, inspection and testing							■						
Change management									■				
Emergency response										■			
Incident/hazard/non-conformances reporting and investigating											■		
Managing materials, waste and discharges								■					
Decommissioning and abandonment							■						
Audit, management review												■	■

and pipeline construction, injection well construction and CO_2 production and injection procedures were developed in-line with the established processes from the oil and gas and chemical industries.

Standards were established for HSE and a bridging document (Table 1.9) was prepared with the principal contractor to ensure that their HSE management systems were in compliance with those developed by CPPL and required to deliver the project.

The subsurface storage risk assessment process used at Otway was a less well defined process and an ongoing area of research (Chapter 8). The Otway Project built on a process developed by the APCRC, using a technique based on developing risk categories through expert solicitation and subjecting the assessed probabilities and consequences through a probabilistic Monte-Carlo process (Bowden et al. 2001; Bowden and Rigg 2004). This allowed the ranking of the Otway site compared to others in Australia and also estimated the project risk against a target of being able to retain 99% of the injected CO_2 in the target reservoir for 1000 years. The quantified risk assessment (QRA) exercise confirmed that the selected site in the Waarre reservoir was adequate to meet these metrics (Chapter 8).

Once the risks were identified and control actions defined, mitigation was managed through a few principle mechanisms:

> operational control by CPPL: this ensured that a rigorous process was in place from planning through to operation to minimise risks. Peer reviews and audits were instituted to ensure robust plans and operational readiness was thoroughly tested through a pre-start audit prior to injection. HSE was allocated the highest priority and, as such, site-specific journey management plans and emergency response plans were created and tested

> insurance for all key operational elements to ensure that well control, property loss and third party liability-related risks were adequately covered

> financial control processes to ensure that, in a limited budget, project flags were raised early enough if cost escalation were to occur.

The project processes and systems worked as planned and the Project was delivered safely without a single lost time incident.

Considerable attention was also paid to reputational issues especially in dealings with governments, landowners, and the local community. While all of these matters were pursued in some detail, a high level risk assessment conforming to the Australian Standard was also developed at Board level to highlight and manage the key risks for the Project. The objective was to identify any matters that might be "showstoppers" and focus attention on managing those risks so that project integrity was not threatened.

1.7.4   The process prior to injection

While the management of operational risks was an ongoing matter, the risk associated with injection received special attention because of its novel nature and the requirement to have a unanimous Board decision.

Prior to the formal CPPL Board decision to carry out injection of the gas at Otway, a pre-start-up review was held in December 2007. This was conducted at a time when almost all construction activities were complete and pre-commissioning activities were in progress. The primary outcome of this meeting was to agree a commissioning plan that would enable the hard target of the official opening on 2 April 2008. Along with this was the need to assess the readiness of the CO2CRC plant for the injection of Buttress gas and check that systems, processes and procedures were in place to ensure that plant integrity could be maintained at all times and that the scientific objectives could be met. This required that there was confirmation of the following:

> facilities were constructed as per the design requirements

> design integrity had been maintained during the construction phase

> flow assurance controls (design, procedures, competencies) were in place

> pre-commissioning checks demonstrated system integrity

> operations staff were sufficiently competent to commence normal running of the facility

> HSE management systems were in place including an HSE case and emergency response capability

> information management systems were in place, including "as built" drawings and vendor information

> operations philosophy had been complied with

> any imperatives had been complied with.

Overall, the audit confirmed that all of these were met although it did identify the need for immediate attention be given to (1) HAZOP close out, (2) operations recruitment and training, (3) high pressure sampling at the Naylor monitoring well and (4) development of a schedule for start up.

As a separate but related exercise, a comprehensive assurance checklist was compiled to ensure that all key risk areas had been addressed. This was summarised with document links to archived files that validated each of the principal check items. This list defined the items, requirements, status, and links to documents that validated these.

The categories covered on this list included:

> regulator/government approvals/licenses

> long-term liability (legal advice and review)

> EPA approvals

> subsurface assurance

> quantitative risk assessment

> risk mitigation

> Otway site closure plan mapping out CPPL's approach to risk minimisation

> decommissioning and rehabilitation

> operations

> HSE requirements

> Project obligations

> Project due diligence

> heritage survey

> Project schedule update/key decision points

> key performance indicators

> cost estimate and budget

> community consultation.

Supporting this in some detail were documents defining the project phases, key performance indicators as required by the EPA as part of the approvals process, the Board responsibilities, any safety incidents, the HAZID/HAZOP register and a legal obligations register. In addition, a QRA register was continually updated throughout the life of the Project. Together these processes enabled Otway 1, Otway 2A and Otway 2B to be safely and successfully undertaken.

1.8   Conclusions

The idea of undertaking a small-scale Australian injection project was developed in 2002–03 and proceeded along a non-linear path for several years as various challenges were met and overcome, before injection finally commenced in 2008. The Project that was successfully delivered in 2008–09 was in fact close to the original concept in terms of the science. But the manner in which it was achieved underwent many changes. Notable amongst these was the need to:

> establish a special purpose operating company

> work very closely with a range of state government bodies because of the lack of a specific regulatory regime for CO_2 storage

> develop innovative arrangements for dealing with long-term liability

> establish clear protocols for scientists, contract staff and visitors to ensure a safe working environment

> recognise the necessity to press ahead with planning, despite the many financial uncertainties that arise in a research environment

> ensure risk and uncertainty were understood, documented and managed as an integral part of the project

> plan to maximise the long-term research opportunities offered by the Otway site

> accept that in a research environment it is often necessary to start planning for the next stage of research before the current stage had been completed

> compromise where necessary to ensure scientific activities can be undertaken despite budgetary limitation, provided this does not impact on science quality and credibility or on key scientific objectives.

Chapter 1 documents how many of these issues were successfully handled over the life of Otway Stage 1 and into Stage 2. The remainder of this volume provides the detail of what science was done, how it was done and what was learned.

1.9   References

Bowden, A., Lane, M. and Martin, J. 2001. Triple bottom line risk management: Enhancing profit, environmental performance, and community benefits, Wiley, USA.

Bowden, A. and Rigg, A. 2004. Assessing risk in CO_2 storage projects. APPEA Journal 44.

Cook, P.J., Cinar, Y., Allinson, G., Jenkins, C., Sharma, S. and Soroka, M. 2011. Enhanced gas recovery, CO_2 storage and implications from the CO2CRC Otway Project. APPEA Journal 51.

Cook, P.J., Rigg, A.J. and Bradshaw, J. 2000. Putting it back where it came from: Is geological disposal of carbon dioxide an option for Australia? APPEA Journal 40(1), 654–666.

Cook, P.J., Causebrook, R., Michel, K. and Watson, M. 2013. Developing a small scale CO_2 test injection: Experience to date and best practice. IEAGHG Report.

Dance, T., Spencer, L. and Xu, J.-Q. 2009. Geological characterisation of the Otway Project pilot site: What a difference a well makes. Energy Procedia, 2871–2878.

Dodds, K., Daley, T., Freifeld, B., Urosevic, M., Kepic, A. and Sharma, S. 2009. Developing a monitoring and verification plan with reference to the Australian Otway CO_2 pilot project. The Leading Edge 28(7), 812–818.

Frederiksen, L. 2002. Buttress-1 Well completion report. Prepared for Santos-Beach joint venture. (Unpublished)

Hennig, A., Etheridge, D., de Caritat, P., Watson, M., Leuning, R., Boreham, C. and Sharma, S. 2008. Assurance monitoring in the CO2CRC Otway Project to demonstrate geological storage of CO_2: Review of the environmental monitoring systems and results prior to the injection of CO_2. APPEA Journal 48.

IPCC 2005. Special report on carbon dioxide capture and storage. Prepared by Working Group 3 of the Intergovernmental Panel on Climate Change [Metz, B., O. Davidson, H.C. de Coninck, M. Loos, and L.A. Meyer (Eds.)]. Cambridge University Press, Cambridge, United Kingdom and New York, NY, USA.

Jenkins, C.R., Cook, P. J., Ennis-King, J., Underschultz, J., Boreham, C., Dance, T., de Caritat, P., Etheridge, D. M., Freifeld, B.M., Hortle, A., Kirste, D., Paterson, L., Pevzner, R., Schacht, U., Sharma, S., Stalker, L., and Urosevic, M. 2011. Safe storage and effective monitoring of CO_2 in depleted gas fields. Proceedings of the National Academy of Sciences of the United States of America (PNAS) 109, E35–E41.

Rigg, A.J., Allinson, G., Bradshaw, J., Ennis-King, J. Gibson-Poole, C. Hillis, R.R., Lang, S. and Streit, J.E. 2001. The search for sites for geological sequestration of CO_2 in Australia: A progress report on GEODISC. APPEA Journal 2001, 711–726.

Sharma, S., Cook, P.J., Berly, T. and Anderson, C. 2007. Australia's first geosequestration demonstration project—the CO2CRC Otway Basin Pilot Project. APPEA J 47:257–268.

Singh, R., and Steeper, T. 2011. The CO2CRC Otway Project: Completion of CRC-2. PESA News Resources, 26–28.

Underschultz, J., Boreham, C., Dance, T., Stalker, L., Freifeld, B., Kirste, D. and Ennis-King, J. 2010. CO_2 storage in a depleted gas field: an overview of the CO2CRC Otway Project and initial results. International Journal of Greenhouse Gas Control 5(4), 922–932.

Zhang, Y., Freifeld, B., Finsterle, S., Leahy, M., Ennis-King, J., Paterson, L., and Dance, T. 2011. Estimating CO_2 residual trapping from a single-well test: Experimental design calculations. Energy Procedia 4, 5044–5049.

2

Tony Steeper

2. COMMUNICATIONS AND THE OTWAY PROJECT

2.1 Introduction

Communications have been very important to the success of the Otway Project. CO2CRC was acutely aware that, as the first geosequestration project in Australia and one of the few active projects in the world, public perceptions of CCS technology were likely to be greatly influenced by the outcomes of the Project. In addition, it was considered essential to win and retain the support of the local community. Any local community opposition would foster negative media coverage of the Project, as well as potentially affect Project approvals. To address these concerns, communication strategies have been in place throughout the life of the Project with the aims of enabling the Project to proceed, fostering good relations with the local community and landowners, and communicating the outcomes of the research to the wider community, in Australia and around the world.

CO2CRC undertook community consultation for the Otway Project following the principles that communication of risks and benefits should be open and transparent, should provide objective information based on good science, and should also provide channels for two way communications with the community and other stakeholders. Over time, the project has become CO2CRC's most important communications asset and has given the organisation invaluable opportunities for showing CCS in action, dispelling myths and communicating CCS technology to the local community, government, industry, researchers and the general public. It has also acquired significance at the national and international level over the past 6 years as the only accessible and public CCS storage project in Australia.

2.2 Strategic communications and the Otway Project

Developing a practical, adaptable and well-planned communication and consultation strategy is key to any successful major project. The Otway Project communication strategies have been adapted and revised over three main project stages.

A communications strategy was first developed in early 2004 (Anderson 2004), covering the first informal meetings by the Chief Executive and the Communications Manager with landowners in the immediate vicinity of the Otway Project to inform them of very preliminary plans to undertake the Project and listen to their views on what was proposed. The first, more formal stage occurred once it was clear that the Project was likely to go ahead, and involved research into the makeup of the local community, the main stakeholders in the project and the regulators. Consultation continued during 2004–05 with key landowners and involved negotiation of agreements, the first public announcement of the Project in 2005 and a large public information meeting to introduce the concept of the research to the local community later that year.

A second and more comprehensive strategy was developed in March 2006 (Anderson 2006a) covering:

> steps leading to the launch of the Project and commencement of injection, including

  – considerable media engagement

  – formation of the project community reference group

  – regular public community reference group meetings

  – face-to-face briefings with landholders and other key stakeholders

  – development of a regular project update newsletter

  – incorporation of Otway Project communications into CO2CRC's wider communications strategy.

In July 2006 social research (Anderson 2006b) was conducted to inform the community consultation programme. This research was very important for gauging community understanding of CCS, understanding any concerns about the Project and clarifying preferred channels for future communications. This research informed ongoing communication activities. Following the provision of additional funding for CO2CRC in 2010, a revised communication strategy (Steeper 2012a) was developed

to inform the community of the new status of the site as a facility for ongoing CCS research. This strategy has been regularly updated to reflect new communication demands on the Otway Project arising from the successful research of Stages 1 and 2, ongoing community engagement activities such as newsletters, public meetings and open days, and the visits programme.

New elements included in the past three years have been social media initiatives, crisis communications planning and issues management, growing resources including an extensive image, diagram and video library (http://www.co2crc.com.au/imagelibrary) and publicly available research papers and reports. In addition, a second round of social research was conducted in 2011 (Steeper 2012b) to inform the ongoing community consultation programme, evaluate previous communication tactics and assess any changes in community perceptions since 2006.

Several aspects of Otway Project communications provide useful lessons for similar projects.

2.2.1   Clear goals and objectives

While CO2CRC pursued wider key messages on CCS through organisational communications, the main Otway Project communication goal has been to build and maintain good long-term relationships with key landholders and neighbouring communities. With this focus in mind, CO2CRC set objectives that were considered to be achievable, effective and measureable. Communication objectives are not always measurable, but social research provided an important evaluation tool.

2.2.2   A good understanding of the community

Nirranda South in south-west Victoria is primarily an area of dairy farming, with a tourism industry associated with the Great Ocean Road which lies a kilometre to the west of the site (Figure 1.2). It is an area that has been the subject of a lot of gas exploration over the years; this brought some unique challenges to community consultation. Understanding this community, its organisations, networks and concerns was critical, particularly in the early stages of the Project.

Figure 2.1: The CO2CRC Chief Executive explains a technical point to a community consultative meeting at the Otway Project visitors' centre.

Issues to contend with in these early stages included a dispersed community, poor existing relationships between some landholders and the oil and gas industry, the need to negotiate landholder agreements, over-expectation of some landowners regarding compensation payments for use of land, the potential for conflict between project and farming operations, local media management, an over-consulted community, resourcing of consultation processes and opposition to development of any kind.

Social research early in the Project played a key role in understanding the community, identifying issues and informing communications tactics. Social research also provided a useful baseline for measuring change in community attitudes to the Project and to CCS more generally.

2.2.3 The right messages for each audience

At the Otway Project, understanding went hand in hand with developing targeted key messages and appropriate communication channels. Once audiences were well understood, key messages were developed for each group.

This approach ensured that messages were relevant, succinct and positive, and addressed strategic objectives. Understanding audiences and developing messages helped define the communication channels and tactics that were needed. Again, social research was useful in this process.

The first that landowners knew of the Project was in early 2004, when the Chief Executive accompanied by the Communications Manager visited all landowners in the immediate vicinity of the site to advise them of the proposal and the likely time frame. At that stage, the purchase of the assets necessary for the Project and issues such as regulation still needed to be resolved, but it was considered essential to ensure that the landowners received the first information on the potential project from CO2CRC and not from some other source. It was also considered important for them to have the opportunity to meet the CEO in order to emphasise the importance of the Project and the fact that the approach being taken was not only open but also non-hierarchical. Subsequently, the approach to the community focused on information sharing—informing key stakeholders not only about the Project and CO2CRC more broadly, but also about how CCS worked and why it was needed.

This was, and continues to be, an ongoing challenge for CCS proponents, researchers and governments but was especially so in the early years of the Otway Project. This kind of early information sharing included:

> one on one consultations with key landholders, the local indigenous community, the police and the Country Fire Authority

> public meetings with the Nirranda community including question and answer sessions

> questionnaires for the Nirranda community

> regular meetings with Moyne Shire Council (the relevant local government body)

> briefings for the Corangamite and Warrnambool Councils (adjacent districts)

> discussions with the Victorian State Member of Parliament for the district

> briefing the Federal Minister for Industry and the Federal Opposition's Industry spokesman

> informing, via email and offered briefings to:

  – local State Upper and Lower House MPs

  – NGOs (e.g. Environment Victoria, ACF, CANA, WWF)

  – Victorian Senators

  – the Federal Member for the area

> development of fact sheets, diagrams, web content and an Otway Project DVD (http://www.co2crc. com.au/imagelibrary2/vid_otway.html) distributed to stakeholders and the media.

Media liaison during this period was frequent, culminating in the official, nation-wide announcement of the Project's official opening and the commencement of CO_2 injection in April 2008. This in turn led to extensive coverage in Australia and worldwide, including numerous radio, newspaper and TV articles and interviews (Anderson 2008).

Once injection was underway, it was important to integrate the community consultation programme with operational concerns. The work of the community liaison officer was, and continues to be highly valuable in this regard.

2.2.4   High quality publications

Making information available on CCS and the Otway Project has been an ongoing and important aspect of communications for CO2CRC. Peer-reviewed scientific journal articles were and are a very important interface with the scientific community in Australia and internationally. However, there are significant challenges in communicating complex science to the public that required a range of formats, channels and levels of detail. CO2CRC has run an active and comprehensive website (http://www.co2crc. com.au/) on CCS since 2003 and on the Otway Project specifically since 2004. The site has been a valuable resource for information and includes publications such as fact sheets, brochures, posters and newsletters (http://www. co2crc.com.au/publications/brochures.html). An image library (http://www.co2crc.com.au/imagelibrary/) was also established to give access to high quality photographs, and diagrams explaining CCS processes and the Otway Project; video footage was also developed for the media.

Images have been vital for explaining CCS to audiences and much effort has gone into presenting the science in clear and easy-to-understand formats including diagrams, still and video photography and animation. It is important that such information is scientifically accurate and this is a real strength of CO2CRC, where graphic designers work closely with expert researchers to ensure accuracy. CO2CRC has put considerable effort into two Otway Project video productions (http://www.co2crc.com.au/ imagelibrary2/videos.html) to set out the science of CCS and the operations undertaken at the site. The animation aspects of these videos in particular have been effective in explaining the science of CO_2 storage, trapping and monitoring and have been widely used by the media and at scientific conferences.

CO2CRC publishes an Otway Project Research Update (http://www.co2crc.com.au/publications/newsletters. html) for the Nirranda and surrounding community. This newsletter is written in Plain English, covers local events and operations at the site and is used to promote community reference group meetings. The newsletter is published at least twice a year and is letterboxed to over 1500 residents in the vicinity of the project site. CO2CRC also produces regular newsletters including CO2 Futures, (http://www.co2crc.com.au/publications/newsletters.html) aimed at CO2CRC stakeholders and the wider public (currently over 1200 subscribers), which updates readers on progress across the CRC, including the Otway Project.

2.2.5   Community reference group

An outcome of early communications was the development of a community reference group in 2006. This group was established to:

> encourage engagement through two-way feedback

> identify and address perceived risks

> monitor community attitudes

> avoid conflict by developing areas of collective agreement

> provide ongoing information on the science of the project.

The group included representatives from each key stakeholder group:

> key landholders

> local government (Moyne Shire) and neighbouring councils

> regulators (EPA, Southern Rural Water)

> NGOs, including the indigenous community

> State Government

> Country Fire Authority.

The group met regularly during the most active operational phase to hear updates from project researchers and operations personnel, and to have the opportunity to ask questions on project progress. Meetings were advertised widely in local papers and in regular project newsletters, and were run at times most conducive to landowner

attendance. This same approach continued during Stage 2 of the Otway Project. The meetings provided an important channel for two way communication as well as giving the community the chance to talk to researchers directly involved in the Project. The Chief Executive frequently attended these meetings, showing the community that CO2CRC is taking consultation very seriously at all levels.

2.2.6   Otway Project liaison officer

Initial communication with the community was undertaken from Canberra but once the Project was confirmed and the level of field activity increased, it became very apparent that there was a need for a locally known liaison officer based in the community. The aim was to provide a go-to person for the community—someone who understood the local issues. The Project was fortunate to be able to appoint a person who lived locally, who had previously been a schoolteacher and who knew the families in the area—an ideal person for the job. This approach has subsequently been vindicated by social research carried out in 2011 which showed that a dedicated Otway Project liaison officer, based in the community, was a key factor in establishing and maintaining good relations with the landowners/occupiers immediately surrounding the project site (see later discussion on the role of the liaison officer).

2.2.7   Site visit programme

A major advantage of running an active demonstration site in an accessible area, less than 3 hours from a major city and an international airport, has been the ability to conduct tours of the Project for key audiences and decision-makers. These visits took various forms but in general were aimed at providing the opportunity to engage one on one with visitors, allowing researchers to explain the science and the Project, and showing CCS in action. This face-to-face engagement with stakeholders has been invaluable for communicating how CO_2 storage works and for dispelling misunderstandings, including being able to show that the footprint of the Project is quite modest. While demand for tours from national and international groups has required considerable resources it has been money well spent and the Otway Project remains CO2CRC's (and Australia's) most valuable and effective communications tool for CCS. The location of the Project near some of Australia's most spectacular coastal scenery (the Port Campbell National Park is approximately 30 km from the site) was an added bonus in that it provided cliff sections through some of the sediments encountered in the uppermost part of CRC-1, plus the opportunity to explain some of the geological processes involved in the storing and trapping of CO_2. Exposures of the weathering profile were also used to explain the difficulties encountered in undertaking onshore seismic surveys at the site. More recently, where time allowed, visits were made to the nearby Iona gas storage plant to explain gas storage from the perspective of an analogue for CO_2 storage. Finally, while not strictly part of the storage "story", the natural occurrence of CO_2 in the Otway Basin was explained to visitors through visits to the volcanic cones in the area (the CO_2 is of volcanic origin) and to the nearby Boggy Creek CO_2 plant. In other words the opportunity was used to provide the broadest possible perspective of the geological setting of the site. In this respect the Project was ideally located compared to most other active CCS project sites which are either offshore or in geologically featureless areas or

in remote locations. It has become apparent that the Otway Project has a further advantage over most other projects, in that the site is leased by and run by the research organisation and that, apart from operational constraints and the necessary limitations imposed to ensure health and safety, CO2CRC encourages and is able to handle visits by many researchers and members of the community, in a way that would be difficult if not impossible where there are commercial constraints and drivers.

2.3 Social research and the Otway Project

One of the advantages of the Otway Project has been the ability to assess community consultation through social research over a 5 year period. Social research in 2006 established a baseline, allowing a second round of research in 2011 to measure changes in attitudes, assess areas for improvement and acknowledge successes.

Independent social research (Steeper 2011) was commissioned by CO2CRC and undertaken by Quantum Market Research between March and July 2006 to inform the consultation process, monitor community attitudes to the Project and geosequestration, and to provide the community with additional opportunities to comment on the Project. The research used both qualitative and quantitative research methods. The quantitative section of the research was a telephone-based Community Perceptions Survey that surveyed 300 people in the region (Corangamite, Warrnambool and Moyne shires, plus Glenelg, Ararat Rural City, Southern Grampians and Colac-Otway Shires). While results showed that climate change was an important issue to most people (93%), only 32% had heard of geosequestration and most knew very little about it. A need for more information was the most obvious outcome. About 50% of respondents felt comfortable about the Project taking place in their region. 53% were worried or concerned about geosequestration. Safety and leakage of CO_2 were the main concerns (71%), with potential impact on groundwater a particular question for many of the farmers in the vicinity.

Figure 2.2: The community liaison officer, Josie McInerney, meets with one of the local dairy farmers to outline future plans for the Project.

The qualitative research was undertaken to explore the issues and attitudes of the community in relation to geological CO_2

storage and the Otway Project. The study consisted of two 1.5–2 hour focus groups in Nirranda and Warrnambool, and five 45 minute in-depth interviews among Nirranda landholders comprising four landowners and an indigenous community leader. Early attitudes toward the Otway Project varied across the three audiences and ranged from being apathetic to being engaged and wanting to know more. There were those who were positive or advocates for the research project and others who were opposed to or rejecters of the Project with grounds for opposition ranging from environmental concerns to an unwillingness to accept the level of compensation available for access to land (based on the formula applied to the oil and gas industry). Most, however, were open to learning more.

Overall, the community in Nirranda, Warrnambool, and in 2006 the surrounding areas, were not convinced that geological storage of CO_2 was a viable option for greenhouse gas mitigation. While some were positive and hopeful, they appeared to be outnumbered by those who needed more information before they could make up their minds about the technology. The participants, particularly those who took part in the focus group discussions, had a multitude of queries and concerns related to geosequestration and all facets of the Project which prevented them from forming a solid opinion. In summary the results from the research were as follows:

> Participants in the social research, particularly those who were involved in the Nirranda and Warrnambool discussion groups, were keen for further community consultation.

> The community had been heavily consulted on other projects (mostly related to oil and gas) in the past—results indicated consultation should be conducted on an as-needed basis only; community meetings and quarterly flyers/newsletters were preferred forms of consultation.

> It was deemed important for CO2CRC to make information about the research project clearly available and transparent to the community but to allow the community to initiate engagement rather than being too intrusive.

> In communicating messages to the Nirranda and surrounding communities, it was thought important that CO2CRC be clear, concise and factual—residents wanted to hear facts, not spin.

In 2011, CO2CRC commissioned a second round of social research (Steeper 2012b) on the Otway Project, also undertaken by Quantum Market Research, using a similar approach and methods to the initial study. The overall aim was to see how the consultation programme was tracking and evaluate changes over time in attitudes and perception. While the study followed a similar format to that used in 2006, using both qualitative and quantitative research methods, the 2011 research took a more locally focused approach, primarily assessing the views of those who live close to the site—Nirranda South, Port Campbell and Timboon—and in particular the landowners immediately around the Project. These results were contrasted with surveys in a regional centre, Warrnambool.

Results from the research showed a markedly increased awareness of CCS since 2006 (increasing from 32% to 69% of respondents over the 5 years), especially in the vicinity of the Otway Project, where awareness increased to 74%. 80% of respondents were positive or neutral about using CCS to reduce CO_2 emissions. Residents in the area felt more comfortable with the presence of the Project over time (48% in 2011 versus 33% in 2006), with locals more comfortable than regional communities. Those who felt better informed felt much more comfortable with the Project (70% for the well informed versus 41% who didn't feel well informed). The research also found that project/landowner relationships had greatly improved over time. CO2CRC was considered to be proactive with information and in dealing with issues that arose. Landowners felt that researchers had developed greater respect for their land and actively sought to minimise disruption to farming operations. A large part of this result can be attributed to the dedicated liaison officer. The research also found several areas where project communications could be improved and these were incorporated into the 2012 communications strategy (Steeper 2012c).

2.4 Operational issues relating to communications and the community

In 2007 CO2CRC recognised that a locally-based liaison officer was essential for effective communications with the community, in particular the local landowners and occupiers.

The officer was required to be:

> aware of local issues and networks

> the first point of contact for the community/landowners

> able to build good relationships/rapport with all stakeholders

> perceived as trustworthy and unbiased

> easily accessible and visible around the community.

The role entailed:

> contact with landowners on a regular basis, through face to face meetings, by telephone and by correspondence

> providing a single point of contact and a consistent approach when dealing with issues

> learning about farming and the ways the Project impacts on farming operations

> understanding the Project and its aims, and communicating Otway Project research and operations in a way that could be understood by the local community and landowners

> keeping stakeholders informed of the need and reasons to access property

> working closely with the site operator (AGR/ Oceaneering) on site visits, issues and the work programme

> providing tours of the Project for a wide variety of interested groups and stakeholders

> acting as a guest speaker for community groups.

A key aspect of the role was to act as the intermediary at the landowner/researcher interface. Regular monitoring surveys on and adjacent to the site required regular access to neighbouring properties. The liaison officer worked with project operations and management on implementing the monitoring programme, arranging approvals from landowners for access to properties for the various surveys. Issues with landowner relationships and seismic surveys led to the need for the liaison officer to develop a protocol for ensuring researchers had minimal impact on farming operations and that mutual respect was a major factor in how the regular monitoring surveys were conducted. Amongst the Otway monitoring activities, seismic surveys involved the greatest level of disruption to farming activities, with access to land sometimes requiring the removal of fences, some impact to the surface of the pasture and a requirement to move cattle from the paddock where the survey was being undertaken. As a result, it was necessary to ensure that seismic operations were only conducted during the months when there was no harvesting underway and in a manner that ensured minimum impact on farm operations. The Otway Project experience clearly showed the barriers that ongoing and frequent 2D and 3D seismic surveys are likely to encounter in many onshore CCS operations—and the need to develop less intrusive monitoring methods.

Researcher Guidelines for Surveys (McInerney and Steeper 2013a; 2013b) were developed with input from

Figure 2.3: The community liaison officer, Josie McInerney, escorts some international visitors around the Otway Project site.

landowners. The protocol was completed by all researchers prior to beginning any field operations and continues to be an essential component of ongoing operations. It has also been an important method of educating researchers regarding required behaviour when on a landowner's property, serving to emphasise that good relations with surrounding landowners and occupiers is vital to ongoing project success.

Visits, tours and events at the Otway Project have been extremely popular from day one and conducting official visits has been a significant part of the liaison officer's role. The Project has generated a great deal of interest around the world, with over 1000 official visitors touring the site from 2008 to 2013. Visitors have included researchers, industry groups, Australian and international government groups, students, community groups and the media. The liaison officer also plays an important role during periods of very busy operations, such as drilling or research operations, by coordinating site access and providing local knowledge in areas such as locating accommodation for visiting researchers.

As a team, the communications manager, liaison officer and project staff have run regular events, including community reference group meetings and open days. While a formal schools programme was beyond the resources of CO2CRC, the liaison officer contributed to the environmental curricula of regional schools, providing information and arranging tours for local teachers.

Figure 2.4: International students attending the 2010 IEAGHG CCS Summer School visiting the Otway site are listening to the sound of CO_2 travelling along the pipeline to the injection well at CRC-1.

2.5 Conclusions

The familiarity of residents near the Otway Project with petroleum exploration and production had positives—drilling and gas storage had been conducted safely for many years—and some negatives, in that not all experiences had been good and larger operators were in some cases able to provide a higher level of monetary compensation for land access for seismic survey acquisition than the Project was able to provide. The community consultation process set up by the Project followed the values and best practices of the International Association of Public Participation and was successful (Ashworth et al. 2010). There was no one element that was responsible for that success. However, factors that undoubtedly contributed to the success included commencing community consultation very early in the life of the Project, having a locally-based liaison officer, taking an open approach to communications including providing direct access to scientists and senior management, establishing a community reference group and also adequately resourcing the community consultation component of the Project, viewing it as a vital and integral part of the project rather than an optional add-on.

2.6 References

Anderson, C. 2004. Geosequestration community consultation plan: I, Cooperative Research Centre for Greenhouse Gas Technologies, Canberra, Australia, CO2CRC Publication Number RPT13-436.

Anderson, C. 2006a. Overarching community consultation strategy: Otway Basin Pilot Project, Cooperative Research Centre for Greenhouse Gas Technologies, Canberra, Australia, CO2CRC Publication Number RPT13-4352.

Anderson, C. 2006b. Public perceptions of the CO_2 Geosequestration Research Project CO2CRC Otway Project April 2006. Cooperative Research Centre for Greenhouse Gas Technologies, Canberra, Australia, CO2CRC Publication Number RPT11-2905.

Anderson, C. 2008. Otway Project launch media report. Cooperative Research Centre for Greenhouse Gas Technologies, Canberra, Australia, CO2CRC Publication Number RPT13-4353.

Ashworth, P., Rodriguez, S. and Miller, A. 2010. Case study of the CO2CRC Otway Project. CSIRO report, Pullenvale, Australia, 10.5341/RPT10-2362 EP 103388.

McInerney, J. and Steeper, T. 2013a. Researcher guidelines for seismic surveys. Cooperative Research Centre for Greenhouse Gas Technologies, Canberra, Australia, CO2CRC Publication Number: RPT13-4370.

McInerney, J. and Steeper, T. 2013b. Researcher guidelines for soil gas surveys. Cooperative Research Centre for Greenhouse Gas Technologies, Canberra, Australia, CO2CRC Publication Number: RPT13-4369.

Steeper, T. 2012a. CO2CRC Otway Project community consultation and communication strategy. Cooperative Research Centre for Greenhouse Gas Technologies, Canberra, Australia, CO2CRC Publication Number RPT13-4368.

Steeper, T. 2012b. CO2CRC Otway Project community perceptions research—qualitative/quantitative findings. Cooperative Research Centre for Greenhouse Gas Technologies, Canberra, Australia, CO2CRC Publication Number RPT12-3862.

Steeper, T. 2012c. CO2CRC Otway Project social research: assessing CCS community consultation, Energy Procedia 37(2013), 7454–7461, Japan, 18–22 November 2012.

3. GOVERNMENT APPROVALS

3.1 Introduction

As the inspirational writer Paulo Coelho (1988, p. 22) said in his famous book The Alchemist, if you really desire to achieve something, the entire universe mysteriously unfolds to help you achieve it. This was the case for the process of gaining approvals for the CO2CRC Otway Project, and what was initially challenging gradually started to be resolved, as people realised the importance of this Project to the State of Victoria. As a result of lateral thinking on the regulators' part and the perseverance of project facilitators in the Victorian Government coupled with a close working relationship with the participating scientists and engineers, this Project became a reality. The context of why this Project was a "Project of State Significance" is that Victoria hosts a significant proportion of the world's brown coal resources (430 billion t). However, brown coal produces large quantities of carbon dioxide in the energy generation process (DPI 2010, p. 1). For Victoria to continue to enjoy the benefits of this vast resource, it is vital that the issue of greenhouse gas emissions

from brown coal be addressed as Australia moves towards a low carbon economy.

The Otway Project was initiated by CO2CRC in 2004, with preliminary discussions during that year with individual landowners and officers of the Moyne Shire. The first community meeting was held in February 2006 with approximately 70 people in attendance. Because CO2CRC had purchased the rights to the surrounding petroleum tenements, it had full access to all prior petroleum-related work and this was used as the basis for the preliminary investigations of the site. Initial work at the site commenced in mid-2006 with the flow testing of the Buttress-1 well, approved by the Department of Primary Industries (DPI). The Buttress-1 well was planned to serve as the CO_2 source for the Project, which gives this Project a unique advantage over most other research and demonstration projects throughout the world. In August 2006, the State authorities that had an interest in regulating this Project liaised with each other in developing the process for approving this "first of a kind" project. The key authorities involved at this stage were the Environment Protection Authority (EPA), Southern Rural Water (SRW) and DPI. Throughout this period there were ongoing discussions between CO2CRC management and officers of DPI, to

Figure 3.1: Official opening of the CO2CRC Otway Project, 2 April 2008 (from right to left: Federal Minister for Resources and Energy, Martin Ferguson; Chief Executive of CO2CRC, Peter Cook; Chairman CO2CRC Board, Tim Besley; State Minister for Energy and Resources, Peter Batchelor; CO2CRC Project Manager, Sandeep Sharma).

not only develop the approvals process for the Project, but also to secure financial assistance from the State for the Project. In October 2006, the Victorian Government signed an agreement with the CO2CRC to progressively release $4 million for Stage 1 of the Project. Financial support was also secured from the Commonwealth Government and industry and this enabled the drilling of Australia's first CO_2 storage well, CRC-1, to progress in February 2007, with regulatory approval from DPI.

In June 2007, the Project was elevated to the highest level of priority, by the gazettal of the declaration by the Minister for Planning as a "Project of State Significance". This included the gazettal of the planning scheme amendment to accommodate this "first of a kind" Project on land zoned for farming. In July 2007, the approvals for Stage 1 of the Project from the EPA and SRW were finalised. In December 2007, the EPA accepted the Environment Improvement Plan (EIP), a requirement of the EPA Stage 1 approval, and allowed the commencement of injection at the Otway site. In April 2008, the Project was launched by the Commonwealth Minister for Resources, Energy and Tourism, Martin Ferguson and the State Minister for Energy and Resources, Peter Batchelor, marking the commencement of injection into Australia's first carbon storage well, CRC-1 (Figure 3.1).

In August 2008 (within the time frame predicted by the CO2CRC), the CO_2 injected was first detected at the monitoring well, Naylor-1, a production well inherited from the former tenement holder and equipped with a suite of monitoring equipment as part of the Project's world class monitoring and verification programme. In September 2009, the researchers concluded that the objectives of Stage 1 in terms of monitoring the migration of the CO_2 in the depleted gas field had been achieved with the injection of approximately 60,000 t of CO_2 and therefore injection at CRC-1 was suspended.

In October 2009, Stage 2 of the Otway Project was approved by the EPA and in late January 2010, a second injection well, CRC-2, was drilled for injection testing in a saline geological formation. The drilling of CRC-1 was approved by DPI as it was accessing the Waarre Sandstone, a depleted gas field. However, the drilling of CRC-2 was approved by SRW as the well was accessing the shallower Paaratte Formation, a saline aquifer. Stage 2 was also partly funded by the Victorian Government in an agreement to provide $2 million which was signed in October 2009. By this time, Victoria's new legislation for carbon storage was in place and about to come into operation. However, it was recognised that the new legislation did not fully cover the continuation of the

Otway Project. Therefore, project-specific regulations were made to enable Stage 2 and subsequent research activities to proceed at the site. These regulations came into operation in December 2009. With all the necessary approvals now in place, injection testing at CRC-2 was able to commence in June 2011 and was completed in September 2011. Throughout this period, a number of publications summarised aspects of the Project (Jenkins et al. 2012). Further work at the Otway site is currently under consideration.

3.2 Challenges of regulating a pilot project

One of the most significant challenges put before this Project was how it was going to be regulated to the satisfaction of the community and ensure that safety and environmental requirements were met in the absence of legislation specific to carbon storage. After investigating all the applicable legislation, it was resolved that the best way forward would be to regulate the Project primarily under the research, development and demonstration approval provisions of the *Environment Protection Act 1970*. While this approval was the overarching approval for this Project, there were other approvals required for various sub-activities, such as the drilling of injection wells, which were regulated under the *Petroleum Act 1998* and the *Water Act 1989*.

In regulating the Otway Project, which was a new type of activity for Victoria and indeed for Australia, it was of utmost importance to ensure that the approval process used was valid and enforceable. The research development and demonstration approval provisions of the EPA offered this security. The various approvals used to regulate this Project are discussed in more detail in the sections below. In the absence of legislation dedicated to regulating CCS activities, there was considerable uncertainty as to who should be regulating which aspects of the operation. After a number of meetings between the CO2CRC and the regulators, it was resolved that in principle the EPA would issue the main approval for the Project, but with parts of the operation regulated by DPI and SRW (Table 3.1).

Table 3.1: Coverage of main Project activities by approving authorities (Source: Ranasinghe 2008, p. 29).

Activity	EPA	DPI	SRW
Construction			
Preparation of Buttress-1 for production of CO_2	✓	✓	
Installation of surface plant at Buttress-1	✓	✓	
Construction of the pipeline (gathering line)	✓	✓	
Drilling and preparation of CRC-1 for injection		✓	✓
Installation of monitoring equipment at Naylor-1	✓	✓	
Operation			
Extraction of gas from Buttress-1		✓	
Treatment of gas mixture (CO_2-CH_4) in surface plant	✓	✓	
Transport of CO_2-CH_4 along pipeline	✓	✓	
Injection of CO_2-CH_4 into CRC-1	✓		✓
Monitoring of site before, during and after injection	✓	✓	✓
Closure			
Plugging and abandonment of wells	✓	✓	✓
Removal of infrastructure	✓	✓	
Monitoring after infrastructure removal	✓	✓	✓

✓ Covered by the authority listed above.

It is clear from Table 3.1 that every aspect of the Stage 1 operation was covered by one or more approvals, ensuring that there was full coverage of the Project through project approvals. Furthermore, when there was more than one approval for an activity, a cooperative approach was used by the regulators to ensure that there was no duplication of approvals which had the potential to cause inconsistencies in requirements and/or be unduly onerous on the Project operators.

When regulating a first of a kind project such as this Project, it was important to go through a detailed analysis to ensure that no aspect of the operation was excluded from regulatory approval coverage. This also helped to identify where the boundaries were, in terms of regulatory responsibility in setting conditions on approvals. It also helped to reduce the regulatory burden where overlaps were observed (Figure 3.2).

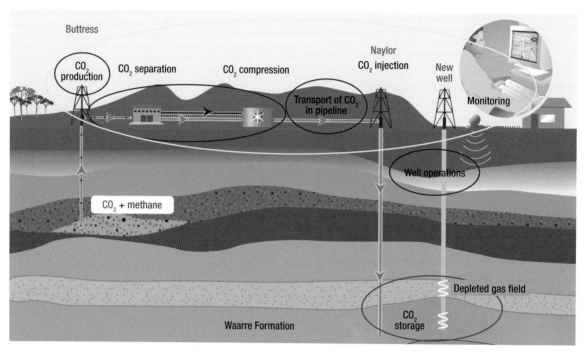

Figure 3.2: Coverage of Project components by regulatory approvals (Source: Sharma 2008, p. 9). Green – EPA and SRW, Red – DPI, Yellow – Planning and Land Use.

3.3   Impact assessment and planning approvals

In Australia, Commonwealth, State and local governments assess if a proposal is likely to impact on its surrounding environment. Generally, if a proposal is identified as potentially having a significant level of environmental or other impact, then a comprehensive process is followed to ensure that any impacts are identified, minimised and managed, by the preparation of impact assessment and management reports and by consulting the stakeholders associated with the Project.

The Otway Project was considered to be unlikely to result in a significant environmental impact; nonetheless, because it was a "first of a kind" project, CO2CRC decided to refer it to the Commonwealth Department of Environment and Heritage in March 2006 under the terms of the *Environment Protection and Biodiversity Act 1999*. In essence, this Act applies a national and international significance test giving consideration to environmental and biodiversity factors, e.g. flora, fauna, ecological

communities and cultural heritage values. In April 2006, the Department published the decision that the Project was not a "controlled action". As the Project was located on farmland, it was considered to pose no risks to environmental and biodiversity factors at the national and international levels and consequently there was no need for the Project to be a "controlled action" or go through a detailed impact assessment process.

A similar process exists at State level but as this Project was a small-scale trial which posed no impacts of State significance, the Project did not trigger the need to be referred to the Victorian Minister for Planning under the *Environment Effects Act 1978* for a decision on whether an Environment Effects Statement was required. In terms of local planning requirements, initially the planning scheme did not recognise carbon storage as an appropriate land use under the existing land requirements. Consequently, the planning scheme was modified to enable the Project to proceed and in June 2007 the decision to amend the Moyne Shire Planning Scheme to accommodate the Otway Project was published in the Victorian Gazette.

3.4   Environmental authority approvals

In principle, any activity which has the potential to significantly impact on the environment should be referred to the environmental protection portfolio area for consideration. Victoria lists these environmentally relevant activities in Schedule 1 of the *Environment Protection (Scheduled Premises and Exemptions) Regulations 2007* which require approval from the EPA. At the time of the development of this Project, Schedule 1 did not include CCS.

Schedule 1 was subsequently amended to include CCS as follows: "carbon sequestration—premises which capture, separate, process or store waste CO_2 for the purpose of geological disposal". However, this amendment was made in 2009 after the approval of Stage 1. Furthermore, the carbon sequestration item in Schedule 1 was then modified to exclude activities approved under the newly introduced *Greenhouse Gas Geological Sequestration Act 2008* which came into operation in December 2009. It is worth noting here that including this item in the regulations did not necessarily solve the problem of how this activity was to be regulated, because environmental considerations did not include other considerations such as safety, land access or liability.

The environmental protection legislation in Victoria allows for the collection of financial assurances for waste disposal facilities. Although the regulations were amended to include carbon sequestration, the financial assurance provisions of the environment protection regulations were not activated for carbon sequestration operations.

As the approval process advanced, it was apparent to the EPA that its regulators needed technical advice from the petroleum regulators at DPI in order to assess the Project proposals and the various reports submitted as part of the EPA approvals process. It was decided the Project would be regulated and approved by the EPA. However, the EPA regulators would seek expert advice from the DPI in assessing and approving the activities proposed as part of the Project. This collaboration between the EPA and the DPI continued to work well throughout the life of the Project. The application for EPA approval for Stage 1 was submitted in November 2006 and approval was given (for a period of 9 years) in July 2007.

The EPA approval for Stage 1 included conditions which related to:

> general requirements such as the need to adhere to environmental policies

> prescriptive requirements such as those relating to noise and liquid storage

> general management requirements

> reporting requirements.

The environmental approval required an Environmental Improvement Plan (EIP) to be submitted prior to injection and the EIP was approved in December 2007. The application for EPA approval for Stage 2 was submitted in May 2009 and the EPA approval was granted (for a period of 6 years) in October 2009. An important step in approving the Project involved the setting of key performance indicators (KPIs) with the EPA. In essence, KPIs were agreed outcomes which the proponent would demonstrate that it had met through the monitoring and verification programme, thereby providing evidence of compliance with the Project approval requirements (Sharma et al. 2007).

3.5   Petroleum authority approvals

The application of the *Petroleum Act 1998* to CCS activities was probably the most complex matter to be resolved. This was because of the perceived ambiguity on whether CCS activities should be treated as petroleum exploration or development activities.

For the purpose of the approvals, the Project construction for Stage 1 involved:

> drilling and installation of the injection well, CRC-1

> construction of the processing plant at the existing production well, Buttress-1

> construction of the pipeline from the extraction well to the injection well

> installation of equipment at the monitoring well, Naylor-1.

Buttress-1 and Naylor-1 were existing wells which were taken over by the CO2CRC in the transfer of tenements from petroleum industry operators. Consequently, the first essential step was for CO2CRC to apply for a transfer of tenements, PPL-11 and PPL-13.

The drilling and installation of CRC-1 was regulated under the *Petroleum Act 1998* as this activity could be considered to be "petroleum exploration" under Section 7 of the *Petroleum Act 1998* by the fact that the gas stream from Buttress-1 contained some 20% methane (CH_4), a hydrocarbon. This activated an exemption from planning requirements under Section 119 of the *Petroleum Act 1998* which enabled the approval for drilling from DPI to be obtained expeditiously. In February 2007, CRC-1 was successfully drilled with approval from the petroleum regulators at DPI. Prior to drilling CRC-1 (i.e. prior to commencing "petroleum operations"), the CO2CRC was required to submit an Operations Plan, in accordance with Section 161 of the *Petroleum Act 1998*, which included an Environmental Management Plan, Safety Management Plan and Geological Prognosis.

Similarly, prior to extracting gas from the Buttress-1 well, CO2CRC was required to submit a Petroleum Production Development Plan, in accordance with Part 5, Division 6 of the *Petroleum Act 1998*, for carrying out petroleum production (in this case, producing CO_2-rich gas) in a petroleum production licence area. However, it was questionable if the activity carried out by CO2CRC actually fell within the definition of "petroleum production" under Section 8 of the *Petroleum Act 1998*. In November 2007, the DPI approved an Operations Plan submitted by CO2CRC covering the extraction and transport of CO_2 but excluding the injection component, which was to be covered by the EPA.

The CO2CRC was exempted from the requirement to prepare a Storage Development Plan, in accordance with Part 5, Division 7 of the *Petroleum Act 1998*, which is a requirement prior to injecting any petroleum into a reservoir in a tenement area, for two reasons: first, it was questionable as to whether the storage proposed in this Project could constitute "petroleum production" as defined in Section 8 of the *Petroleum Act 1998* which specifically stated: "Petroleum production is . . . (b) the

injection and storage of petroleum in reservoirs for the purpose of later recovering it." As there was no intention of recovering the injected material, it was considered safe to conclude that CO2CRC did not need to prepare a Storage Development Plan. Second, it had been confirmed by this stage that the injection activity of the operation would be covered under the EPA approval, addressing environmental and safety aspects with expert advice from the petroleum regulators with experience in approving Storage Development Plans for the petroleum sector.

When the Buttress-1 gas mixture was tested at the early stages of the Project, it was expected that the gas would mainly be CO_2. But upon testing, it was found that the Buttress gas contained some 20% CH_4, which was seen as a potential disadvantage to the Project because the Project was about storing CO_2. However, when it came to regulating the Project, it was an advantage because as with the drilling of CRC-1, the presence of some CH_4 meant that this activity could be readily regulated under the *Petroleum Act 1998*, thus making the approval process more efficient because it was processed by experienced petroleum regulators and also offered access to exemptions.

The construction and operation of the processing plant was regulated by both the EPA and DPI regulators. Although this may appear as an overlap in regulatory roles, this was needed because the EPA considerations focused on the environmental aspects and the DPI considerations focused mainly on the safety aspects of the plant.

The construction of the pipeline from Buttress-1 to CRC-1 could have been regulated in two ways—as a petroleum pipeline under the *Pipelines Act 2005*, or as a gathering line under the *Petroleum Act 1998*. The pipeline could have been seen as falling under the *Pipelines Act 2005* because of the CH_4 content in the gas mixture, given that petroleum is defined in the *Pipelines Act 2005* as: "any naturally occurring or processed mixture of one or more hydrocarbons and one or more of the following: hydrogen sulphide, nitrogen, helium, carbon dioxide or water." In this case, the gas mixture was CO_2 with some CH_4 (a hydrocarbon).

As the approval process under the *Pipelines Act 2005* would have been onerous, given that it was only a short (2 km) pipeline, it was resolved to treat the pipeline as a "gathering line" under the *Petroleum Act 1998*, which is defined (in Section 82) as: "a pipeline which is situated wholly within a production licence area and that is used (or intended to be used) or designed to convey petroleum (or a petroleum product) from one place to another in that area."

The question was then raised whether it was possible to treat this pipeline as a gathering line, considering the above definition refers to "within a production licence area" and the pipeline crossed from one production licence area to another. It was resolved that the pipeline did serve as a gathering line although it straddled two licence areas which lay adjacent to each other and were held by the same operator. Consequently the pipeline was approved as a gathering line under the *Petroleum Act 1998*, which included environmental and safety considerations.

Once again as the petroleum regulators were familiar with approving such pipelines, this approval was processed quickly, taking some of the burden of regulation from the EPA. Energy Safe Victoria was involved with reviewing safety considerations in approving this pipeline.

The other parts of the operation which were approved by DPI regulators via Operations Plans included the installation of new equipment at Naylor-1 and the seismic survey work. As mentioned previously, in Stage 2, CRC-2 was drilled under the *Water Act 1989* but with expert advice from the petroleum regulators in DPI. Once again, the pipeline connecting into the new injection well was treated as a gathering line under the *Petroleum Act 1998*, making the approval process less onerous.

3.6 Water authority approvals

Under Section 76 of the *Water Act 1989*, approval is required to dispose of any matter underground by means of a bore. In the case of the Otway Project, this approval was issued by SRW, which is the authority responsible for protecting the groundwater resources. In July 2007, CO2CRC was issued with the licence for Stage 1 which was valid to 30 June 2010.

In summary, this licence contained conditions relating to:

> monitoring, record keeping and management of quantities being injected

> prevention of pollution

> annual reporting.

This approval could be seen as having an overlap with the EPA approval. However, this approval focused on the protection of groundwater resources and SRW was the organisation which had the most intimate knowledge of the quality and quantity of groundwater resources in the area, thereby making it best placed to consider the implications of injection. SRW considerations were an important part of approving this Project.

The annual fee specified in the SRW licence is calculated on the amount of material injected. But as the *Water Act 1989* did not cater for storage of large quantities of CO_2 (this was a technology not considered during the drafting of the *Water Act 1989*) these charges were waived for this non-commercial research demonstration Project.

In Stage 2, the drilling of CRC-2 was approved under Section 67 of the *Water Act 1989*. The conditions of approval were primarily related to well construction and installation requirements, whereas the injection testing was carried out under the original Stage 1 licence, together with the EPA approval which was the main instrument of approval for this Project.

3.7 Land access and acquisition

Land access matters are amongst the most sensitive issues to document but it is important to capture any lessons from such experiences because land access issues can significantly impact on projects. For the most part, the CO2CRC was able to establish very good relations with landowners in the vicinity of the Project. However, in one particular instance, this was not the case and it was necessary for CO2CRC to obtain project facilitation involvement from government to help resolve the matter.

The problem was that a critical part of the Project (i.e. Buttress-1) was located within land which was privately owned by a landholder who refused to give access to the relevant part of his property (a track approximately 10 m wide) to install a new small diameter pipeline. Under the terms of its petroleum tenements, CO2CRC had control of Buttress-1 and the rights to produce gas from that well. However, it needed to transport the gas by pipeline to the public road, a distance of approximately 200 metres. Landowners in the Project area were conscious of the level of compensation offered by the petroleum sector when their land was accessed under Section 128 of the *Petroleum Act 1998*. However, petroleum operations are commercial operations, whereas this Project was clearly non-commercial and had to operate under a tight budget. While CO2CRC was prepared to pay an appropriate level of compensation, there were budgetary constraints that meant it could not go over and above a "fair" amount, due to it being a non-commercial research project. The overwhelming majority of local people recognised this and were realistic in their expectations. Compensation offers were made to the landholder in question here and negotiations stretched over an extended period without success. Compounding this situation, pressure was building on the Project from its funding providers to adhere to timelines and milestones set in contractual agreements but timelines were slipping as the matter was being investigated.

As the land access options available through the *Petroleum Act 1998* were not appropriate in this instance, the only option left in the absence of landholder cooperation was one where the matter had to be dealt with by the Victorian Civil and Administrative Tribunal (VCAT); this could be a lengthy process and significantly delay the Project as it awaited a decision from VCAT on the amount of compensation payable to the landholder under Section 128(1)(c) of the *Petroleum Act 1998*. Consequently it was decided to investigate other options to resolve this issue.

Covering the Project (or at least part of it) under the *Pipelines Act 2005* was considered, because of the land acquisition provisions in Division 2 of that Act which allowed the Victorian Minister for Energy and Resources to compulsorily acquire land for the purpose of pipeline construction, should a situation such as that being encountered by the

Project were to arise with a landholder. However, the public notification and consultation process under this Act were seen as also likely to be very time consuming for this small Project and consequently the *Pipelines Act 2005* did not seem to offer a satisfactory solution.

After detailed consideration, the only workable option appeared to be under Part 9A of the Victorian *Planning and Environment Act 1987*. Section 201F of the *Planning and Environment Act 1987* allows the Minister for Planning to declare a proposed development to be of "State or Regional Significance" by notice published in the Government Gazette. This would allow the Secretary of the Department of Infrastructure (DOI) to compulsorily acquire land for the purposes of a declared project under Section 201I of the Planning Act. The Minister for Planning may declare specified land required for a declared project to be "Special Project Land" for the purposes of the Victorian *Land Acquisition and Compensation Act 1986*. The process in the *Land Acquisition and Compensation Act 1986* could then be followed by the Secretary of DOI in compulsorily acquiring the land required for the Project where an agreement could not be reached. Although this offered a speedy resolution, it was a highly sensitive matter and was not the preferred approach. Nonetheless, given the inability of the parties to resolve the impasse through a cooperative process, it was felt necessary to resort to a more coercive approach as a last resort. The CO2CRC and its collaborating organisations were firmly of the view that the Project was indeed a project of considerable significance to the State of Victoria and therefore, somewhat reluctantly, felt it had no alternative other than to follow a coercive course of action although it would have much preferred a cooperative approach.

At this stage, a communications plan was developed, including a detailed schedule on how the above process would be followed, and all of the senior officials involved were well briefed on the subject to ensure that they were able to answer any questions on the subject, especially regarding why such a process was needed. In the event, the process went smoothly and was completed in a relatively short time, putting the Project back on track. Subsequently the landowner also received a fair level of compensation, without undue delay to the Project.

In summary, the Project was declared as being of "State significance" and a small portion of the landholder's property was compulsorily acquired by the Victorian Government to cover the pipeline route and infrastructure at Buttress-1, and ownership was temporarily transferred to the Secretary of DOI. Once the necessary portion of the property was acquired, DOI was able to issue a licence to CO2CRC to undertake its research activities, including the production and transport of Buttress gas, but requiring that CO2CRC would conduct its operations taking the landholder's concerns into account. An officer from DOI was allocated to liaise between the landholder and the CO2CRC; this was a valuable step to ensure that any issues between the parties were closely monitored and managed. After the commencement of construction, a compensation amount was determined and paid to the landholder, based on an independent valuer's recommendations.

Owing to the urgency of resolving this issue, Sections 7(1)(c), 106(1) and 26(4) in the *Land Acquisition and Compensation Act 1986* were utilised to expedite the land acquisition process. On this occasion, the matter was resolved speedily without compromising the integrity of the approach. The Project was then able to proceed. In summary, the lesson here was that land access issues need to be taken fully into account right at the start of a project. Furthermore, sufficient time needs to be allocated for any difficult negotiations and indeed, perhaps realistically assuming that there will be at least one difficult issue such as this, to be addressed with each project. It should also be recognised that land access issues are likely to apply not only to the project area, but also to the envelope within which monitoring and verification work will be carried out as shown by some land access issues which arose during the seismic survey work at the Otway site.

3.8   Miscellaneous approvals

In addition to the major approvals discussed in the above sections, there were other authorities involved at various stages of this Project including:

> Energy Safe Victoria, e.g. safety aspects with the gathering line

> Worksafe Victoria, e.g. use of explosives in seismic monitoring

> Country Fire Authority, e.g. the requirement to notify of any potentially fire-generating activities

> Department of Sustainability and Environment, e.g. Crown land and State water monitoring bore access.

Other considerations included Native Title and heritage values.

3.9   Transitional arrangements

The Otway Project commenced in the absence of legislation specific to carbon storage but by the time that Stage 2 was being developed, the *Greenhouse Gas Geological Sequestration Act 2008* had been finalised and was about to come into operation. Upon reviewing this Act, CO2CRC through its lawyers raised the matter that the new Act did not fully cater for the continuation of research activities at the Otway Project. Therefore, project-specific regulations, *Greenhouse Gas Geological Sequestration (Exemption) Regulations 2009*, were made to exempt the Otway Project from the *Greenhouse Gas Geological Sequestration Act 2008*, provided it had approval for its activities under the environmental, petroleum or water legislation. This was an important exemption for the Project; without it, it would have been very difficult for the Project to proceed beyond Stage 1.

3.10   Liability and responsibility

One of the important early considerations for CO2CRC in a pilot project in which there were many collaborating parties was the question of who would take on the responsibility for being the operator of the site for the purpose of obtaining approvals. The unincorporated joint venture was not a suitable entity to hold the approvals which were to be issued to this Project and a company needed to be created for this purpose (see Chapter 1).

In November 2005, for the purpose of creating an entity that could hold approvals to operate the Project, CO2CRC Pilot Project Limited (CPPL) was formed, consisting of a subset of 10 industry members from the original joint venture. The members of this company, CPPL undertook the task of being the operator of a pilot project in a regulatory environment which was not entirely clear and at a time when there was limited clarity on the issue of liability. The members came from oil, gas and coal sectors, were familiar with the risks and liabilities in their own sectors and possessed knowledge which was transferable to this new and emerging CCS sector. This helped them in considering and managing any risks associated with the Otway Project (see Chapter 1).

The members sought to clarify the responsibility matrix early in project planning in order to understand the liability regime which would apply to this Project, before allowing the commencement of any operations. After gaining an in-principle understanding of how liability would apply as the Project progressed, they agreed to proceed with the Project. In summary it was agreed that the operator would hold liability for the Project during construction and operation, and for a period of time when monitoring was carried out for the purpose of demonstrating that the CO_2 has been stored safely and sustainably. After requirements agreed between the operator and the regulator were met and when the tenements were handed back to government, long-term liability would then transfer to the State, but this did not rule out any common law liabilities for the operator.

The following considerations were made in allowing the Project to progress:

> The site was selected carefully to guarantee success and every effort was made to use best practice control measures including a world class monitoring and verification programme.

> The members involved were familiar with the risks associated with their respective sectors and were well placed to assist with minimising and managing any risks associated with this Project.

> The members involved through CPPL would have the operational control of activities at the site to ensure good processes are implemented and any risks managed.

> In the unlikely event that there was an incident, it would not be catastrophic, in that the Otway Project was a small-scale, highly controlled and carefully monitored experiment.

Another important consideration was that the Otway Project would have contractors working on site. Therefore the operators of such projects needed to be mindful that if a contractor caused an incident owing to negligence on the operator's part to educate the contractor of their obligations, then the operator could be held responsible, unless of course the contractor held the licence for the activity it was undertaking which caused the incident. In other words the operator (CPPL) needed to give some thought to how contractors would be managed on site to minimise the possibility of an incident. Specifically, the operator would need to inform the contractors of the regulatory requirements including obligations associated with other stakeholders such as landowners. In the event, the Otway Project was undertaken in an exemplary manner with no serious issues.

3.11   Stakeholder engagement

One of the matters of critical importance to government, especially with a pilot project such as this where perceptions of risks were high, was to ensure that any politically sensitive issues are managed appropriately. The engagement aspects of the Otway Project have been documented by Sharma et al. (2007) and Ashworth et al. (2010), and the Project provided many valuable lessons for other CCS projects through its successes and challenges. The community engagement and communications aspects of the Project are covered in detail in Chapter 2. This section is focused on what was particularly relevant to government.

One of the things which worked especially well in the engagement strategy was the community reference group meetings which were attended by the local community, regulators and project experts. Having the regulators and the project experts in the room with the community was valuable in allaying any concerns the community had in relation to the Project and CCS more generally, as the community could direct their questions to the people immediately involved in the Project.

The local community transitioned from initially being quite sceptical about the Project to later being supportive of the Project and feeling comfortable enough to even answer questions posed by some members of the community.

This was clear evidence that the acceptance of this new technology can be improved through a proper education campaign and access to experts. This has been confirmed in a recent study by Ashworth et al. (2012) for the Victorian Government on understanding stakeholder attitudes to CCS in Victoria.

From a government perspective, one of the highlights of this Project was how the relationship with the media was managed. Both the CO2CRC and DPI monitored the media to capture any issues early and promptly addressed these issues, particularly at critical times such as during the process of compulsory acquisition and the Project launch when project critiques were likely to surface. Regular interactions between the Project and government helped with keeping government spokespeople well informed of any issues which might be raised with them. The CO2CRC also effectively captured any media opportunities, such as meeting project milestones and funding announcements, to convey positive and accurate messages about the Project and more broadly the CCS sector. Effectively managing the media is undoubtedly a crucial part of any such project.

One of the most successful initiatives of the CO2CRC in stakeholder engagement was hiring a locally-based community liaison officer. This officer has made a significant contribution to this Project by keeping the local community and landholders onside and by individually hearing and addressing their issues.

3.12 Conclusions

From a governmental regulatory perspective, the lessons from this Project were:

1. Adequate time and resources should be allocated for the project approvals processes, particularly for pilot projects where the regulatory framework for the project is unclear.

2. Project operators and regulators need to collaborate at the concept phase of the project to clarify the process for approving the project.

3. The project approvals process needs to be appropriate to the scale and the likely impact from the project.

4. Although a CCS project may be approved by environmental regulators, the petroleum regulators have a valuable contribution to make in carbon storage, based on their experience with the petroleum sector.

5. The petroleum industry regulation model has a lot to offer for the regulation of the carbon storage industry and the expertise required for regulating the petroleum industry is transferable to the carbon storage sector.

6. Water authority approvals are an important part of regulating carbon storage projects, particularly where there are freshwater aquifers in the vicinity of the project.

7. Adequate time and resources need to be allocated for any potential land access issues even if a project is a small-scale, non-commercial research project.

8. In developing new legislation alongside projects in operation, the transitional provisions need to provide for projects already in existence.

9. Although, liability is a challenging topic to resolve, discussions must take place early in project planning, to clarify the distribution of liability over time especially for the long term.

10. Stakeholder engagement is a critical part of any pilot project and needs to be planned and managed carefully, including a proactive approach to managing media matters.

In conclusion, the Otway Project presented a number of challenges from a government perspective but the government team involved was able to work through each of these challenges with the project team and together they delivered an iconic project for Victoria which has clearly put Victoria on the map in advancing the research into greenhouse gas technologies.

3.13   References

Ashworth, P., Jeanneret, T., Romanach, L. and James, M. 2012. Understanding stakeholder attitudes to carbon capture and storage in Victoria. CSIRO, Pullenvale, Australia, EP125331.

Ashworth, P., Rodriguez, S. and Miller, A. 2010. Case study of the CO2CRC Otway Project. CSIRO, Pullenvale, Australia, EP 103388.

Coelho, P. 1988. The Alchemist. Harper Torch, New York.

DPI 2010. A principal brown coal province. Department of Primary Industries, Victorian Government, Australia, pp. 1–4.

Jenkins, C., Cook, P., Ennis-King, J., Underschultz, J., Boreham, C., de Caritat, P., Dance, T., Etheridge, D., Hortle, A., Freifeld, B., Kirste, D., Paterson, L., Pevzner, R., Schacht, U., Sharma, S., Stalker, L. and Urosevic, M. 2012. Safe storage and effective monitoring of CO_2 in depleted gas fields, Proceedings of the National Academy of Science of the USA, 109(2), 353–354.

Ranasinghe, N. 2008. Regulating carbon capture and storage in Australia. Masters Thesis, Griffith University, Nathan, Queensland, Australia.

Ranasinghe, N. 2013. Regulating a pilot project in the absence of legislation specific to carbon storage. Energy Procedia 37, 6202–6215.

Sharma, S. 2008. Demonstrating CO_2 storage in the Otway Basin. Presentation at CO2CRC Symposium, Queenstown, New Zealand, 1–28.

Sharma, S., Cook, P., Berly, T. and Lees, M. 2009. The CO2CRC Otway Project: Overcoming challenges from planning to execution of Australia's first CCS project. Energy Procedia 1, 1965–1972.

Sharma S, Cook P, Jenkins C, Steeper T, Lees M and Ranasinghe N (2011), The CO2CRC Otway Project: Leveraging experience and exploiting opportunities at Australia's first CCS site. Energy Procedia 4, 5447–5454.

Sharma, S., Cook, P., Robinson, S. and Anderson, C. 2007. Regulatory challenges and managing public perception in planning a geological storage pilot project in Australia. International Journal of Greenhouse Gas Control 1(2), 247–252.

Generally, applications for approval and approval documents are public documents and can be accessed by getting in touch with the proponent (CO2CRC) and approving authority (e.g. EPA, SRW, DPI), respectively. All Victorian Government Acts and Regulations are available on the Victorian Government legislation website on http://www.legislation.vic.gov.au/ and all gazettal documents are available from the Victorian Government Gazette website on http://www.gazette.vic.gov.au/.

The contents of this chapter are based on publicly available information and compiled solely for the purpose of capturing the learnings from the experience of regulating a pilot project for the benefit of future project operators and regulators. The author disclaims all liability for any error, loss or consequence which may arise from relying on the information in this document. The contents of this chapter are not based on legal advice.

4. DESIGN AND OPERATIONAL CONSIDERATIONS

Craig Dugan, Ian Black, Sandeep Sharma

4.1 Introduction

Developing the most appropriate surface plant and facilities configuration to enable the Otway Project to go ahead in a timely and cost-effective manner, and at the scale proposed, required a great deal of consideration. This chapter summarises the options that were developed and outlines the final configuration that was seen as the best way forward.

CO2CRC Pilot Project Ltd (CPPL), the formal owner and operator, engaged Process Group Ltd, a small company specialising in gas processing (and a founding participant in the CO2CRC), to assist with the evaluation of the CO_2 processing plant design options. Upstream Petroleum (subsequently AGR Ltd; now Upstream Production Solutions) was engaged by CPPL to design the field facilities and undertake the field operations. The design was intended to deliver a fit-for-purpose facility that was

not over-designed, that was at a commercially significant scale (defined by CO2CRC as up to 100,000 t of CO_2) in order to provide a reasonable analogue for future full-scale commercial CCS activities and which would also deliver the scientific objectives of the Project. All this required a close working relationship between AGR, Process Group, CPPL directors, CO2CRC management and researchers.

The decision to use Buttress CO_2 resulted in a number of challenges during the conceptual design phase. In developing the original Otway Project concept, it was anticipated the injected CO_2 would be relatively pure CO_2, analogous to what would be expected from a commercial carbon capture plant. The Buttress reservoir offered a large source of inexpensive CO_2, but it was not pure CO_2 as it contained significant quantities of hydrocarbons, mainly methane.

Before the process plant could be engineered, it was necessary to confirm the compositional analysis of the CO_2 source at Buttress-1 and establish the characteristics of the well. When Santos drilled the Buttress-1 mono-bore well in 2002, it was never perforated for production, as the gas was found to contain a very high percentage of CO_2. The casing was 13% chrome steel but the wellhead was carbon steel which was not suitable for production of wet CO_2 owing to potential corrosion and embrittlement issues. Therefore in May 2006 the existing wellhead was changed to a wellhead with Xylan-coated casing and stainless steel trim. The completion (perforation) was undertaken using a Prism gun fired remotely on a standard wire line (slick line) with stainless steel cable, rather than using the alternative approach of an E-line with an electrical connection to the surface.

To better understand the gas reserves and the character of CO_2 stored in Buttress-1, an isochronal test programme designed by the CO2CRC were carried out to determine at what rate the well could be flowed and how quickly the formation would recover. The isochronal test comprised a series of flow and shut-in tests with flow being recorded, together with wellhead temperatures and pressures.

One of the issues considered was the noise that might be created during the test, in dropping the pressure from approximately 100 bars to atmospheric pressure. This problem was overcome by making a large pit in the ground, running the vent pipe to the pit, connecting it to a sparge and burying it in the pit. This proved to be very successful in that there was no noise problem.

The only complication that arose from the gas pressure drop was the associated temperature drop and the resultant formation of solid carbon dioxide or dry ice (Figure 4.1). During the flow tests, multiple safety precautions were implemented to eliminate the possibility of exposure of personnel to levels of CO_2 above safe limits. The precautions

Figure 4.1: Flow testing of the Buttress-1 well produced large quantities of solid carbon dioxide at the site due to the pressure drop and resultant temperature drop.

included safety barriers and multiple industrial fans to disperse the CO_2 produced during the flow testing. The level of CO_2 in the area surrounding the pit was also carefully monitored to ensure safe operations; a significant change in the CO_2 level could only be detected within 25 mm of the dry ice mound. The test interpretation clearly showed that there was sufficient gas present in the Buttress structure to meet the condition of producing to 100,000 t of CO_2.

The well test performed for Buttress-1 indicated presence of 78.66 mol % CO_2, 18.91 mol % CH_4 and 2.43 mol % of other hydrocarbons heavier than methane plus some nitrogen. The post-well completion isochronal tests saw the accumulation of solid CO_2 plus water ice in the purpose-designed acoustic discharge pit (Figure 4.1) This melted extremely slowly and after some 6 weeks it was decided to excavate and remove it (in industrial skips) to an approved disposal site. At that time there was no indication of Buttress-1 producing significant amounts of wax from the wellbore (see later discussion on operational issues).

4.2   Options for gas processing

In order to try to match expected conditions for a commercial large-scale carbon capture plant with Otway, the option was first considered that the Buttress gas would be purified to approximately the level of purity of commercial grade CO_2. This would have involved removing the naturally occurring hydrocarbon contaminants, which at the early stage of development were initially thought to represent approximately 10–15% by volume of the Buttress gas (but ultimately proved to be 21%). The initial aim was to take the Buttress gas, dehydrate and purify it, and then further pressurise the remaining concentrated CO_2 gas for injection into the Naylor reservoir, approximately 2 km from the proposed plant location.

Four alternative processes for treatment of the Buttress gas were developed and investigated.

Each design had to address specific challenges posed by the gas chemistry. Naturally occurring gas will always be present in equilibrium with water; gas is associated with water present in the pore structure of the reservoir.

These reservoirs are generally at high pressures and with temperatures of 80°C or more. As the gas is produced, the pressure is reduced, the gas cools and hydrates can form, depending on the temperature and pressure conditions. A hydrate is a naturally occurring solid which forms when water and CO_2 or components of natural gas are cooled to below the hydrate formation temperature. In the case of the Buttress gas the hydrate formation temperature was 17°C.

Therefore any Otway processing facility had to ensure hydrates did not form during the process. Moreover as the ground temperature at Otway was below 15°C, it was deemed that transporting the Buttress gas to the Naylor injection well by underground pipeline would risk the gas cooling below 17°C, potentially resulting in the pipeline becoming blocked by hydrates.

The other significant problem posed by the presence of water was corrosion due to the formation of carbonic acid. Carbonic acid is extremely corrosive to standard grades of steels and, unless the water is removed, requires the plant to be built from corrosion-resistant stainless steel.

With all these various considerations in mind, several plant design options were considered and these are outlined below.

4.2.1 Option 1: Free liquid knock out and compression or pumping

This process design option considered removal of free liquids from the inlet feed gas by knock out, and direct compression of the vapour leaving the inlet slug catcher (Figure 4.2). In this case the gas would not have been dried.

The implication of having a water-saturated carbon dioxide and light hydrocarbon feed is that all processing equipment would be required to be manufactured from stainless steel, including the pipeline to the Naylor reservoir, because the presence of water in the predominantly carbon dioxide feed would have been highly corrosive if free liquids were formed. The most likely section where free liquids could form would be in the 2 km pipeline. There would also be the additional possibility of hydrate formation. To minimise this, the vapour leaving the injection compressor would not be cooled by an after-cooler. This would enable hot compressed vapour to pass through the pipeline and down the Naylor well without dropping the fluid temperature below the hydrate formation temperature of 17°C.

The benefits of this option include:

> no requirement for molecular sieve driers (MSD)

> no refrigeration cooling required

> small emission quantities

> cost effectivess of the plant package.

However, there were a number of disadvantages including:

> requires multi-stage stainless steel reciprocating compressor, knock out, inter-cooling etc.

> injection fluid contains 21% hydrocarbon impurities

> potential for hydrate formation

> stainless steel process package and pipeline required.

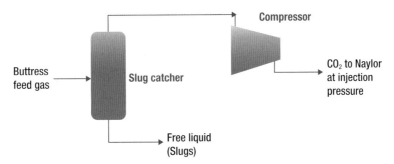

Figure 4.2: Option 1: Free liquid knock out and compression or pumping.

4.2.2 Option 2: Dehydration and compression

The second process design option considered water removal from the natural gas, followed by compression to an injection pressure of 20,000 kPag (Figure 4.3). In this design option, the inlet feed gas is passed through a plant choke valve and inlet slug catch and then dehydrated using an MSD (molecular sieve dryer). The dry gas leaving the MSD is then sent to a reciprocating multi-stage compressor for pressurising to 20,000 kPag. Due to the high discharge pressure, the compressor would also require inter-stage and after-cooling by air cooled heat exchangers and inter- and after-stage knock out pots. The gas leaving the compressor would be cooled to 45°C before finally being sent to the pipeline. This process will also require a take-off of some of the product gas for regenerating the MSD. Hence, there is also the issue of what to do with the spent regeneration gas.

The benefits of this configuration include:

> no refrigeration cooling required

> low emissions.

However, there were a number of disadvantages including:

> requires multi-stage reciprocating compressor, knock out, cooling etc.

> injection fluid contains hydrocarbon impurities

> requires regeneration gas booster compressor, knock out, cooling etc. (if applicable)

> high suction pressure on compressors.

4.2.3 Option 3: Dehydration and liquefaction of natural gas

The third process design option looked at water removal from the natural gas followed by liquefaction only of the feed gas and finally pumping the liquid to the injection pressure of 20,000 kPag (Figure 4.4).

As in the previous processes, the inlet feed gas is passed through a plant choke valve and an inlet slug catch, and then dehydrated using molecular sieve driers. The dry gas leaving the molecular sieve driers is then condensed in the inlet chiller at a temperature of –12°C using propane refrigerant. The liquids leaving the inlet chiller are then passed through a pressure control valve where the operating pressure is reduced to 5600 kPag. The fluid is then sent to the outlet separator where any non-condensable materials are removed from the liquid and vented. The liquids leaving the outlet separator are then compressed to reservoir pressure for injection. The final composition of fluid to be injected into the Naylor reservoir is 85% carbon dioxide and 15% light hydrocarbons. 2500 kg/h of the dry gas leaving the MSD would also be used for regeneration of the desiccant beds. This regeneration gas would first be heated to the regeneration temperature of 290°C and then passed through the offline desiccant bed for removal of water. Hence, the regeneration gas leaving the MSD bed would be resaturated with water. There are then three possibilities for directing this gas:

> Vent to atmosphere, which obviously is not a viable option given the greenhouse gas objectives of the Project.

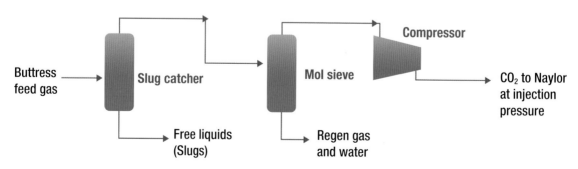

Figure 4.3: Option 2: Dehydration and compression.

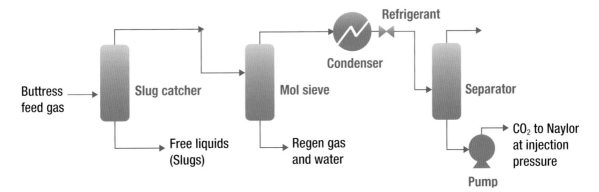

Figure 4.4: Option 3: Dehydration and liquefaction of natural gas.

> Repressurise the gas with a booster compressor and then reintroduce with the inlet feed—the additional cost associated with this option would be very high.

> Condense the regeneration gas and then pump it back to the inlet feed pressure in order to condense the gas stream. However, the temperature would drop below the hydrate formation temperature.

The benefits arising from this configuration include:

> pumping liquids in lieu of compression

> low emissions.

The disadvantages include:

> injection liquid contains hydrocarbon impurities

> requires low temperature refrigeration package

> require regeneration gas booster compressor, knock out, cooling etc. (if applicable)

> high suction pressure on pumps.

4.2.4   Option 4: Dehydration and purification by cryogenic distillation

The fourth process design option considered was water removal from the natural gas followed by liquefaction and distillation of the carbon dioxide component, and finally pumping the liquid carbon dioxide at the injection pressure of 20,000 kPag (Figure 4.5).

Under this option, the inlet feed gas would first pass through a plant choke valve and inlet slug catcher. The vapour leaving the slug catcher would then pass through an inlet filter coalescer for removal of free liquid carry-over and then be sent to the MSD for water removal. This system of drying was selected for water removal in lieu of glycol dehydration, as the small flow of gas did not warrant the expense of a glycol dehydration package.

The dry gas leaving the molecular sieve bed is passed through a dust filter to remove any desiccant carried over by the dry gas and then partially condensed using heat exchanged with the hydrocarbon off-gas from the distillation column. Prior to the dry gas/liquid entering the distillation column, the pressure and temperature of the fluid are reduced over a control valve to 6200 kPag and 13.2°C respectively. These operating conditions correspond with a suitable feed point in the fluid phase envelope for distillation of the carbon dioxide/hydrocarbon components.

The purpose of the distillation column would be to purify liquid carbon dioxide to a concentration of 97 mol % while also limiting the amount of carbon dioxide in the off-gas to 3 mol %. The carbon dioxide concentration in the bottom liquid and the overhead off-gas are maintained by an external kettle reboiler and an external reflux condenser, which are operated at the temperatures of 20°C and 35°C respectively.

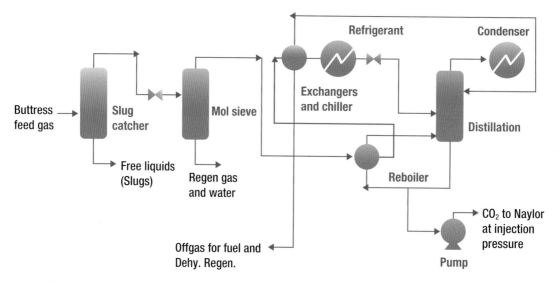

Figure 4.5: Option 4: Dehydration and purification by cryogenic distillation.

The liquid carbon dioxide from the distillation column is then pumped to an injection pressure of 20,000 kPag and passed through a flow transmitter before entering the pipeline. The off-gas from the top of the distillation column is passed through the inlet chiller and sent to the molecular sieve regeneration heater for regenerating the off-line molecular sieve bed.

The benefits from this configuration would be:

> produces carbon dioxide liquid for injection with purity of 97 mol %

> removes light hydrocarbon impurities from the carbon dioxide feed and using the off-gas for regenerating the MSD. The spent regenerated gas could then be either flared or used as a fuel source

> pumping liquids in lieu of compression.

There were however, a number of disadvantages:

> requires low temperature refrigeration package

> high suction pressure on pumps

> highest capital cost of all the options

> expensive to run.

4.2.5 Option 5: Dehydration and purification by membranes

Under this option, gas from Buttress would be passed through a slug catcher and molecular sieve driers remove free liquids and water respectively as previously discussed (Figure 4.6). The gas would then enter a polymeric membrane which allows the separation of methane and carbon dioxide via the use of a membrane material specifically selected with a high affinity or selectivity for carbon dioxide. The gas would be reduced to medium pressures (6 to 7 MPa) for processing through a polymeric membrane. CO_2 would permeate through the membrane and emerge at a much lower pressure (< 20 kPag), while methane would be recovered at full pressure. The CO_2 would then need to be compressed to the required injection pressure.

The benefits of this option include:

> production of carbon dioxide liquid for injection with a purity of 97 mol %

> removal of light hydrocarbon impurities from the carbon dioxide feed and allowing the use of off-gas for regenerating the MSD. The spent regeneration gas would then be either flared or used as a potential fuel source

> the system is simple.

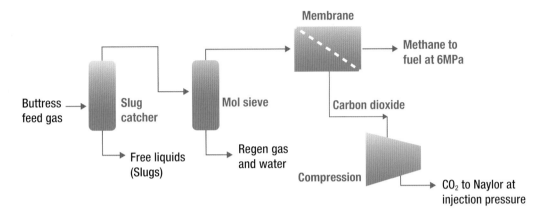

Figure 4.6: Option 5: Dehydration and purification by membranes.

However, there were disadvantages with the option, including:

> all CO_2 is recovered at close to atmospheric pressures, thus requiring a considerable amount of compression (up to the injection pressure), or compression to 2 MPa followed by liquefaction and pumping (a complex operation)

> the possibility of membrane contamination.

4.2.6 Gas processing selection

Following detailed evaluation of these options it appeared that Option 4 offered the best way forward for the Project. This selection was primarily based on the desire to inject near-pure CO_2 into the target injection zones. But as this option was further developed and the Buttress and Naylor wells were refurbished, it became apparent that cost constraints were going to require the process selection to be revisited as Option 4 was too expensive.

Further consultation with the project geologists and reservoir engineers assessed the implications of not purifying the carbon dioxide and the effect this would have on the overall injection trials. The conclusion of this evaluation was that while pure carbon dioxide would be "nice to have", given that there was a significant amount of residual methane in the target reservoir, it was not critical to the success of the Project and that the gas mixture could be injected without jeopardising the Project. On this basis, the process review focused on Option 1, which involved simple removal of free liquids for the Buttress well followed by compression. The major challenge to making this option viable was to develop a mechanism to ensure the carbon dioxide pipeline to Naylor did not become blocked with hydrate. To avoid this, detailed modelling of the temperature profile of the pipeline was carried out. It was found that the heat put into the carbon dioxide as a result of compression could be used to keep the pipeline warm and ensure the carbon dioxide arriving at CRC-1 was above its hydrate formation temperature. The issue of corrosion in this pipeline could be solved by manufacturing it from stainless steel. While this increased the cost of the pipeline, this added cost was substantially less than the cost of installing a system to dry the carbon dioxide.

The only remaining issue to be solved was how to stop the formation of hydrate when the pipeline was not flowing and a drop in temperature occurred. To solve this, a novel system was added so that during prolonged shutdowns the pipeline was automatically purged with dry high pressure nitrogen from a small high pressure nitrogen storage system installed at Buttress. Being an inert gas, nitrogen would sweep the pipeline of CO_2 and water, thus preventing the formation of hydrates. A small amount of nitrogen would ultimately be injected into the Naylor well but, given the small quantity involved, it was concluded that this would have no significant effect on the injection trials.

There were a number "firsts" arising from the Otway gas processing system:

> As far as can be ascertained, Otway is the only site in the world where wet (water-saturated) carbon dioxide for sequestration has been successfully processed without any problems arising from hydrate formation or corrosion. This innovation may offer significant potential savings to future large-scale CCS projects where the high cost of drying the carbon dioxide could potentially be avoided. Clearly, a tradeoff with having a more expensive stainless steel pipeline exists and will need to be evaluated on a case by case basis.

> The simplicity of the process makes it possible to operate the facility with minimum operator intervention. In the case of Otway, the initial idea was to be able to remotely manage the plant with only a few visits per week and having an operator on standby for emergencies. However, with the wax and repeated shutdown issues this was not possible and active operator intervention was required.

> The process selected not only minimised capital cost but was also the lowest operating cost option, thus maximising availability of funds for the research.

> The simplicity of the plant design meant it had the smallest foot print of all the designs considered and therefore minimum visual and noise impact on the surrounding area.

The processing plant was installed and began operations in March 2008.

4.3 Facilities and pipeline design considerations

At the outset, no infrastructure existed at the proposed site for the Project. The main requirements were earthworks for the access road and site for CRC-1 and the facilities comprising the amenities, visitors' centre and control room. Electrical power was installed to all locations including the air monitoring station. This included transformers and power lines from the nearby 22 kV supply.

4.3.1 Process envelope

The process envelope was based on data obtained from the Buttress-1 and Naylor-1 wells. Naylor-1 was the depleted natural gas well that was to be used for monitoring and to all intents and purposes had the same downhole characteristics as the new injection well CRC-1, which was drilled into the same formations, albeit downdip into the water zone. The downhole pressures and temperatures were measured during drilling and completion.

A model was developed to predict how the temperatures and pressures at the surface at each well (Buttress-1 and CRC -1) would change throughout the life of the Project. In general terms, it was predicted that during flowing of Buttress-1, as the temperature at the surface increased, the surface pressure would also increase. Similarly, at CRC-1 the formation downhole would slowly cool, with the result that the density of the column of CO_2 would increase and hence the surface pressure would decrease.

Originally the surface pressure at Buttress-1 was higher than at CRC-1 and the model predicted that Buttress-1 would flow freely into CRC-1. This proved to be the case, but there was only a flow of approximately 50 t per day due to pressure losses along the pipeline, rather than the 140 t per day required.

The process envelope was used to identify the optimum selection of pipeline size, to determine compression requirements and the timing of when compression would be required. The compressor was sized and selected in accordance with the process scenarios identified on the process envelope.

4.3.2 Pipeline design

Water-saturated gas from Buttress-1 was a challenge from the point of view of selecting the appropriate material for construction of pipe and facilities and second from the point of view of flow assurance due to the potential presence of hydrates and also the formation of carbonic acid in the presence of free water. Accordingly, for the entire surface facility and pipeline, SS 316L, a CO_2 resistant steel, was chosen as the most suitable construction material.

Because of the use of stainless steel for the pipeline, cost was a major consideration when deciding on the diameter.

Using pressures determined from the model and pressure loss considerations, the pipeline design pressure was set at 150 bars. This was to suit the maximum pressure likely to be seen at the maximum flow rate. The option was either a 50 mm nominal bore or an 80 mm nominal bore. The larger diameter would have had a lower point-to-point pressure drop; however, apart from the additional cost to increase the diameter, there was also the requirement to increase the pipe wall thickness, further adding to the cost of material and construction.

Multiple simulation runs were carried out using Hysis software. This had to take into account all process factors such as temperature, pressure and fluid characteristics. It was finally determined that the flow rate required could be achieved without exceeding the maximum allowable stress under AS-2885. The discharge pipeline from the compressor outlet to the CRC-1 wellhead was buried for a distance of approximately 2.25 km and fabricated from dual certified 316/316 L stainless steel (SS), seamless, Sch. 40S, 50 mm NB piping. The same specification was used for the compressor suction above-ground piping.

The maximum design temperature for the pipeline (at the outlet of the compressor) was fixed at 50°C. At this temperature, the control system would (1) alarm and set the flow from the compressor to low and (2) if the high temperature remained, then it would have automatically shut the compressor down (alternatively, or in conjunction, pipeline protection could be based on pressure). The minimum design temperature of the SS pipeline was –10°C. The design pressure for the pipeline was 15 MPa.

The below-ground pipeline was in accordance with AS-2885. The above-ground piping was in accordance with ASME B31.3.

Other than for the well-heads (API-5000 rated), the agreed pipe flanging for the facilities was 1500#. The pipeline design pressure of 15 MPa was limited by the sch40s pipe walls and not by the 1500# flanging. Thus the pipeline rating was "de-rated" 1500#. All 1500# flanging were standardised on 1500#RF as directed by CPPL.

The two wellhead shutdown valves were slow-closing to constrain surge pressures to code limits. Pipeline stress analyses were performed to ensure that thermal line growth stress was acceptable and that line restraint from the surrounding backfilled spoil was acceptable. Underground sections of pipeline were protected against pitting, stress corrosion cracking and other forms of corrosion, by yellow jacket coating of straight pipe and Denso wrapping of fittings and bends. In addition, the pipeline was also protected by a simple sacrificial anode cathodic protection system and insulating above-ground sections of piping at CRC-1. Drain points (assembly comprising ball valve, globe valve and blind flange) were installed at low points segments in the underground pipeline to facilitate line clearing in an upset process scenario.

As part of the design review, it was decided that no provision would be made for a chemical injection system for hydrate suppressant. However, 25 mm NB blind flanged injection points were provided at selected locations to allow hydrate suppressant injection, should it prove necessary at any stage.

For unplanned system shutdowns the following approaches were documented as operating procedures to control hydrates:

1. The pipeline contents were displaced with nitrogen prior to potential pipeline temperature decreases to the upper limit of the hydrate formation range (17°C). This required a nitrogen purge of the pipeline (depending on the time of pipeline cool-down after a shutdown). The pipeline contents would be discharged to the atmosphere from the CRC-1 cold vent stack or displaced downhole into CRC-1.

2. The "fall-back" option was to depressurise the pipeline from both ends at the same time and at roughly equal depressurisation rates and then wait for the hydrates to dissociate. All depressurising activities would be performed manually.

3. The entire 2 km transfer pipeline was also provided with vent pits at equal distances to enable depressuring of the pipeline, if hydrates were suspected at any location along the underground section of the pipeline. These vent pits consisted of a double-block and bleed valve arrangement with a flange end to connect a removable pipe vent with a top mounted restriction orifice, in order to ensure controlled depressurisation.

A number of other approaches were rejected for hydrate control, including displacement of the hydrates with pressure (rejected due to safety concerns from a projectile plug) and heating of hydrate plugs (rejected due to safety concerns from a projectile plug, given the possibility of differential expansion of the hydrate plug).

Start up required filling the pipeline with nitrogen and then heating the process stream to the required higher temperature (via a water bath heater) so that the pipe at the CRC-1 end of the pipeline would not be below the hydrate temperature. 3000 psi bullets (Man-packs) were used for nitrogen purging of the pipeline. Pipeline contents were displaced via the CRC-1 cold vent or discharged downhole in CRC-1 in the event of an unplanned shutdown. For the initial start and for starts after line depressurisation, the line was pressurised with nitrogen up to 8–10 bars prior to commencing process injection, with the nitrogen injected into the pipeline upstream from the compressor scrubber and filters.

A tracer chemical injection facility was set up at CRC-1. Suitable valving arrangement were established and procedures framed to evacuate and load chemical tracers into the main injection pipeline, using injection gas to carry the slug of chemical tracer downhole into the CRC-1 well.

Provisions were made for the future pigging of the transfer pipeline. These provisions were limited to 5D pipe bends and blind flanged off-takes for any future connection of the launcher and receiver as required.

4.4   Facilities design

4.4.1   Environmental considerations

It was important to the Otway Project to ensure that environmental issues were fully taken into consideration in designing the facilities.

These issues included:

> All sources of noise from the facility were attenuated so that their night-time impact at the nearest existing residence were less than 34 db(A) as required under the EPA guidelines.

> The performance of all depressurisation and relief vents were modelled using dispersion analysis techniques, to ensure resultant ground level gaseous concentrations would not exceed prescribed safety levels.

> The facility was designed in such a way that there would be no loss of liquid containment (produced water, operational fluids, or lubricants) from plant and equipment.

> Visual impact was minimised through low level construction wherever possible and a colour scheme that blended with the natural surroundings.

> The Buttress-1 and CRC-1 well-head areas were installed with CO_2 and CH_4 gas leakage detection alarm and shutdown functions.

The facilities were designed for the following conditions:

> minimum ambient temperature   –2°C

> maximum ambient temperature   45°C
  (for materials, corrosion, robustness)

> maximum ambient temperature   35°C
  (for air cooler performance)

> design maximum temperature (generally   60°C
  for above-ground exposed equipment)

> design maximum temperature for pipelines   50°C

> design maximum temperature   100°C
  (compressor and water bath heater)

> design maximum temperature   90°C
  (above ground LV cables)

> minimum in-ground temperature   14°C

4.4.2   Equipment design criteria

The Buttress-1 well had been drilled in 2002 to a monobore design with 7⅝″ surface casing set at 378 mRT and 3½″ 13Cr95 Fox Production Casing at TD of approx. 1717 mRT. This resulted in some design constraints, but the following design criteria were used for the Project:

> A 9″ 5000 psi × 3 1/16″ 5000 psi mono-bore wellhead/Xmas Tree was installed. This was upgraded in April 2006 to a Xmas Tree with the internal wetted surfaces Xylan coated for corrosion resistance.

> The prime objective was to produce CO_2 from the Waarre "C" reservoir at a rate of 3 MMscfd for up to 2 years.

> Downhole completion and wellhead were designed for high CO_2 service for the life of well (> 10 years).

> The well did not have downhole installation of a subsurface safety valve (SSSV).

> A surface safety valve SDV-1003 (ESD = SSV or Hi-Lo) was installed at the downstream side of the wing valve on the Xmas Tree. Any higher or lower pressure than a preset level seen in the well downhole side the SSV would shut in the well and prevent transmission of higher well pressures and product to the downstream flow-line.

> A monitor line allowed monitoring of the well annulus pressure. The maximum expected shut-in wellhead pressure was about 1500 psi.

> Similarly, any higher or lower temperature than the preset level seen in the well or in the pipeline also generated automatic well shutdown. The calculated maximum wellhead temperature was not expected to exceed 55°C and the maximum reservoir temperature was expected to be 68°C.

> Non-return valves on the flow-line were installed to prevent backflow from downstream in the event of a wellhead failure.

4.4.3   Gas supply

A suction scrubber vessel (V-001) was provided to remove free liquids from the process stream prior to compression. Accumulated water automatically discharged to the low pressure flash pot by means of an on-off valve. This pressure vessel complied with AS-1210 requirements in regard to design, materials, welding, fabrication and testing.

The low pressure flash pot vessel (V-002) was used to separate any entrained liquid hydrocarbon/CO_2 that vaporised from the bulk liquid as its operating pressure was reduced to approximately 50 kPag. Any liquid remaining in the low pressure flash pot would then be discharged automatically into the produced water storage tank (T-001) for removal from site, while the vapour would be vented at a safe location.

The process stream leaving the suction scrubber was then sent to the cartridge filter (F-001) for removal of solid particles. Filtration was required to protect the compressor from solids entrained with the gas. This pressure vessel complied with AS-1210 requirements in regard to design, materials, welding, fabrication and testing.

The gas leaving the filter was sent to an electric water bath heater (HE-001) for pre-heating. The purpose for pre-heating the gas was for the start-up operating conditions, where the Buttress surface temperature and consequently gas compressibility was outside the allowable range of the compressor. Heating the gas brought the gas compressibility back within the compressor's allowable range and had the added benefit of reducing the probability of hydrate formation in the transfer pipeline under normal operating conditions. The start-up flow-rates were increased to the specified design of 3 MMscfd, as the Buttress surface temperature increased to steady state normal operating conditions.

4.4.4   Compression

A single stage reciprocating compressor (K-001) was installed at Buttress-1 (Figure 4.7), to boost the Buttress surface pressure, to overcome transfer pipeline pressure losses and provide any additional pressure required for injection at CRC-1. It was housed in a suitable enclosure for noise attenuation. The compressor was controlled via suction and discharge pressure control, through a variable speed electric drive and recycle-valve control. The control panel was also designed to receive input from a flow transmitter for flow control, with suction and discharge pressure checks, or suction and discharge pressure control with flow checks. Remote and local controls were also defined and made available. The compressor was sized and selected to accommodate all process scenarios identified on the operating envelope, with the exception of the start-up after shutdown at 24 months with 100% nitrogen in the transfer pipeline. A compressor after-cooler (AC-001) was installed to limit the temperature of the compressor discharge gas stream to a maximum temperature of 50°C. The after-cooler was sized on a maximum ambient air temperature of 35°C at full recycle. The after-cooler was not operated following the start of the injection, as it was found to overcool the CO_2. Action was taken to provide screens around the cooler to prevent natural draft and minimise overcooling.

Figure 4.7: Buttress gas production and processing facilities.

4.4.5   Siting the plant and equipment

Many options were evaluated on siting of the plant and the pipeline route. Finally it was decided to place the compressor plant at the Buttress-1 well site and locate some ancillary equipment and the visitors' centre at CRC-1. It was also determined that the best route for the pipeline was east along Callaghans Road and then south alongside Sodas Lane (see Chapter 1).

CRC-1 was drilled to a mono-bore design with 7⅝″ surface casing set at 500 mRT and 4½″ 13Cr80 bear production casing at TD of approx. 2200 mRT. An 11″ 5000 psi × 4¹⁄₁₆ the internal wetted surfaces were Xylan coated for corrosion resistance.

Various seismic surveys were carried out to better define the geometry of the reservoir formation and determine the optimum position in the Waarre C reservoir for injection of 3 MMscfd of CO_2 into the water leg of the Naylor structure.

The downhole completion and the wellhead were designed for high CO_2 service for the life of the well (assumed to be 10 years or more). The cased hole wellbore diameter was designed for running of production logs when required and included provision for downhole pressure and temperature gauge installation.

A Surface Safety Valve SDV-1008 (ESD = SSV or Hi–Lo) was installed at the upstream side of the wing valve on the

CRC-1 Xmas tree. Any higher or lower pressure than a preset level would be seen in the CRC-1 downhole side and the SSV would have shut in the well. A monitor line also provided the opportunity to permit monitoring of the well annulus pressure. The design pressure for the Xmas tree and completion equipment was 5000 psi WP. Similarly, any higher or lower temperature than the preset level seen in the well, or in the pipeline, also generated automatic well shutdown. The calculated maximum wellhead temperature was not expected to exceed 90°C. Non-return valves on the flow-line prevented backflow from downstream in the event of a wellhead failure.

Produced water was transferred from the LP flash pot (V-002) to an above-ground 10 m³ plastic storage tank (T-001) on level control. The accumulated contents of the storage tank were transported by road tanker for disposal at a site approved by the regulator.

4.4.6   Control and instrument design criteria

A fully instrumented control system (an Emerson Delta V DCS) was installed, incorporating an independent safety shutdown system. The plant could be operated from Buttress-1 or from CRC-1 or remotely over the internet. Apart from enabling operations and monitoring the plant, the DCS had the capability to store data and plot trends over time. Special requirements, notably critical process conditions, were monitored by SIL-rated devices which made use of the different physical instrument principles (pressure, temperature, flow transmitters, etc.) as applicable. The response time requirement meant that safety shutdown valves (at wellheads and on skids) were set to shut in an appropriate closure time to keep pipeline surge pressures to acceptable limits.

Consideration was given to safety instrumented system (SIS) field devices and the general field device requirements when developing the safety requirements. Specific details considered included:

> Standard configurations for inputs into the SIS and for outputs from the SIS were adopted for the Project. This encompassed the types of testing regimes that were required to meet the SIL and reliability targets.

> The defined SIL required included the actual final control element to isolate the process and the selection of the power unit to activate shutdown. The plant protection system was designed as "fail-safe" for all valves.

Final shutdown devices (trip devices) such as solenoid valves were designed and installed so that if de-energised or depressurised, this would cause a shutdown of the equipment they were designed to protect. Each solenoid was driven from a separate output module of the SIS and each solenoid coil was continuously monitored for open circuits. Solenoids were generally de-energised to trip and were direct acting rather than pilot-operated.

Trip devices driven from the SIS such as solenoids required manual reset to re-establish the system after a shutdown. The latching was performed by logic in the SIS. Reset was performed via a reset switch into the SIS, either initiated from the field or from the control room. Solenoid valves with latched operation and manual reset were not used. All manual resets were readily accessible by operators and located so that safe operation of the plant could be ascertained before activation.

4.5 Unanticipated operational problems

4.5.1 Wax removal

Upon commencement of injection in April 2008, it was observed that the well fluid started producing wax which was dark brown in colour and sticky in nature. This required a complete shutdown of the surface facility and manual removal of the wax before restarting the plant. This necessary shutdown due to wax ingression started to become a daily feature. To overcome this problem, the LP flash pot, originally designed to operate at a slightly positive pressure of 350 kPag, was converted into an atmospheric vessel with a cover on the top, to prevent ingress of rain. Along with this modification, a hand-hole in the shape of a rectangular window was cut into the lower half of the LP flash pot. This hand-hole was closed with a lid so that the wax could be removed easily. This alleviated the issue of forced shutdowns of the surface facilities. It was noted that the nature and colour of the wax changed over the period, but the removal procedure was found to be satisfactory throughout the operations.

Surprisingly, all the wax drop-outs were contained by the LP flash pot; later investigations of downstream piping and equipment revealed that the wax did not travel any further down on the injection gas side.

4.5.2 Water production

It was expected that water fall-out from the Buttress-1 gas due to the Joule-Thompson effect would be progressive and the gas conditioning skid had been designed on this basis, but it soon became apparent that this would not work. Initially, no produced water was seen in the scrubber for a number of hours. When water was subsequently produced, it came in large slugs and the process was shut down because of the high water level in the scrubber. The exact reason for this slug flow could not be confirmed but may have been due to the produced water droplets falling back down the well and accumulating until the volume became sufficiently large that it formed a slug that was then gas-lifted to the surface. Action was taken to rectify this, by raising the high level alarm and changing the discharge orifice on the outlet to the LP scrubber.

4.5.3   Compressor vibration

Vibration issues caused by reciprocating compressors are a common problem which can result in fatigue failures in adjacent piping. Immediately upon start up at Otway, it was considered that the level of vibrations transmitted by the compressor was unacceptable and was likely to rapidly lead to a failure of the high pressure stainless steel piping. At that time, free flow of gas between Buttress and CRC-1 was possible, so compressor operation was minimised until the root cause of the vibration could be found and changes were made to the piping system to bring any vibration within acceptable limits. Vibration specialists engaged to carry out a necessary on-site analysis recommended shortening some of the dead legs and providing more rigid anchorage for the off-skid piping without compromising thermal expansion.

During the injection period between early 2008 and most of 2009, two weld failures were experienced. Both were at socket welds on the gas conditioning skid. Subsequent laboratory analysis proved these were, as expected, fatigue failures.

Throughout the operation there were numerous failures of the stainless steel valves. Steel of a different grade was trialled but this proved to be worse than the original valves and consequently valves with the original specification were put back in. The issue was managed by routinely changing the valves. Several failed valves were sent away for routine analysis; the problems were attributed to pitting corrosion and fatigue.

4.6   Conclusions

The plant functioned safely and allowed the Project to meet its objectives. In over a year of operation approximately 66,000 t of CO_2-rich gas was injected into the Naylor structure without a single lost time incident or major environmental incident.

As has been discussed, the plant design was a compromise between the available budget and the minimum specifications needed to meet the broader project objectives. The plant was designed as an unmanned (or not regularly manned) system which could be controlled by the distributed control system (DCS) and required only intermittent intervention. The design parameters were guided by the well test which was of a short duration and thus was not fully representing the way in which the Buttress reservoir behaved while in production. Some of the key issues faced, as discussed, were related to the excessive wax and water production, weld failures and unintended shutdowns of the plant due to the triggering of various control parameters in the DCS.

The plant operating parameters had to be adjusted during the early run and several modifications were made to the plant and the process system to redefine the operating philosophy. A system of regular intervention was instituted so that the operator could manage time efficiently. Fortunately for the Project, the plant operator was also a local farmer and was able to divide time between the part-time requirements of plant operation and farming activities.

Compared to commercial projects, research projects are more likely to face budget constraints and have less flexibility in being able to design potential failure options. In addition, there was not a lot of local experience in Australia with wet gas compressors and therefore a period of learning was involved.

The Otway Project took a risk-adjusted approach towards its design and operating philosophy which worked well, except in a few cases where modifications had to be made. If the budget had allowed a longer production test and more extensive heat tracing, operating costs associated with the water bath heater would have minimised unintended shutdowns. Notwithstanding all this, pragmatic solutions within the constraints were developed by the project team and the plant worked as predicted.

Finally, as far as is known, Otway is the only site in the world where wet (water-saturated) carbon dioxide for sequestration has been successfully processed without any problems arising from hydrate formation or corrosion. This approach may offer significant potential savings to future CCS projects where the high cost of drying the carbon dioxide could potentially be avoided. Clearly, a tradeoff with having a more expensive stainless steel pipeline exists and this will need to be evaluated on a case by case basis.

Tess Dance

5. CHARACTERISING THE STORAGE SITE

5.1 Introduction

Site characterisation is defined by the CO2CRC as "The collection, analysis, and interpretation of subsurface, surface and atmospheric data (geoscientific, spatial, engineering, social, economic, environmental) and the application of the knowledge to judge, with a degree of confidence, if an identified site will geologically store a specific quantity of CO_2 for a defined period and meet all required health, safety, environmental, and regulatory standards". Depleted petroleum reservoirs, such as the Naylor Field, are regarded as desirable CO_2 storage sites, due to the perception that much of the data gathering and characterisation was done in the exploration and development phase of the field's life and that they are proven traps, having held hydrocarbons in the past (Stevens et al. 2000). However, it cannot be assumed that the extent of site characterisation needed for a depleted petroleum reservoir site will be any less stringent than that needed for any other site when assessing injectivity, capacity and containment of CO_2 (Chadwick et al. 2007; Jenkins et al. 2012). A field or structure that was charged naturally with hydrocarbons over perhaps millions of years may not have the same physico-chemical response when injected with CO_2 at high rates over a short space of time. Similarly, the CO_2 storage capacity of depleted oil or gas fields will not necessarily equate to the original volume of gas produced, particularly in reservoirs with strong aquifer drive. Finally the geochemical reaction potential of CO_2, once it is dissolved in water, may compromise seal integrity at a site where the original gas (e.g. methane in the reservoir) had a relatively low reaction potential.

While the data available for the Naylor depleted gas field were sufficient for CO2CRC to determine the structure had held hydrocarbons within a porous and permeable sandstone, at a depth of about 2000 m overlain by impermeable mud rock and that it might be suitable as a storage site, in many respects the available data could not meet the requirements for a comprehensive CO_2 storage site characterisation. For example, there was no conventional core, or side wall cores from either the reservoir or seal, and there was only a very basic suite of pre-production logs. Although the production pressure data proved useful in the early stages of flow simulation history matching, it only provided half the picture when trying to assess the post-production aquifer recharge potential.

The CO2CRC Otway Project addressed each of these issues with targeted data acquisition and specialist analysis. Core and logs were gathered during drilling of the injector well, and the existing production well was re-logged to understand hydrodynamic conditions prior to injection. As a result, the Project provides a valuable example of the process of site characterisation.

5.1.1 Workflow

Site characterisation activities at the Otway site were divided into four distinct phases (Figure 5.1). Phase 1 began in 2004 when CO2CRC undertook initial site screening of sedimentary basins close to major Australian emission centres. The three main regions under investigation were the Bowen Basin in south-east Queensland (Sayers et al. 2007), the Perth Basin in Western Australia (Causebrook et al. 2006) and the Otway Basin in Victoria. As discussed in Chapter 1, the Otway Basin was chosen (Figure 1.2) when the depleted Naylor natural gas field and the Buttress CO_2 field became available.

The second phase of modelling then commenced in 2005 to assess the feasibility of the site against the project objectives. Initially injection into the shallower formations (Figure 1.5), above the gas reservoir, was considered, and for a while these became the subject of seismic mapping, looking for structural trapping above the Waarre Formation using the Naylor-1 well as a "conceptual" injector. It was soon recognised that there were few structural traps at this level. In addition this proposal introduced key risks associated with injection into freshwater aquifers.

Therefore it was decided to concentrate on the Waarre Formation as the injection reservoir at the Naylor structure (Figure 1.6). It was accepted that due to the production well's small diameter (3.5″ or 88.9 mm), Naylor-1 could not be used for both injection and monitoring, so planning and modelling was undertaken to site a new injector, CRC-1 (Spencer et al. 2006).

Following the drilling of CRC-1, phase 3 then commenced; this provided the opportunity to incorporate the much needed core data and high resolution wire-line log information into a new set of static models (Dance et al. 2009). These new data were combined with the pre-existing data from nearby wells and fields, as well as with a good quality 3D seismic survey covering most of the Port Campbell Embayment, in order to better characterise the reservoir and overlying formations. Table 5.1 summarises the improvements made to the existing database that addressed the key uncertainties at the field.

Finally, the last phase in characterisation of the field was achieved during the "demonstration" stage of the CO2CRC Otway Project and encompassed the post-injection model calibration. History matching of the dynamic model against the injection and monitoring data provided insights into the reservoir's bulk permeability, pressure response, and reservoir heterogeneity (see later details of dynamic modelling in Chapter 16).

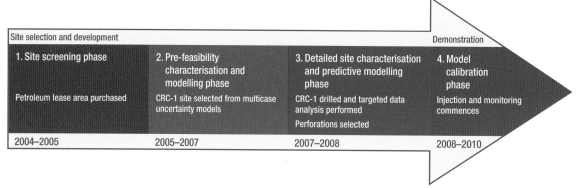

Figure 5.1: Timeline and four phases of site characterisation during site selection, development, and demonstration of the CO2CRC Otway Project.

Table 5.1: Database comparison before and after targeted acquisition programme.

Initial Site Screening Database	Uncertainty	Updated CO2CRC Database	Improvements
No full hole or side wall cores.	Two depositional models, no palynological or reservoir analyses available.	49 m of core, 24 m through the reservoir.	Conventional and special core analysis provided quantitative data on porosity/permeability. Core logging improved facies model.
Minimum wire-line log suites.	Only basic petrophysics: porosity and water saturation.	Sonic, FMI, petrophysical logs, MDT, and injection test.	Stress estimation for mechanical rock strength and fault model. Petrophysical interpretation and fluid saturations incorporated in static model.
No VSP.	Poor understanding of velocity gradient.	Naylor-1 and CRC-1 VSP data and baseline 3D seismic survey.	Full earth velocity model constructed. Confirmation of depth conversion model.
Poorly constrained gas-water contact/local water gradient.	Reserves estimation imprecise and simulation models less constrained.	Naylor-1/CRC-1 RST logs and hydrodynamic data.	Reduced uncertainty in the post-production gas water contact. Improved dynamic modelling history matching.

In the entire Project scheduling, phases 2 and 3 of the site characterisation required by far the most resources and time to complete, and comprised the bulk of the geological characterisation workload. This will most likely be true for many other projects because it is in these phases where a detailed assessment is made against the criteria for a suitable storage site, namely injectivity, capacity, and containment. Therefore, this chapter focuses mainly on these important phases in the site selection and development stage of the CO2CRC Otway Project. A summary of the earth science and reservoir engineering objectives is given below, and the following sections describe in more detail the methodology used and lessons learnt.

5.1.2   Objectives

The schematic in Figure 5.2 summaries the objectives of the comprehensive site assessment and lists examples of the types of data and analysis that were necessary for the Project to proceed. In summary the steps were to:

1. assess the **site details** in a regional sequence stratigraphic setting, establish the regional hydrodynamics and field history and map the Naylor structure to plan for optimal injector location

2. determine from core, logs, and well tests that the reservoir has sufficient **injectivity** to allow for

Figure 5.2: Datasets and analysis used for the five criteria of CO_2 storage site assessment.

up to 100,000 t of CO_2 to be injected over the 2 years allocated for the experiment

3. build a static 3D geological model of the structure and determine that there is sufficient **capacity** based on the spill points of the reservoir, porosity distribution and current fluid/gas saturations to accommodate the planned maximum 100,000 t of CO_2

4. characterise the **reservoir heterogeneity** to understand how sedimentary features are likely to affect vertical and horizontal fluid flow within the reservoir sands

5. assess any risks to **containment** in the overlying Belfast Mudstone seal or from adjacent faults, by mapping seal continuity and 3D fault geometry on the seismic section, and testing cores for CO_2 retention potential and geomechanical rock strength properties (see Chapters 6 and 7).

5.2   Site details

Details of the geology of a site, reservoir seal pairs, existing natural resources which may be impacted by CO_2 injection (as well as production history in the case of a depleted oil or gas field) are critical in the first stages of a site assessment.

5.2.1   Regional geology and stratigraphy

Regardless of the size of a CCS project, understanding the regional, basin-wide, geological setting is an important step in characterising the targeted reservoir-seal pairs. A sequence stratigraphic approach is preferred as it combines well correlations, biostratigraphy and seismic mapping to provide a predictive model for distribution of reservoir and seal lithologies. A regional review of the Otway Basin tectonic and stratigraphic development is given in Krassay et al. (2004). The western part of the Otway Basin is structurally restricted by the Otway Ranges to the east and bounded by structural highs to the north and west. Its development, and that of the adjacent Shipwreck Trough, which extends off-shore, was coeval with the eastern Gondwanan breakup along the Australian Southern Margin and with the Tasman Sea seafloor spreading to the east (Woollands and Wong 2001). The Waarre Formation is the basal unit of the Sherbrook Group (Turonian–Maastrichtian ~91–65 Ma) sitting directly on top of the Otway Unconformity, which marks a period of compression, folding, uplift and erosion in the Mid Cretaceous (Krassay et al. 2004). The Waarre Formation is overlain and sealed by the Flaxman Formation and the Belfast Mudstone.

Turonian deposition was associated with the initial syn-depositional faulting during a phase of basin extension. Deposition of the overlying Flaxman Formation and Belfast Mudstone occurred during the subsequent sea level rise. Following on from this Mesozoic extension, which ended in break-up and subsequent seafloor spreading along the southern margin, there was a long interval of margin subsidence. This was punctuated in the mid Eocene by local inversion and, since the mid Miocene, by regional compression and fault reactivation. The resulting structural style in the vicinity of the study site (Figure 5.3) comprises large north dipping half-grabens separated by the linkage of transfer fault zones (Hill and Durrand 1993).

There are several broadly similar sequence stratigraphic chronostratigraphic systems and descriptions of lithostratigraphy in use in the Otway Basin (Buffin 1989; Laing et al. 1989; Kopsen and Scholefield 1990; Morton et al. 1995; Geary and Reid 1998; Boult et al. 2002). Here, the system published by Partridge (2001) (see Figure 5.4) has been adopted because it focuses on the Sherbrook Group in wells close to the study site. Partridge (2001) subdivides the Waarre Formation into units A, B and C. The basal unit, A, is a fine-grained lithic sandstone with low to moderate porosity. The middle Unit B consists of hard, grey to black carbonaceous mudstone. The upper unit (Unit C) is the main gas producing reservoir in the area and consist of poorly

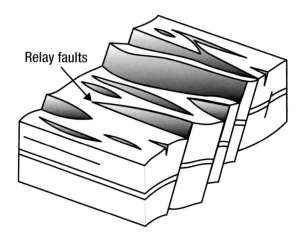

Figure 5.3: Structural model for the Otway Basin showing the development of half-grabens through the linkage of transfer faults.

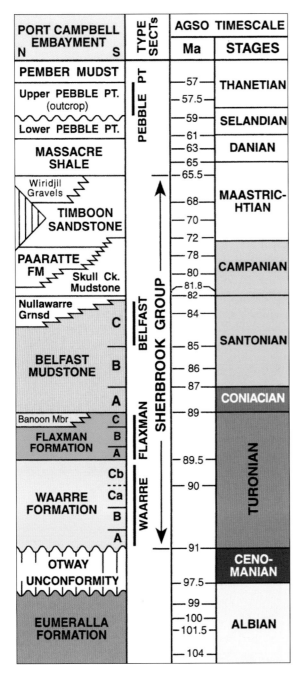

Figure 5.4: Stratigraphic column of sedimentary units in the Port Campbell Embayment (after Partridge 2001).

sorted very fine to coarse quartz sands and occasional gravels, 2 to 14 m thick, separated by minor mudstones which vary from 0.5 m to 3 m in thickness.

The first step in site selection was to perform well correlations of these reservoirs and seals over the area (Figure 5.5). In the onshore area, the Waarre C Formation is relatively thin particularly in the area of Naylor and surrounding fields (approximately 25 m to 40 m thick). By comparison, the Belfast Mudstone (seal) is up to 400 m thick. The local Waarre C lithostratigraphic correlation around the Naylor area is relatively straightforward. The Buttress-1 and Boggy Creek-1 wells form a natural grouping that has a fairly uniform but thin Waarre C sand (approx. 10 m to 20 m). The CRC-1, Naylor-1, Naylor South-1, and Croft-1 wells (Figure 1.3) also form a natural group, but with a thicker Waarre C section (approx. 30 m to 40 m); the thickness difference between the two groups is interpreted as due to thickening across the approximately north north-west syn-depositional growth fault to the north of Naylor-1 (Figure 5.6). This interpretation of a sequence of stacked episodic relatively thin deposits separated by poorly defined sequence boundaries is consistent with deposition in a rift margin environment subject to episodic faulting and extension.

Seismic interpretation was carried out on the existing Nirranda-Hetysbury 3D Survey. This survey had excellent resolution, with 24 fold data to a depth of 4 seconds, a bin size of 20 m and an extensive total area of 83.5 km². Supplementing this was the CO2CRC 2008 baseline 3D seismic, ZVSP, walk-away VSP and 3D VSP data which provided even greater detail directly over the study area. These were all acquired prior to injection as part of the Otway Project time-lapse monitoring programme (Dodds et al. 2009). The new data also allowed a thorough well-to-seismic tie for the CRC-1 well. The polarity convention used was SEG negative, i.e. an increase in impedance is represented by a trough. The gas-bearing Waarre C reservoir reflector is a relatively "bright" (high amplitude) peak (Figure 5.7). The Naylor structure is bound on three sides by faults; variance and instantaneous frequency volumes were used to help image these faults with clarity in 3D.

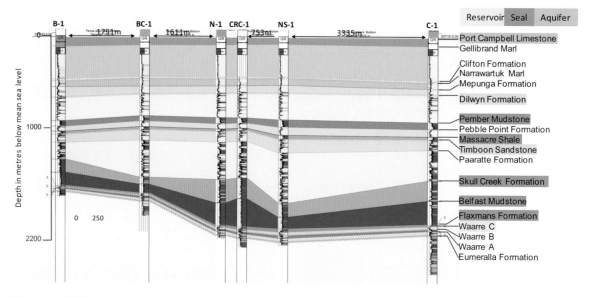

Figure 5.5: Well log (gamma ray) correlation of stratigraphic formation tops in the study area. Refer to Figures 1.3 and 5.7 for well locations: B-1 (Buttress-1), BC-1 (Boggy Creek-1), N-1 (Naylor-1), NS-1, (Naylor South-1), C-1 (Croft-1).

The top and base of the reservoir were mapped, in addition to the top of the Flaxman Formation and top Belfast Mudstone seal. Mapping of these horizons was extended away from the Naylor Field to confirm their regional continuity throughout the Port Campbell Embayment. The resulting surfaces were depth converted using velocity derived from check shot data at the wells.

The Waarre C Formation lies at a depth of between 1980 m TVDSS and 2180 m TVDSS (Figures 1.6 and 5.5). The Belfast Mudstone is between 1340 m TVDSS and 2010 m TVDSS and is 280 m thick on average throughout the site. At this depth the pressure and temperature of the reservoir are in excess of the critical point where the CO_2/methane gas mixture enters the supercritical state. This is important because in this form it is much denser than gaseous CO_2 and therefore a greater volume of CO_2 can be stored in the pore space available (Holloway and van der Straaten 1995; Cook et al. 2000).

In addition to focusing on the injection interval and cap rock, a full earth model was constructed to map the faults and the overlying stratigraphy that may be impacted by injection. Figure 5.7 includes two seismic sections: (1) an approximate east-west section over the field and (2) a north-south section, with mapped formations delineated by coloured lines.

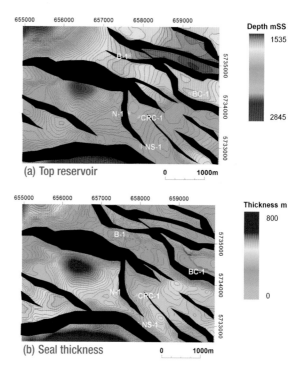

Figure 5.6: (a) Depth structure map of the top of the reservoir in metres sub-mean-sea-level; and (b) seal thickness map in metres. Well name abbreviations: B-1(Buttress-1), BC-1 (Boggy Creek-1), N-1 (Naylor-1), NS-1, (Naylor South-1), C-1 (Croft-1). Black polygons denote faults.

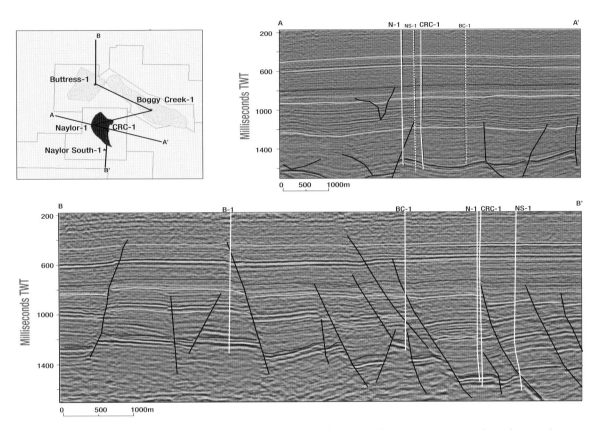

Figure 5.7: Seismic sections A-A' and B-B' (approximately east-west and north-south orientations respectively as shown in location diagram for the study area). Key formation horizons from the base up are Waarre C (red), Belfast Mudstone (green), Skull Creek Mudstone (light green), Timboon Sandstone (yellow), Pebble Point Formation (pink), Pember Mudstone, (purple), Dilwyn Formation (orange), Clifton Formation (blue).

Immediately overlying the Belfast Mudstone is the Skull Creek Mudstone, deposited in the Early Campanian. The Skull Creek Mudstone consists of dark grey to black, carbonaceous mudstones, with minor interbeded siltstones and sandstones that become more frequent towards the top. The section probably represents the outermost prograding toe of a delta system deposited in an open marine environment. Because the Skull Creek Mudstone is mostly fine grained, it consequently has low hydraulic conductivity and it contributes to the primary seal capacity of the underlying Belfast Mudstone across the study area.

Overlying the Skull Creek Mudstone is the Paaratte Formation (the focus of the Otway Stage 2 experiments), which is Campanian to Maastrichtian in age. It comprises laminated fine-grained sandstones with moderate to good

porosity and fair to excellent permeability, interbedded with siltstones and mudstones. The Formation is up to 400 m thick in the study area. The overall coarsening upward nature of the sequence suggests a prograding deltaic to shallow marine depositional environment.

The Timboon Sandstone overlies the Paaratte Fomation, from which it is distinguished by a "blocky" electric log signature (Figure 5.7). It consists of predominately poorly consolidated, fine grained, micaceous sand, believed to have been deposited by fluivial processes in an upper delta plain (Gallagher et al. 2005). Water samples obtained from the Timboon Sandstone generally have total dissolved solids (TDS) values around 500 ppm, suggesting this unit may have significant potential for future use as a town water supply (Duran 1986). It is categorised by the Victorian state EPA as potable water, which is water with between 501 ppm and 1000 ppm TDS. However, the potential of the Timboon Sandstone as a water supply is limited, as it does not outcrop and is only recharged by downward vertical flow from the overlying sediments. To date, this aquifer has not been exploited because of its depth and the abundance of freshwater in shallower aquifers. Nevertheless, it has been flagged as a future resource and, as such, its integrity must be assured.

Above the Timboon Sandstone are the formations of the Wangerrip Group including the Massacre Shale and the Pember Mudstone, which were characterised in the context of their potential to provide secondary seals at the site, in the unlikely event that CO_2 were to breach the primary container. The Massacre Shale, which lies between 931 m and 1026 m TVDSS, is a glauconitic mudstone deposited during a widespread transgressive event and although it is relatively thin (approx. 20 m to 30 m thick) it can be mapped with continuity across much of the Otway Basin. The Pember Mudstone is a pro-deltaic, silty mudstone approximately 50 m thick in the study area. Facies changes that could compromise the seal continuity are recognised on the regional scale; however, sealing potential appears good in the study area.

Above the Pember Mudstone is the Dilwyn Formation. It comprises a thick (approximately 250 m) sequence of shallow marine to coastal plain sandstones and mudstones. The Dilwyn Formation is a major freshwater aquifer (< 1000 ppm TDS), supplying water for urban use to surrounding towns in times of drought.

Overlying the Dilwyn Formation is the Heytesbury Group. The main aquifer in this Group is the Port Campbell Limestone. This karstic limestone outcrops extensively in the area and forms spectacular cliff exposures at the coast. The aquifer is the primary groundwater supply in the region and is currently exploited for urban use, agriculture and irrigation. Understanding the baseline hydrological conditions of these two major aquifers as well as water chemistry and pH was critical in the overall Otway Project site characterisation, and more information on this can be found in de Caritat et al. (2009), Hortle et al. (2011) and Chapter 13.

5.2.2   Field history

The Naylor Field was discovered by SANTOS with the drilling of the Naylor-1 well in May, 2001. It was drilled on the basis of a direct hydrocarbon seismic indicator at the level of the Waarre C Formation and reached total depth in the Eumeralla Formation at 2105 m TVDSS. Initial proven plus probable (2-P) reserve estimates for this small structural closure were 1.47×10^8 m^3 (or approximately 5.4 Bscf) original gas in place. The discovery pressure was 19.5928 MPa at 1993.34 m TVDSS (around the middle of the Waarre C). Because the field was expected to be small prior to drilling, economic considerations required exceptional cost minimisation during development. The operator completed the well as a mono-bore with 3½″ (88.9 mm) casing and there was no additional sampling or testing. That is, there was no conventional core, no side wall cores, and only a basic set of wire-line logs. The well was perforated over the upper 4 m of the Waarre C and produced approximately 9.5×10^7 m^3 (or ~3.3 Bscf) of natural gas (~86% methane) from the Waarre C between June 2002 and October 2003 at which time the well started taking in water. As the cost of water handling equipment was prohibitive, production from the Waarre C was no longer economically viable. Reservoir pressure at this time was down to 11.8612 MPa (converted from the reported flowing tubing head pressures). A casing patch was installed and the well was reperforated at the Waarre A stratigraphic level and a further 0.8×10^7 m^3 (or ~0.3 Bscf) was produced briefly between November 2003 and July 2004 until the well was again killed due to the influx of formation water; it was subsequently shut in. The nearest well, Naylor South-1 (~860 m to the south east), was drilled because the post Naylor-1 assessment suggested a possible field extension across to the Naylor South structure (Figure 5.6(a)). The well did not intersect producible hydrocarbons (only residual methane), and was subsequently abandoned. As in the case of Naylor-1, Naylor South-1 had a minimal test programme, with no cores or any side wall cores.

In 2006, Naylor-1 was logged by CO2CRC using the Schlumberger reservoir saturation tool (RST) which confirmed the position of a post-production gas-water contact (GWC) at 1988.4 m TVDSS, approximately

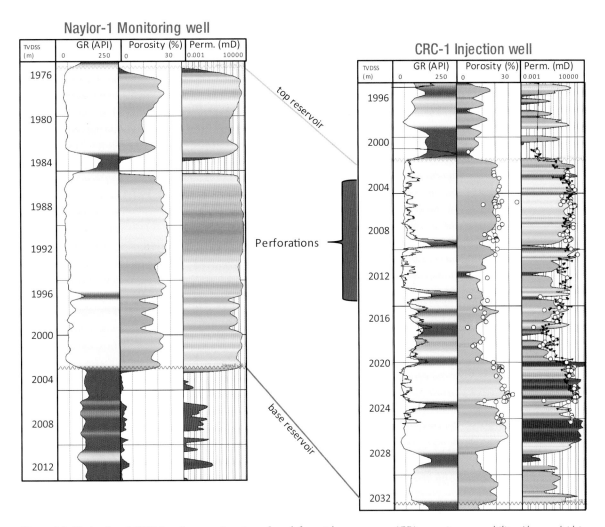

Figure 5.8: Naylor-1 and CRC-1 well composites. Logs from left to right gamma ray (GR), porosity, permeability. Also overlaid in CRC-1 tracks are core gamma ray (black curve), core porosity, core permeability (circles), and mini-perm (black triangles).

11 m below the top of the Waarre C. There was an average residual gas saturation of 20% throughout the remaining 14.5 m down to the boundary with the Waarre B. Pressure information was also obtained at this time; it indicated repressurisation to 17.4 MPa after depletion from production. This pressure recovery has been attributed to the strong regional aquifer drive of the greater Waarre C (Hortle 2008).

The injection well, CRC-1, was spudded on 15 February 2007 and reached a total depth of 2249 mRT (2199.3 m

TDVSS, 109 m in the Eumeralla Formation) on 8 March 2007. It was set and cased as a 4½″ (114.3 mm) vertical mono-bore. Along with vital core and log information, it also provided reservoir temperature (82°C), and pressure (17.8 MPa) information , adding to the dynamic history profile of the field. The pre-injection reservoir simulations were then constrained using a history matching process that honours flow rate and cumulative production data, bottom hole pressure during production and post-production aquifer recharge (Xu et al. 2006).

5.3   Injectivity

Focusing on the reservoir itself, it was necessary to establish that the target had sufficient injectivity to cope with planned volumes and rates. For the Otway Project, the target was up to 100,000 t of CO_2 over 2 years. This translates roughly to 3 MMscf/d (~150 t per day). For this to be feasible with only a single injection well operation, in a reservoir 25 m thick, it was desirable to have absolute reservoir permeability values in the order of > 100 millidarcies. Even though Naylor lacked any permeability information from core, production data and cores from nearby fields (Boggy Creek-1) suggested that highly permeable sands were likely be encountered downdip where the injection was planned. This was confirmed when CRC-1 was drilled. The core programme included recovery of over 49 m of core, including 24 m of continuous core through the Waarre C. The suite of wire-line log information gathered comprised gamma ray, nuclear magnetic resonance (NMR), elemental capture spectroscopy (ECS) and formation micro imager (FMI), which were recorded to complement the standard resistivity-density-porosity logs. In addition several modular formation dynamic tester (MDT) samples allowed multiple pressure measurements and the recovery of multiple fluid samples from the Waarre C Formation, as well as from shallower reservoir sections (Figure 5.8).

Porosity and permeability measurements were performed on vertical and horizontal core plugs at in-situ stress conditions, and supplemented by profile permeametry (mini-perm) measurements recorded on the whole core surface every 5 to 10 cm. The downhole depths from the 96 core measurements and mini-perm were corrected using core gamma ray correlated against downhole gamma-ray logs, so the core porosity and permeability could be matched to the log curves. The core-derived bulk density was compared to the log-derived bulk density and the match was found to be satisfactory. Results shown in Figure 5.8 indicate the porosity of the Waarre C ranged from 2% to 25% while the permeability averaged 1 darcy, with up to 5 darcies measured in some of the cleaner sandstone intervals of the formation.

Relative permeability information was derived from laboratory work on a core sample from the Waarre C in order to understand CO_2/water two-phase flow at reservoir pressure and temperature conditions. The analysis, performed at Stanford University (Perrin et al. 2009), involved flooding the core sample with mixed CO_2 and brine and measuring the pressure at the inlet and outlet with two high accuracy pressure transducers. The difference of the two pressures was used to calculate the relative permeability. X-ray CT scanning was also used to determine CO_2 saturation at a fine scale after the flooding and provided 3D porosity and saturation maps of the sample. The results gave a residual water saturation S_{lr} of 44.4% and a relative permeability to gas at this saturation ($kr_{g_{max}}$) of 0.608. The study also revealed that microscopic grain size heterogeneities and clay lamina impact on porosity distribution and consequently on the distribution of CO_2 saturation in the reservoir.

Injectivity testing at the well, using water, provided information on the bulk permeability of the reservoir. Initially injectivity was poor and pressure built up rapidly. This was assessed as being due to a combination of formation invasion by the drill fluids and mud build-up on the surface of the high permeability reservoir sands and was in keeping with high permeability (1–5 darcies) recorded by conventional core analysis of some of the sands within this interval. Subsequently the perforated interval was extended and the reservoir allowed to back-flow into the well. This flushed the mud filtrate and consequently injectivity was much improved.

Other factors can impact on the rates of CO_2 injection, including near-wellbore dry-out, salt precipitation, fines mobilisation and mineralisation (Burton et al. 2009). Various petrological analyses, including X-ray diffraction (XRD) and scanning electron microscopy (SEM), were performed on 34 samples from the Waarre C Formation and two from the Flaxman Formation, in order to examine these potential effects (Schacht 2008). Samples were taken as 1½″ (38 mm) core plugs from existing CRC-1 cores and thin sections of the samples were cut perpendicular to the bedding plane. In general, the petrographic analyses focused on the mineral content and textural relationships of the rocks. Likely CO_2 chemical interaction within the Waarre C Formation was predicted to involve the in-place potassium feldspar and mica, as well as the dissolution of patchy carbonate cements. CO_2-induced diagenetic products were expected to be minor, due to the absence of cations suitable for mineral trapping of CO_2 in this formation. As a result CO_2-water-rock interactions were not expected to interfere with the ability to inject CO_2 at CRC-1.

Geomechanical assessments were also conducted (van Ruth and Rogers 2006; Vidal-Gilbert et al. 2010) in order to estimate the maximum pore pressure increase the reservoir could sustain during injection. These studies concluded that the maximum sustainable pore pressure increase for the reservoir was 9.6 MPa (~1395 psi) and that the seal could sustain an increase of up to 16.5 MPa above the pre-injection conditions (Chapter 7). The study by Vidal-Gilbert (2010) used results from triaxial rock mechanical tests on CRC-1 cores to constrain models of the minimum pore pressure increase required to cause fault reactivation. For faults oriented optimally with respect to the regional Otway Basin stress regime, the values ranged from 1 MPa to 15.7 MPa, given the initial pore pressure at the top of the reservoir was 17.5 MPa just prior to injection. These limits were used in a series of dynamic models (Xu et al. 2006; Undershultz et al. 2011). In each modelled case, the maximum injection pressure (bottom-hole pressure) was below the initial discovery pressure of the reservoir (19.5 MPa) at the end of the injection and therefore the proposed injection rate of 150 t/d (about 3 MMscf/d) was considered feasible.

5.4 Capacity

In the preliminary stages of site selection for the CO2CRC Otway Project (2004–06) up to a maximum of 100,000 t of CO_2 was proposed to be injected and stored. At the time other projects around the world were running tests with much lower tonnages (see Figure 18.1) so it was considered that this demonstration project was relatively "large scale", making it more relevant to a commercial scale injection project. During the site characterisation study this relatively large reservoir storage capacity requirement was a key assessment issue in the context of the question: "Will there be sufficient storage space available for up to 100,000 t"? More specifically it needed to be technically feasible to achieve this tonnage given the size of the structure down to the spill point, the thickness of the reservoir, and the net-to-gross. Reservoir storage capacity is a complex

function of the density of the CO_2 at subsurface reservoir conditions, the pressure and temperature at the time of injection, and the effective pore space available minus the space occupied by the existing gas cap and the residual methane, while still remaining below the maximum allowable pore pressure increase. There is an important difference in this estimate compared to asking: "How much space can be available at the site"?

Estimating the total effective capacity of a depleted field for storing CO_2 requires a calculation that accounts for dynamic effects such as the increasing pressure due to the hydrodynamic aquifer drive and consequent change in size of the free gas cap. Similarly, in theory the site may be engineered to make more useable space available and minimise the pressure build-up by producing natural gas and water using increased pressure caused by the injection of CO_2, making capacity more of a function of reservoir dynamics and economic feasibility.

A simplistic production-based calculation of storage capacity at the Naylor Field was first undertaken assuming the volume of gas produced equated to the equivalent intended injection volume. The Naylor Gas Field originally contained an estimated 1.47×10^8 m^3 or ~5.2 BSCF (billion standard cubic feet) of initial gas in place (measured at standard temperature and pressure). The cumulative production from the Waarre C reservoir was 9.5×10^7 m^3 (~3.3 BSCF), which was about 64% of the initial gas in place. This volume of produced gas was equivalent to approximately 1.5×10^5 t of the Buttress Field gas mixture of 80% CO_2 and 20% CH_4 by mole fraction. On this basis there was 150% of the required storage capacity at the depleted Naylor Field.

However, this volume-for-volume basis for capacity estimates is only useful in depleted fields where there is weak aquifer drive and injection is performed soon after depletion. If there is only minor invasion of formation water post-production, the same pore space in the reservoir is still available for gas, and so in returning to the original reservoir pressure, the same subsurface volume can be stored as originally produced. Hydrodynamic assessment of the greater Waarre Aquifer by Hortle (2006) concluded that the regional Waarre Formation aquifer is a well connected aquifer in regional hydraulic communication across the Port

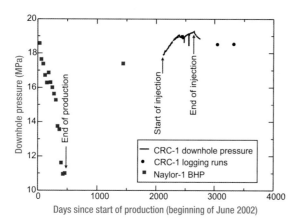

Figure 5.9: Pressure versus time recorded at the Naylor-1 and CRC-1 wells during production (pressure depletion), post-production (recovery), CO$_2$ injection, and post-injection.

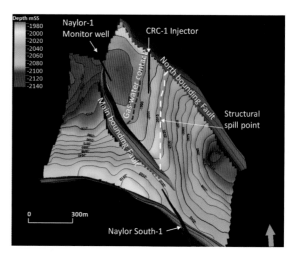

Figure 5.10: The Naylor Field 3D structural model, including top reservoir horizon, faults, spill point, and post production gas-water contact.

Campbell Embayment. The flow rate within the Waarre Formation is quite fast at about 0.39 m/year; estimated assuming an average permeability of 500 md. Although there is strong evidence of regional draw-down, due to a long history of production across the Port Campbell Embayment, the Naylor Field still maintained a relatively rapid pressure recovery. Production at Naylor-1 ceased at the end of October 2003, when the formation pressure was around 10 MPa. When injection began in March 2008, the reservoir pressure had recovered to around 17.8 MPa (Figure 5.9). This indicated a substantial influx of formation water from the aquifer system, and a consequent reduction in capacity given the desire not to exceed the discovery pressure of 19.5 MPa.

Having mapped the structure in detail and derived average porosity from cores and logs, a 3D static geocellular model of the reservoir was constructed (Figure 5.10) from which volumetric-based capacity could be estimated using the equation proposed by the United States Department of Energy (DOE 2006):

$$G_{CO_2} = A \, h_n \, g \, \phi_e \, \rho \, E$$

The bulk rock volume was calculated from the static model by multiplying the reservoir area (A), net gas column height (h_n), and geometry of the structural spill of the three way closure (g). An average effective porosity (ϕ_e), in combination with the bulk rock volume ($A \, h_n \, g$) provides an estimate of total pore space available for storage. The

storage efficiency factor (E) provides a measure of the fraction of this total pore volume from the gas that has been produced and that can be filled by CO$_2$. The storage efficiency factor accounts for irreducible water saturation, as well as an estimate of the irreducible gas saturation. As the structure was interpreted to have been filled to spill point the irreducible gas saturation estimate could be "blanket-applied" to the whole bulk rock volume

A methane gas cap remained at the top of the structure down to 2039.5 mRT at Naylor-1 (equivalent to ~1989 m TVDSS). Below the post-production gas-water contact, prior to injection, the pore space contained an average 20% residual methane saturation, with the remaining 80% being formation water. This was confirmed by the Reservoir Saturation Tool logging at CRC-1 and Naylor-1. On the time scale of the injection period (1–2 years), it was considered that the injected CO$_2$/methane mixed gas could displace some of the formation water but would not access the entire 80% of the pore space previously occupied by formation water. Numerical simulation, run prior to injection, suggested that the water saturation within the reservoir at the end of the injection period would be 40–50%, leaving only 30–40% of the pore space accessible for storage of injected gas. The density of the injected CO$_2$-CH$_4$ mixed gas was 360 kg/m^3, giving an estimated storage capacity within the Naylor Field of between 113,000 and 151,000 t, which exceeded the proposed injection volume, and therefore it was concluded from both capacity estimation methods that the site had more than sufficient capacity to meet the Project aims.

5.5 Reservoir heterogeneity

Reservoir heterogeneity will impact on the migration behaviour of injected CO_2 as well as on storage effectiveness. The spatial distribution of sand and shale bodies as well as their grain size, sorting, roundness, clay content and mineral digenesis can control vertical and horizontal connectivity within a reservoir (Ambrose et al. 2007). Modelling by Flett et al. (2004) found the increased number of baffles in heterogeneous formations (shale layers etc.), results in tortuous migration pathways which slow the movement of CO_2. Hovorka et al. (2004) suggested these tortuous flow paths increase the volume of rock contacted by the CO_2, resulting in a larger net storage capacity for heterogeneous formations. The reservoir quality (porosity and permeability distribution) of the clastic Waarre C Formation is strongly linked to the original depositional facies (see Table 5.2), as there has been only very minor mineral digenesis—mainly kaolinite replacement of feldspar with no net change in porosity. This was advantageous for any reservoir characterisation, because once a depositional model had been established, and the size and spatial distribution of sand bodies and shale baffles ascertained from well-studied reservoir analogues, then the length, width and anisotropy ranges of these depositional facies could be used to predict the flow pathways away from the wells without significant cause for concern that diagenetic

overprinting would complicate the model. Consequently, at the Naylor Field, it was important to characterise the depositional setting and resulting sedimentary features that could have any impact on migration of CO_2 and timing of plume arrival at the Naylor-1 monitoring well.

5.5.1 Sedimentary facies scale

The cored interval from CRC-1 intersected the lower portion of the Flaxman Formation and the Waarre C, terminating a few metres above the Waarre B. High resolution X-ray CT scanning was performed on the CRC-1 cores to visualise the internal sedimentary and structural features at millimetre scale. Few fractures were noted, but fine laminae in the form of thin carbonaceous layers were recorded throughout. A few millimetres of the core were cut from the entire length to provide a clean flat surface along the length of the core; the core was then described in detail and selected photographs taken. Sedimentological interpretations of the cores showed a complex stratigraphy that included incised valley fill deposits within the Waarre C Formation, overlain by transgressive to offshore open marine deposits in the Flaxman Formation (Dance and Vakarelov 2008); a weak unconformity sequence boundary was noted, separating the two units. The interpretation that the two formations were not contemporaneous was supported by biostratigraphic evidence (Partridge 2006), as well as by the nature of the different depositional environments. These included a transgressive tidally-influenced fluvial succession transitioning into a marine-dominated succession in the lower part of the core; a fluvially-dominated stacked channel interval in the middle portion of the core; and a transgressive to

Table 5.2: Results of conventional core analysis and microtomographic derived reservoir quality for each of the depositional facies.

Facies	Porosity %	k mD	Kv/kh	Pore Size	Throat Size	Pore/Throat Ratio	Connectivity
Transgressive sand	10–19	62–2795	0.63	11.2	4.1	2.4	3.4
Channel sands	9–34	8–2428	0.38	–	–	–	–
Gravel dominated	6–28	3–3750	0.4	10.2	5.3	2.4	5.1
Abandoned channel fill	1–3	0.002–0.3	0.9	–	–	–	–
Wave reworked sands	9–14	1–281	0.3	11.6	3.8	3.0	3.3
Tidal sands	18–21	440–6000	0.8	23	10.3	3.5	4.9

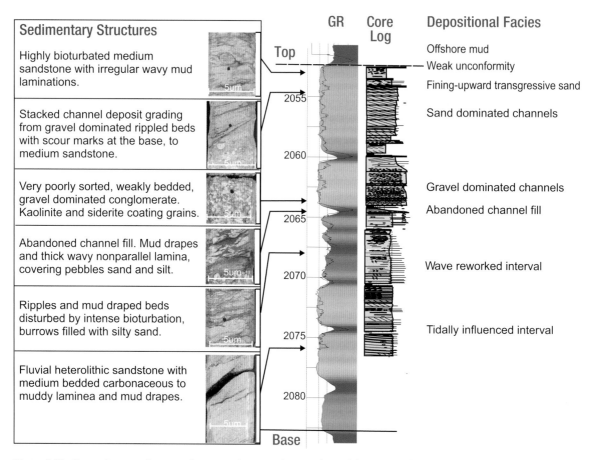

Figure 5.11: Reservoir core sedimentary description log, core photographs, and depositional facies.

offshore marine interval in the Flaxman Formation. The tidally-influenced fluvial interval in the Waarre C was interpreted to have been deposited in an incised valley during a lowstand to transgressive system tract. The fluvial interval was deposited during a drop and then rise of relative sea level, and probably related to a pulse of valley incision followed by valley fill. The Flaxman Formation was deposited during a subsequent transgression (under open marine conditions), floored by a transgressive surface of erosion topping the fluvially-dominated interval of the Waarre C. Tidally-influenced fluvial intervals overlain by restricted marine facies are commonly associated with transgressive estuarine settings occupying former incised valleys (Shanley and McCabe 1993).

Six depositional facies which contribute to the overall heterogeneous nature of the Waarre C Formation (Figure 5.11) were identified. These were defined by their grain fabric and sedimentary structures, which in turn reflected the environment in which they were deposited.

A strong relationship was identified between these six depositional facies and reservoir quality; the facies (sand channels and shales) undoubtedly constrain the spatial arrangement of permeability streaks and low flow baffles between the injector and monitoring wells. A paleo-environmental model was therefore essential to understanding the dimensions and orientations of the six interpreted facies. The stacked nature of the sediments interpreted from core observations suggested the environment was dominated by river courses forced to conform to the north-west/south-east trending topographic troughs.

Regionally, deposition of the Waarre C Formation was probably affected by contemporaneous structural control, which would have had an important influence over orientation of feeder systems, valley incision and marine incursion, suggesting the river courses would be forced to conform to the north-west/south-east trending topographic troughs. Subsequently the study drew on depositional models and analogues for fluvially-dominated low sinuosity channels, feeding shallow marine inlets, frequently influenced by tidal processes and marine storm surges such as those encountered at present-day Hervey Bay, Queensland, Australia (Figure 5.12). Resulting sand and shale distribution appropriate to this type of setting mean the permeability conduits within channels can be expected to be highly connected in excess of the distance between the injection and monitoring wells (> 300 m). Both Naylor-1 and CRC-1 intersected at least two 1–3 m thick shale baffles. The main uncertainty was whether they were continuous or truncated between the wells. This had implications for interpreting vertical connectivity between the injection perforations and the sampling points at Naylor-1 which span these shales. Unlike the channel sands, the distribution of shale-dominated abandoned channel fill was expected to be more restricted due to down-cutting channels eroding the fine grained sediment. Thus the resulting permeability baffles were not expected to be greater than 80 m to 200 m wide.

5.5.2   Pore scale

Recent advances in digital core analysis now mean that pore scale properties can be studied from X-ray microtomographic images (Figure 5.13). The pore and mineral phase structure of the reservoir core material from CRC-1 was enumerated in 3D using X-ray microtomographic technology (Knackstedt et al. 2010). Quantification of the pore space interconnectivity, pore to throat ratio, and pore shape allowed for analysis of the permeability heterogeneity and anisotropy of each sand type present in the reservoir. The results supported the conclusion that reservoir quality and resulting CO_2 flooding processes are related to the different depositional facies.

The X-ray microtomographic study involved sampling four of the facies: (1) the poorly consolidated, poorly sorted

Modern day analogue

Conceptual model

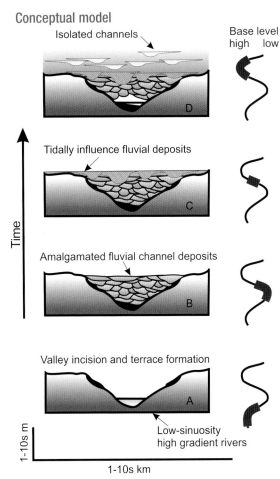

Figure 5.12: Hervey Bay in Australia, a modern day analogue for the paleo-depositional environment for the Waarre C Formation (photograph courtesy of Simon Lang); and a conceptual depositional model for an incised valley fill sequence (modified from Shanley and McCabe 1993).

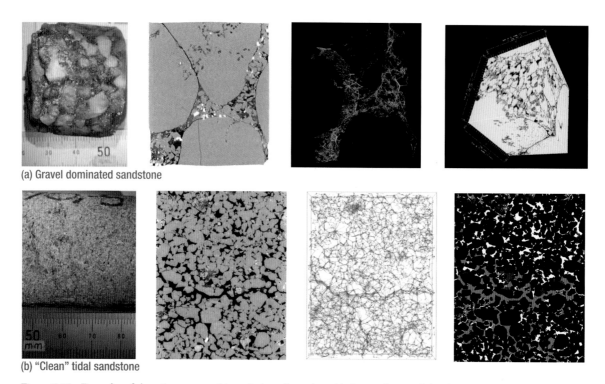

Figure 5.13: Examples of the microtomographic analysis performed on (a) the poorly consolidated gravel; and (b) the well sorted, quartz-rich tidal sandstone; from left to right the images are of the core specimen, an image slice parallel to bedding, the 3D pore network connectivity (green), and the simulated residual non-wetting phase CO_2 (red).

gravel dominated channel sandstone; (2) the fine laminated wave reworked sandstone; (3) the highly bioturbated transgressive sandstone; and (4) the relatively clean, well sorted, well rounded quartz sands of the tidal channel sandstone. 1½″ (38 mm) plugs were imaged with micro-CT at a resolution of ~20 microns. 2D backscattered scanning electron microscopy (SEM) and automated mineralogical identification (QEMSCAN®) data were acquired and registered on the 3D image so that virtual slices of the sample grains, pores and minerals could be viewed from any angle. The samples were flooded with an analogue fluid (n-hexane) that mimics CO_2 behaviour at ambient conditions, and were scanned again at various states of saturation. This provided insight into fluid distribution in the pore spaces (Figure 5.13).

Results are summarised in Table 5.2. Porosity and permeability from conventional core analysis are compared for each facies along with the microtomographic-derived mean pore size, mean throat size, pore to throat aspect

ratio, and connectivity factor for the four facies tested. These parameters were shown to be related to the residual (non wetting phase) fluid saturation. For example low connectivity (< 4) and high aspect ratios have been correlated to high trapped non-wetting phase saturations (Chatzis et al. 1983). The wave-reworked sample appears to be an example of this: it has distinct anisotropy in the pore network due to the strong laminations and relatively good permeability horizontal to bedding (k_x = 130 md and k_y =165 md), but low permeability perpendicular to bedding (< 1 md in the z direction). The low connectivity value of 3.3 suggested quite high trapped non-wetting phase residual saturations. Similarly, the heavily bioturbated sample exhibited lower connectivity (3.4), due to extensive clay-filled burrows and pyrite-rich laminations reducing the porosity. This indicated higher residual (non-wetting phase) trapping was possible in these rocks. For the gravel-dominated sample the horizontal and vertical permeability obtained were 2.8 and 2.1 darcies respectively. The mean connectivity for the sample was relatively high (5.1), due to the large angular grains; therefore the facies had relatively lower residual (non-wetting phase) trapping potential. The clean tidal sandstone had well connected porosity throughout and, at this scale, little heterogeneity was observed. Permeability values were isotropic—in the order of 500 md. The higher connectivity of 4.9 indicated relatively lower potential for high residual (non-wetting phase) saturations.

5.6   Containment

5.6.1   Primary containment mechanisms

Structural trapping was considered to be the dominant mechanism for containment of the CO_2 at the Naylor Field. The CO_2 would rise due to buoyancy, towards the top of the fault-bound trap (as a result of the CO_2 being the non-wetting phase). The CO_2 would settle beneath the methane gas cap at the top of the structure, as it was slightly denser than the methane; nevertheless some mixing of the two gas volumes was expected to occur. This mixed composition, but continuous gas column, was contained by both the overlying seal and the seal juxtaposed across the bounding fault.

Coring at CRC-1 allowed sampling of the primary seals in the Belfast Mudstone and Flaxman Formation. The Belfast Mudstone is an exceptionally good seal, which is known to have diapiric and shale flow features throughout the Otway Basin (Stilwell and Gallagher 2009). Mercury injection capillary pressure tests were conducted on samples of the seal by Daniel (2007). These tests determined threshold or breakthrough pressures which were subsequently used to calculate the carbon dioxide retention height of the sealing rocks (see Chapter 6). Pore throat size distributions were also determined for the analysed samples and the laboratory mercury/air values were converted to equivalent subsurface supercritical carbon dioxide (scCO$_2$) values to determine subsurface water saturation versus height relationships. Experimental evidence by Broseta (2005) and Chiquet et al. (2007) showed that scCO$_2$ may be partially wetting (depending on contact angle), with respect to quartz and mica-rich rocks under subsurface conditions. As a consequence of this evidence, CO_2 column

heights were calculated with contact angle sensitivities from 0° to 60° in 20° increments to indicate the possible minimum column height. For example, at a contact angle of 0° the Belfast Mudstone sample minimum column heights ranged from 607 m to 851 m with an average scCO$_2$ column height of 754 m. However, using a contact angle of 60° the minimum column heights for the same samples ranged from 303 m to 426 m. The maximum possible column height of the plume was expected to be in the order of 43 m, given the top of the structure is at 1972 m TVDSS and the spill point is at 2015 m TVDSS.

Samples from the seal were also analysed using scanning electron microscopy (SEM), which confirmed the geological interpretation and inferred that the formations' exceptionally good sealing capacity was due to burial depth and depositional environment. Significant compaction reduced the mudstone's microporosity; the high percentage of clay minerals, specifically smectite and illite, suggested it was deposited in a distal marine depositional environment. Because it was unconsolidated to semi-consolidated during the syn-depositional phases of tectonic deformation, it is possible that many of the fault planes interpreted to go through the Belfast Mudstone lithology are in fact effectively "annealed", excepting perhaps those reactivated during the late Tertiary. Backing up the assertion that the Belfast Mudstone forms an exceptionally good seal is the fact that there are virtually no significant gas fields in formations shallower than the Waarre Formation, nor any gas effects (e.g. small bright spots on seismic sections indicative of migration of gas through this otherwise immature section) in any reservoirs above the Waarre Formation.

The faults flanking the Naylor Field terminate within the Belfast Mudstone and do not appear to have been reactivated during the Tertiary. Their reactivation potential was the subject of a study by Vidal-Gilbert et al. (2010) that investigated fault activation propensity in both strike slip and normal stress regimes. The minimum pore pressure increase required to cause fault reactivation for optimally-oriented faults ranged from 1 MPa to 37 MPa, with an initial pore pressure at the top of the reservoir of 17.5 MPa, depending on assumptions made about stress regime, fault strength, reservoir stress paths and Biot's coefficient (see Chapter 7).

5.6.2 Secondary containment mechanisms

Residual gas trapping was likely to be a containment mechanism for some of the CO_2 as it migrated updip from the injector to the top of the structure. The CO_2 becomes trapped in the pore space as a residual immobile phase by capillary forces. The injected gas displaces methane and formation water (gas-water relative permeability hysteresis). At the tail of the migrating CO_2 plume, imbibition processes are dominant, as the formation water (wetting phase) re-enters the pore space behind the migrating CO_2 (non-wetting phase). When the saturation level of the CO_2 falls below a certain level, it becomes trapped in the intragranular pore space by capillary pressure forces (snap-off), and ceases to flow (Ennis-King and Paterson 2001; Holtz 2002; Flett et al. 2003; Flett et al. 2004). A trail of residual immobilised CO_2 is left behind the plume as it migrates upward (Juanes et al. 2006). Estimates for the residual gas saturation (S_{gr}) were derived for the Waarre C from the special core analysis (SCAL) tests, RST logging, and digital core analysis. The values for S_{gr} vary between 20–40% and are a function of the ratios of pore throats to pore bodies in the sandstone. As mentioned in the previous section, these characteristics were imaged in detail using X-ray microtomography (Knackstedt et al. 2010). By characterising the formation at multiple scales (in-situ downhole, core plug scale and at the pore scale), this type of secondary containment can be better understood. It is apparent that reservoir heterogeneity in the form of small-scale sedimentary features has a strong relationship with residual trapping potential of the reservoir sands just as it does with K_v/K_h.

Mineral trapping results from the precipitation of new carbonate minerals (Gunter et al. 1993). Long-term trapping of CO_2 in carbonate phases is limited in the Waarre C due to the low abundance of necessary reactive minerals (Schacht 2008). Additionally, the limited contact the injected gas had with moving formation water inhibited the gas reacting with any cations. Similarly, trapping by dissolution of CO_2 into the formation water was limited (Boreham et al. 2011). Conversely, a study of the greensand units of the Flaxman Formation by Watson and Gibson-Poole (2005) found that the mineral trapping potential of this overlying formation provided increased security to CO_2 storage in the Waarre C. Not only does the lower porosity of the greensands slow down the vertical migration of the CO_2 plume, but the higher proportion of labile minerals (carbonate, glauconite, and chlorite) provided the cations necessary for mineral storage of CO_2.

5.7 Site analogue

The Iona Field is a produced natural gas field in south eastern Victoria approximately 20 km east of Naylor Field. 0.532 Bcm of the initial recoverable gas reserve was produced from the field and the depleted field is currently being used as a peak demand underground gas storage site, supplying to the domestic market during the winter months. As the injection reservoir is also the Waarre C, it provides a valuable analogue for the Otway Project. The Iona storage operation commenced in December 2000 with the injection of 0.28 Bcm gas by April 2001 in its first year of operation (Mehin and Kamel 2002). The site has proven capability to be able to inject up to nearly 2000 t of natural gas per day and withdraw around 5000 per day without incident (see Chapter 7).

Eight wells have been drilled in the Iona Field area, in a dense array, where the wells are often less than a few hundred metres apart. This well density offers an insight into the potential variability and heterogeneity of the Waarre C reservoir that is not available at any other location. Significant core, dip meter and reservoir studies have been undertaken in an effort to better characterise the reservoir at this site and optimise engineering practice. The minimum thickness of the Waarre C at the site is 33.1 m and the maximum is 40.2 m. None of the Iona wells have particularly thick shales within the upper sandstone section of the Waarre C unit. The thickest shale occurs in the upper Waarre C section in Iona Observation-1, but is absent in Iona-4, only 500 m away. From dip meter and downhole log interpretation, the general depositional channel orientation is 35–40° (i.e. north-north-east to north-east), and the effective shale-out distance is estimated to be as little as 200 m. Sedimentary event units (e.g. channel bodies) are 2–3 m thick within the sandstone facies. Unlike at Naylor, there is no indication of bioturbation within either the sands or the shales. The most likely paleo-environment, based on the dip patterns, is that the sands of the reservoir interval at Iona were deposited by a high energy braided stream system, with a general north-east/south-west trend, that flowed towards the north-east.

Three of the Iona Field wells have core data for the Waarre C interval, and recent interpretations of the available cores have strengthened the view that the depositional environment was predominantly fluvial (Tenthorey et al. 2013). The dominant lithology for the Waarre C Formation is quartz arenite with minor feldspars. The reservoir quality is exceptional and available core reports show that porosities range from 13 to 30% with the average in the high 20%. The overburden permeability is in the range 244 md to 20 darcies with most samples between 5 darcies and 20 darcies. It was established that, because of the exceptionally high permeability, water flow through the reservoir is strongly channelised. Vertical to horizontal permeability ratios vary from 0.1 to greater than 1. Eight core plugs from Iona-4 underwent special core analysis (SCAL) tests. The average S_{lr} was 9.5%, the average S_{gr} was 34.2%, the average relative permeability k_{rgmax} at S_{gr} was 0.86, and the average relative permeability to water at residual gas saturation was 0.11.

The Iona Field Waarre C unit is not exactly analogous to the reservoir encountered at Naylor Field. The interpretation that sediments at Iona are almost entirely fluvial is in contrast to the tidal channels and marine influence observed at Naylor, including marine biota in the biostratigraphic studies. However, reservoir quality at both Naylor and Iona is exceptionally good; porosities within the sands are generally in the range 25–30% and the associated permeability (at reservoir conditions) is in the multi-darcy range. Consequently injectivity is excellent. This geological information was most relevant in the early stages of CO2CRC Otway Project site selection as it implied, along with data from other wells, that good reservoir quality was regional and that the Waarre C reservoir thickness would be relatively uniform in the immediate area of Naylor (Spencer et al. 2006).

Iona was also useful as an analogue to complement the geomechanical modelling (Chapter 7). There was a wealth of engineering data from the constant injection/ withdrawal cycles that could be used to improve the

understanding of the mechanical stresses on the reservoir and seal. Tenthorey et al. (2013) analysed the results of dynamic simulations of the injectivity, pressure evolution, storage capacity and maximum fluid pressures sustained by the faults. The geomechanical simulations for the Iona Field were rerun using CO_2 gas instead of methane in order to evaluate the effects of the different physical properties on fault seal retention column heights (i.e. wetting behaviour). Modelling the worst case scenario, where the faults have no cohesion, it was found the faults could sustain 2 MPa of pore-pressure increase without reactivation. In the more than 10 years of operation, the Iona Field has experienced pressure oscillations in the order of 1–2 MPa with no observable seismicity (Tenthorey et al. 2010; 2013). Similarly modelling at the Naylor Field suggested that faults cutting across the Waarre C interval possess some cohesion and the bounding faults at the Naylor Field have shown no signs of reactivation under the injection pressures (see Chapter 7).

5.8 The evolution of the static models

5.8.1 Phase 2 pre-feasibility uncertainty models

As already mentioned, in the initial pre-feasibility phase, there were only limited data available to guide the reservoir modelling. There was no core from the field and therefore no evidence for sedimentological facies modelling and no direct measurement for porosity and permeability. Vital checkshot data were missing in order to constrain the time-to-depth conversion of seismic horizons that defined the structural model. Well information was supplemented from regional data and extrapolated from adjacent fields (Spencer et al. 2006). Adding to this uncertainty, there were two plausible regional paleo-depositional models for the Waarre C Formation: (1) a transgressive shoreline model, whereby the main depositional trends are perpendicular to the expected CO_2 flow direction (Buffin 1989); and (2) a braided fluvial model (Faulkner 2000) where reservoir sands are highly connected and deposited parallel to the direction of flow. A series of static models were created to investigate the optimal location for the injector well

(CRC-1). Along with the two depositional model cases mentioned above, two additional extreme cases for reservoir connectivity were also investigated: (1) a fast migration model with high permeability and no shale baffles, and (2) a slow migration model with low permeability with major barriers to flow. These were considered geologically improbable but remotely possible. Without cores it was necessary to include these models for completeness, in order to understand the end member limits to the geological uncertainty.

Uncertainty in the seismic interpretation was also investigated by modelling several different possibilities for the structural dip of the reservoir that would result from shifts in the depth conversion of the horizons. This was important in order to risk the potential impacts of breakthrough occurring too early to record meaningful results, or the risk of CO_2 not arriving at the monitoring bore in the Project time frame. All of the cases were simulated against the production data from Naylor-1 using the ECLIPSE™ modelling package (Xu et al. 2006). The extreme cases could not match the pressure or production data and so these cases were discounted. The braided fluvial model provided a suitable history match with the least adjustments required to the bulk permeability. Results from a biostratigraphic study (Partridge 2006) became available during the course of these initial simulations and further discounted the transgressive shoreline model, lending more weight to the braided fluvial model (Spencer et al. 2006). This model characterised the reservoir as having a net to gross of 90%, (i.e. 90% sands to 10% shale) and an average of 700 mD for bulk permeability. The optimal placement of the well, which was derived from these results, was between 280 m and 300 m in the downdip direction from Naylor-1, with breakthrough of injected CO_2 at the Naylor well, predicted to be between 6 and 14 months from the commencement of injection.

5.8.2 Phase 3 detailed pre-injection static models

A new suite of static models were created; these incorporated geological details such as porosity, permeability, pressure, and the geometry of the reservoir including faults, sedimentary layers, and facies (rock types) distribution. These are the primary characteristics controlling the behaviour of stored

CO_2. The underlying PETREL™ geo-cellular grid is based on a UTM projection of a 20 m × 20 m optimally-oriented (for the direction of flow) grid. Layering was 0.5 to 2 m thick and upscaling of the petrophysical logs to this resolution appeared to capture the vertical variation in the data and adequately represent vertical permeability. The PETREL™ model and a subsequent ECLIPSE™ model used for pre-injection modelling (Underschultz et al. 2011) uses an irregular cell geometry which honours the geometry of stratigraphic bedding, i.e. on-lap and erosional surfaces between the incised valley fill and the overlying transgressive sands. The ECLIPSE™ grid maintained this geometry and also incorporated local grid refinement to investigate near-wellbore effects. The TOUGH2 simulation grid converted the geological model to a regular grid (see Chapter 16).

Because of the strong relationship identified between the six depositional facies and reservoir quality, the new models used facies objects (sand channels and shales) to constrain the spatial arrangement of permeability streaks and low flow baffles between the wells. Similarly a method of incorporating facies-based permeability anisotropy was developed, and was set directly in ECLIPSE™. K_v/K_h ratios from core were applied as multipliers for each facies code; this was an improvement on just using one ratio for the whole reservoir because the impact of low vertical permeability in the shale-rich facies could now be discretely assessed. Depositional analogues, appropriate to this type of setting, indicated the permeability conduits within channels were likely to be highly connected over the distance of 300 m between the injector and monitoring well.

Due to the fact that the lengths predicted for sand bodies exceeded the correlation distance between wells, there was less uncertainty in the stochastic modelling. There was also uncertainty in the distribution of shale barriers as analogues suggested they were not expected to be greater than 80–100 m wide due to down-cutting channels eroding the fine grained sediment. Both Naylor-1 and CRC-1 intersected at least two 1–3 m shale intervals. The main uncertainty was whether these intervals were continuous or truncated between the wells; this had implications for interpreting vertical connectivity between the injection perforations and the U-tubes which spanned these shales in both wells. Two cases were then created to address this (Figure 5.14). Case 1 used a small correlation length for the shales (60–80 m), and Case 2 used a long correlation

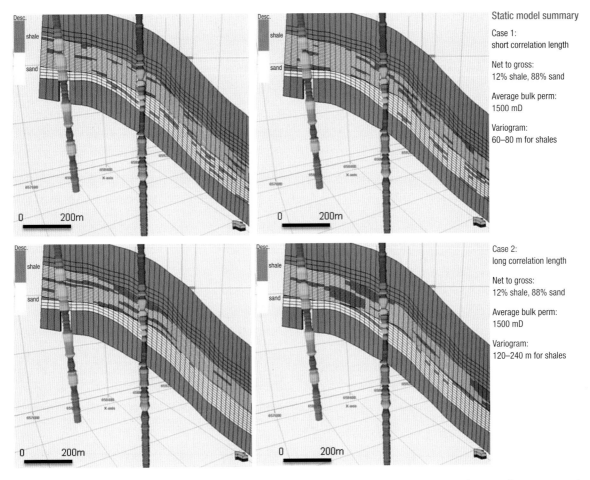

Static model summary

Case 1:
short correlation length

Net to gross:
12% shale, 88% sand

Average bulk perm:
1500 mD

Variogram:
60–80 m for shales

Case 2:
long correlation length

Net to gross:
12% shale, 88% sand

Average bulk perm:
1500 mD

Variogram:
120–240 m for shales

Figure 5.14: Examples of static model realisations (2 of 5 realisations shown for each) in cross-section between the injection and monitoring boreholes. These were used in the pre-injection simulation and characterisation phase.

length (120–240 m). Five equi-probable realisations were generated for each case, giving a total of 10 models. These are summarised in Figure 5.14. Prior to injection, dynamic simulation was performed on four of the 10 static models using ECLIPSE™; the resulting predictions suggested an expected arrival time of CO_2 at the monitoring bore of between 4 and 8 months.

5.9   Conclusions

The site characterisation of the Naylor Field for the CO2CRC Otway Project was essential to establish the injection rates and CO_2 volumes that could be accommodated and that long-term containment would

be assured. The workflow comprised four distinct phases: (1) an initial site screening phase; (2) a pre-feasibility phase, which assessed the regional geology and utilised existing data; (3) the targeted characterisation phase, which involved data acquisition specifically to address uncertainty unique to the CO_2 storage concept; and finally (4) the post-injection model calibration phase which was used to assess the storage performance. The geo-engineering workflow identified five key criteria that should be addressed: site details, injectivity, capacity, reservoir heterogeneity and containment. Cores, well logs (in particular formation microimaging (FMI), nuclear magnetic resonance (NMR), and modular dynamic testing (MDT) samples), and seismic data acquired by

CO2CRC, were necessary to better understand reservoir and seal properties. Specialised analysis such as SCAL core flooding, X-ray microtomography and tri-axial rock mechanics were employed to better understand reservoir potential and fracture limits. Other outcomes from the study were:

> Reservoir characterisation needed to extend to the regional scale, in order to understand the hydrodynamic relationship a proposed injection site may have on adjacent producing fields and overlying aquifers, particularly if the latter contained important freshwater resources.

> Injectivity may be hindered by low permeability, mobilisation of fines, mineral precipitation, or excessive pressure build up. Cores, well testing, and time-lapse pressure measurements were required to assess these effects. These types of data may not always be available from the field production phase.

> When calculating pre-injection storage capacity, a storage efficiency factor is applied to account for both irreducible water saturation and the irreducible gas saturation. Managing the volumes injected within the pressure limits of the field meant the capacity estimate would change over time, particularly in the case of a depleted field with strong water recharge. The influx of water and partial repressurisation due to a strong aquifer drive meant the volume of gas produced did not equate to CO_2 equivalent capacity.

> Reservoir heterogeneity at the facies scale impacted on plume migration and preferential channels flow through high permeability streaks. Pore scale heterogeneity impacts on residual trapping potential and vertical versus horizontal permeability within reservoir sands.

> Combined petrographic analysis, mercury injection capillary pressure tests, and stratigraphic mapping were necessary to characterise the overlying mudstone seal. The potential for CO_2-mineral reactions within the overlying seal meant containment may be compromised by dissolution or further enhanced by precipitation.

> Underground gas storage facilities can provide useful analogues to CO_2 storage sites, particularly if they are in close proximity, or have similar reservoir settings to the planned injection sites as was the case with the Iona facility and the Otway site. Residual gas estimates, injectivity rates, rock mechanics, geo-body connectivity and heterogeneity (for reservoir modelling) all provided substitute information in situations where data were otherwise lacking.

Site characterisation will always be specific to project objectives such as how much CO_2 is to be injected and at what rate. The extent of data acquisition and analysis will impact on the level of risk at the site. Above all, site characterisation should aim to address the uncertainty surrounding risks and should be tested and updated as necessary during the storage performance monitoring of the project.

5.10   References

Ambrose, W.A., Lakshminarasimhan, S., Holtz, M.H., Núñez-López, V., Hovorka, S.D. and Duncan, I. 2007. Geologic factors controlling CO_2 storage capacity and permanence: case studies based on experience with heterogeneity in oil and gas reservoirs applied to CO_2 storage. Environmental Geology 54, 1619–1633.

Boreham, C., Underschultz, J., Stalker, L., Kirste, D., Freifeld, B., Jenkins, C. and Ennis-King, J. 2011. Monitoring of CO_2 storage in a depleted natural gas reservoir: gas geochemistry from the CO2CRC Otway Project, Australia. International Journal of Greenhouse Gas Control 5(4), 1039–1054.

Boult, P.J., White, M.R., Pollock, R., Morton, J.G.G., Alexander, E.M. and Hill, A.J. 2002. Lithostratigraphy and environments of deposition. In: Boult, P.J. and Hibburt, J.E. (Eds.), The petroleum geology of South Australia. Volume 1: Otway Basin, South Australia. Department of Primary Industries and Resources. Petroleum Geology of South Australia Series 1, 2nd edition.

Buffin, A.J. 1989. Waarre Sandstone development within the Port Campbell Embayment. Australian Petroleum Exploration Association Journal 29(1), 299–311.

Burton, M., Kumar, N. and Bryant, S.L. 2009. CO_2 injectivity into brine aquifers: Why relative permeability matters as much as absolute permeability. Energy Procedia 1(1), 3091–3098.

Causebrook, R., Dance, T. and Bale, K. 2006. Southern Perth Basin site investigation and geological model for storage of carbon dioxide. CO2CRC Report Number: RPT06-0162, Canberra, Australia. Available at http://www.co2crc.com.au/publications.

Chadwick, A., Arts, R., Bernstone, C., May, F., Thibeau, S. and Zweigel, P. (Eds.), 2007. Best practice for the storage of CO_2 in saline aquifers – Observations and guidelines from the SACS and CO2STORE Projects: Nottingham. British Geological Survey.

Chatzis, I., Morrow, N.R. and Lim, H.T. 1983. Magnitude and detailed structure of residual oil saturation. SPE Journal 23, 311–326.

Chiquet, P. and Broseta, D. 2005. Capillary alteration of shaly caprock minerals by carbon dioxide. Society of Petroleum Engineers, Paper No. 94183.

Chiquet, P., Broseta, D., and Thibeau, S. 2007. Wettability alteration of caprock minerals by carbon dioxide. Geofluids 7 (2), 112–122.

Cook, P.J., Rigg, A.J. and Bradshaw, J. 2000. Putting it back from where it came from: Is geological disposal of carbon dioxide an option for Australia? APPEA Journal 40(1), 654–666.

Dance, T. and Vakarelov, B. 2008. Sedimentology of the Flaxman Formation and Waarre C Formation from the CRC-1 cores 5&6. CO2CRC Report No: RPT08-1293, Canberra, Australia. Available at http://www.co2crc.com.au/publications.

Dance, T., Spencer, L. and Xu, J. 2009. Geological characterisation of the Otway Project pilot site: What a difference a well makes. Energy Procedia 1(1), 2871–2878.

Dance, T. 2013. Assessment and geological characterisation of the CO2CRC Otway Project CO_2 storage demonstration site: From prefeasibility to injection. Marine and Petroleum Geology 46, 251–269.

Daniel, R.F. 2007. Carbon dioxide seal capacity study, CRC-1, CO2CRC Otway Project, Otway Basin, Victoria. CO2CRC Report No. RPT07-0629, 67, Canberra, Australia. Available at http://www.co2crc.com.au/publications.

Dodds, K., Daley, T., Freifeld, B., Urosevic, M., Kepic, A. and Sharma, S. 2009. Developing a monitoring and verification plan with reference to the Australian Otway CO_2 pilot project. The Leading Edge, vol. 28 (7), 812-818. DOI: 10.1190/?1.3167783.

DOE 2006. Carbon sequestration atlas of the United States and Canada: Appendix A – Methodology for development of carbon sequestration capacity estimates. National Energy Technology Laboratory, Department of Energy. Available from: http://www.netl.doe.gov/publications/carbon_seq/atlas/Appendix%20A.pdf.

Duran, J. 1986. Geology, hydrogeology and groundwater modelling of the Port Campbell hydrogeological sub-basin, Otway Basin, S.W. Victoria. Report 1986/24. Geological Survey of Victoria, Victoria.

Ennis-King, J. and Paterson, L. 2001. Reservoir engineering issues in the geological disposal of carbon dioxide. In: Williams, D.J., Durie, R.A., McMullan, P., Paulson, C.A.J. and Smith, A.Y. (Eds.), Proceedings of the Fifth International Conference on Greenhouse Gas Control Technologies (GHGT-5), 13–16 August, Cairns, Australia. CSIRO Publishing, 290–295.

Faulkner, A. 2000. Sequence stratigraphy of the late Cretaceous Sherbrook Group, Shipwreck Trough, Otway Basin; B.Sc. (Hon) Thesis. National Centre for Petroleum Geology and Geophysics, University of Adelaide.

Flett, M.A., Taggart, I.J., Lewis, J. and Gurton, R. M. 2003. Subsurface sensitivity study of geologic CO_2 sequestration in saline formations. In: The Second Annual Conference on Carbon Sequestration, 5–8 May, Alexandria, USA. National Energy Technology Laboratory, United States Department of Energy.

Flett, M., Gurton, R.M. and Taggart, I.J. 2004. Heterogeneous saline formations: long-term benefits for geo-sequestration of greenhouse gases. GHGT7: Proceedings of the 7th International Conference on Greenhouse Gas Technologies, 5–9 September, Vancouver, Canada.

Flett, M., Beacher, G., Brantjes, J., Burt, A., Dauth, C., Koelmeyer, F., Lawrence, R., Leigh, S., McKenna, J., Gurton, R., Robinson IV, W.F. and Tankersley, T. 2008. Gorgon Project: Subsurface evaluation of carbon dioxide disposal under Barrow Island. SPE Asia Pacific Oil and Gas Conference and Exhibition, 20–22 October, Perth, Australia, p. 21.

Förster, A., Norden, B., Zinck-Jørgensen, K., Frykman, P., Kulenkampff, J., Spangenberg, E., Erzinger, J., Zimmer, M., Kopp, J., Borm, G., Juhlin, C., Cosma, C. G. and Hurter, S. 2006. Baseline characterization of the CO2SINK geological storage site at Ketzin, Germany. Environmental Geoscience 13, 145–161.

Förster, A., Schoner, R., Forster, H.-J., Norden, B., Blaschke, A.-W., Luckert, J., Beutler, G., Gaupp, R. and Rhede, D. 2010. Reservoir characterisation of a CO_2 storage aquifer: the Upper Triassic Stuttgart Formation in the Northeast German Basin. Marine and Petroleum Geology 27, 2156–2172.

Gallagher, S.J., Taylor, D., Apthorpe, M., Stilwell, J.D., Boreham, C.J., Holdgate, G.R., Wallace, M.W. and Quilty, P.G. 2005. Late Cretaceous dysoxia in a southern high latitude siliclastic succession, the Otway Basin, southeastern Australia. Palaeogeography, Palaeoclimatology, Palaeoecology 223, 317–348.

Geary, G.C. and Reid, I.S.A. 1998. Hydrocarbon prospectivity of the offshore eastern Otway Basin, Victoria for the 1998 Acreage Release. Victorian Initiative for Minerals and Petroleum Report 55, Department of Natural Resources and Environment.

Gunter, W.D., Perkins, E.H. and McCann, T.J. 1993. Aquifer disposal of CO_2-rich gases: reaction design for added capacity. Energy Conversion and Management 34(9–11), 941–948.

Hill, K.C. and Durrand, C. 1993. The western Otway Basin: an overview of the rift and drift history using serial composite seismic profiles. PESA Journal 21, 67–78.

Holloway, S. and van der Straaten, R. 1995. The Joule II Project: the underground disposal of carbon dioxide. Energy Conversion and Management 36(6–9), 519–522.

Holtz, M.H. 2002. Residual gas saturation to aquifer influx: a calculation method for 3-D computer reservoir model construction. SPE Gas Technology Symposium, 30 April – 2 May, Calgary, Alberta, Canada. Society of Petroleum Engineers, SPE Paper 75502.

Hortle, A., de Caritat, P., Stalvies, C. and Jenkins, C. 2011. Groundwater monitoring at the Otway project site, Australia. Energy Procedia 4, 5495–5503.

Hortle, A., Xu, J. and Dance, T. 2008. Hydrodynamic interpretation of the Waarre Formation Aquifer in the onshore Otway Basin: Implications for the CO2CRC Otway Basin Project. Proceedings of the 9th International Conference on Greenhouse Gas Technologies (GHGT-9), 16–20 November, Washington DC.

Hortle, A., Xu, J. and Dance, T. 2009. Hydrodynamic interpretation of the Waarre Fm Aquifer in the onshore Otway Basin: Implications for the CO2CRC Otway Project. Energy Procedia 1(1), 2895–2902.

Hovorka, S.D., Doughty, C., Benson, S.M., Pruess, K. and Knox, P.R. 2004. The impact of geological heterogeneity on CO_2 storage in brine formations; a case study from the Texas Gulf Coast. In: Baines, S.J. and Worden, R.H. (Eds.), Geological storage of carbon dioxide. Geological Society Special Publication 233, 147–163.

Jenkins, C., Cook, P., Ennis-King, J., Underschultz, J., Boreham, C., de Caritat, P., Dance, T., Etheridge, D., Hortle, A., Frefeld, B., Kirste, D., Paterson, L., Pevzner, R., Schacht, U., Sharma, S., Stalker, L. and Urosevic, M. 2012. Safe storage and effective monitoring of CO_2 in depleted gas fields. Proceedings of the National Academy of Science of the USA 109(2), 353–354.

Juanes, R., Spiteri, E.J., Orr, F.M. and Blunt, M.J. 2006. Impact of relative permeability hysteresis on geological CO_2 storage. Water Resources Research 42(12), 1–13.

Knackstedt, M., Dance, T., Kumar, M., Averdunk, H. and Paterson, L. 2010. Enumerating permeability, surface areas, and residual capillary trapping of CO_2 in 3D: Digital analysis of CO2CRC Project core. Presented at SPE Annual Technical Conference and Exhibition, Florence, Italy, 19–22 September. SPE paper 134625.

Kopsen, E. and Scholfield, T. 1990. Prospectivity of the Otway Supergroup in the central and western Otway Basin: Australian Petroleum Exploration Association Journal 30(1), 263–279.

Krassay, A.A., Cathro, D.L. and Ryan, D.J. 2004. A regional tectonostratigraphic framework for the Otway Basin. PESA Eastern Australasian Basins Symposium II, 97–116.

Laing, S., Dee, C. N. and Best, P.W. 1989. The Otway Basin. Australian Petroleum Exploration Association Journal 29(1), 417–429.

Mehin, K. and Kamel, M. 2002. Gas resources of the Otway Basin in Victoria. Department of Natural Resources and Environment (State of Victoria), Petroleum Development Branch, ISBN 0 7306 9481 X. Available at http://www.nre.vic.gov.au/minpet/index.htm.

Morton, J.G.G., Alexander, E.M., Hill, A.J. and White, M.R. 1995. Lithostratigraphy and environments of deposition. In: Morton, J.G.G. and Drexel, J.F. (Eds.), The petroleum geology of South Australia, Volume 1: Otway Basin. Mines and Energy South Australia, Report 95/12, 127–139.

Partridge, A.D. 2001. Revised stratigraphy of the Sherbrook Group, Otway Basin. In: Hill, K.C. and Bernecker, T. (Eds.), PESA Eastern Australasian Basins Symposium, 455–464.

Partridge, A.D. 2006. Depositional environments of the Waarre C unit – the palynological constraints. International Biostrata Pty Ltd Report 2005/24.

Perrin, J.C., Krause, M., Kuo, C.W., Miljkovic, L., Charoba, E. and Benson, S.M. 2009. Core-scale experimental study of relative permeability properties of CO_2 and brine in reservoir rocks. Energy Procedia 1, 3515–3522.

Sayers, J., Marsh, C., Scott, A., Cinar, Y., Bradshaw, J., Hennig, A., Barclay, S. and Daniel, R. 2007. Assessment of a potential storage site for carbon dioxide: a case study, southeast Queensland, Australia. Environmental Geosciences 13(2), 123–142.

Schacht, U. 2008. Petrological investigations of CRC-1 core plug samples: Waarre C, Flaxman, and Paaratte Formation. CO2CRC report number RPT08-1234, Canberra Australia. Available at http://www.co2crc.com.au/publications.

Shanley, K.W. and McCabe, P.J. 1993. Alluvial architecture in a sequence stratigraphic framework: a case history from the Upper Cretaceous of southern Utah, USA. In: Flint, S.S. and Bryant, I.D. (Eds.), The geological modeling of hydrocarbon reservoirs and outcrop analogues. International Association of Sedimentologists Special Publication 15, 21–56.

Spencer, L., Xu, Q., LaPedalina, F. and Weir, G. 2006. Site characterization of the Otway Basin Storage Pilot in Australia. Proceedings of the 8th International Conference on Greenhouse Gas Control Technologies (GHGT-8), 19–22 June, Trondheim, Norway.

Stevens, S.H., Kuuskraa, V.A, Gale, J.J. and Beecy, D. 2000. CO_2 injection and sequestration in depleted oil and gas fields and deep coal seams; worldwide potential and costs. AAPG Bulletin 84, 1497–1498.

Stilwell, J.D. and Gallagher, S.J. 2009. Biotratigraphy and macroinvertebrate palaeontology of the petroleum-rich Belfast Mudstone (Sherbrook Group, uppermost Turonian to mid-Santonian), Otway Basin, southeastern Australia. Cretaceous Research 30, 873–884.

Underschultz, J., Boreham, C., Dance, T., Stalker, L., Freifeld, B., Kirste, D. and Ennis-King, J. 2011. CO_2 storage in a depleted gas field: an overview of the CO2CRC Otway Project and initial results. International Journal of Greenhouse Gas Control 5(4), 922–932.

Tenthorey, E., Nguyen, D. and Vidal-Gilbert, S. 2010. Applying underground gas storage experience to geological carbon dioxide storage: A case study from Australia's Otway Basin. Proceedings of the 10th International Conference on Greenhouse Gas Control Technologies (GHGT-10), Amsterdam, The Netherlands, 19–23 September.

Tenthorey, E., Vidal-Gilbert, S., Backe, G., Puspitasari, R., John, Z., Maney, B. and Dewhurst, D. 2013. Modelling the geomechanics of gas storage: A case study from the Iona Gas Field, Australia. International Journal of Greenhouse Gas Control 13, 138–148.

van Ruth, P. and Rogers, C. 2006. Geomechanical analysis of the Naylor Field, Otway Basin, Australia. CO2CRC Report number RPT06-0039, Canberra Australia. Available at http://www.co2crc.com.au/publications.

Vidal-Gilbert, S., Tenthorey, E., Dewhurst, D., Ennis-King, J., Van Ruth, P. and Hillis, R. 2010. Geomechanical analysis of the Naylor Field, Otway Basin, Australia: implications for CO_2 injection and storage. International Journal of Greenhouse Gas Control 4(5), 827–839.

Watson, M. and Gibson-Poole, C. 2005. Reservoir selection for optimised geological injection and storage of carbon dioxide: a combined geochemical and stratigraphic perspective. In Conference Proceedings of the Fourth Annual Conference on Carbon Capture and Sequestration, DOE/NETL, May, Virginia, United States.

Woollands, M.A. and Wong, D. (Eds). 2001. Petroleum atlas of Victoria. Victorian Department of Natural Resources and Environment. Available at http://dpistore.efirst.com.au/product.asp?pID=9&cID=6.

WRI. 2008. CCS Guidelines: Guidelines for carbon dioxide capture, transport, and storage. Washington, DC, World Resources Institute, 144.

Xu, J., Weir, G., Ennis-King, J. and Paterson L. 2006. CO_2 migration modelling in a depleted gas field for the Otway Basin CO_2 storage pilot in Australia. Proceedings of the 8th International Conference for Green House Gas Control Technologies (GHGT-8), 19–22 June, Trondheim, Norway.

Richard Daniel, John Kaldi

6. EVALUATING CO$_2$ COLUMN HEIGHT RETENTION OF CAP ROCKS

6.1 Introduction

The evaluation of regional and local top seals has been a significant component of petroleum exploration. Seal evaluation was used for the Otway Project in order to determine whether a confining seal would support an injected volume of CO$_2$ for a significant period of time.

Seal capacity or column height determination, using mercury injection capillary pressure (MICP) analysis, has been utilised in the petroleum industry since the technique was developed by Purcell (1949) and refined by Picknell et al. (1966) and Wardlaw and Taylor (1976).

The sealing capacity of a rock is a function of pore throat size, contact angle (wettability) and interfacial tension. The column height of CO$_2$ in a reservoir increases as (1) pore throat size in the seal decreases; (2) the contact angle between CO$_2$/water/rock decreases and (3) the interfacial tension between CO$_2$ and water increases. Wettability is a function of both interfacial tension (σ) and contact angle (cos θ) (though the contact angle alone is commonly used interchangeably with wettability).

The pressure at which mercury first enters the sample (after the mercury has filled any surface irregularities in the sample) is termed the entry or displacement pressure (P_d). The capillary threshold pressure (P_{th}) is the pressure at which the non-wetting phase (mercury or CO$_2$) begins to flow through the rock as a continuous phase. This is further discussed in Section 6.6. The pressure is determined from the mercury injection capillary pressure graph, which has the incremental pore volume on the secondary axis; it is the combination of these two graphs which gives the most repeatable result for P_{th}. Buoyancy pressure is the pressure exerted due to the difference in density between the wetting phase (commonly water) and the non-wetting phase (commonly hydrocarbons or CO$_2$). The difference in buoyancy pressure between the two fluids (CO$_2$ and water), together with the threshold pressure of the CO$_2$, makes it possible to determine column height and hence storage or reservoir volume.

6.2 Mercury injection capillary pressure

Mercury injection capillary pressure (MICP) analysis is based on a technique which is rate-limited, as noted in the Darcy equation. This equation, after Purcell (1949), describes the general function of flowrate vs pressure drop:

$$V = k \frac{g_c}{\mu} \frac{\Delta P}{L} \qquad (1)$$

where V is the superficial velocity of fluid, k is permeability, g_c is a dimensional gravity constant, μ is the fluid viscosity, ΔP is the pressure differential and L is the pore length.

The velocity of flow in a viscous liquid, such as mercury, is proportional to the pressure drop and inversely proportional to the length and surface area of the pore. Hence, given a specific limited flow velocity, the complete filling of a porous network will be a function of time. The larger the volume of pores, the more time is required to fill the total pore volume. Therefore mercury porosimetry is most accurate when mercury is given sufficient time to fill all the available pores at the same pressure (equilibration).

6.3 Methodology

The mercury injection porosimetry analyses for rocks evaluated in this study were carried out using a Micromeritics Autopore 9410 instrument. This instrument comprises two separate systems: one for low pressures and the second for high pressures. The low pressure run must always be done first, followed quickly by the high pressure run, to preclude the possibility of extra mercury intrusion into the sample by capillary action while the sample is held at atmospheric pressure at the conclusion of the low pressure run.

The system operates using the equilibration by time method—after the required pressure for a reading is attained, it is held for 25 seconds to allow the amount of mercury entering the pores to stabilise. This is done because the process of mercury filling the pores is not an instantaneous one. Mercury begins entering the pores as soon as the pressure exceeds the value required for the pore throat diameter, but the time required to fill the

pores depends on the volume and shape of the pores. The equilibration over time allows the pores to fill. If equilibration is not attained, then the filling may not be complete when the reading is taken, which leads to estimation of lower pore volumes and smaller pore sizes than is actually the case. Readings of mercury intrusion are taken by measuring the electrical capacitance of the penetrometer. This varies as the mercury is intruded from the precision bore stem into the pore space of the sample during the analysis.

Each sample is usually dried at 60°C for at least 24 hours, weighed and placed into a penetrometer (a glass chamber attached to a precision bore glass tube, which has been nickel-plated) and the entire assembly is weighed. This is placed in the low pressure port and evacuated to 0.05 torr. This vacuum is held for 30 minutes to ensure that no vapour remains in the sample. After this time the penetrometer is filled with mercury and the low pressure run is carried out. The pressure is increased incrementally from 13.8 kPa (2 psia) to 199.5 kPa (28.94 psia), with a reading taken after 25 seconds of equilibration at each pressure. At the end of the low pressure run, the penetrometer returns to atmospheric pressure. The sample is removed from the instrument and weighed to obtain its weight plus that of the mercury.

The penetrometer is then placed in the high pressure chamber, which uses hydraulic oil to take the pressure incrementally from 199.5 kPa (28.94 psia) to 413.7 MPa (60,000 psia). Again readings are taken after a 25 second equilibration period. The pressure is then decreased incrementally from 413.7 MPa (60,000 psia) to 139 kPa (20 psia), with readings taken after the equilibration period. The sample is removed from the penetrometer and weighed. The specific gravity and porosity of the sample can be calculated once it is removed from the penetrometer and weighed.

6.4 Pore throat size determination

MICP analysis uses the physical principle that a non-reactive, non-wetting liquid will only penetrate a porous medium once sufficient pressure is applied to force its entrance into the pore system. The relationship between the applied pressure and the pore throat radius into which mercury will intrude is given by the modified Washburn (1921) equation, as suggested by Purcell (1949) and Schowalter (1979):

$$r = 2\sigma\cos\theta/P_c \qquad (2)$$

where P_c is the applied capillary pressure, r is the pore throat radius, σ is the surface tension of mercury and θ is the contact angle between mercury and the pore wall (Figure 6.1).

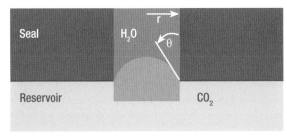

$$r = 2\sigma\cos\theta/P_c$$

r = pore throat radius θ = contact angle, CO$_2$/water/rock
σ = interfacial tension, CO$_2$-water P_c = capillary pressure

Figure 6.1: Diagram showing the relationship between pore throat radius and capillary pressure, interfacial tension, contact angle, using Equation 2, i.e. the smaller the pore throat, the higher the pressure required to intrude the throat (after Purcell 1949).

These equations assume that all pores are circular cylinders. As pressure increases during analysis, the MICP instrument senses the intrusion volume of mercury by the change in capacitance between the mercury column and a metal sheath surrounding the stem of the penetrometer (Vavra et al. 1992). The pressure and volume data are continuously acquired as the mercury column shortens in the stem and penetrates the sample.

The following values for the air-mercury system can be used to convert these capillary pressure data to effective pore throat size (Vavra et al. 1992):

Air/mercury contact angle ($\theta_{a/m}$) = 140°

Interfacial tension ($\sigma_{a/m}$) = 486 mN/m (micronewton/metre or dyne/cm)

Wettability ($\sigma_{a/m} \times \cos\theta_{a/m}$) ≈ 372

These result in the following general relationship of capillary pressure to pore throat radius:

1 psi ≈ 100μm	100 psi ≈ 1 μm
10 psi ≈ 10 μm	1000 psi ≈ 0.1 μm

6.5 CO$_2$ contact angle

Until recently gas in the subsurface, whether hydrocarbon gas or CO$_2$/water/rock, was assumed to have a contact angle (θ) of 0°, as water was thought to be the wetting phase (Schowalter 1979). Experimental work by Morrow et al. (1990) to determine the wettability factor used the general assumption that most rocks are preferentially water wet, and can therefore be expressed as:

$$wettability = \sigma_{b/CO_2}.\cos\theta_{b/CO_2}$$

where σ is the interfacial tension, θ is the contact angle and b/CO_2 is the brine/CO$_2$ system.

Pore-level injection modelling, incorporating a distribution of varying pore throat radii and formation wettability (including contact angle, interfacial tension (IFT) and fluid viscosities) developed by Ferer et al. (2002), also assumed that the injected CO$_2$ was immiscible in a water-wet porous medium (CA = 0°). However Span and Wagner (1996) have found that brine can contain up 58 kg of scCO$_2$ per 1000 kg of formation brine, which has the affect of modifying the supercritical CO$_2$ (scCO$_2$)/water/rock contact angle.

CO$_2$ column heights have been calculated for the Muderong Shale (a major top seal in the Carnarvon Basin, NW Shelf, Australia) by Dewhurst et al. (2002), who used the following data: interfacial tension of 25 mN/m, CO$_2$ density of 0.65 g/cm^3 and water density of 1.05 g/cm^3. A contact angle sensitivity analysis of between 0° and 45° was used, as there were few data available on scCO$_2$/water/rock contact angles at subsurface conditions. The maximum sensitivity resulted in a calculated reduction of CO$_2$ column height from 789 m to 558 m.

Experimental data by Chiquet and Broseta (2005) and Chiquet et al. (2007), using scCO$_2$ droplets immersed in brine, showed that quartz and mica substrates (as proxy minerals for fine grained rocks) under low pressures (< 1.0 MPa or 145 psi) become less water-wet in the

Table 6.1: CO_2/water/substrate contact angle (θ) variation with increasing salinity and pressure (from Chiquet and Broseta 2005).

Substrate/ Salinity	Contact Angle (θ) Low Pressure < 0.5 MPa	Contact Angle (θ) High Pressure 10 MPa
Clay – 0.01 M (584 ppm)	0–20°	60–80°
Clay – 0.1 (5840 ppm)	0°	40–60°
Clay – 1 M (56,000 ppm)	0–20°	60–80°
Quartz – 0.01 M (584 ppm)	20–30°	40–55°
Quartz – 0.1 (5840 ppm)	0–10°	20–45°
Quartz – 1 M (56,000 ppm)	20–30°	40–55°

presence of $scCO_2$, i.e. contact angles (θ) vary from 0° to 20° for mica (a clay proxy) and 20° to 30° for quartz. Under higher pressures (10 MPa or 1450 psi), the contact angle (θ) increases to 60°–80° for mica and 40°–55° for quartz (Table 6.1). The contact angles were measured through the CO_2 droplet (φ, see Figure 6.2) and subtracted from 180° to give the wetting phase contact angle (θ). These experiments were carried out above the substrate in a pressure cell (Figure 6.2). In a second part of the experiment, the CO_2 droplet was introduced beneath the mineral substrate in the pressure cell and the experiment repeated. The results were determined to be 10–15° less at low pressure with negligible differences at high pressures (Figure 6.2). The purpose of the experiment below the substrate was to test the contact angle results against a buoyancy effect (Chiquet and Broseta 2005). Chiquet et al. (2007) also demonstrated that the difference in storage capacity between CO_2 as a non-wetting and as a partially wetting phase (contact angle: 74°) would be approximately 69% less at 1200 m as a result of the commensurate lowering of the capillary membrane seal pressure (i.e. a reduced column height) due the changes in wettability.

The contact angle (CO_2/water/mica or quartz) is also affected by brine concentrations. For instance, θ decreases

by ~20° from 0.01 M NaCl to 0.1 M NaCl and then increases by ~20° from 0.1 M NaCl to 1 M NaCl brine solutions (Table 6.1). A decrease in CO_2 solubility also occurs with increasing salinity (Chiquet and Broseta 2005; Chiquet et al. 2007).

These experiments were carried out on single mineral plates rather than a shale rock surface. Experimental $scCO_2$/water/rock contact angle studies on shale surfaces have encountered difficulties due to rapid dispersion of clay platelets in aqueous solutions. Research by Shah et al. (2008) determined that dense acid gases (CO_2 and H_2S) lower the IFT when compared with water /hydrocarbon systems and are responsible for a loss in capillary sealing potential. The results obtained by the methodologies which Chiquet et al. (2007) used on $scCO_2$/water/rock contact angles were repeated and found to be lower than originally reported. The pressures used in these experiments were not high enough for the CO_2 to be in a saturated super critical state.

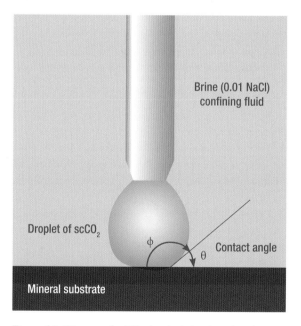

Figure 6.2: Diagram of $scCO_2$ droplet being introduced onto a mineral substrate, at subsurface conditions, to determine the contact angle (modified from Chiquet and Broseta 2005). Yang and Gu (2005) used a similar method but introduced a water droplet via a needle into various CO_2 states to determine the interfacial tension at subsurface conditions.

Wettability studies by Yang et al. (2008) have also shown that the contact angle advances under the influence of high pressure. Experimental research, using samples of the Midale Vuggy (dolomitic limestone reservoir rock) from the Weyburn Field CO_2 EOR site in Canada, illustrated that the contact angle advanced from 91.23° at 27°C and 0.1 MPa (14.5 psi) to 116° at 27°C and 12.01 MPa (1742 psi) and 130° at 27°C and 25 MPa (3626 psi). This effectively changed the scCO₂/water/rock system to a hydrophobic system, i.e. the water became predominantly the non-wetting phase. At a higher temperature (58°C) the angle only advanced to 100°, which was attributed to each phase (CO_2 and water) permeating the other (Yang et al. 2008). This research demonstrated that CO_2 wettability in scCO₂/brine/calcareous or dolomitic cap rocks can be significantly higher than in quartz/clay dominated cap rocks, with contact angles above 90°.

Shah et al. (2008) also experimentally determined the advancing and receding contact angles for a CO_2/brine/rock system on a muddy limestone reservoir rock from the south of France. These results generally showed lower contact angles than Yang et al. (2008) determined for the Vuggy Limestone in Canada. However, it is important to note that both experimental results did demonstrate that scCO₂ can be partially wetting when injected into the subsurface under certain conditions.

Yang et al. (2008) concluded that when modelling the properties of scCO₂ during injection, the injected CO_2 first displaces the brine from the reservoir rock and the water changes from wetting to partially non-wetting as the pressure increases. This change in wettability can increase storage capacity by increasing residual trapping as scCO₂ becomes a partially wetting phase. However, as detailed in the following section, when extrapolating these conclusions to cap rocks, the effect is that the CO_2 column height retention will be reduced.

6.6   Determination of seal capacity or column height

MICP analytical data are also used to determine the maximum column height and the water saturation of a sedimentary rock as a function of height above the free water level (FWL). These data must be converted to a subsurface CO_2/water system before the mercury injection data can be used to determine seal capacity (column height). The following equation can be used (after Schowalter 1979 and Vavra et al. 1992):

$$P_{b/CO_2} = P_{a/m}\ \frac{(\sigma_{b/CO_2} \cos\theta_{b/CO_2})}{(\sigma_{a/m} \cos\theta_{a/m})} \qquad (3)$$

where P_{b/CO_2} is the capillary pressure in the CO_2/water system, $P_{a/m}$ is the capillary pressure in the air/mercury system, σ_{b/CO_2} and $\sigma_{a/m}$ are the interfacial tensions of the water/CO_2 and the air/mercury systems respectively, θ_{b/CO_2} and $\theta_{a/m}$ are the contact angles of the CO_2/water/substrate and air/mercury/substrate systems respectively. As highlighted in Equation 3, the role of wettability (contact angle) and interfacial tension (IFT) in determining column height is significant. In the petroleum industry these parameters are known experimentally through extensive research using synthetic and proxy hydrocarbons, which agree with subsurface conditions as demonstrated by Smith (1966), Schowalter (1979), Anderson (1986), Zhang et al. (1997), Bi et al. (1999), Al-Siyabi et al. (1999) and Morrow (1990). In the geological storage of carbon dioxide, the role of wettability and more specifically, contact angle is not well known. Research on interfacial tension in scCO₂/water systems at subsurface conditions has been shown to be as low as 23 mN/m by several authors (Hildenbrand et al. 2004; Chalbaud et al. 2006; Bennion and Bachu 2006; Chiquet et al. 2007). While there is still discussion as to the wettability of scCO₂ (Daniel and Kaldi 2008; 2009; Yang et al. 2008; Chiquet and Broseta 2005; Chiquet et al. 2007) most workers consider scCO₂ to be non-wetting. The conclusion reached from reviewing these studies is that a range of contact angles (sensitivities) need to be applied when interpreting retention heights for the geological storage of CO_2 (see Daniel and Kaldi 2008; 2009). Buoyancy pressure drives CO_2 (non-wetting phase) movement in the subsurface and forces it into the pore throats of a rock, subsequently displacing water (wetting phase). Buoyancy is the density difference (g/cc) between CO_2, and water by the column height (ft) and the pressure gradient of water (0.433 psi/ft). In other words the greater the column height of CO_2, the greater is the buoyancy pressure forcing the CO_2 into the pore

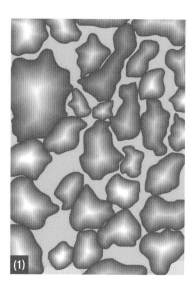
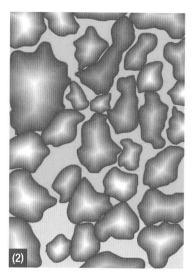
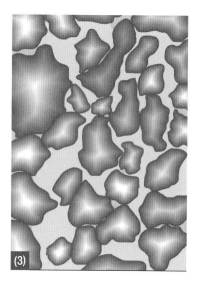

Figure 6.3: The sequence of non-wetting phase displacement, from entry pressure through threshold pressure to continuous flow. As buoyancy pressure increases, non-wetting phase starts to exceed surface tension forces (entry pressure) and displaces the wetting phase from the larger pore throats (1). Increasing pressure further forces the non-wetting phase through smaller pore throats, until the threshold pressure is reached (2). The non-wetting phase commences to move freely through the seal rock with increasing pressure above the threshold, leaving residual water saturation (3).

network. Threshold pressure (P_{th}) is the pressure at which the non-wetting phase (mercury or CO_2) begins to flow through the reservoir or seal rock as a continuous phase (Figure 6.3). The pressure at which this flow occurs is the point where the non-wetting phase commences to migrate through the seal rock (Smith 1966; Schowalter 1979; Katz and Thompson 1986; Sneider 1997).

A reservoir for potential CO_2 storage is made up of lithologies with different pore throat sizes. These pore throats will have different displacement and threshold pressures, and varying CO_2 saturations as a function of height (h) above the free water level (FWL). In any given CO_2 charged reservoir, the lowest indication (or maximum height) of CO_2 in a particular rock type approximates the threshold pressure (P_{th}) for that rock. The P_{th} (equivalent column height) can be considered as the CO_2/water contact for that particular rock type. It should be noted that a reservoir with multiple rock types may have several corresponding CO_2/water contacts, but will have only one FWL. It is therefore important to determine the FWL, which is required to determine the maximum column height ($hmax$) of CO_2 (or any non-wetting fluid) in the reservoir (Schowalter 1979).

In order to determine $hmax$, capillary pressure data must first be converted to height above free water level by using the equation:

$$Pc_{b/CO_2} = h(\rho_b - \rho_{CO_2})\ 0.433 \qquad (4)$$

then $hmax$ is calculated using Equation 5 (Smith 1966; Vavra et al. 1992)

$$hmax = (P_{ths} - P_{thr}) \div (\rho_b - \rho_{CO_2})\ 0.433 \qquad (5)$$

where Pc_{b-CO_2} is the capillary pressure (psi) reservoir CO_2-water system, h is the containment height (ft), ρ_b is the subsurface water density (g/cc), ρ_{CO_2} is the subsurface CO_2 density (g/cc) at ambient conditions, 0.433 (psi/ft) is the gradient of pure water at ambient conditions, P_{ths} is the threshold pressure of the seal (psi) and P_{thr} the threshold pressure of the reservoir (psi).

6.7   Interpreting threshold (breakthrough) pressure

There are several methods of determining threshold pressure, which are outlined below and illustrated in Figures 6.4 to 6.7:

> Schowalter (1979) found that the capillary leakage saturation point, breakthrough pressure or threshold pressure (P_{th}) commonly ranged between 4.5% and 17%. He therefore suggested 10% saturation as "a rule of thumb".

> Sneider (1987) and Sneider et al. (1997) examined the column heights of examples from over 200 reservoirs and noted that seal leakage occurs at between 5% and 10% of the pore volume (< 1.0% bulk volume) of the non-wetting phase saturation He therefore determined threshold pressure of reservoir rocks, referred to as breakthrough or leakage saturation pressure, as the 7.5% non-wetting phase saturation displacement pressure.

> Boult (1996) and Boult et al. (1997) utilised the intersection of the tangent to the slope of the upper inflection (plateau) with the injection pressure axis to determine threshold pressures of seal rocks (Figures 6.4 to 6.7).

> Kivior (2000) determined the threshold pressure of seals by using the first derivative of the injection curve—equivalent to the point where the derivative curve spiked upwards, i.e. after the entry pressure (Figures 6.4 to 6.7).

> Boult (1996) and Dewhurst et al. (2002) utilised a combination of the upward inflection point of the injection curve tied to a significant upward rise in incremental volume curve to determine threshold pressure of seal rocks (Figures 6.4 to 6.7).

The samples used by Schowalter and Sneider (op. cit.) were from reservoirs and as a consequence the entry (displacement) pressure and threshold pressure are nearly the same. However, in the subsequent studies listed above, the data were derived from sealing lithologies. They all required that the curve be corrected for "closure" or conformance to be accurate. Closure or conformance is the process whereby mercury fills surface irregularities such as nicks, gouges small fractures and vugs on the sample (see curve 1 in Figures 6.4 to 6.7). Figures 6.6 and 6.7 highlight the determination of threshold pressures of a cuttings sample, both without and with correcting for conformance and highlights the differences in capillary pressure that can be estimated by this method. If the entry pressure can be determined accurately from the injection curve, then the P_{th} will be very close to that picked using the Boult (1996) and Dewhurst et al. (2002) method.

Two representative seal (cap rock) Otway samples analysed using mercury injection capillary pressure methodology are described below.

Sample 1 (core) is a very fine silty claystone with approximately 5 by 2 mm lenticular bodies of siltstone throughout the claystone. Bulk mineralogy from X-ray diffraction (XRD) analysis indicates co-dominant clays and quartz with accessory muscovite and feldspar, where further clay analysis highlighted a co-dominant mixed layer of smectite/illite and chlorite/berthierine with accessory kaolin.

Sample 2 (cuttings) is a fine to very fine silty calcareous claystone. Bulk mineralogy from XRD analysis indicates co-dominant clays, calcite and quartz with accessory glauconite and feldspar. Further analysis of the clays present shows co-dominant smectite and chlorite with minor berthierine and illite.

Both of these samples are representative of seals and intraformational barriers found in Otway Basin wells, which include CRC-1, CRC-2, Boggy Creek-1, Flaxman-1, Buttress-1 and Croft-1 (Figure 1.3).

6.7.1 Interpretation

The slope extrapolating technique for interpreting threshold pressure, as described and used by Boult (1996), extrapolates the poorly sloping section of the injection curve that represents the maximum rate of mercury injection into the rock (curve 2, Figures 6.4 to 6.7). The line is extended down to the zero non-wetting phase saturation line where the pressure (P_{th}) is read. This works well for rocks that have a well defined slope, but is not accurate for rocks with poorly defined slopes such as sandy mudstones, fractured sidewall cores (SWC) and cuttings in general. Again this method is more accurate when corrected for conformance (Boult 1996) (see curve 2, Figures 6.4 to 6.7).

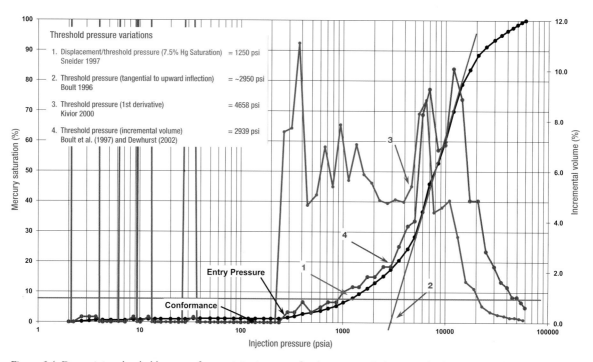

Figure 6.4: Determining threshold pressure from an injection curve for the core sample (core sample 1) uncorrected for conformance.

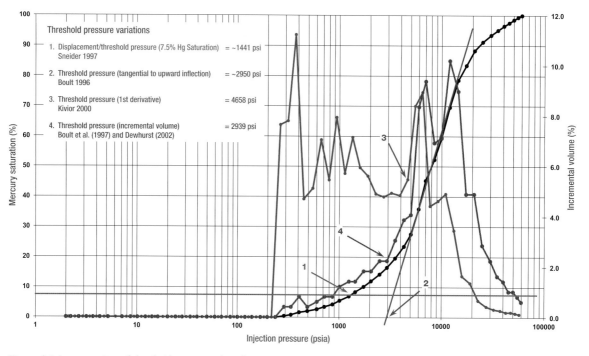

Figure 6.5: Interpretation of threshold pressure values from an injection curve for the same core sample (sample 1 as in Figure 6.4) corrected for conformance; note the range of values for P_{th} for both the uncorrected and corrected curve.

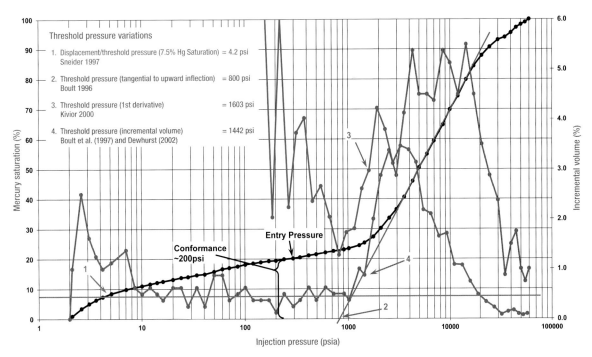

Figure 6.6: Interpretation of threshold pressure values from an injection curve of the cuttings sample 2 uncorrected for conformance. Note low P_{th} values for methods 1 and 2.

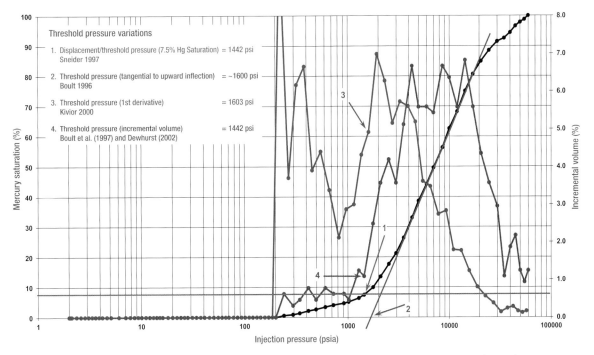

Figure 6.7: Interpretation of threshold pressure values from an injection curve of cuttings sample 2 corrected for conformance. Note similar values once the entry pressure has been determined and the curve corrected for conformance.

The method used by Kivior (2000) for determining threshold pressure utilises the first derivative of the injection curve. This produces a graph where a peak results from the rapid change between applied pressure and volume injected. The point of maximum inflection, where the mercury has commenced to flow through the rock, corresponds to the pressure at which the non-wetting phase (mercury in the lab, carbon dioxide in the field) migrates rapidly through the rock (curve 3, Figures 6.4 to 6.7).

Boult (1996) and Boult et al. (1997) also employed an alternate method utilising significant intrusion on the incremental curve combined with the cumulative injection curve to determine threshold pressure for seals rocks which have poorly defined slopes. Boult (1996) also noted that this technique can be used to accurately determine the threshold pressure of sidewall cores (SWC) and cuttings. This method was first described by Leverett (1941), where reference was made to the determination of displacement pressure through the interpretation of significant intrusion on the incremental volume curve. Dewhurst et al. (2002) also utilised this technique and determined the threshold pressure by using the point of upward inflection of the injection curve in conjunction with the steep gradient increase in incremental injection volume curve as shown in Figures 6.4 to 6.7 (curve 4). The threshold pressure is determined graphically by (1) combining the injection and incremental volume curves against pressure; (2) determining the point at which the incremental volume curve approaches the critical pore throat size (or modal pore throat size); and (3) the point at which the injection curve has its maximum upward inflection (curve 4 in Figures 6.4 to 6.7).

As can be seen from Table 6.2 there is a significant disparity between the values obtained by the various methods, some of which have significant variance from the uncorrected to conformance corrected curves. This is especially highlighted when attempting to determine the entry pressure and conformance on cuttings samples and the concomitant problem of interpreting the threshold pressure (P_{th}).

Threshold pressure determination utilising the techniques by Kivior (2000) and Boult (1996)/Dewhurst et al. (2002) did not require the determination of entry pressure and could be consistently applied across core, cuttings and SWC (Figures 6.4 to 6.7). These two methods also produced consistent results for core and especially cuttings (and fractured SWC) whether the injection curve has been corrected for conformance or not. As noted from the results obtained from these two methods (Table 6.2) the Kivior (2000) method always has a significantly higher threshold pressure than the Boult (1996)/Dewhurst et al. (2002) method. This point (curve 3) is situated on the section of curve where more than proportionate non-wetting phase is moving through the rock (the injection curves in Figures 6.4 to 6.7). The threshold pressure as determined by Boult (1996)/Dewhurst et al. (2002) occurs on injection curve 4 before significant flow through the rock has started, so it is at the threshold of leakage on the injection curve in Figures 6.4 to 6.7. This threshold point has also been demonstrated by Katz and Thompson (1986) in their experiments on increasing electrical conductivity at the point where mercury begins to flow through the rock.

Table 6.2: Threshold pressure comparisons of uncorrected and conformance corrected mercury injection curves for a core sample and a sample of cuttings (samples represent different seal rocks).

	Sample 1, Core: Uncorrected, psi	Sample 1, Core: Corrected, psi	Sample 2, Cuttings: Uncorrected, psi	Sample 2, Cuttings: Corrected, psi
Sneider 1997, 7.5% non-wetting saturation, break-through pressure	1250	1441	4.2	1442
Boult 1996, tangent to upward inflection, P_{th}	2950	2950	800	1600
Kivior 2000, 1st derivative, P_{th}	4658	4658	1603	1603
Boult 1996, Dewhurst et al. 2002, incremental and injection curves P_{th}	2939	2939	1442	1442

Subsurface properties for supercritical CO$_2$ are variable, as densities range from 0.42 to 0.74 g/cc and water densities range from 0.97 to 1.05 g/cc with salinities of ~1000 to ~65,000 ppm. Higher salinities (with commensurate higher densities) are known from some formations (Rowe and Chou 1970; Span and Wagner 1996).

Interfacial tensions, CO$_2$ concentration and water densities are determined using calculations after Span and Wagner (1996) and Rowe and Chou (1970). Generally the interfacial tension for water/CO$_2$ varies from 23 to 27 mN/m and sensitivities for contact angle vary between 0° and 60° as recommended by Daniel and Kaldi (2008; 2009). Once the capillary pressure values have been converted to h (height, ft or converted metres), height versus mercury (non-wetting phase) saturations can be plotted. Conversion of mercury (non-wetting phase) to CO$_2$ (non-wetting

phase) yields a height versus CO$_2$ saturation plot. The non-wetting phase saturation can be converted to water (wetting-phase) saturation (Schowalter 1979), using the conversion:

$$S_w = 1 - S_{nw} \qquad (6)$$

where S_w is the wetting phase (water) saturation and S_{nw} the non-wetting phase (scCO$_2$) saturation.

Using Equations 5 and 6, a graph of containment height versus water saturation can be used to estimate potential CO$_2$ storage volume at various water saturations. Figure 6.8 shows the typical column height versus saturation relationship, highlighting the maximum containment heights for CO$_2$ storage using the various techniques for determining threshold pressure.

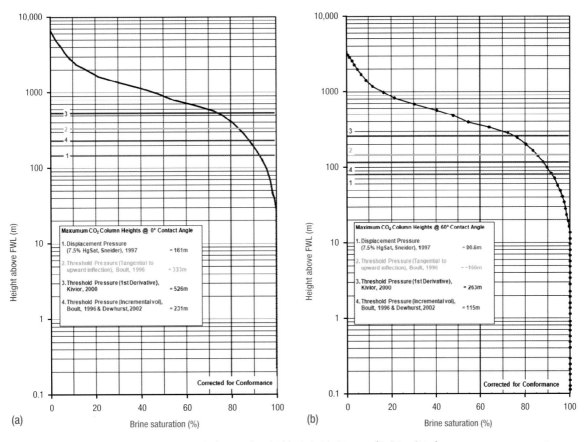

Figure 6.8: Graphs of CO$_2$ containment height for sample 1 (Table 6.2), (a) CA = 0°, (b) CA = 60°, showing variations in containment height using methodologies outlined by Sneider (1987) (purple), Boult (1996) (green), Kivior (2000) (blue) and Boult (1996) and Dewhurst et al. (2002) (red), (CA: contact angle).

Table 6.3: Description and petrophysical summary of sealing rock samples analysed by MICP from CRC-1 and CRC-2, Otway Project; CRC-1 and CRC-2 Pember Mudstone, Paaratte Formation and Belfast Mudstone Seals and Intraformational Barriers petrophysical data.

Sample Depth	Formation	Macro-Lithology Description	Seal Threshold Pressure (Air-Hg system) (psia/MPa)
CRC-1 917.00	Pember Mudstone	m-dk gry sb-fssle silty mudstone, sli carb and mic	70 / 0.48
CRC-2 930.63	Pember Mudstone	m-dk gy, slty, vf-v crs sandy mudstone, occ well rounded granules (~5 mm) throughout, bedding slightly graded	1410 / 9.72
CRC-2 933.18	Pember Mudstone	m-dk gy, slty, vf-v crs sandy mudstone, minor (< 10%) sb rnd-sb ang granules (~5 mm) through out, indistinct bedding	1436 / 9.90
CRC-1 1262.95	Paaratte Formation	m gry/grn, pr srt, sb fssle, sli vf-crs sdy, mudstone, sli calc	2055 / 14.17
CRC-1 1264.95	Paaratte Formation	m gry/grn, pr srt, sli sdy, v slty mudstone	1422 / 9.8
CRC-1 1268.80	Paaratte Formation	m gry/grn, pr srt, sb fssle, sdy, slty, mudstone	199 / 1.4
CRC-2 1321.25	Paaratte Formation	interbedded lt-m gy, very fine-fine muddy sandstone and m gy very fine sandy mudstone, irregulat wavy contacts, some biotubation	698 / 4.81
CRC-2 1322.73	Paaratte Formation	undulose interbedded lt-m gy, mudstone and biotubated very fine sandy to muddy siltstone, bioturbated	1432 / 9.87
CRC-2 1323.89	Paaratte Formation	lt-m gy, very fine sandy mudstone with m gy subhorizontal mud clasts (8 × 3 mm), isolated specs of white clay in the matrix	2472 / 17.04
CRC-2 1328.89	Paaratte Formation	m-dk gy, quartzose slightly coarse to fine silty claystone	5976 / 41.20
CRC-2 1330.56	Paaratte Formation	lt-m gy, slightly very sandy mudstone with occ mudstone clasts and bioturbation	2932 / 20.22
CRC-2 1434.70	Paaratte Formation	lt-m gy, vf-f sandy mudstone with lenticular/ wavy silt laminae (< 5 mm) and clay clasts (10 × 3 mm LxH)	2922 / 20.15
CRC-2 1435.31	Paaratte Formation	lt-m gy, f-vf muddy quartzose sandstone with ~1 mm slightly wavy quartz siltstone laminae 2–4 mm apart, significant bioturbation throughout	85 / 0.59
CRC-2 1439.72	Paaratte Formation	lt-m gy, f-m qtz sandstone with poikilotopic dolomitic and minor calcareous cement; clay filling uncemented pores	51 / 0.35
CRC-2 1440.21	Paaratte Formation	lt gy, vf-f grn, sb ang, qtz sandstone with pervasive slightly ferroan dolomitic cmt and minor vermicular kaolinite and trace chlorite fill; F- m grn flakes of relict biotite present throughout	1003 / 6.92
CRC-2 1448.58	Paaratte Formation	lt-m gy, f-m qtz sandstone with rare clays, fully cemented with dolomitic cement	1433 / 9.88
CRC-2 1449.01	Paaratte Formation	lt gy to blk, crs-vf miceaous qtz sandstone, relict biotite laths (up to 3 mm) common throughout with rare angular mudstone clasts up to 4 mm	8.6 / 0.06
CRC-2 1473.27	Paaratte Formation	lt gy, f-c quartzose sandstone with poikilotopic dolomitic and minor calcareous cementing; clay filling uncemented pores	16 / 0.11
CRC-2 1481.46	Paaratte Formation	lt-dk gy, silty - vf sandy mudstone with fine discontinuous beds (< 3 mm) of vf-f quartzose sand	4986 / 34.38
CRC-2 1485.39	Paaratte Formation	lt-dk gy, silty - vf sandy mudstone, bioturbated with small sublenticular clasts (~5 mm) of vf quartzose sandstone with fine laminae (2–5 mm) of dk grey mudstone (~2 cm apart)	85 / 0.59
CRC-2 1523.10	Skull Creek Mudstone	m-dk gy, wavy crs laminated slty mudstone with ellipsoidal vf-f sand clasts (long axis parallel to bedding), bioturbated	1426 / 9.83
CRC-1 1900.70	Belfast Mudstone	dk-gry, sli slty, tr glauconitic claystone, highly fractured	6965 / 48
CRC-1 1900.99	Belfast Mudstone	dk-gry, sb fssle, sli slty glauconitic claystone	4979 / 34.3
CRC-1 1901.50	Belfast Mudstone	dk-gry, sb fssle, claystone	6972 / 48.1
CRC-1 1901.85	Belfast Mudstone	dk-gry, sb fssle, claystone, tr carbonaceous	4987 / 34.4
CRC-1 1902.25	Belfast Mudstone	dk-gry, sb fssle, claystone, tr carbonaceous	6969 / 48

Seal Threshold Pressure (Brine-CO$_2$ system) (psia/MPa) CA 0°	Column Height Capacity (Metres)				Porosity % (MICP)	Residual Saturation % (MICP)	Critical Pore Throat Diameter μm (MICP)
	CA 0°	CA 20°	CA 40°	CA 60°			
5.84 / 0.04	4.9	4.6	3.8	2.5	16.80	81.7	1.98
100 / 0.68	129.44	121.6	99.2	64.7	28.70	59.7	0.0471
102 / 0.70	131.20	123.3	100.5	65.6	23.10	60.5	0.0282
149.4 / 1.03	221.80	208.4	169.9	110.9	26.30	54.2	0.04
1.3.3 / 0.09	153.50	144.3	117.6	76.8	28.10	58.1	0.056
14.5 / 0.1	20.40	19.2	15.6	10.2	20.00	60.0	0.068
49.4 / 0.34	90.58	85.1	69.4	45.29	14.80	58.6	0.0561
101 / 0.7	187.51	176.2	143.6	93.76	12.00	59.7	0.0469
175 / 1.21	324.74	305.2	248.8	162.37	13.30	57.1	0.0198
423 / 2.92	788.89	741.3	604.3	394.44	10.10	66.7	0.0164
208 / 1.43	386.58	363.3	296.1	193.29	12.10	59.9	0.0114
208 / 1.43	400.27	376.1	306.6	200.14	10.50	52.9	0.0198
6.1 / 0.42	10.31	9.7	7.9	5.15	13.26	57.1	0.82
3.63 / .025	5.64	5.3	4.3	2.82	7.80	52.7	1.4058
71.33	136.72	128.5	104.7	68.36	2.80	54.7	0.0962
51 / 0.35	196.54	184.7	150.6	98.27	1.75	57.1	0.0396
0.6 / 0.004	0.150	0.14	0.12	0.08	32.05	77	16.63
1.14 / 0.008	0.83	0.8	0.6	0.4	10.80	51.5	6.636
355 / 2.45	694.18	652.3	531.8	347.1	13.30	50.9	0.0165
6.1 / 0.042	10.44	9.8	8.0	5.2	17.70	60.2	1.673
101.7 / 0.70	199.73	187.7	153.0	99.9	11.00	48.5	0.0165
509.2 / 3.51	850.5	799.2	651.5	425.3	10.70	58.5	0.017
364 / 2.51	607.7	571	465.5	303.8	10.70	57.3	0.018
509.7 / 3.51	851.4	800	652.2	425.7	8.20	62.8	0.0096
364.6 / 2.51	608.9	572.2	466.5	304.5	7.97	69.6	0.0094
509.5 / 3.51	851.4	800.1	652.2	425.7	10.20	56.9	0.0185

6.8   Results for CRC-1 and CRC-2

Results of MICP analysis on samples from the Pember Mudstone showed that the carbon dioxide seal retention height varied from 4.5 m to 131 m (contact angle: 0°). At a contact angle sensitivity of 60° these containment heights vary from 2.5 m to 65.6 m, based on experimental evidence as described in Table 6.3.

Sealing lithology (intraformational barriers) containment heights for the Paaratte Formation range from 20.4 m to 694 m and these heights vary from 10.2 m to 347 m with a contact angle sensitivity of 60°. Reservoir facies that were analysed from the Paaratte Fm indicate threshold pressures (brine/CO_2 system) between 0.6 psi (0.004 MPa) and 6.1 psi (0.042 MPa) (Table 6.3).

A single analysis was performed on a sample of the Skull Creek Mudstone from CRC-2, which indicated a containment height of 199 m (CA 0°), with a height of 100 m at a 60° contact angle (Table 6.3).

The Belfast Mudstone MICP analyses (CRC-1) ranges in CO_2 containment height from 607 m to 851 m (CA 0°), with a maximum sensitivity contact angle of 60° indicating column heights between 304 m and 426 m (Table 6.3).

The techniques for determining threshold pressure employed by Sneider et al. (1997) (average bulk displacement) and Boult (1996) (tangential method) rely significantly on being able to determine the entry pressure from the mercury injection curve. The entry pressure is relatively easy to determine when analysing the curves from core samples, but when cuttings or SWC samples are analysed, there is potentially a more than proportionate surface area to the sample and therefore a commensurate increase in surface rugosity. The resulting conformance of this comparatively large surface area can be represented at times by up to 60% of the mercury injected, which can cause difficulties in determining entry pressures. Another possible artefact (though not fully established at this point) occurs primarily in cuttings samples and is the effect of water absorption during the drilling and sample washing processes, which causes a shaly seal rock to expand and results in an enlargement of the surficial pore throats. These two artefacts tend to mask the true entry pressure as the sample surface may have degraded over time and

the pore throat size is in fact an artefact expressed as a lower threshold pressure shown on the graph (Figure 6.6).

Threshold pressure determination utilising the techniques by Kivior (2000) and Boult (1996)/Dewhurst et al. (2002) is independent of the need to determine entry pressure and can be consistently applied across core, cuttings and SWC (Figures 6.4 to 6.7). As can be seen from the results obtained from these two methods (Table 6.2) the Kivior (2000) method always has a significantly higher threshold pressure than the Boult/Dewhurst method. The threshold pressure as determined by Boult (1997)/Dewhurst et al. (2002) occurs on the injection curve before significant flow has started to migrate through the rock and therefore represents safe repeatable pressures for determining volumes of CO_2 that seal rocks can potentially contain.

6.9   Conclusions

Mercury injection capillary pressure (MICP) analysis technology can be applied to establish the suitability of a top seal for the containment of CO_2. MICP was used for determining the containment height of seals and intraformational barriers in the Otway CRC-1 and CRC-2 wells. Intraformational barriers and seals from the Pember Mudstone, Paaratte Formation, Skull Creek Mudstone and Belfast Mudstone were analysed to determine their capillary containment height potential.

In light of the wide array of sealing lithologies analysed during the Otway Project, it was concluded that MICP analyses of cap rocks were important for the evaluation of CO_2 storage capacity and containment and that the containment provided by the seals in the Naylor structure was adequate for the scale of CO_2 injection in Stage 1 and that proposed for Stage 2C. Coring of the seal/barrier was an integral part of the sampling programme and ensured accurate analyses, provided the core was correctly preserved and analysed soon after recovery.

The method which produced the most reliable results (the Boult/Dewhurst method) was determined by graphing a combination of mercury saturation and incremental pore volume data versus mercury injection pressure derived from MICP analysis. This method is relatively simple and effective and could be used on a wide range of rocks (seal and reservoir) as well as both core and cuttings for the Otway Project and potentially other CCS projects.

6.10 References

Al-Siyabi, Z., Danesh, A., Tohidi, B., and Todd, A.C. 1999. Variation of gas-oil-solid contact angle with interfacial tension: Petroleum Geoscience 5, 37–40.

Anderson, W. 1986. Wettability literature survey – Part 2; Wettability measurement: Journal of Petroleum Technology, November, 1986, SPE No. 13933, Society of Petroleum Technology, p 1246–1262.

Bennion, D.B., and Bachu, S. 2006. Dependence on temperature, pressure and salinity of the IFT and relative permeability displacement characteristics of CO$_2$ injected in deep saline aquifers. SPE No. 102138.

Bi, Z., Zhang, Z., Xu, F., Qian, Y., and Yu, J. 1999. Wettability, oil recovery, and interfacial tension with an SBS-Dodecane-Kaolin System. Journal of Colloid and Interface Science 214, 368–372.

Boult, P. J. 1996. An investigation of reservoir/seal couplets in the Eromanga Basin; Implications for petroleum entrapment and production. Development of secondary migration and seal potential theory and investigation techniques. PhD Thesis, Department of Applied Geology, University of South Australia 1 and 2.

Boult, P.J., Theologou, P.N. and Foden, J. 1997. Capillary seals within the Eromanga Basin, Australia: Implications for exploration and production. In: R.C. Surdam (Ed.), Seals, traps and the petroleum system. AAPG Memoir 67, 143–167.

Chalbaud, C., Robin, M. and Egermann, P. 2006. Interfacial tension data and correlations of brine/CO$_2$ systems under reservoir conditions. SPE No. 102918.

Chiquet, P. and Broseta, D. 2005. Capillary alteration of shaly cap rocks by carbon dioxide. Society of Petroleum Engineers, Paper No. 94183.

Chiquet, P., Broseta, D. and Thibeau, S. 2007. Wettability alteration of caprock minerals by carbon dioxide. Geofluids 7, 112–122.

Daniel, R.F. and Kaldi, J.G. 2009. Evaluating seal capacity of cap rocks and intraformational barriers for CO$_2$ containment. In: M. Grobe, J.C. Pashin, and R.I. Dodge (Eds.), Carbon dioxide sequestration in geological media – State of the Science AAPG Studies in Geology 59, 335–345.

Daniel, R.F. and Kaldi, J G. 2008. Evaluating seal capacity of caprocks and intraformational barriers for the geosequestration of CO$_2$. Eastern Australian Basins Symposium III Conference Proceedings, 2008, Sydney.

Dewhurst, D.N., Jones, R.M. and Raven, M.D. 2002. Microstructural and petrophysical characterisation of Muderong Shale: application to top seal risking: Petroleum Geoscience 8, 371–383.

Ferer, M., Bromhal, G.S. and Smith, D.H. 2002. Pore-level modelling of carbon dioxide sequestration in brine fields: Journal of Energy and Environmental Research 2, 120–132.

Hildenbrand, A., Schlomer, S., Krooss, B.M. and Littke, R. 2004. Gas breakthrough experiments on pelitic rocks: comparative study with N$_2$, CO$_2$ and CH$_4$. Geofluids 4, 91–80.

Katz, A.J. and Thompson, A.H. 1986. Quantitative prediction of permeability in porous rock. The American Physical Society 34, 8179–8181.

Kivior, T. 2000. Late Jurassic and Cretaceous seals of the Vulcan Sub-basin. Proceedings of the Technical Workshop – Hydrocarbon sealing potential of faults and cap rocks, Australian Petroleum cooperative Research Centre, McLaren Vale, 2000.

Leverett, M.C. 1941. Capillary behaviour in porous solids. American Institute of Mining and Metallurgy, Petroleum Engineering, Transactions 142, 152–169.

Morrow, N.R. 1990. Wettability and its effect on oil recovery: Journal of Petroleum Technology 42, 1476–1484.

Picknell, J.J., Swanson, B.F. and Hickman, W.B. 1966. Application of air-mercury capillary pressure data in the study of pore structure and fluid distribution: SPE Journal 6, 55–61.

Purcell, W. R. 1949. Capillary pressures – their measurement using mercury and the calculation of permeability therefrom: Petroleum Transactions – American Institute of Mining, Metallurgical and Petroleum Engineers 186, 39–48.

Rowe, A.M.J. and Chou, J.C.S. 1970. Pressure-volume-temperature-concentration relation of aqueous NaCl solutions. J Chem Eng Data 15, 61–66.

Schowalter, T.T. 1979. Mechanics of secondary hydrocarbon migration and entrapment. AAPG Bulletin 63, 723–760.

Shah, V., Broseta, D. and Mouronval, G. 2008. Capillary alteration of caprocks by acid gases. Society of Petroleum Engineers, Paper No. 113353.

Smith, D.A. 1966. Theoretical considerations of sealing and non-sealing faults. AAPG Bulletin 50, 363–374.

Sneider, R.M. 1997. Practical petrophysics for exploration and development. Course Notes. AAPG Education Department, Tulsa, 396p.

Sneider, R.M., Sneider, J.S., Bolger, G.W. and Neasham, J.W. 1997. Comparison of seal capacity determinations: Core versus cuttings. In: R.C. Surdam, Seals, traps and the petroleum system. AAPG Memoir 67. AAPG Tulsa, Oklahoma, 1–12.

Span, R. and Wagner, W. 1996. A new equation of state for carbon dioxide covering the fluid region from the triple-point temperature to 1100K at pressures up to 800 MPa. J Phys Chem Ref Data 25, 1509–1596.

Vavra, C. L., Kaldi, J. G., and Sneider, R. M. 1992. Geological applications of capillary pressure. AAPG Bulletin 76, 840–850.

Wardlaw, N.C. and Taylor, R.P. 1976. Mercury capillary pressure curves and the interpretation of pore structure and capillary behaviour in reservoir rocks. Bulletin of Canadian Petroleum Geology 24, 225–262.

Washburn, E.W. 1921. A note on the method of determining the distribution of pore sizes in a porous material. Proceedings of the National Academy 7, 155–116.

Yang, D., Gu, Y. and Tontiwachwuthikul, P. 2008. Wettability determination of the reservoir brine-reservoir rock system with dissolution of CO_2 at high pressures and elevated temperatures. Energy and Fuels 22, 504–509.

Yang, D. and Gu, Y. 2005. Interfacial interactions of crude oil-brine-CO_2 systems under reservoir conditions: SPE No. 90198, 12p.

Zhang, L., Ren, L. and Hartland, S. 1997. Detailed analysis of determination of contact angle using sphere tensiometry: Journal of Colloid and Interface Science 192, 306–318.

Eric Tenthorey

7. GEOMECHANICAL INVESTIGATIONS

7.1 Introduction

7.1.1 Pressure management

Geomechanics is concerned with understanding the mechanical response of rock and soil under various conditions. In terms of the storage of carbon dioxide, the bulk of the mechanical concerns centre around the injection-driven fluctuations in fluid pressure, although mineral reactions and pre-injection drilling of wells also have geomechanical implications that can potentially be significant. In this chapter, the focus is solely on the important geomechanical phenomena at the Otway Project that are related to the injection of CO_2.

Injection of CO_2 invariably results in some degree of fluid over-pressurisation with respect to the prevailing pressure profile. If the system is pressurised sufficiently, then a number of deleterious effects may occur, such as fracturing of the reservoir or cap rock or reactivation of previously existing faults. These physical weaknesses in the system arise due to the pore fluid pressure counteracting the forces from the in-situ stress field and are a manifestation of a reduction in the system's effective stress state. The effective stress concept was first developed by Terzaghi (1943) for soil systems, where an increase in pore fluid pressure results in an equally reduced effective stress on the rock mass (Engelder 1992). The implication of reduced effective stresses on the three principal stress axes is that the shear/normal stress ratio increases on virtually all imaginary planes in the system, which essentially means that the system is closer to failing in shear mode (Figure 7.1). Similarly, elevated fluid pressures also shift the system closer to the tensile failure condition, where the fluid pressure is equal to or greater than the minimum principal stress.

The influence of fluid pressure on the effective stress state of the system is the main reason why injectivity is such an important issue for a CCS project. Reservoirs with a large storage capacity and high permeability help to ensure that CO_2 can potentially be injected at high rates, while minimising the increase in fluid pressure and preserving the mechanical integrity of the reservoir, cap rock and any faults in the vicinity of the injected CO_2. However, reservoir rocks with these favourable characteristics are not always available as a storage opportunity, making it particularly important to understand the geomechanics of lower porosity/lower permeability reservoirs.

7.1.2 Fault reactivation and dilation tendency

Ensuring that pre-existing faults remain stable during CO_2 injection is one of the more important aspects of the geomechanical workflow during a CCS operation. It is well known that fault growth and the reactivation of faults leads to or facilitates fluid flow in and around the fault system. Obviously, this is not desirable during CO_2 injection as any permeability enhancement may facilitate leakage of CO_2 to shallower levels. While there are no known examples associated with CO_2 storage, in extreme cases, fault reactivation might also result in significant seismic events that could potentially cause damage at the surface, as was the case for the "hot dry rock" geothermal project in Basel in 2006. In that case, injection of water 5 km below ground triggered a significant seismic event, apparently via shear reactivation of a major fault (Deichmann and Giardini 2009), which eventually resulted in great public concern and the cancellation of the project. It is crucial that the possibility of such events occurring as a result of the injection of CO_2 is minimised, if CCS is to be widely deployed for reducing CO_2 emissions to the atmosphere.

The likelihood of fault reactivation is controlled by the shear/normal stress ratio on the fault plane (a function of the fault's orientation with respect to the in-situ stress field), and the frictional and cohesive properties of the fault plane. In general, faults oriented at about 30° to the maximum principal stress and 60° to the minimum principal stress directions possess highest shear to normal stress ratios and are theoretically most prone to reactivating. As fluid pressure increases, the normal stress will reduce, while the shear stress remains unchanged. When the shear/normal stress ratio reaches the friction coefficient value, then shear reactivation should in theory occur (assuming the fault has no cohesion). For most rocks, the friction coefficient is around 0.6. A practical example of how fault stability is gauged using this technique is presented in a later section of this chapter discussing the geomechanical workflow at Otway.

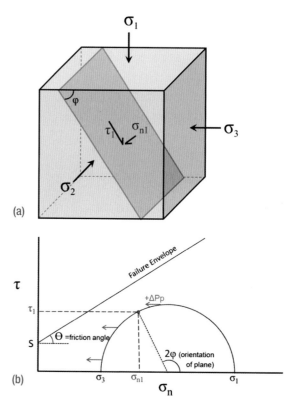

Figure 7.1: (a) Illustration showing the three principal stresses acting on a body of rock. Every imaginary plane within the volume has a shear stress (τ) and normal stress (σ_n) that can be resolved. The τ/σ_n ratio is one of the main parameters controlling fault stability. (b) Mohr-Coulomb diagram showing the relative stability of variably oriented planes within a volume of rock. For the two dimensional case shown, shear and normal stress for different planes will lie on the perimeter of the circle. Increasing pore fluid pressure (P_p) results in a reduction of the effective normal stresses thereby shifting the Mohr circle toward the failure envelope. S is the cohesion of the fault or body of rock.

7.1.3 Cap rock integrity

A cap rock is a geological formation capable of hydraulic sealing over time in the presence of natural geomechanical, geochemical and hydrogeological changes, such as those likely to occur when CO_2 is stored in an underlying reservoir rock. There are a number of specific geomechanical processes that can affect the hydraulic integrity of cap rocks, leading to leakage and/or failure of the seal. Massive injection into the formation may induce capillary leakage, hydraulic fracturing and shear deformation or fracturing of cap rocks. Capillary leakage under hydrostatic conditions takes place when the pressure difference between the injected fluid phase and the water phase in the pores exceeds the capillary entry pressure of the seal.

If the capillary pressure is very high, seal integrity loss may occur due to fracturing of the cap rock. These fractures are driven by a reduction of the effective stress and generally form when the minimum effective stress becomes zero (assuming that the cap rock has little or no tensile strength). Such hydraulic fracturing seems to be a common natural leakage mechanism for fluids through cap rocks in overpressured basins (Jimenez and Chalaturnik 2002). It is therefore important that the fracture gradient is adequately characterised during assessment of the in-situ stress and that injection pressures are kept below this gradient. However, as discussed below, pressurisation of the reservoir can result in poroelastic driven stress transfer that can shift the fracture gradient in the cap rock to lower values, thereby facilitating unwanted tensile failure.

7.1.4 Wellbore stability issues

Wellbores provide access between the surface and the reservoir, making them a preferential flow path for leakage from the reservoir. Wellbores are composed of several components: the borehole itself, the annular space between the borehole and the casing, and the casing itself. The propensity of boreholes to leak depends mainly on the quality of the wellbore system and its operational history (Nauroy 2011).

The first concern arises during drilling. The stability of a borehole during the drilling phase is a function of in-situ stresses, geomechanical properties of the different geological formations, direction and azimuth of the borehole, drilling techniques, drilling mud-formation interactions and maximum bottom hole pressure (BHP). It is therefore preferable that as much geomechanical information as possible is acquired before drilling a well, especially in relation to the in-situ stress field. Determining the appropriate mud weight while drilling is particularly critical to ensuring wellbore integrity, especially when deviated wells or significant overpressures are involved.

The operational history of a borehole relates to the production/injection and stimulation processes that

have occurred since the well was drilled. Moreover, the interaction between the borehole and the reservoir and its overburden must be taken into account. Problematic wellbore issues relate to failure within the reservoir due to expansion (injection) and compaction (withdrawal), tensile failure of the overburden, shearing of the boreholes due to reservoir and/or overburden shearing, and shearing along pre-existing fractures. Sand production in the well over time can also be a problem.

Finally, well abandonment must also be considered, as the long-term well integrity may not be fully understood. The injection of CO_2 can give rise to a variety of coupled physical and chemical processes which may affect the hydraulic integrity of wellbores. These may include casing failures, sealing deficiency of abandoned wells, ageing and long-term degradation, and wellbore failures due to geomechanical events in the reservoir and its overburden. These coupled processes can also result in poor operational performance during the active years of the well.

7.2 Key data for geomechanical assessment of the Otway site

7.2.1 In-situ stress field

To conduct any meaningful geomechanical assessment, it is critical to have an adequate understanding of the in-situ stress field, which encompasses the magnitudes and orientations of the three principal stresses. The stress field will largely determine how and when a rock will fail during pressurisation and will also control which faults are most prone to reactivation. The in-situ stress field is usually assumed to be Andersonian, meaning that one of the principal stresses is oriented vertically and the other two are oriented horizontally. A number of different sources and techniques must be used to determine the in-situ stress field and these are described in this section.

Determining the orientation of the three principal stresses is usually straightforward, assuming some well data are available for the area. Since one stress direction is vertical, then either the minimum or maximum horizontal stress direction must be determined in order to have all three orthogonal stress directions. The orientation of maximum horizontal stress can be measured from the occurrence of borehole deformation as observed by various methods. Borehole breakouts (BHBOs) are compressive failure features which can form within the well and are oriented at 90° to the maximum horizontal stress. Dip meter logs or four-arm callipers can be used to determine the orientation of breakouts, although microresistivity images (e.g. FMI, FMS) are preferred as they offer better resolution. Drilling-induced tensile fractures (DITFs), by contrast, form parallel to the maximum horizontal stress and are also best characterised using microresistivity images. In some cases further information on the horizontal stress orientations can also be acquired from earthquake focal mechanisms and from over-coring stress analysis.

The vertical stress is the stress applied at any given point due to the weight of the overlying rock mass and fluids. This stress is generally calculated by integrating the bulk density of the overlying rock mass and fluids over depth. The most accurate density measurements are obtained using logging while drilling or wire-line logs (e.g. RHOZ). If no direct density measurements are available for parts of the section then it is possible to use check shot velocities to calculate average density over a given interval, by applying empirical relationships between velocity and density (Gardner et al. 1974).

The magnitude of the minimum horizontal stress (S_{hmin}) can be estimated from leak-off test and mini-fracture test data. While leak-off pressures in vertical wells reflect S_{hmin}, the same pressure tests from deviated wells are a function of the vertical and horizontal stresses and wellbore trajectory with respect to those stresses (Aadnoy 1990; Brudy and Zoback 1993). Limited information on S_{hmin} can be obtained from a formation integrity test (FIT). The formation is not fractured during an FIT, so the yield pressure cannot be inferred. FIT pressures can therefore be used to gauge the lower bound to S_{hmin}.

The magnitude of the maximum horizontal stress (S_{Hmax}) is generally the most difficult stress to characterise and is usually derived from borehole failure features in conjunction with mini-frac tests and theoretical relationships involving the tensile strength of the rock. Indirect geological indicators and inversion of earthquake focal mechanisms can also help constrain the stress regime, and therefore the magnitude

of S_{Hmax}. A detailed description of possible workflows involved with determining S_{Hmax} and the in-situ stress field in general, is provided by Zoback (2007).

Figure 7.1 is a schematic illustration showing the three principal stresses and their importance for fault reactivation as shown by a Mohr-Coulomb failure diagram. In Figure 7.1(a), σ_1, σ_2 and σ_3 correspond to the maximum, intermediate and minimum stresses, respectively. In this figure, the stress field corresponds to a normal faulting regime, which is believed to be the case for the Otway site. Within the stressed volume, every plane possesses a different shear/normal stress ratio, which is what drives the propensity for faults to reactivate in an injection scenario. Generally, planes oriented at about 30° to σ_1 ($\varphi = 60°$) will be most unstable (assuming a normal coefficient of friction). The Mohr circle (semicircle) in Figure 7.1(b) shows the shear stress (τ) and effective normal stress (σ_n) for all hypothetical planes within a volume, together with the failure envelope representing the point that shear reactivation will occur. Planes oriented parallel and perpendicular to the maximum and minimum stress directions possess zero shear stress and therefore cannot reactivate in shear mode. When a gas or liquid is injected into the subsurface and is accompanied by an increase in fluid pressure, the net effect is to reduce the effective normal stress on all planes, while having no effect on shear stress. Barring any complex poroelastic effects, this causes the Mohr circle to migrate to the left on Figure 7.1(b) and approach the failure criterion. Once the failure envelope is crossed, faults will either reactivate or shear failure of intact rock will occur.

7.2.2   3D fault surfaces

During the seismic interpretation process for a CO_2 storage operation, it is important that faults in close proximity to the injection site are characterised in some detail. Faults are important geological features that can have profound geomechanical implications on several fronts. As discussed above, faults may reactivate due to elevated fluid pressures, thereby creating permeable pathways for unwanted fluid migration. Faults are also key components for basin compartmentalisation, which can lead to heterogeneous pore pressure profiles over short distances. The degree to which faults contribute to such compartmentalisation depends on the hydraulic properties of the fault zone itself; these are controlled by displacement, lithology and juxtapositional relationships across the fault.

The standard petroleum industry techniques used to image reservoirs and faults involve interpretation of seismic reflection lines, time slices and coherence cubes. In general, interpretation of closely spaced 3D seismic cubes will allow resolution of faults with displacements as small as about 5–10 m. Since faults are known to exhibit fractal patterns in nature, it is certain that many subseismic faults also exist; these could exert some influence on fluid migration. However, in terms of compartmentalisation, juxtaposition relationships and fault rupture implications, it is the faults with larger offsets that are most important and it is therefore critical that these faults are well characterised. During the interpretation and fault mapping process, it is also important to accurately characterise fault bends, relays and intersections, as these are likely to be sites with high fracture densities which could contribute to abnormally high fluid flow. These features usually also have significantly different orientations when compared to most of the fault surfaces, meaning that they will probably have different fault stability metrics. Once the fault networks have been mapped out in detail and converted to depth, they can be exported to various software packages for geomechanical modelling purposes.

7.2.3 Rock mechanical tests

Conducting or acquiring the results of any rock mechanical tests that have been conducted at a site can be very useful when trying to understand the likely geomechanical response of the system during CO_2 injection. Some mechanical tests can also help to predict if and where wellbore stability issues may be a problem. One of the main benefits of having an adequate rock mechanical test database to draw from is that the tests allow calibration of the sonic logs so that a continuous record of mechanical properties can be generated for each well. These records can then be used to generate a 3D geomechanical model of the field which can in turn be used to examine various phenomena from brittle failure to other complex poroelastic phenomena.

When embarking upon a mechanical testing programme designed to characterise and calibrate mechanical properties of a section, it is critical to first develop a well planned sampling programme. If the samples selected for testing do not adequately characterise the various end-members encountered throughout the section, then the transforms developed for the sonic logs are likely to be very inaccurate. Furthermore, core testing should be conducted on specimens that are as homogeneous as possible and devoid of any obvious fracturing, damage or alteration, as these will affect the mechanical test results. When testing mudstone samples, it is preferable that the core be preserved in a manner that minimises dehydration of the core, as loss of bound and unbound water would reduce integrity.

Rock mechanical properties that require quantification prior to developing a geomechanical model may reflect either the elastic or brittle behaviour of the rock under stress. The following are the principal properties of interest that were considered during the Otway Project rock mechanical testing programme:

> Young's modulus (E). This property is a measure of a rock's stiffness. It is defined as the ratio of the uniaxial stress to uniaxial strain and is acquired in mechanical tests during loading of the rock in the elastic regime. For rocks, the general unit of measurement is in GPa.

> Poisson's ratio (ν). This is the ratio of the lateral expansion to axial shortening under a given stress. The value is determined using strain gauges during the loading period in a mechanical test. This property is unitless.

> Friction angle (θ). The friction angle is represented by the failure envelope on a Mohr diagram, as determined by a series of compressional tests. The friction angle is often described by the coefficient of internal friction (μ), where $\mu = \tan^{-1}\theta$. The friction angle determines how rock strength will vary as the confining pressure is increased.

> Unconfined compressive strength (UCS). UCS is the axial stress (or differential stress) that can be supported by a rock when the effective confining pressure is equal to zero. Directly testing for this parameter is problematic, due to splitting of

the core which frequently occurs under uniaxial conditions. The value for UCS is therefore calculated once the Mohr failure envelope has been determined from a series of tests. UCS is measured in MPa.

> Cohesion (S_o) is the shear stress that can be supported by a rock when the normal force on all internal planes is equal to zero (zero effective confining pressure). It is determined from the intersection of the Mohr failure envelope with the shear stress axis (at zero effective stress) on a Mohr diagram. Cohesion is closely correlated with the tensile strength. Units are in MPa.

> Tensile strength (T) is a measure of rock strength in the extensional domain. In this study T was determined via a Brazil tension test, in which compressive stresses are applied to a circular disc, with the sample failing due to the formation of a tensile splitting fracture. Units are in MPa.

7.3 Geomechanical workflow at the Otway site

7.3.1 Initial modelling using the FAST methodology

During Stage 1 of the Otway Project, the geomechanical work conducted on the Naylor Field was somewhat limited due to the low CO_2 injection pressures. The reservoir models predicted that less than 1 MPa of excess pressure would be generated in the reservoir, which was confirmed by pressure observations during CO_2 injection. The main work conducted was in relation to the stability of the faults, as even small pressure perturbations can destabilise a fault if it is critically stressed.

The methodology used to gauge stability of the three key faults at the Naylor Field was the Fault Analysis Seal Technology (FAST) of Mildren et al. (2002). This technique uses a Mohr-Coulomb approach to assess fault reactivation propensity by calculating the increase in pore pressure required to cause reactivation. While the actual fault modelling was performed using Traptester software,

a significant amount of work related to acquiring specific numbers for input into the model. In addition to the actual fault surface characterisation, values were needed for the magnitude and orientation of the principal stress directions and also for the frictional and cohesive properties of the fault surfaces themselves.

At the Otway site, the orientation of S_{Hmax} was determined to be 142°N based on borehole breakouts in the CRC-1 borehole (van Ruth and Nicol 2007). This S_{Hmax} orientation is broadly consistent with the regional orientation for the onshore Victorian Otway Basin (Hillis et al. 1998; www.asp.adelaide.edu.au/asm). The magnitude of the vertical stress was calculated by integrating the density logs from Naylor-1, CRC-1 and five other wells in close proximity. The results were all consistent and indicated that at 2025 m depth, the vertical stress gradient was 21.4 MPa/km. To calculate S_{hmin}, 17 leak-off tests and 17 formation integrity tests from the onshore Victorian Otway Basin were compiled and analysed. In addition, an extended leak-off test at 512 m depth was undertaken during the drilling of the CRC-1 well (van Ruth 2007). The regional LOT data indicated a S_{hmin} of 14.62 MPa/km, which was consistent with the interpreted average fracture closure pressure from the CRC-1 extended leak-off test.

The greatest uncertainty regarding the in-situ stress tensor was in regard to S_{Hmax}, which in turn had very large implications for the fault modelling, as the faulting regime (i.e. strike slip versus normal) could not be determined reliably. The magnitude of S_{Hmax} was calculated using two methods: The first was based on the open hole hydraulic fracturing technique (Amadei and Stephansson 1997) in which:

$$S_{Hmax} = 3S_{hmin} - P_b - P_p + T$$

where P_b is the breakdown pressure, P_p is the pore pressure and T is the tensile strength of the formation. Tensile strength was assumed to be the difference between breakdown pressure and fracture reopening pressure during the extended leak-off test, which implied that the S_{Hmax} to S_{hmin} ratio was 1.3. Using this technique the S_{Hmax} gradient for the Otway site was 19 MPa/km, which suggested a normal faulting environment (i.e. the vertical stress being the largest stress).

S_{Hmax} was also calculated using the frictional limit theory, which states that the ratio of the maximum to minimum effective stress cannot exceed the magnitude required to cause faulting on an optimally oriented, pre-existing, cohesionless fault plane (Sibson 1974). The frictional limit to stress is given by:

$$\frac{S_1 - P_p}{S_3 - P_p} \leq \left\{ \sqrt{(\mu^2 + 1)} + \mu \right\}^2$$

where μ is the coefficient of friction, S_1 is the maximum principal stress and S_3 is the least principal stress. The maximum principal stress is assumed to be maximum horizontal stress. Assuming that S_{hmin} = 14.62 MPa/km as described above and P_p = 8.64 MPa/km at the Waarre C level, then the magnitude of S_{Hmax} was constrained to 27.3 MPa/km at reservoir level. This suggests that the Otway site is in a strike slip faulting regime, which is in contrast to the S_{Hmax} calculations using the extended leak-off test data.

For the FAST modelling, two scenarios for fault rheology were explored: cohesionless and healed. For the cohesionless scenario, the friction coefficient on the fault surfaces was assumed to be 0.6 and the cohesion was assumed to be 0. The healed scenario used a value of 0.78 for the friction coefficient and 5.4 for the cohesion. These values were chosen as end-members for the range of possible fault rheologies. In reality, the actual cohesion and friction of the Otway faults probably lie somewhere in the middle of this range.

The main conclusion of the CO2CRC pre-injection geomechanics report (van Ruth et al. 2007) was that the absolute values of fault reactivation propensity differed markedly due to the great uncertainties relating to the in-situ stress tensor and fault properties. The reactivation propensity (for optimally oriented faults) was expressed in terms of the pressure pulse that could be supported by the faults. The results ranged between 0 MPa and 28.6 MPa depending on the parameters that were input into the model. This spread did little to constrain the fault stability at the Otway field and operations proceeded in a very cautious manner to ensure that fault integrity was maintained. It was recognised that information acquired during Stage 2 of the Otway Project (discussed later) would allow the in-situ stress field, and therefore the fault model, to be better constrained.

7.3.2 Incorporation of reservoir stress path effects

The reservoir stress path (sometimes referred to as pore pressure-stress coupling) is an important geomechanical phenomenon that can greatly impact on fault stability by imparting changes to the in-situ stress field via poroelastic interactions. The reservoir stress path generally describes the degree to which the minimum horizontal stress changes in response to perturbations of pore fluid pressure, either in an injection or withdrawal scenario (e.g. Teufel et al. 1991; Addis 1997; Hillis 2000). Characterising such changes to the stress tensor is very important from the operational point of view, as it will determine the fracture gradient and also influence the pressures at which a fault will theoretically become unstable, assuming Mohr-Coulomb failure criteria.

The reservoir stress path and its effect on the reservoir may be understood conceptually as follows: when a liquid or gas is injected into a reservoir, pore fluid pressure builds up and the reservoir has a tendency to expand in all directions, as the vertical direction is bound by a free surface (the Earth's surface), there is no change to the vertical stress. Conversely, as the reservoir tends expand laterally, there is a counteracting force that is imparted into the reservoir which causes the minimum horizontal stress to increase.

Following the initial fault modelling work (described in the previous section), an analytical modelling study was undertaken to characterise the reservoir stress path at the Otway site and determine how it would affect the fault stability modelling that had been previously conducted. The results of the work are described by Vidal-Gilbert et al. (2010). This study uses the formulations described in Rudnicki (1999) to calculate the effects of geometry and elastic properties on the local in-situ stress field. In the analytical treatment, the functions describing the changes to the horizontal stress are incorporated into the governing equations for the normal and strike-slip faulting criteria. Although there was still significant uncertainty in the model due to a poor constraint on the stress tensor and the mechanical properties of the fault zones, this study did narrow the range of pore fluid pressures that could theoretically cause fault reactivation.

The key results of this study are shown graphically in Figure 7.2. The pressure required to cause fault reactivation varies significantly, depending on whether a normal or strike-slip faulting regime is assumed. For the strike slip scenario (Figure 7.2(a)), the Naylor fault in the vicinity of the Naylor-1 well has the greatest propensity to reactivate, with a pressure increase of only about 2 MPa needed to bring the fault to a critically stressed state. In contrast, when a normal faulting regime is modelled, the faults most prone to reactivation can sustain greater fluid pressures (5 MPa or more) and trend toward the northwest. The implication of both of these results for the Otway Project was that the Stage 1 operations were operating at pressures below those which could have caused reactivation of

proximal faults. Based on the pressure data shown by Vidal-Gilbert et al. (2010), the pressure increases during Stage 1 were only about 1.5 MPa.

7.3.3 Additional stress information gained from the Otway Stage 2, CRC-2 well

The drilling of the CRC-2 well occurred after the completion of Stage 1, but provided much needed additional information on the in-situ stress field and on the geomechanical characterisation of the Otway site in general. Such information would preferably have been available for Stage 1, but this was not possible. Another extended leak-off test was conducted as was an MDT mini-frac testing programme designed to give a detailed picture of the horizontal stresses at different levels within the Paaratte Formation (Chapter 5). These further tests allowed the stress field to be characterised with greater confidence, thus allowing future geomechanical work to be conducted with greater accuracy. Furthermore, an extensive rock mechanical testing programme was conducted, in which five specimens were tested for various brittle and elastic properties. These five tests, together with the two conducted following the drilling of the CRC-1 well, allowed calibration of the sonic logs, thus providing continuous records for mechanical properties down through the CRC-1 and CRC-2 wells.

In the early stages of CRC-2 drilling, an extended leak-off test (ELOT) was conducted at 500 m depth in the Narrawartuk Marl (shoe depth 494 m, hole depth 504 m). The primary purpose of the test was to determine the magnitude of the minimum horizontal stress (σ_{hmin}) with the practical application of determining borehole stability during drilling. The results of the ELOT also provided much needed information on the in-situ stress state in the Naylor Field. In addition to providing an estimate of the minimum horizontal stress, ELOTs also allowed estimates to be made of the tensile strength and maximum horizontal stress—parameters that are essential for modelling fault stability and the geomechanical response of the reservoir and cap rock.

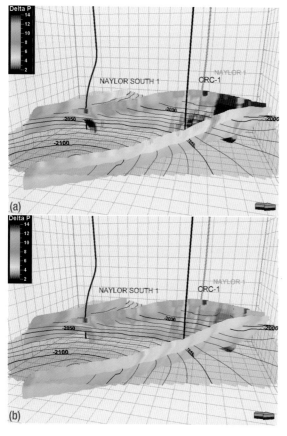

Figure 7.2: Fault reactivation propensity for all faults at the Otway site (a) using scenario 1: SSFR assumption, (b) using scenario 2: NFR assumption. Based on the modelling, a normal faulting regime would lead to a more stable configuration for faults in the area. See Vidal-Gilbert et al. 2010 for details.

The results showed that the three leak-off cycles had fracture closure pressures of around 300 psi. It should be noted that these were excess pressures measured at the surface and must be added to the pressure of the static mud column in order to extract an absolute fracture closure pressure. The leak-off pressures from the three cycles were averaged and yielded a σ_{hmin} gradient of 14.77 MPa/km (Tenthorey et al. 2010). ELOTs can also be used to gain some information on the tensile strength of the formation at the testing depth, allowing some estimate to be made of the maximum horizontal stress magnitude (σ_{Hmax}). Tensile strength can be estimated from the difference between the fracture breakdown pressure and the fracture propagation pressure, which in the case of the CRC-2 ELOT was estimated to be 174 psi (1.12 MPa). The σ_{Hmax} could then be calculated. Using the relevant values described by Tenthorey et al. (2010), together with σ_{hmin} = 14.77, MPa/km yielded a σ_{Hmax} value of 9.22 MPa at test depth or a σ_{Hmax} gradient of 18.43 MPa/km.

Mini-frac testing using Schlumberger's Modular Formation Dynamics Testing (MDT) tool is similar in principle to an ELOT, albeit on a much smaller scale and with much better spatial control on fracture propagation. Using the MDT tool configured with inflatable packers, it is possible to pressurise an interval of 1 m with drilling mud until a tensile fracture develops. Following such breakdown of the formation and propagation of the fracture, pumping is stopped and the pressure allowed to decay, with the fracture closure pressure determined using the double tangent method. It is necessary to conduct multiple pressurisation and propagation cycles during mini-frac testing, to ensure that the fracture propagates a distance of at least four borehole radii (and preferably six or seven radii) so that wellbore stress distortion effects are eliminated (Carnegie et al. 2000). The results from two successful mini-frac tests indicated that S_{hmin} is approximately 14 MPa/km, which is consistent with the ELOT. A value for S_{Hmax} of 16 MPa/km was determined, and was broadly consistent with the results of the ELOT.

The results of the CRC-2 ELOT and MDT mini-frac tests were very important additions to the Otway dataset. They provided information regarding the magnitudes of the horizontal stresses in the vicinity of the Naylor structure, and therefore allowed refinement of previous geomechanical models that contained significant uncertainties. For the purposes of updating the contemporary stress field at the Otway Project site, the vertical stress gradient, as calculated by van Ruth et al. (2007), was used. The vertical stress was likely to be unchanged at CRC-2 due to the close proximity to CRC-1. As was the case for the CRC-1 geomechanical models, the maximum horizontal stress orientation, based mainly on borehole breakouts, was determined to be 142° (van Ruth and Nicol 2007). The updated details of the contemporary stress field at the Otway Project are presented in Table 7.1. These results suggested that the Naylor structure is in a normal faulting regime, consistent with the study of Berard et al. (2008).

7.3.4 Acquisition of rock mechanical test data

Following the drilling of the CRC-2 well, five specimens were selected from cored intervals, so that an extensive rock mechanical testing programme could be carried out. All mechanical tests yielded values for Young's modulus, Poisson's ratio, cohesion, friction angle and unconfined compressive strength. Two samples were also tested for tensile strength. The results of all the tests are discussed in detail by Tenthorey et al. (2011).

When embarking upon a mechanical testing programme designed to characterise and calibrate mechanical properties of a section, it is critical to first develop a well planned sampling programme. If the samples selected for testing do not adequately characterise the various end-members encountered throughout the section, then the transforms

Table 7.1: Refined parameters for in-situ stress tensor following the drilling of the CRC-2 well.

σ_{hmin} (ELOT)	14.77 MPa/km	(Tenthorey et al. 2010)
σ_{hmin} (Mini-fracs)	13.86 MPa/km	(Tenthorey et al. 2010)
σ_{Hmax} (ELOT)	18.43 MPa/km	(Tenthorey et al. 2010)
σ_{Hmax} (Mini-fracs)	16.16 MPa/km	(Tenthorey et al. 2010)
σ_{v} (density logs)	21.4 MPa/km	(van Ruth et al. 2007)
Maximum horizontal stress direction	142°N	(van Ruth and Nicol 2007)

developed for the sonic logs are likely to be very inaccurate. Furthermore, core testing should be conducted on specimens that are as homogeneous as possible and devoid of any obvious fracturing, damage or alteration, as these will affect the mechanical test results. As previously mentioned, if testing mudstone samples, it is preferable that the core is preserved so that dehydration of the core is minimised, as loss of bound and unbound water would reduce integrity.

The rock mechanical test results for Otway were used to calibrate the compressive and shear sonic logs for rock mechanical properties, so that a continuous record for the various properties could be acquired for the Paaratte and adjacent formations. The transforms developed using the CRC-2 core proved to be very satisfactory for accurately calibrating the sonic logs through the Paaratte Formation, especially for Young's modulus. These transforms were then extended deeper in the section, to the level of the Belfast and Waarre Formations. The transform results compared well with the rock mechanical test results from the CRC-1 well, thereby validating the broad applicability of the transforms at the Otway site.

The continuous proxy logs developed for the various mechanical properties turned out to be very important for the geomechanical modelling conducted on the Iona Field (Section 7.5) and will be the cornerstone of any future geomechanical modelling of the Otway site. Without the rock mechanical tests and the transforms developed from the tests, it would be impossible to accurately estimate or quantify Otway rock mechanical properties using standard wire-line log data.

7.4   3D geomechanical modelling

For major CCS projects in which large volumes of CO_2 are to be injected over long periods of time, a detailed geomechanical assessment and understanding of the storage site will be critical. As discussed previously, a detailed characterisation of the in-situ stress tensor is key to developing a geomechanical model with any significant meaning. In the first instance, this information can be used to determine the stability of the key faults within a region or a site, as was done at the Otway Project. However, when significant pressures are expected to occur in the subsurface, the development of a detailed and coupled 3D geomechanical model would be highly

desirable. Such models allow prediction of reservoir and surface deformation during gas injection (e.g. Rutqvist et al. 2010; Vidal-Gilbert et al. 2009) and also provide information on poroelastic phenomena that can lead to variations of the in-situ stress tensor in and around the site. The impacts of a changing stress tensor on the fracture gradient and on fault stability have been discussed previously. Understanding these stress transfer processes and combining this with other geomechanical phenomena, such as stress arching, requires complex analytical or finite element modelling.

To date there has been little work published on geomechanical modelling for projects involving CO_2 injection. Much of the published work is either analytical (Soltanzadeh and Hawkes 2009; Vidal-Gilbert et al. 2010) or of a 2D nature (Rouania et al. 2006; Orlic 2009; Vilarrasa et al. 2010). However, some notable studies have conducted 3D finite element modelling which incorporated complex poroelastic phenomena during CO_2 injection (Minkoff et al. 2003; Lucier et al. 2006; Lucier and Zoback 2008; Shi and Durucan 2009; Vidal-Gilbert et al. 2009; Rutqvist et al. 2010). The main benefit of constructing 3D finite element models is that the complex lithological and structural variability can be incorporated and explored. Most of the 3D geomechanical modelling studies described above are coupled to some degree with the reservoir models. In the case of the fully coupled models, the changes to porosity and permeability driven by depletion or pressurisation will be fed back into the reservoir model as the dynamic simulation proceeds.

The development of detailed geomechanical models, which describe the geomechanical state of a 3D system through time, have only recently become commonplace. This is partly due to technological advancements in computing, but also because operators have become increasingly aware of the negative impact that can be associated with poor geomechanical management. It should be emphasised that successful development of such models requires a detailed geomechanical characterisation of not only the reservoir interval, but also of the rock volume enclosing it (Vidal-Gilbert et al. 2009; Rutqvist et al. 2010). In most cases, rock mechanical tests are used to calibrate the sonic logs so that continuous records for poroelastic properties are generated down any given well, as was done for the

Paaratte Formation. This is made possible by the fact that compressive and shear wave velocities are controlled by the elastic properties of the rock itself. However, it should be noted that calibration is not straightforward and that some empirical relationships are required to convert the dynamic moduli as calculated from the sonic logs to the static moduli as acquired from the rock mechanical tests (Wang 2000). It is also critical to have a quantitative understanding of the brittle failure parameters, such as unconfined compressive strength and friction angle, as these properties will control whether the reservoir and cap rock will be able to support the changes in pore fluid pressure, or whether new fault and fracture networks form.

The gridding size and level of detail which is input into the geomechanical model depends mainly on the desired outputs from the model, the data that are available and the computing power available to the modeller. Clearly, it is generally desirable to have greater geomechanical detail within the reservoir interval as this is where fluid pressures are most variable, both spatially and temporally. For the purposes of modelling CO_2 storage systems and understanding containment risk, it may also be necessary to extend this detail into the cap rock. A number of geomechanical software packages are available to perform such modelling, including ABAQUS, FLAC3D and VISAGE. The ease with which the various packages interface with the different reservoir modelling packages varies significantly and should be taken into consideration before modelling is commenced. Outputs from the geomechanical modelling will include information on changes to the local in-situ stress field caused by injection pressures, fault and cap rock integrity and strain within the reservoir and surrounding formations, including vertical displacement at the Earth's surface. An example of such outputs from a coupled 3D geomechanical model is given in Section 7.5.

7.5 The Iona gas storage facility as an analogue for CO$_2$ storage

The Iona Field is an underground gas storage facility located in southwest Victoria, approximately 20 km east of the CO2CRC Otway Project. The formation being used for gas storage is the Waarre Formation, the same geological unit used for CO_2 storage during Stage 1 of

the Otway Project. Therefore, in addition to providing important geomechanical constraints which can be applied broadly (there are many hundreds of natural gas storage sites around the world and only a few CO_2 storage sites), Iona is also useful in that it may provide information that can be used to better understand the effects of CO_2 injection at the Otway Project. The Iona facility has a long injection/withdrawal history together with accompanying pressure data, from which history-matched static and dynamic flow models have been generated. This information was used as the foundation for a 3D geomechanical model describing the physical evolution of the field since its conversion to a gas storage facility in 2000 (Tenthorey et al. 2013).

The Iona Field was converted from a producing gas field to a gas storage facility in 2000. At Iona, methane gas is piped from offshore gas fields and stored in the Upper Cretaceous Waarre C Formation, approximately 1300 m below ground level. Gas injection and withdrawal are carried out via five wells, with two observation wells on the edge of the field to ensure that the reservoir is not filled to spill. Structurally, the Iona Field is a tilted anticline which is bound by two main faults, a similar configuration to that at the Otway Project site. The south fault behaves as a juxtaposition seal to the gas column, while the seal capacity of the north fault is uncertain, as it has barely been exposed to gas.

A significant portion of the Iona modelling work pertained to building the static geomechanical model. To populate the static geomechanical model with rock mechanical properties, the transforms developed from the CRC-2 rock mechanical testing programme were used and applied to the sonic logs available at Iona. Unfortunately, rock mechanical testing of Iona core itself proved to be impossible, mainly due to the poor quality and friable nature of the core plugs. Once the Iona wells were calibrated for mechanical properties, the 3D volume was populated, using empirical relationships in each well between the mechanical properties and parameters such as porosity.

Once the static modelling was completed, the dynamic geomechanical modelling on the field was performed. In the case of Iona, the modelling conducted was one-way coupled, meaning that the pressure profile from the reservoir model was selected at key times in the field's history and fed into the geomechanical model. In this way it was possible to assess the geomechanical effects of gas injection or withdrawal, relative to the pre-production condition. Unlike fully coupled models, a one-way coupled model does not allow any geomechanical influences on permeability or porosity to be fed back into the reservoir model. Figure 7.3 shows the injection/withdrawal history together with the time steps chosen for geomechanical analysis. Most of the attention focused on time steps S2 and S3 as these were the periods of peak withdrawal and injection, respectively.

The dynamic geomechanical modelling of the Iona Field yielded several interesting results. First, during the periods of large pressure change due to gas injection or withdrawal, there was a strong reservoir stress path which resulted in a significant modification of the horizontal stresses at the reservoir level, but also above the reservoir in the cap rock. The reservoir stress path both during pressurisation

and depletion had a predicted value of about 0.75, which meant that for each unit change in fluid pressure, the horizontal stress magnitudes would change by 75% of that value. Compared to previous studies on reservoir stress path effects in producing fields, this value is at the upper end of what has been observed (Hillis 2001; Goulty 2003; Teufel et al. 1991; Whitehead et al. 1987). There are significant implications arising from a strong stress path, including modification of the fracture gradient in the reservoir and cap rock and also complex effects pertaining to fault stability and potential reactivation during pressurisation or depressurisation. These effects are discussed in more detail by Tenthorey et al. (2013).

Another output of interest from the model was the expected surface movement due to pressurisation or depletion of the reservoir. During time steps S2 and S3, the maximum predicted subsidence and heave were 9 mm and 2.5 mm, respectively (Figure 7.4). While the

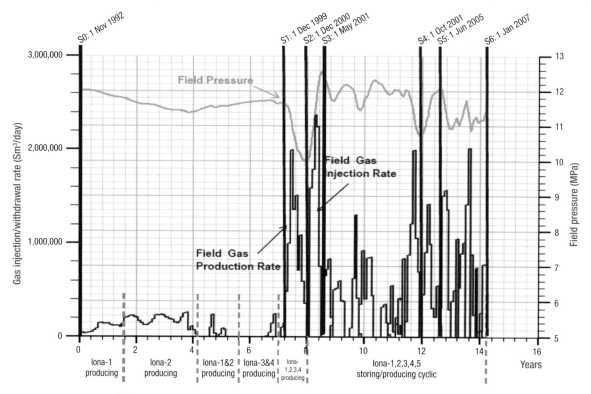

Figure 7.3: Injection/withdrawal history, together with the field pressure for the Iona gas storage facility, between 1992 and 2007. Time-step selections for the dynamic geomechanical modelling are also shown. Geomechanical response of the reservoir and surrounding formations was assessed mainly at times of peak injection or withdrawal, as these are the period of greatest interest geomechanically.

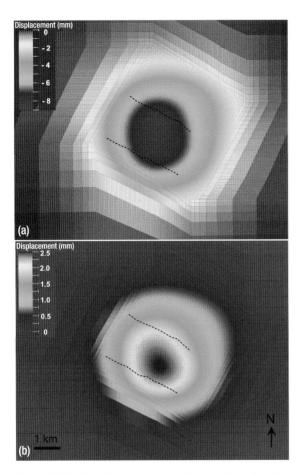

Figure 7.4: Surface displacements predicted at Iona for time steps S2 (a) and S3 (b) relative to S0. The dashed lines represent the projection of the two bounding faults to the surface. Colour scale is in metres, with a maximum displacement of 9 mm in (a) 2.5 mm in (b).

Finally fault stability and cap rock integrity during gas injection and withdrawal were assessed by looking at the plastic strain accumulation as determined by the model results. A value of 1% strain is generally used to delineate the point at which the risk of rock failure becomes elevated. Unlike elastic strain, plastic strain is irreversible and only increases during the life of a field. Plastic strain is predicted to be greatest in the northern bounding fault, with values up to 0.09% in some cells, at the end of the model run. Even the cells that exhibit these maxima in plastic strain are well below the critical value at which fault reactivation or rupture becomes a significant risk. The same plastic strain metric was used to evaluate cap rock integrity. The maximum change in plastic strain within the Belfast Mudstone at the end of the history match is 0.135%, and occurs in the vicinity of the northern bounding fault. This value is well below the 1% benchmark and therefore suggests that there is minimum risk of cap rock integrity failure. However, it should be noted that the fault model data were only available in the reservoir section. Should the faults extend into the overburden, then a higher potential for cap rock integrity failure would have been possible.

The Iona geomechanical modelling study provided valuable information on two fronts which can be applied to our understanding of gas storage in general. First, it provided a step by step workflow for building a detailed 3D dynamic geomechanical model for a potential gas storage facility. This included calibration of sonic well data for rock mechanical properties using mechanical test data, population of the static model with mechanical properties, pre-production stress initialisation and execution of the dynamic geomechanical model using history matched flow models. Although many geomechanical models such as these have in all likelihood been constructed, there is a surprising paucity of information in the public domain regarding the construction and execution of such models. Second, the information gained from the modelling study can be applied not only to operational aspects of the Otway project but also to other potential gas storage facilities around the world.

greatest surface displacements were expected directly above the reservoir, some displacement was expected as far as 2.5 km away from the centre of the reservoir projection. The surface displacements predicted by the model were smaller, but on the same order as those observed at the In Salah CO_2 Project, using satellite-based interferometry (Rutqvist et al. 2010). Having the ability to accurately predict such ground movements is a critical aspect of confidently communicating to the public the surface manifestations of CO_2 injection in the subsurface.

7.6 Conclusions

The geomechanical data collected for Otway Stage 1 and Otway Stage 2 proved to be an extremely important component of these injection tests in that they provided a comprehensive geomechanical assessment and understanding of the storage site. A detailed characterisation of the in-situ stress tensor was important for developing the geomechanical model used to determine the stability of the key faults within the Otway site. Successful development of the model at Otway required not only a detailed geomechanical characterisation of the reservoir interval, but also of the rock volume enclosing it. It was also critical to have a quantitative understanding of the brittle failure parameters, such as unconfined compressive strength and friction angle, as these properties controlled whether the reservoir and cap rock would support any changes in pore fluid pressure associated with the injection of CO_2.

Despite the importance of constructing 3D finite element models in which complex lithological and structural variability can be incorporated and explored, to date there has been little work published on geomechanical modelling for projects involving CO_2 injection. Therefore the Otway geomechanical study has an importance that extends beyond the obvious relevance to the Otway site, particularly as operators become increasingly aware of the negative impact that can potentially be associated with poor geomechanical management of storage sites. The information obtained from the nearby Iona gas storage site proved to be an important adjunct to the data collected at the actual Otway site and suggests that greater use should be made of data from natural gas storage sites when considering CO_2 storage.

7.7 References

Addis, M.A. 1997. The stress-depletion response of reservoirs. SPE Annual Technical Conference and Exhibition, SPE 38720, San Antonio, Texas, 5–8 October.

Amadei, B. and Stephansson, O. 1997. Rock stress and its measurement. London, Chapman and Hall.

Aadnoy, B.S. 1990. Inversion technique to determine the in-situ stress field from fracturing data. Journal of Petroleum Science and Engineering 4, 127–141.

Berard, T., Sinha, B.K., van Ruth, P., Dance, T., John, Z. and Tan, C. 2008. Stress estimation at the Otway CO_2 storage site, Australia. SPE Paper 116422.

Brudy, M. and Zoback, M.D. 1993. Compressive and tensile failure of boreholes arbitrarily inclined to principal stress axes: Application to the KTB boreholes, Germany. International Journal of Rock Mechanics and Mining Science 30, 1035–1038.

Carnegie, A., Thomas, M., Efnik, M.S., Hamawi, M., Akbar, M. and Burton, M. 2000. An advanced method of determining insitu reservoir stresses: Wireline conveyed micro-fracturing. SPE Paper 78486.

Deichmann, N. and Giardini, D. 2009. Earthquakes induced by the stimulation of an enhanced geothermal system below Basel (Switzerland). Seismological Research Letters 80/5, 784–798, doi:10.1785/gssrl.80.5.784.

Engelder, T. 1992. Stress regimes in the lithosphere. Princeton University Press.

Gardner, G.H.F., Gardner, L.W. and Gregory, A.R. 1974. Formation velocity and density—the diagnostic basics for stratigraphic traps. Geophysics 39: 770–780.

Goulty, N.R. 2003. Reservoir stress path during depletion of Norwegian chalk oilfields. Petroleum Geoscience 9(3), 233–241.

Hillis, R. 2000. Pore pressure/stress coupling and its implications for seismicity. Exploration Geophysics 31, 448–454.

Hillis, R.R. 2001. Coupled changes in pore pressure and stress in oil fields and sedimentary basins. Petroleum Geoscience 7(4) 419–425.

Hillis, R.R., Meyer, J.J. and Reynolds, S.D. 1998. The Australian stress map. Exploration Geophysics 29, 420–427.

Jimenez, J.A. and Chalaturnik, R.J. 2002. Integrity of bounding seals for geological storage of greenhouse gases. SPE/ISRM 78196.

Lucier, A. and Zoback, M. 2008. Assessing the economic feasibility of regional deep saline aquifer CO_2 injection and storage: A geomechanics-based workflow applied to the Rose Run Sandstone in Eastern Ohio, USA. International Journal of Greenhouse Gas Control 2(2), 230–247.

Lucier, A., Zoback, M., Gupta, N. and Ramakrishnan, T.S. 2006. Geomechanical aspects of CO_2 sequestration in a deep saline reservoir in the Ohio River Valley region. Environmental Geosciences 13(2), 85–103.

Mildren, S.D., Hillis, R.R. and Kaldi, J. 2002. Calibrating predictions of fault seal reactivation in the Timor Sea. APPEA Journal 42, 187–202.

Minkoff, S.E., Stone, C.M., Bryant, S., Peszynska, M. and Wheeler, M.F. 2003. Coupled fluid flow and geomechanical deformation modeling. Journal of Petroleum Science and Engineering 38(1–2), 37–56.

Nauroy, J-F. 2011. Geomechanics applied to the petroleum industry. Ed. Technip, Paris.

Orlic, B. 2009. Some geomechanical aspects of geological CO_2 sequestration. KSCE Journal of Civil Engineering 13(4), 225–232.

Rudnicki, J.W. 1999. Alteration of regional stress by reservoirs and other inhomogeneities: stabilizing or destabilizing? Proceedings of the Ninth International Congress on Rock Mechanics, Paris, France, August 25–28, edited by G. Vouille and P. Berest, Vol. 3, 1629–1637.

Rouainia, M. et al. 2006. Hydro-geomechanical modelling of seal behaviour in overpressured basins using discontinuous deformation analysis. Engineering Geology 82(4), 222–233.

Rutqvist, J., Vasco, D.W. and Myer, L. 2010. Coupled reservoir-geomechanical analysis of CO_2 injection and ground deformations at In Salah, Algeria. International Journal of Greenhouse Gas Control 4, 225–230.

Shi, J.-Q. and Durucan, S. 2009. A coupled reservoir-geomechanical simulation study of CO_2 storage in a nearly depleted natural gas reservoir. Energy Procedia 1, 3039–3046.

Sibson, R.H. 1974. Frictional constraints on thrust, wrench and normal faults. Nature, 249, 542–544.

Soltanzadeh, H. and Hawkes, C.D. 2009. Assessing fault reactivation tendency within and surrounding porous reservoirs during fluid production or injection. International Journal of Rock Mechanics and Mining Sciences 46(1), 1–7.

Tenthorey, E., John, Z. and Nguyen, D. 2010. CRC-2 Extended Leak-Off and Mini-Frac Tests: Results and implications. Cooperative Research Centre for Greenhouse Gas Technologies, Canberra, Australia, CO2CRC Publication Number RPT10-222.

Tenthorey, E., Maney, B. and Dewhurst, D. 2011. Description of geomechanical properties for the Paaratte and surrounding formations. Cooperative Research Centre for Greenhouse Gas Technologies, Canberra, Australia, CO2CRC Publication Number RPT11-3124.

Tenthorey, E., Vidal-Gilbert, S., Backe, G., Puspitasari, R., Pallikathekathil, Z.J., Maney, B. and Dewhurst, D. 2013. Modelling the geomechanics of gas storage: A case study from the Iona gas field, Australia. International Journal of Greenhouse Gas Control 13, 138–148.

Terzaghi, K. 1943. Theoretical soil mechanics. John Wiley and Sons, New York.

Teufel, L.W., Rhett, D.W., Farrel, H.E. 1991. Effect of reservoir depletion and pore pressure drawdown on in situ stress and deformation in the Ekofisk field, North Sea. Rock mechanics as a multidisciplinary science. In: Roegiers, J.C. (Ed.), Proceedings of the 32nd US Symposium, Balkema, Rotterdam.

van Ruth, P. and Nicol, A. 2007. Structural interpretation of the CRC-1 resistivity image log. Cooperative Research Centre for Greenhouse Gas Technologies, Canberra, Australia, CO2CRC Publication Number RPT07-0962.

van Ruth, P. 2007. CRC-1 extended leak-off test. Cooperative Research Centre for Greenhouse Gas Technologies, Canberra, Australia, CO2CRC Report Number RPT07-0608.

van Ruth, P., Tenthorey, E. and Vidal-Gilbert, S. 2007. Geomechanical analysis of the Naylor structure, Otway Basin, Australia. Cooperative Research Centre for Greenhouse Gas Technologies, Canberra, Australia. CO2CRC Publication Number RPT07-0966.

Vidal-Gilbert, S., Nauroy, J-F. and Brosse, E. 2009. 3D geomechanical modelling for CO_2 geologic storage in the Dogger carbonates of the Paris Basin. International Journal of Greenhouse Gas Control 3, 288–299.

Vidal-Gilbert, S., Tenthorey, E., Dewhurst, D., Ennis-King, J., van Ruth, P. and Hillis R. 2010. Geomechanical analysis of the Naylor Field, Otway Basin, Australia: Implications for CO_2 injection and storage. International Journal of Greenhouse Gas Control 4, 827–839.

Vilarrasa, V., Bolster, D., Olivella, S. and Carrera, J. 2010. Coupled hydromechanical modeling of CO_2 sequestration in deep saline aquifers. International Journal of Greenhouse Gas Control 4(6), 910–919.

Wang, Z. 2000. Dynamic versus elastic properties. In: Seismic and acoustic velocities in reservoir rocks. SEG Geophysics Reprint Series, No. 19.

Whitehead, W.S., Hunt, E.R. and Holditch, S.A. 1987. Effects of lithology and reservoir pressure on the in-situ stresses in the Waskom (Travis Peak) Field. Society of Petroleum Engineers of AIME, (Paper) SPE, 139–152.

Zoback, M. 2007. Reservoir geomechanics. Cambridge University Press, Cambridge.

Maxwell Watson

8

8. CONTAINMENT RISK ASSESSMENT

8.1 Introduction

A key objective of the CO2CRC Otway Project was to demonstrate that underground storage of CO_2 is an effective option for reducing greenhouse gas emissions without any adverse effects on community safety and environment. An important part of this objective was to show that containment risk could be assessed through a process of risk and uncertainty analysis that was able to take fully into account the possibility of unlikely events inducing leakage. Due to the technical nature of subsurface leakage, it was decided that a specific risk analysis technique could be applied to containment risk. Any such assessment strongly relies on the characterisation of the subsurface fluid flow and understanding of the changes resulting from the injection. The resulting independent containment risk assessment could then be used to manage stakeholder and public perception that the key CCS risk is associated with leakage. The resulting risk assessment could also be used to determine the optimal risk mitigation and monitoring programme required to assure CO_2 containment.

At the time of the Otway Project development, no CCS-specific commercial tool existed that provided suitably specific risk assessment for subsurface containment of CO_2. The Project therefore utilised its own risk assessment research within the CO2CRC in combination with the generic proprietary risk assessment method RISQUE (Bowden et al. 2001). The Otway Project built on a process developed under the precursor Australian Petroleum Cooperative Research Centre (APCRC), using a technique where specific risk categories were populated with quantitative risk parameters (Bowden and Rigg 2004). While CCS was considered a new application of RISQUE, importantly the tool and methodology still met the industry standard of risk assessment, and was very transparent in its application.

8.2 Methodology

The assessment of storage risk at the Otway site involved an understanding of uncertainty in the subsurface storage complex. At the start of any subsurface project, this uncertainty is quite broad and is not expected to be fully resolved. For the Otway Project, first, the mechanisms that might provide conduits for CO_2 leakage, were characterised; these included the Belfast Mudstone top seal, the bounding faults, the CRC-1 and Naylor-1 wells, and the lateral bounds of the storage complex. Statistically, these mechanisms were expected to be competent to a determined threshold before a leakage event would initiate. These thresholds, including their inherent uncertainty, needed to be determined. Second, the changes caused by CO_2 injection and storage (including changes in pressure, temperature, stress or chemistry) needed to be modelled. These changes could possibly lead to CO_2 leakage through processes such as fault reactivation, cap rock fracturing, chemical erosion of wellbore or cap rock, and other storage-related events such as induced seismicity and ground deformation. Understanding the range of static properties and associated uncertainties, and then comparing them to the modelled dynamic changes invoked in the subsurface due to CO_2 injection (which also hold uncertainty), was seen as the key to effective risk assessment.

Risk in the RISQUE method is defined simply as the relationship between the likelihood of an individual risk event, and the impact of a risk event (in this case, CO_2 leakage along an identified pathway). The approach and calculation of leakage risk was quite simple using this method. A facilitated workshop environment, at which all the interpreted data and models were discussed and uncertainty considered, enabled rational consideration of risk and uncertainty with site-specific experts. The overall question that was assessed by the site-specific experts was the risk that some of the injected CO_2 would leak out of the defined storage container. To add quantification to

the assessment, the project team established leakage limits at less than the likely retention suggested by the IPCC (IPCC 2005). Therefore the acceptable limit was set at 1% total leakage over 1000 years. This allowed the ranking of the Otway Project to be compared to other projects. At this point in the Otway Project development, it was planned to inject up to 100,000 t of CO_2, so the acceptable total leakage limit over 1000 years was 1000 t. This risk assessment technique considered leakage of CO_2 outside of the defined storage complex as the risk impact. It did not assess the consequence of a leakage event, which was instead assessed using full earth modelling and large-scale modelling of leakage paths.

The process of quantification of containment risks was to systematically define each risk on the following basis:

> likelihood of occurrence (0–1 represented at a log scale), within a confidence limit (CL) range of CL05 and CL95

> impact in terms of leakage rate (tonnes CO_2 per year) with a most likely (CL50) and conservative estimate (CL95) (example given in Figure 8.1)

> number of these risks (e.g. number of wells)

> duration of loss (time that the event would be active).

A qualitative probabilistic framework was used to link to a quantitative meaning of the likelihood input, assuring consistency in estimation (Table 8.1). The basis of this table was developed originally for mature industries and not the currently immature CCS industry. However, if the basis of other projects analogous to CCS (including acid gas storage, natural gas storage, enhanced oil recovery (EOR) and hazardous waste storage) were considered, the table was considered acceptable for use in the Otway and other CCS projects.

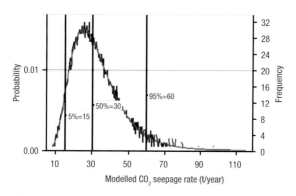

Figure 8.1: Example of a probability plot with log normal distribution representing the uncertainty within a subject matter specialist's estimation. This probabilistic distribution was used in the calculation of the risk quotient. For leakage, high estimate: CL 95% 60; best estimate: CL 50% 30.

Assessment of leakage rate impacts utilised either specific modelling outputs or literature. In particular a CO2CRC study used observed outflow rates of fluids from a range of scenarios to constrain potential CO_2 leakage rates (Streit and Watson 2005). These scenarios included outflow from springs and wells after earthquakes, examples from natural gas storage, and flux measurements of surface seepage above natural subsurface CO_2 accumulations. These empirical data, in addition to modelled predictions of flow rates, provided the team judging the risk with a broad range of possible values on leakage rates, which in turn served as an input into the risk assessment tool.

All risk assessment inputs provided by the expert panel were subjected to a probabilistic analysis using the RISQUE tool's inbuilt Monte-Carlo method. This modelling performed numerous simulations to handle the variables that were input as probability distributions. Various levels of risk confidence were available to accommodate the various risk appetites for the forthcoming project decisions.

Two containment risk assessments were performed for the Otway Project. The first risk assessment, conducted in 2005, was performed before the drilling of the CRC-1 well, with results required to provide assurance that the Project was viable and the cost to drill CRC-1 was justifiable. The second risk assessment, conducted in July 2007, was performed after the CRC-1 well was drilled to incorporate the additional data and interpretations and to prepare the Project for final approvals.

Table 8.1: Qualitative-quantitative basis table. This table was used by the subject matter expert to assign a consistent quantification to expert judgement within an order of magnitude range.

Qualitative description	Likelihood	Basis
A. Certain	1 (or 99.9)	Certain, or as near to as makes no difference
B. Almost certain	0.2–0.9	One or more incidents of a similar nature has occurred here
C. Highly probable	0.1	A previous incident of a similar nature has occurred here
D. Possible	0.01	Could have occurred already without intervention
E. Unlikely	0.001	Recorded recently elsewhere
F. Very unlikely	1×10^{-4}	Has happened elsewhere
G. Highly improbable	1×10^{-5}	Published information but in a slightly different context
H. Almost impossible	1×10^{-6}	No published information of similar case

8.3 Risk assessment context

To frame the assessment of a given risk event, the exact context of the project activity regarding storage had to be clearly established. As the Otway Project evolved, the context was adjusted to best match the Project operation plan. For the final risk assessment that was performed pre-injection, the Project context was described as follows:

> CO_2 would be sourced from the nearby CO_2-rich Buttress Field and transported via pipeline to the injection point (CRC-1) located to the east and downdip from the depleted Naylor gas field in the Port Campbell region of the onshore Otway Basin.

> The injection volume was fixed at 3 mmscf/d for a period of 2 years for a total of 100,000 t stored.

> The single well to be used as the injector would be the CRC-1 well, located approximately 300 m from the crest of the structure. The existing Naylor-1 well was to be used as the monitoring well, with both wells in contact with the CO_2 plume throughout the risked 1000 year period.

> Downhole pressure would be measured in both wells and injection would be stopped if pressure deviated significantly from modelled forecast.

8.4 Storage complex

The Otway Project storage interval was the approximately 30 m thick Mid Cretaceous Waarre Sandstone Unit C reservoir, at a depth of 2000 m with porosity of 20–25% and permeability of 600–700 mD (see Chapter 5). It was anticipated that the CO_2 would migrate updip into the crest of the structure. The sandstone is overlain by a series of mudstone up to approximately 500 m thick including the Flaxman Formation, the Late Cretaceous Belfast Mudstone, and the Skull Creek Formation. The storage complex for the Otway Project was defined as the entire PPL-13 tenement block, and below the top of the Skull Creek Formation. Overlying Skull Creek was the secondary reservoir, the non-potable aquifer Paaratte Formation, overlain by the Late Cretaceous Timboon Sandstone, the deepest potable water aquifer

Table 8.2: Risk input data from the July 2007 update.

Initiating Events	Probability of Event	Number of Items	Leakage Rate (CL50%) (t/year/item)	Leakage Rate (CL95%) (t/year/item)	Duration (Years)	Mass Lost (t) from Primary Container	Normalised Risk Quotient (CL80)
Leakage through permeable zones in seal	Almost impossible	1	0.01	0.05	1000	10	0.00001
Leakage: along bounding fault (Naylor Fault)	Highly improbable–very unlikely	1	30	60	1000	30,000	0.9
Leakage: along fault junction to Buttress Fault	Very unlikely–unlikely	1	30	60	1000	30,000	9
Leakage: along bounding fault (Naylor South Fault)	Almost impossible	1	30	6	990	2000	0.002
Well leakage: Naylor-1	Very unlikely–possible	1	30	0	15–1000	450	0.45
Well leakage: CRC-1	Highly improbable–unlikely	1	30	60	15	450	0.045
Regional overpressurisation: short-term leakage	Highly improbable–unlikely	1	300	3000	0.08	25	0.0025
Regional overpressurisation: long-term leakage	Highly improbable–unlikely	1	10	20	1000	10,000	1
Local overpressurisation	Almost impossible	1	30	60	0.02	1	0.000001
Exceeding spill-point	Almost impossible	1	1000	5000	2	2000	0.002
Earthquake induced fracturing: short-term leakage	Very unlikely	1	1000	10,000	0.08	83	0.0083
Earthquake induced fracturing: long-term leakage	Very unlikely	1	30	60	1000	30,000	3
Migration direction	Almost impossible	1	1000	5000	2	2000	0.002

of the region (see Chapter 5). The secondary reservoir unit, the Paaratte Formation, was not incorporated into the storage complex for the purpose of this Otway Stage 1 risk assessment, but was a very important component of Stage 2 (Chapter 17).

8.5   Risk items

All the risk items described below relate to geological aspects of CO_2 containment. Output from the expert panel from the most relevant 2007 and 2005 risk assessments are provided in Table 8.2 (2007 risk assessment) and Table 8.3 (2005 risk assessment).

8.5.1   Leakage from permeable zones in seals

In characterising the seal quality of the primary regional seal (approximately 500 m thick mudstone, including the upper Flaxman Formation, Belfast Mudstone and Skull Creek Formation), factors that were taken into account included the high degree of well control, clear seismic evidence of lateral continuity, excellent seal quality, as well as recognising that CO_2 had the additional barrier of the lower permeability reservoir component of the lower Flaxman Formation (see Chapter 5). Seal capillary analysis determined that the maximum modelled gas column (approximately 30 m) below the seal was an order of magnitude less than what would be required for buoyancy pressure to overcome the capillary pressure at the wells (lowest seal capacity result determined at 303 m; Daniel 2007; Chapter 6). The seal interpretation was

Table 8.3: Risk input data from the July 2005 assessment.

Initiating Events	Probability of Event	Number of Items	Leakage Rate (CL50%) (t/year/item)	Leakage Rate (CL95%) (t/year/item)	Duration (Years)	Mass Lost (t) from Primary Container	Normalised Risk Quotient (CL80)
Leakage through permeable zones in seal	Highly improbable	1	0.01	0.05	1000	10	0.0001
Leakage: along bounding fault (Naylor Fault)	Highly improbable–very unlikely	1	30	60	1000	30,000	0.9
Leakage: along fault junction to Buttress Fault	Very unlikely–unlikely	1	30	60	1000	30,000	9
Leakage: along bounding fault (Naylor South Fault)	Not assessed						
Well leakage: Naylor-1	Very unlikely–possible	1	30	60	15–1000	450	0.45
Well leakage: CRC-1	Not assessed						
Regional overpressurisation: short-term leakage	Unlikely	1	30	60	0.08	3	0.003
Regional overpressurisation: long-term leakage	Unlikely	1	6	10	100	600	0.6
Local overpressurisation	Almost impossible	1	30	60	0.02	1	0.000001
Exceeding spill-point	Almost impossible	1	50,000		0.08	4166	0.004166
Earthquake induced fracturing: short-term leakage	Very unlikely	1	1000	10,000	0.08	83	0.0083
Earthquake induced fracturing: long-term leakage	Very unlikely	1	30	30	1000	30,000	3
Migration direction	Highly improbable	1	50,000		0.25	12,500	0.125

extrapolated across the whole cap rock for the storage area. Therefore, this risk refered to the possibility that there was a substantial decrease in the seal quality away from the wells, due to the seal's heterogeneity providing a sufficient pathway for CO_2 leakage. This included the potential for existing fracture-based or other intrusion-based pathways with sufficiently high permeability for leakage, or for CO_2 reactions with the seal lithology resulting in the creation of a leakage pathway through the seal.

The likelihood of leakage from permeable zones in the seal was judged as *Almost Impossible* (reduced from *Highly Improbable* in the 2005 assessment). A loss rate for this event was considered to be 0.01–0.05 t per year (CL50–CL95) and, if it were to occur, was judged conservatively to leak for the full duration of the risked period (maximum of 1000 years).

8.5.2 Leakage from faults

Assessment of this risk event relied on the ability to identify faults from seismic data, in particular to establish whether it was likely that the faults had:

> insufficient throw to create a clay smear seal and/or a lack of a direct juxtaposition to the Belfast Mudstone, which would prevent across-fault leakage

> sufficient fault length to consider along-flow leakage out of the storage complex.

For the Naylor structure the Naylor, Naylor East (at the junction to Naylor fault), and Naylor South faults were all considered as potential pathways for leakage from faults to the secondary Paaratte Formation reservoir (Figure 8.2). Dynamic modelling showed that CO_2 would initially

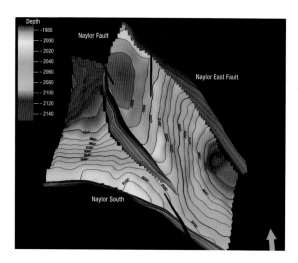

Figure 8.2: Location of the three faults investigated in the greater Naylor structure (arrow points due north).

migrate to the Naylor fault, relying on the fault seal to the west to limit any migration in that direction and, as the CO_2 continued to accumulate, also relying on the fault junction seal between the Naylor and Naylor East faults to further constrain migration. The Naylor South fault was included to capture the risk of leakage, in the event that some of the CO_2 plume migrated around the spill point of the initial Naylor structure into the "greater Naylor structure" (see Section 8.5.5) due to exceeding the spill point of the storage site.

CO_2 was modelled to accumulate against the Naylor fault and the Naylor East fault at the junction with the Naylor fault. A juxtapositional seal was therefore required for these faults at the Waarre C interval. The Naylor fault and Naylor East fault were interpreted to have a strong across-fault seal as they were juxtaposed against Belfast Mudstone at the Waarre Sandstone interval for the storage structure. Hence the chance of across-fault flow out of the storage complex was ruled out.

For faults of sufficient length to cut the storage complex, the geomechanical properties of the faults needed to be considered for the likelihood of connected fracture conduits along the fault plane (see Chapter 7). Leakage via fault reactivation as a result of enhanced natural or anthropogenic mechanisms was handled separately.

The Naylor fault extends vertically through the approximately 500 m thick top seal. Fault reactivation modelling showed that this fault was reasonably well orientated and stressed, meaning reactivation could occur. Work by Lyon et al. (2005), suggested that faults with this geomechanical regime in the Otway Basin could retain open fractures, which then could allow along-fault migration. The Naylor East fault was interpreted to not extend vertically to the Paaratte Formation (the secondary storage container). However, the junction of this fault with the Naylor fault was of concern to the Project and was judged to have a higher potential than the Naylor fault itself for allowing along-fault leakage, to the full vertical extent of the fault, then migrating along the Naylor fault out of the storage complex.

As the Naylor fault and the junction of this to the Naylor East fault were shown to potentially allow along-fault flow out of the primary container, an empirical analysis was performed to gather evidence on fault leakage. This analysis was an evaluation of 3D seismic data to assess if there was any evidence that could be attributed to previous gas migration, including hydrocarbon-related diagenetic zones (HRDZs), or gas chimneys. No features relating to gas migration were detected above the Naylor structure or in the region of the Project. This was consistent with the lack of pressure communication between the Waarre Sandstone and the Paaratte Formation determined from drilling of CRC-1 and the hydrodynamic evaluation of the area (Hennig 2007). No gas shows were seen in the Paaratte Formation and only trace gas (7–20 ppm) readings were recorded from the cuttings (Greaney 2008). Thus, even with the new data there was no evidence to support vertical hydrocarbon loss via a fault conduit for the Naylor fault or fault junction.

The Naylor South fault (Figure 8.2) was interpreted to have a strong across-fault seal as it juxtaposed the Belfast Mudstone against the Waarre Sandstone interval for the greater part of the Naylor structure. This fault extended through to the Pember Mudstone and was interpreted to have leaked hydrocarbons in the past, as shown by the presence of a palaeo gas column. This fault is optimally orientated for reactivation and was therefore considered to be a potential conduit for CO_2 flow. However, there was a low chance, determined through modelling, that any of the CO_2 injected would reach this fault as it was outside of the primary storage structure.

The likelihood of leakage via faults was judged as *almost impossible–unlikely*, with the fault junction viewed as having the highest likelihood for leakage. A loss rate for this event was assessed to be 30–60 t per year (CL50– CL95), and if it were to occur was judged conservatively to leak for the full duration of the risked period (maximum of 1000 years). The risk inputs for the set of fault leakage risks are summarised in Tables 8.2–8.5.

8.5.3   Leakage due to excessive regional pressurisation

This risk event related to the potential for the reservoir pore pressure build-up being higher than modelled across the Naylor region (and of particular concern near existing faults) as a result of injection of CO_2. An unexpectedly large change in pressure (and consequently stress) would present the potential for leakage due to fault reactivation (which might also cause induced seismicity and ground deformation), fracturing through the top seal, or damage to wells (Vidal-Gilbert et al. 2009). Fault reactivation can occur when the maximum shear stress acting on a fault plane exceeds shear strength of the fault (Chapter 7). Therefore this risk also needed to take into account the possibility that any faults which were reactivated during pressurisation would leak rapidly for a short period of

Table 8.4: Pore pressure increase (ΔP_p) required to reactivate critically-orientated faults depending on assumptions made about various geomechanical properties (Vidal-Gilbert et al. 2009).

Scenario	Stress Regime	Fault Strength	ΔP_p range (MPa)
1	Strike slip fault regime w. low horizontal stress	Cohesionless faults	1.0–9.9
		Healed faults	10.8–23.9
2	Normal fault regime	Cohesionless faults	5.3–25.9
		Healed faults	13.9–37
3	Strike slip fault regime w. high horizontal stress	Cohesionless faults	2.3–12.9
		Healed faults	14.3–29.9
Expected pre-injection pressure P_p			17.5MPa

time, but then the leakage would slow to a low rate for the remainder of the risk period.

There was an excellent understanding of the pressure history of the Naylor Field and hence good capability to predict the likely post-injection pressures in the Naylor structure. New data from the CRC-1 well further improved the history match from pre-production of the Naylor gas accumulations, throughout production, and during aquifer recharge (Figure 8.3). After the drilling of CRC-1, aquifer recharge of the Naylor structure was found to be stronger than previously modelled. It was predicted that, by the end of injection, the change in pressure would be 1.5 MPa, which would have brought the absolute reservoir pore pressure to near the pre-production pressure of the gas-filled Naylor system.

The estimated maximum sustainable pore pressure change without brittle deformation occurring in the Waarre Sandstone was 10.9 MPa (van Ruth et al. 2007).

The absolute values of fault reactivation propensity differed markedly, depending on which fault strength and maximum stress scenario was used. The highest reactivation propensity (for optimally-orientated faults) was estimated to range between 1.0 and 28.6 MPa (Table 8.4). Therefore, difference in assumed fault strength and maximum horizontal stress arising from uncertainties in

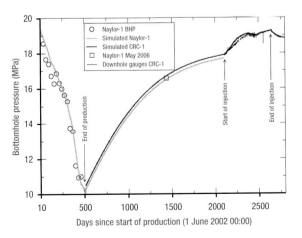

Figure 8.3: Reservoir pressure profile from pre-production pressure to the likely end of injection pressure at the end of Otway Stage 1.

the geomechanical model resulted in large uncertainties in modelled fault reactivation propensity.

The risk of injection activity fracturing the intact Belfast Mudstone seal was very low. A geomechanical assessment of the seal (Chapters 6 and 7) showed that the maximum modelled pore pressure change of 1.5 MPa was far less than the pore pressure required to fracture intact seal rock (6.3 MPa).

At the time of the risk evaluation, the degree of elastic response of the reservoir from the depletion to the re-pressurisation of the system, was uncertain. However, a storage analogue study from the nearby Iona natural gas storage facility (Chapter 7) supported the view that re-pressurisation of Waarre Sandstone systems, including the Naylor Field, would not lead to any structural deformation or fracturing (Tenthorey et al. 2013).

Based on simulations in which pre-production aquifer recharge and likely injection rates were taken into account, after looking at the geomechanical properties of the faults and seals, it was considered unlikely that sufficient pressurisation would take place to reactivate faults or initiate fracturing through the top seal.

While the risk of CO_2 leakage out of the storage complex due to excessive regional pressurisation was considered acceptable, it was decided it would be prudent to assess the impact of this potential risk event. A CO_2 leakage event was designed from the Naylor fault to the west of Naylor-1 where it intersected the overlying Paaratte Formation. The modelling then focused on the migration of CO_2 from the leak point into the Paaratte high permeability reservoir unit. Various cases were assessed based on differing leakage rates, relative permeabilities and leakage durations. The modelling indicated that, in the event that CO_2 migrated up the fault zone into the Paaratte at assumed rates, the CO_2 would not migrate more than about 250 m from the leak point (i.e. still within the Paaratte Formation and within the PPL-13 tenement). After 1000 years almost 90% of the leaked CO_2 would have dissolved.

The likelihood of leakage due to excessive pressurisation was judged as *highly improbable–unlikely*. The short term leakage rate for this event was 300–3000 t/year (CL50–CL95) for a 1 month period, followed by a long-term

leakage rate for this event of 10–20 t/year (CL50–CL95) for a 1000 year period.

8.5.4　Leakage from wells

The wells included for this risk assessment were the Naylor-1 and CRC-1 wells. The assessment of these wells was based on the understanding of the cementing quality and materials used in the wells. Additional consideration was given to the location of the well relative to the CO_2 through the period of plume migration and accumulation.

As the CRC-1 injection well had been purpose-drilled for CO_2 injection, it was considered that the CO_2-resistant materials and engineering reduced the risk of CO_2 leakage from that well. Thus, the chance of leakage through the injection well was considered *highly improbable–unlikely*. The total amount of CO_2 loss that could occur from the injector well would have been significantly less than from the Naylor well as the plume migrated up structure from the injector.

It was expected that the CO_2 would encounter and pool around the Naylor-1 well at the crest of the Naylor structure within approximately 6 months of the start of injections, below the remaining hydrocarbon accumulations. Leakage through the well was considered to be *very unlikely–possible*, due to uncertainty in the pre-project well completion quality. The expert panel viewed leakage behind casing as the main risk in the Naylor-1 well, enhanced through cement dissolution by CO_2. However, given the hydrocarbon accumulation at the crest of the structure, the risk of CO_2-brine interacting with the cement at the critical

Table 8.5: Risk inputs from the three fault leakage-related risks for the Otway Project.

Risk	Probability of Event	Loss Rate (t/year)	Loss Duration (Years)
Naylor Fault	Highly improbable–very unlikely	30–60	1000
Fault Junction (Naylor East)	Very unlikely–unlikely	30–60	1000
Naylor South	Almost impossible	30–60	990

point for sealing (where the well intersects the base of the top seal) was considered to be *unlikely*. Additionally, as this well would be actively used to monitor the reservoir during and after the experimental period, any leakage event would probably be of short duration, as well-leakage management for CO_2 would be implemented.

The likelihood of CO_2 leakage from wells was judged as *highly improbable–unlikely* for CRC-1 and *very unlikely–possible* for Naylor-1. The short-term leakage rate for this event was 30–60 t/year (CL50–CL95) for a 1000 year period.

8.5.5 Leakage due to exceeding the spill point

In addition to assessing the risk of CO_2 leaking vertically beyond the top seal, the risk of leakage laterally out of the storage complex was considered. The mechanisms for lateral leakage of CO_2 considered for this Project were "exceeding the spill point of the storage structure" and "incorrectly predicting the migration pathway". Lateral leakage through the bounding faults was considered as part of the "leakage from faults" section.

Leakage from exceeding the spill point of the storage site refers to the structure not having the capacity predicted and the CO_2 migrating laterally around the defined shallowest structural closure, resulting in CO_2 loss. Spill point leakage out of the primary Naylor structure was reconsidered for this containment risk assessment. The spill point of the Naylor trap was defined by the southern extent of the Naylor Fault (Figure 8.4). Because of the resolution of the data, there was a degree of uncertainty in the exact depth of the spill point due to the transition of the fault juxtaposition from Belfast Mudstone to Flaxman Formation "seal" to Flaxman Formation "reservoir" to Waarre Sandstone. However, if the spill point had been much shallower than predicted, the Naylor South structure would be likely to be filled, with a common gas-water contact (GWC) as the Naylor gas accumulation below the spill point. Therefore both the Naylor and Naylor South structures would become one common accumulation—the "greater Naylor structure". If the spill point was as predicted or

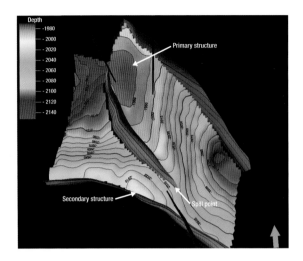

Figure 8.4: Location of the primary and secondary structures of the Greater Naylor structure in the Otway Project model.

deeper, there was sufficient storage capacity so that the likelihood of leakage via the spill point was considered to be *almost impossible*.

Assessment of this risk likelihood after an update to the geological model clearly demonstrated that the risk of leakage from exceeding the spill point was *almost impossible*. The risk assessment addressed the case where approximately 10% of the injected volume (approximately 10,000 t) leaked at the injection rate.

8.5.6 Leakage due to incorrectly predicting the migration direction

Leakage resulting from incorrect migration direction at the Otway site was only possible if the CO_2 plume migrated south, downdip, around the southern tip of the Naylor Fault into the Naylor South structure. This was considered *almost impossible* in reality yet theoretically possible if the injection pressure was greater than buoyancy drive, and if permeability streaks existed in an orientation favourable for this migration. Worst-case scenarios were simulated; the results clearly showed that this risk was *almost impossible* (Figure 8.5) and, if it were to occur, would leak at 1000–5000 t/year for a period of approximately 2 years.

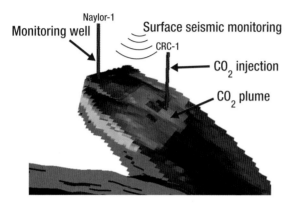

Figure 8.5: Otway Project simulation of plume distribution illustrating a dominantly buoyancy-driven flow, migrating updip and away from spill point.

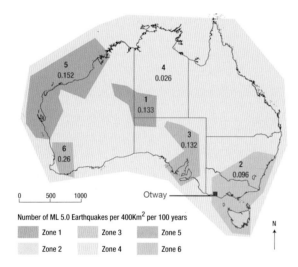

Figure 8.6: Australian earthquake frequency data from Geoscience Australia for potential storage sites evaluated by CO2CRC. The Otway Project site (red) is located in zone 2.

8.5.7 Leakage from natural earthquake induced fracturing

This risk event considered the possibility of leakage arising from a naturally occurring event. In order to assess this risk the approach taken was to evaluate the frequency of magnitude 5 earthquakes in this region and over the area of the intended storage site. Due to the nature of earthquake-induced fracturing, it was anticipated that there would be an initial short-term, high loss rate of CO_2 followed by a long-term, low loss rate (Streit and

Watson 2005). Australian earthquake frequency data from Geoscience Australia were used to establish the likelihood over any area in this region and this was applied to the Otway site (Figure 8.6).

8.6 Risk assessment output

The results of the two containment assessments of the Otway Project containment risk performed in 2005 and 2007 are shown in Figure 8.7 (2007 risk assessment) and Figure 8.8 (2005 risk assessment). As the first risk assessment was performed before the drilling of the CRC-1 well, it was not the assessment used as the basis for developing the Project operation plan. The second risk assessment, with the benefit of the additional CRC-1 well data and interpretations, was used as part of the decision support package for the final project approvals and for development of the monitoring plan.

The histogram display in Figure 8.7 shows the dominant risks assessed in 2007. The y-axis is the risk quotient on a logarithmic scale. The risk quotient was normalised against the acceptable project containment risk (red dashed line), which was 1% of the target injection volume of 100,000 t with an 80% confidence level. For reporting, three levels of confidence were selected as the model outputs for each risk, namely optimistic, planning and pessimistic.

The profile shows that the overall project risk (far right column) was one to two orders of magnitude below the target risk quotient, depending on the confidence level used. No containment risk was found to be at an unacceptable level from this quantitative assessment. The Otway Project was therefore considered to be low risk in terms of injected CO_2 containment for the Waarre Sandstone within the Naylor structure. The dominant containment risk event was fault leakage (at the junction of the Naylor East Fault to the Naylor Fault). Other notable, yet acceptable risks included (in decreasing order of risk quotient) earthquake-related long-term leakage risk, excessive regional pressurisation short-term leakage risk, Naylor Fault leakage and Naylor-1 well leakage risk.

Overall the quantitative assessment of containment risk demonstrated that the Otway Project had low risk. Risks were easily within the threshold targets and were considered acceptable on this basis. Extensive geological characterisation and dynamic modelling were performed and supported this conclusion; the mitigation and monitoring programme was optimised to ensure containment.

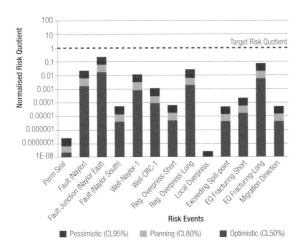

Figure 8.7: July 2007 RISQUE output for the Otway Project. Each risk quotient is plotted against a logarithmic *y*-axis that has been normalised to the Target Risk Quotient. An optimistic, planning and pessimistic risk quotient is given for each risk to represent the uncertainty in the inputs.

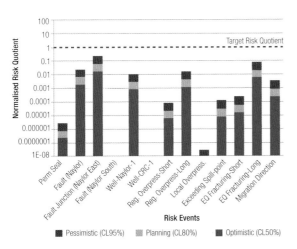

Figure 8.8: September 2005 RISQUE output for the Otway Project.

8.7 Conclusions

The risk assessment process followed for the Otway Project Stage 1 was an important part of taking the Project forward from a number of perspectives. First, it provided the regulators with a level of confidence in the rigour of the evaluation process. Second, it provided the community with confidence in the transparent process being followed to ensure that the Project would be undertaken in a manner that was safe and that there was a very low level of risk. Third, it provided a framework for integrating many aspects of the Project including a diversity of data sources. To date, few other small-scale injection projects have documented the risk assessment process that was followed in taking the project forward. The experience at Otway has shown the importance of ensuring that a rigorous and well documented risk assessment process is followed.

8.8 References

Bowden, A.R. and Rigg, A. 2004. Assessing risk in CO_2 storage projects, Australian Petroleum Production and Exploration Association Journal 42(1), 677–702.

Bowden, A.R., Lane, M.R. and Martin, J.H. 2001. Triple bottom line risk management—enhancing profit, environmental performance and community benefit. John Wiley & Sons.

Daniel, R.F. 2007. Carbon dioxide seal capacity study, CRC-1, CO2CRC Otway Project, Otway Basin, Victoria, Cooperative Research Centre for Greenhouse Gas Technologies (CO2CRC), Canberra, Australia. Report No: RPT07-0629.

Greaney, T. 2008. CRC-1 Well Completions Report PPL 13 Rev 0. AGR Asia Pacific Report No 34632-DR-05-0006.

Hennig, A. 2007. Hydrodynamic interpretation of the formation pressures in CRC-1: vertical and horizontal communication, Cooperative Research Centre for Greenhouse Gas Technologies (CO2CRC), Canberra, Australia. Report No: RPT-0626.

IPCC 2005. Special report on carbon dioxide capture and storage. In: Metz, B., Davidson, O., de Coninck, H.C., Loos, M. and Meyer, L.A. (Eds.), Contributions of Working Group III of the International Panel on Climate Change. Cambridge University Press, Cambridge, United Kingdom and New York, USA.

Lyon, P., Boult, P.J., Watson, M.N. and Hillis, R. 2005. A systematic fault seal evaluation of the Ladbroke Grove and Pyrus traps of the Penold Trough, Otway Basin. Australian Petroleum Production and Exploration Association Journal 45(1), 459–476.

Streit, J.E. and Watson, M.N. 2005. Estimating rates of potential CO_2 loss from geological storage sites for risk and uncertainty analysis. In: Rubin, E., Keith, D. and Gilroy, C. (Eds.). Proceedings of the 7th International Conference of Greenhouse Gas Control Technologies (GHGT-7), Vancouver, 5–9 September 2005, 1307–1314.

Tenthorey, E., Vidal-Gilbert, S., Backe, G., Puspitasare, R., John, Z., Maney, B. and Dewhurst, D. 2013. Modelling the geomechanics of gas storage: a case study from the Iona gas field, Australia. International Journal of Greenhouse Gas Control 13, 138–148.

van Ruth, P., Tenthorey, E. and Vidal-Gilbert, S. 2007. Geomechanical analysis of the Naylor structure, Otway Basin, Australia—pre injection. Cooperative Research Centre for Greenhouse Gas Technologies (CO2CRC), Canberra, Australia. Report No: RPT07-0966.

Vidal-Gilbert, S., Tenthorey, E., Dewhurst, D., Ennis-King, J., van Ruth, P. and Hillis, R. 2009. Geomechanical analysis of the Naylor field, Otway Basin Australia: implications for CO_2 injection and storage. International Journal of Greenhouse Gas Control 4(5), 827–839.

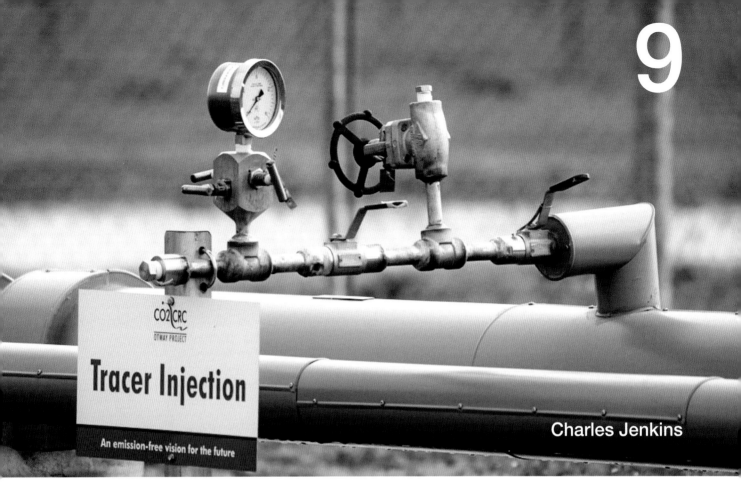

Tracer Injection

CO2CRC OTWAY PROJECT

An emission-free vision for the future

Charles Jenkins

9. MONITORING AND VERIFICATION

9.1 Introduction

The minimum requirement for any CCS monitoring programme is that it should provide the information that satisfies regulators that legislative requirements are being met by a storage project. However, these requirements are usually qualitative and somewhat imprecise. For example in the European Union, operators are required to demonstrate "absence of detectable leakage" and other jurisdictions have a range of similar phrases. In the United States there is a strong concern with the security of underground sources of drinking water (USDWs) and the EPA requirement is that monitoring must "demonstrate the non-endangerment of USDWs". Clearly—and quite properly—the interpretation and application of regulations will take account of a range of current concerns and perceptions of risk. Hence the type of monitoring programme that is required for a given project, and the way in which its data are interpreted against the regulatory requirements, may well be unique. Common to all jurisdictions, however, is the need to demonstrate "no leakage" (although this is qualified with a

variety of adjectives and adverbs), as well as more routine requirements for health and safety associated with high concentrations of CO_2.

A key part of a regulator's considerations is the risk assessment for a storage operation. Logically, monitoring should focus on the highest risks, although in the absence of much experience with actual CCS projects, these risks have to be largely assessed by analogy and by expert opinion. Ideally the risks would be quantified as quite specific and predictive scenarios, so that it is possible to know exactly how to look for the risks and how to rule out their occurrence (see Chapter 8).

The social and political context will affect the scope and stringency of regulatory requirements, and operators may also be involved in separate negotiations with powerful but informal activist groups. As a result of these, measurements which monitor some perceived risk may be promised, where the risk is not only tiny but, to the extent that it exists at all, is poorly understood. Regular monitoring of the biota in streams at beauty spots is an actual example of this in one CCS project.

Related to the political context is "defensive monitoring". It is nowadays quite possible for activist groups to undertake their own monitoring, particularly of environmental variables such as water quality or soil gas composition. Typically such near-surface measurements show large variability with season or location and can be difficult to interpret in isolation. Defensive monitoring is a tactical move to establish a pre-injection baseline in key variables which may later become the subject of controversy.

It is ironic that the only area of monitoring where the target is very clear is the one that is least often mentioned outside the scientific community: the purpose of CCS is of course to keep CO_2 out of the atmosphere. It is possible to calculate the maximum tolerable leakage rate from CO_2 storage, as the rate that does not result in net global warming from the storage operation. These rates are subject to some debate, but are around 0.1–0.01% year^{-1} (Enting et al. 2008; Haugan and Joos 2004; Shaffer 2010). Notice this is not the expected, anticipated or desirable rate, but the rate that is neutral in terms of impact on global temperatures, given the energy penalty of CCS. For an industrial-scale storage site of say 10 Mt of CO_2, this maximum tolerable rate is 10,000 t year^{-1}. In some circumstances this size of leak would be quite easy to detect, but it is not "no leakage".

Finally, monitoring may be necessary for carbon accounting purposes. The European Union has the best-developed regulations in this area. These essentially presume no leakage unless there is evidence (not very closely defined) to the contrary. There is then a complex penalty scheme which withdraws carbon credits in a way that depends on the precision of the monitoring techniques and the frequency with which they were used.

9.2   Designing a monitoring programme

"Leakage" can be defined as the escape of CO_2 from the primary storage container—in other words, movement of CO_2 above the seal. The CO_2 might also move through the seal along an unmapped fault, for example. Beyond this, because the CO_2 is buoyant, it will move upwards until it encounters another seal, or dissolves in groundwater, or escapes to the atmosphere. On encountering a seal, the CO_2 will potentially spread out (again driven by buoyancy) and may form a laterally extensive plume.

Another possibility is that the CO_2 will pass through the seal along a faulty wellbore; and may then move a considerable distance vertically along the wellbore or annulus, before perhaps exiting into a near-surface aquifer or possibly even reaching the surface.

A monitoring programme cannot verify "no leakage" unless the term is more closely defined. It is not possible to measure "no leakage" directly; there is no practical method for measuring the amount of CO_2 in a reservoir with a precision that permits a relevant comparison with the amount that has been injected (which in any case is subject to additional metering errors). It is far more likely (in fact inevitable if the term "no leakage" is taken literally) that there will be discrepancies between "no leakage" models and observations. The "no leakage" scenario only remains when all the plausible leakage possibilities have been ruled out, at some predetermined level of statistical confidence. This is a workable approach because in well-designed storage, by definition, any leakage mechanism is very unlikely. Given definite leakage mechanisms, we can design monitoring techniques that clearly test (to the detection limit) if these types of leakage are occurring. This is the Sherlock Holmes principle: "… when you have eliminated the impossible, whatever remains, however improbable, must be the truth."[1]

Monitoring for leakage is therefore broader than monitoring the reservoir. Measurements there, e.g. high quality 3D seismic imagery, can be very important. They calibrate geological and flow models and give warning of unexpected migration. But measurements above the seal are crucial to establishing "no leakage" (defined in some practical sense). It may be that quite a small accumulation of CO_2 could be imaged by seismic techniques in an aquifer immediately overlying the seal. It might be possible (at Otway) to detect around 10,000 t for instance. If monitoring proves to be convincingly *unable* to detect any such accumulations, then a leakage mechanism can be ruled out at a good level of accuracy; to within 0.1% for the case of the hypothetical 10 Mt storage referred to earlier. To take the other example, if CO_2 were to escape up a wellbore and emerge near enough to the surface to escape to the atmosphere, it would be readily detected by the atmospheric surveillance at Otway if rates exceeded

---

[1]   A. Conan Doyle. 1890. The Sign of Four, Chapter 6.

about 5000 t year^{-1}. Again, since this sensitivity would apply for industrial-scale storage it corresponds to good fractional sensitivity. These arguments are made in detail by Jenkins et al. (2012).

Two properties then characterise a monitoring technique from the point of view of its capacity for leak detection. One is its "conditional sensitivity", which indicates the chance of detecting X t of CO_2 *if* it finds its way into the domain of investigation of the technique in question. In the case of the Otway atmospheric measurements, the conditional sensitivity was affected by the intrinsic sensitivity of the instrument that measured concentration, by dilution depending on wind speed and direction, and by varying background levels of CO_2 driven by ecosystem respiration. The second property of a technique is the "arrival function"—what fraction of a leak of X t from the reservoir will in fact find its way into the domain of investigation? Clearly this may be hard to estimate (in the atmospheric example, it would entail modelling the passage of CO_2 up a wellbore of uncertain characteristics) but the estimate has to be made to enable further meaningful interpretation. The combination of arrival function and conditional sensitivity tells us the utility of a monitoring technique for leakage; if either of these two properties is poor then the technique overall cannot perform well.

The Stage 1 Otway design used the concept of "assurance" measurements. Simply put, these are measurements intended to "assure" stakeholders that some asset of public concern (e.g. groundwater) is unaffected by CO_2 storage activities in the vicinity. It is inherent in the concept that the arrival function of an assurance technique is very low—otherwise it would function as a leakage monitoring technique.

In a "politically" benign environment, an assurance programme needs little interpretation: e.g. the collection of water samples annually from farmers' wellbores (as in Otway Stage 1). These results, year by year, tell the farmers that the water in their wells has only changed its composition by amounts comparable to regional and seasonal variations. Note that this presentation of the data refers to particular places and times. Without considerably more effort in interpretation, it does not say anything about the water quality either next year or in some new location not previously sampled.

The notion of assurance monitoring is attractive as a way of bundling up measurements that are taken to deal with perceived rather than actual risks. However, this approach has some issues unless such data are treated as rigorously as any other monitoring data.

The main issue is the interpretation of data for which the arrival function is only known in a qualitative way to be "small", typically with no definite transport mechanism identified for the passage of CO_2 into the region of investigation. Unless this is better defined, there is no rigorous basis for classifying the measurements as merely "assurance", and if the mechanism exists and is well defined, then the measurements can potentially be used for leakage monitoring. This is the logical objection to classifying some measurements as assurance. The "political" issues are equivalent but perhaps more serious. Assurance-type measurements, by definition, investigate things that stakeholders care about; e.g. they may ask, "Will CO_2 get into this aquifer?" If the answer is "No" this begs the question about why it is then being monitored. If the answer is "Yes", then it is necessary to have a clear idea of the mechanism, its likelihood, and the exact way in which the monitoring programme will detect it.

The contingent sensitivity is also important. Assurance measurements are not only about things people care about, but they are also about typical environmental parameters, such as water or air quality. Because they refer to open systems, subject to many driving forces, such parameters vary, often strongly. The contingent sensitivity has to be known, so that if random changes are seen, we can correctly attribute them to natural variability. The contingent sensitivity is also the key to answering the question, "You say you saw nothing—but what could you have seen?" This is the Nelson question.[2]

It is often thought that the difficulties with assurance data in particular can be circumvented with a "baseline". For example, a time series of water quality data may be collected before a storage project commences injection, and then compared with the same type of post-injection measurements. The hypothesis here is that if there is no

---

[2] At the Battle of Copenhagen in 1801, Lord Nelson famously put his telescope to his blind eye to avoid seeing a signal from his commander to withdraw from the action.

leakage, the data before and after injection will be the same. But they will not be, because at a minimum, they will be different due to random analytical and sampling error, and there may be much larger long-term variations because of droughts, water extraction, and so on. Comparing baseline and post-injection data therefore has exactly the same difficulty as comparing the predictions of a no-leakage model with the data: there will be discrepancies, and without a definite predictive *leakage* model, it is logically impossible to say if these matter or not, and in fact also impossible to say if the design and execution of the measurement is fit for purpose.

The issue of false alarms is an important one. Because environmental data can be very variable, false alarms have to be considered. Typically there is a trade-off between the false alarm rate and the sensitivity. If small excursions from the mean are accepted as significant, then it is likely a small leak will be detected. We are also likely to be fooled by measurement error more often if there is high sensitivity. The false alarm rate of a technique is another useful property to know.

It is hazardous to collect data that no-one understands, even if the mere act of collection seems to keep some people happy initially. Sooner or later, someone will ask about the data in more detail, and a lack of clear answers could be extremely damaging to a storage project. There needs to be a clear plan for the interpretation of data before it is collected, including agreement with stakeholders about the thresholds that would indicate, perhaps not leakage per se, but the need for further, specified investigations.

The discussion so far has been somewhat forensic, and this is because of the conjunction of monitoring and *verification;* if claims are to be made about compliance or safety, then a probing analysis is justified. In the early stages of any CCS project, there will be many aspects which are not fully understood, and there is clearly a good case for exploratory scientific work, collecting data in a general way to build up a better understanding of the systems involved. An example might be induced seismicity: the risks are low, but it is worth monitoring seismicity at an injection site to improve understanding of the specific mechanisms involved.

Obviously, if such observational work gets confused in the public mind with monitoring for compliance and safety, there is potentially a public relations issue. "Why are you measuring this if you say there's no risk?" This is not an insuperable problem, and an important one to overcome, but to do so will require clear thinking about the purpose of any exploratory monitoring, and a good communications plan with stakeholders (see Chapter 2).

9.3   Designing the Otway monitoring programme

The CO2CRC Otway Project Project, Stage 1, was permitted through the Victoria EPA under a research and development approval provision but with specialist technical advice provided to the EPA by DPI (Chapter 3). An extensive monitoring programme was undertaken, but many of the techniques were experimental. The concept and early stages of the Project have been described by Sharma et al. (2007), and a comprehensive assessment of the Project at its close has been given by Jenkins et al. (2012).

The monitoring and verification programme for Stage 1 of the Otway Project had several overlapping objectives (Dodds et al. 2009):

> ensuring safe operations

> demonstrating compliance with regulatory requirements

> evaluating a range of monitoring methods

> providing data to calibrate models.

At the outset of the Project, the scope and nature of regulations were unclear, but were developed during the initial phases of the Project. Basic principles had been set out in the 2004 Regulatory Impact Statement of the Commonwealth Organisation of Australian Governments (COAG) which considered that the regulatory framework for monitoring and verification should:

> provide for the generation of clear, comprehensive, timely, and accurate information that is used to effectively and responsibly manage environmental,

health, safety, and economic risks and to ensure that set performance standards are being met

> determine to an appropriate level of accuracy the quantity, composition, and location of gas (CO_2) captured, transported, injected, and stored and the net abatement of emissions. This should include identification and accounting of fugitive emissions.

These were ambitious, perhaps aspirational words and monitoring at the Otway Project provided the opportunity to establish what they meant in practice. As discussed in Chapter 3, the intention of the Victorian regulatory authorities was that the regulation of the Project would be done by standard mechanisms with the addition of "key performance indicators" (KPIs). In the early planning stages of the Project these were deliberately non-specific as to the monitoring methods or assessment of evidence from these methods. This was entirely appropriate, given the novel technologies that would need to be implemented. In the later stages of the Project, the way in which the KPIs could be met became operationally more specific as to technology and outcomes. This reflected the reality that transition through the earlier KPIs required being more prescriptive while also recognising that the Project should seek to limit the amount of monitoring that should be done to meet the needs of the KPIs rather than trying to measure every possible parameter. As a consequence of this, development of the regulatory regime was obviously informed by what was thought to be possible in monitoring, and later legislation in Victoria and elsewhere has been influenced by results from the Otway Project.

The Project risk assessment (Chapter 8) was an important input to the Otway monitoring and verification (M&V) programme. It was recognised that "risk" should be monitored where possible. The main risks that were identified were leakage along wellbores, and along faults, with the former being much more probable. Since a wellbore can connect several geological zones, or provide a path to the surface, this aspect of the risk assessment mandated a range of monitoring at various depths. Combined with the objective of assessing a range of techniques at Otway, the possibility of wellbore or fault leakage determined many modes of the monitoring programme. These modes included monitoring immediately above the seal (seismic), monitoring overlying aquifers (groundwater), soil gas

(a wide-area technique particularly looking for fault leakage) and atmospheric monitoring (most suited to detecting leakage from small areas near the existing wells). These are shown schematically in Figure 9.1

The M&V programme was not designed from first principles to meet an objective of quantifying the amounts stored, although this aspect became more prominent as Stage 1 progressed (with M&V often expanded to the acronym MVA, meaning monitoring, verification and accounting). However, quantification was and continues to be very challenging, and regulation elsewhere (e.g. in the EU) is still imprecise on this topic. The Otway Stage 1 M&V undertook some careful design work in two areas in this respect. Detailed seismic forward modelling and experimentation established that it was unlikely that injected CO_2 would be detected in the reservoir (Chapter 10), but there was good sensitivity to leakage into overlying aquifers (Li et al. 2006; Urosevic et al. 2011). Likewise, analysis and design of the atmospheric monitoring showed good sensitivity to emissions from the vicinity of the wells (Leuning et al. 2008). In both cases, however, it was uncertain what might have happened to the CO_2 on its way to the places where it could be detected. The idea of a contingent sensitivity

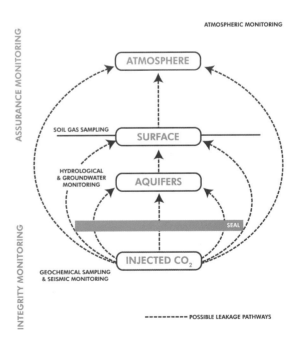

Figure 9.1: Schematic representation of the integrity and assurance monitoring.

was seen as a way to clarify the situation. Where the CO_2 reached a particular location, sensitivity to leakage with a particular monitoring technique then requires careful consideration. The question of how the CO_2 reached the particular location was typically much more complicated to analyse and model, but at least the difficulty could be separated from the measurement issue.

An important aspect of Otway monitoring concerned social licence and securing and maintaining the consent of stakeholders such as local residents, politicians, and NGOs. A comprehensive monitoring programme was seen as being helpful in this regard. Typically the concern of these stakeholders focused on assets such as groundwater or air quality. The groundwater, soil gas and atmospheric measurements were sometimes categorised in Otway Stage 1 as "assurance" measurements (distinguished from "containment" measurements which investigated the reservoir level directly). Although these measurements were the ones which were most useful to assure stakeholders that all was well, they were in fact no different to "containment" measurements. First, they were undertaken because of the existence of definite, albeit imprecise, mechanisms of leakage; and in the case of the atmospheric measurements, there was very well-defined sensitivity to leakage. Second, over the lifetime of Stage 1 it became clear from experiences elsewhere that any measurement taken at a storage site was, as a matter of political reality, about containment. Any measurement taken at a storage site, even if it showed no significant change above background, inevitably prompted the question: "But what level of leakage could you have detected?" Also, an important contemporary reason for a monitoring programme which included a range of environmental variables was that many of these could easily be measured by third parties. If anomalies were then claimed to exist, it was important for someone (perhaps project proponents) to be able to refer to a well-established baseline and measurement history in that variable.

The selection of techniques for the Stage 1 M&V, as discussed, was informed by risk assessment and regulatory discussion, as well as sensitivity to local concerns. Where possible, the sensitivity of the techniques to leakage was investigated, and thorough assessments were made of the circumstances of the particular monitoring domain. For example, information was gathered on the hydrology of the overlying aquifers to establish flow directions and rates as far as possible. Quantitative modelling was made of the effectiveness of several areas where the physics was best understood: these were seismic (Li et al. 2006; Urosevic et al. 2010), electromagnetic (Christensen et al. 2006), gravitational (Sherlock et al. 2006), and atmospheric concentration and flux measurements (Leuning et al. 2008). These topics are considered in more detail in Chapters 10, 11 and 15. Electromagnetic and gravitational measurements seemed likely to be useful but required expensive borehole installations or logging. It was clear from the outset that the sensitivity of land seismic surveys would probably not be good enough to detect CO_2 in the reservoir, but given the importance of imaging at least above the seal, this monitoring method was energetically pursued (Chapter 10). The effectiveness of atmospheric monitoring could be modelled in advance in more detail than any other technique, essentially because the dispersion mechanisms in the atmosphere were well understood (Chapter 15) and the medium was subject to far fewer uncertainties than the deep subsurface. The monitoring programme comprised the following methods:

1. 3D repeated seismic surveys, as well as vertical seismic profiles (VSP) and offset VSP (Chapters 10 and 11). These attempted to image the stored gas using its contrast in acoustic reflectivity. Two 3D surveys were taken at yearly intervals post-injection, and were amongst the most costly and logistically complex of the monitoring operations.

2. Direct sampling at the Naylor monitoring well (Chapter 12). This programme retrieved reservoir fluids from 2 km depth at pressure and successfully recorded the accumulation of injected gas at the crest of the storage reservoir.

3. Pressure and temperature measurements in the reservoir (Chapter 16). These gauges failed on deployment although subsequently, data were taken in the injection well.

4. Microseismic monitoring at reservoir depth (Chapter 11). These receivers failed on deployment and were later replaced by shallow microseismic arrays in water wells. The intent was to monitor any seismic activity caused by injection.

5. Monitoring of groundwater chemistry (Chapter 13). This programme monitored existing wells in the deep (800 m) Dilwyn Aquifer, as well as in a large number of shallow wells in the near-surface Port Campbell Aquifer. Data were collected every 6 months.

6. The composition of soil gas was measured from samples taken at approximately 2 m depth, on an extensive grid about 2 km on a side (Chapter 14). These measurements were taken yearly.

7. Atmospheric composition was measured continuously at a tower 700 m NNE of the injection well (Chapter 15), as was the vertical flux of CO_2 from an area within approximately 100 m from the tower (the flux measures the production of CO_2 from the local ecosystem).

8. A small number of wire-line logs were run in the injection well, to attempt to determine the vertical distribution of CO_2 near the wellbore (Chapter 5).

9. The concentration and flux of tracers added to the injected stream were measured (Chapters 12 and 16).

Methods which were not deployed, but which have attracted some interest elsewhere, included gravimetry, dipmeters, surface deformation (INSAR), various types of electromagnetic sensing, in-situ vegetation surveys, soil flux surveys, and remote sensing of various vegetation indices.

The design of many of the techniques deployed at Otway was strongly influenced by practical issues. The biggest practical constraints probably applied to monitoring methods 5–7 (above). In the case of groundwater, it was impractical to drill new monitoring wells and therefore available water bores were used. Since these were not close to the injection or monitoring well, their usefulness as monitors of leakage was very limited; in fact the deeper water bores were, as far as could be determined, actually upstream of the storage site. Emphasis for these data was then primarily on assurance for residents and water regulators that the groundwater was unaffected in the actual bores that were measured. Extrapolation to statements about containment were of limited usefulness but, as noted, it was nonetheless expected by some stakeholders. The density and frequency of the soil gas sampling grid was also fixed by the practical issue of access to land and the fact that reliable data could not be obtained in the wet seasons. Finally, although a network of atmospheric sensors would have been much more versatile, atmospheric sensing for Otway Stage 1 was restricted to one set of sensors. This was because of the high cost at that time of sensors that could measure the tiny perturbations of CO_2 concentrations expected to result from hypothetical leakages at levels of interest.

An important feature of most of these measurements was the establishment of a baseline. In the case of seismic measurements, the concept was simple—an image of the reservoir before CO_2 was injected. For other techniques, the idea was more complex. Environmental variables such as groundwater composition are not stable in time, but are driven by other environmental factors such as drought and water extraction. In general therefore the purpose of the baseline measurements was to establish both a level and a range for the quantities of interest. In the case of the atmospheric measurements, the baseline data served to calibrate an ecological model that could predict (to some extent) the seasonal and diurnal variations in CO_2 concentration. How long or spatially extensive a baseline dataset should be was difficult to establish in advance, especially if understanding of the processes at work was limited—which was the case at the Otway site. In Otway Stage 1, the decisions were essentially pragmatic, because they were limited by the available time and resources. Potentially the riskiest area for this pragmatic approach was microseismic sensing: because of the nature of earthquakes (a power-law distribution of intensities), a reliable baseline, at least in principle, has to be of very long duration. In the event, however, the baselines proved adequate from the point of view of assurance monitoring for all of the techniques. The Otway experience suggests that design of baseline surveys will probably continue to be an area of worthwhile research.

Related to the baseline was a particular feature of the Otway site, namely the natural occurrence of CO_2 at depth. This was one of the main reasons for adding tracers to the injected Buttress gas. Of the four tracers (SF_6, R134A, Kr and CD_4), only CD_4 was not found in the environment and so baseline levels of the others were important to establish.

Given this, however, the tracers did provide valuable information at both the reservoir and environmental level. In addition, the isotopic distinctiveness of the injected CO_2 compared to atmospheric CO_2 made it a valuable tracer and indeed isotopic analysis was informative even at the level of distinguishing residual CO_2 in the Naylor structure from CO_2 sourced from the Buttress structure.

Some of the monitoring data were used for history matching. In particular, the injection pressure was used to fine-tune the bulk permeability of the reservoir models. This was an exception to the general rule that models were not calibrated with post-injection data. This was to ensure that the models made predictions about post-injection behaviour, the objective being to determine how well the storage process could be characterised in advance. In a more conventional workflow, relevant data (in the Otway case, the information from the Naylor fluid sampling) could have been used to refine the reservoir model as the injection proceeded.

9.4   Evaluation of monitoring techniques

9.4.1   Assessing technology against objectives

A wide range of monitoring techniques is under discussion for land- and sea-based projects. The Canadian QUEST Project, Shell Canada (2010) listed 56 techniques, for example. Later work by Shell (unpublished) attempted to illustrate quantitatively the way in which risk was reduced as more and more monitoring methods were added to a portfolio. Another common way of assessing monitoring techniques is to plot cost against effectiveness in a matrix; this was done for the In Salah Project (Mathieson et al. 2011). There are also a number of best practice manuals and guidelines, which are listed by Soroka (2012) and Cook et al. (2013).

Bearing all this in mind it is appropriate to now consider techniques used for Otway Stage 1, briefly assessing each against purpose, sensitivity, arrival function, false alarm rate, cost and complexity. Adjectival assessments will be used to grade the techniques. Other CCS sites will be

different but evaluating the Otway techniques in this way provides a worked example of evaluation of some commonly-used techniques.

The Stage 1 Risk Assessment was an important point of reference (Watson et al. 2007; Chapter 8). All risks were assessed to be very low, with the largest being reactivation of faults, wellbore leakage, and earthquakes. Although the Otway area is seismically inactive, it is highly faulted, which in turn informs most of the monitoring methods.

9.4.2   Seismic surveys

From the outset, it was known that it would be very difficult and probably impossible to image CO_2 in the Naylor reservoir; because CO_2 was displacing methane, and as the two have similar acoustic properties, there is little differential signal. The various seismic surveys were pursued as a research project, at reservoir level. However, imaging above the seal addressed the risk of CO_2 passage through an unmapped or reactivated fault. Very little was known about these hypothetical faults so the arrival function into the overlying aquifer (the Paaratte Formation) was unknown. By contrast, because the measurement of noise was well understood, the conditional sensitivity for the detection of small accumulations in the Paaratte Formation was quite well known, with around 10,000 t being in principle clearly detectable. The false alarm rate was not quantified but at the 10,000 t level must be much better than 1 in 20.

The seismic surveys were the most expensive of the monitoring techniques employed, and posed many logistical difficulties. They were also the most intrusive on farmers' land and were difficult to manage for this reason. On the other hand, they provided vital well-characterised information just above the reservoir.

Seismic imaging is not invariably successful, especially on land, and is likely to be very expensive for industrial-scale plumes of stored CO_2, which may be kilometres in extent. However, it can provide vital information as the migrating CO_2 "lights up the geology"[3], and constrains leakage dramatically if imaging is successful above the

----

[3] Phrase due to Susan Hovorka, lead scientist at the Cranfield injection project.

storage zone. It is hard to imagine most storage projects being undertaken without regular seismic surveys.

9.4.3 Fluid sampling in the monitoring well

The direct recovery of fluid samples from Naylor was a vital, high signal-to-noise measurement for Stage 1: it was the only successful direct measurement at reservoir level. As such it directly addressed the risks of exit of CO_2 from the reservoir by any mechanism. The ability of the reservoir models to fit the breakthrough data, as CO_2 migrated towards the reservoir crest, showed directly that the expected amount of CO_2 was moving through the reservoir at the predicted rate. The arrival function was *a priori* 100%, the false alarm rate negligible, and the sensitivity in the 5000–10,000 t range.

Three tracers were added to the injected gas (SF_6, CD_4 and Kr) to identify this gas unambiguously on arrival at Naylor, but in the event there was hardly any doubt. It was hoped that the various solubilities of the tracers would result in interestingly different breakthrough curves, but the measurements did not show differences at a level of signal-to-noise that were informative (see Chapter 12).

Overall the Naylor sampling programme was a great success, but it was expensive, both in capital and operating costs. A considerable amount of development went into the sampling and analytical techniques, and success would not have been achieved without the efforts of many capable scientists.

9.4.4 Pressure measurements

As noted, the pressure gauges in Naylor failed on deployment, and only an injection pressure was available from CRC-1. However, even this injection pressure was extremely useful for the calibration of reservoir models (see Chapter 16). Pressure measurements are probably not sensitive to small leakages from the reservoir, but they are a vital sanity check on reservoir models, which in turn map out the migration path of the CO_2 plume and are a key input to risk assessments for that reason.

9.4.5 Microseismic

The earthquake risk identified at the site relates to historical seismic activity, not induced seismicity. The surface array of three seismic stations monitored down to reservoir depth and was sensitive to very small tremors, well below felt intensity. Small numbers of events were detected and the research interest of the data was in seeing if these events clustered on some of the larger faults in the area.

The sensitivity of the technique was high and the false alarm rate was low, because of the sensitive detectors and careful signal processing (note that operating the system requires experts). The real issue is that the arrival function —in this case, the probability that stored CO_2 would in some way affect seismicity—was extremely small, and the magnitude of the risk, if it existed, was unknown. As it happens, the level of seismicity at the site is so low that this does not really raise the issues of interpretation that have been explained for assurance data. However, these would come into sharp focus if there were more seismic activity, and, if events were felt, the focus would be sharp indeed.

It is worth noting that there is a particular technical issue with seismic baselines. The aim would be to be able to say that a particular post-injection seismic event is not unusual compared to the pre-injection baseline. It is a statistical difficulty with the Gutenberg-Richter Law that the longer the observations continue, the larger are the earthquakes that will be detected. If a post-injection microseismic survey lasts much longer than the baseline survey, it is a statistical inevitability that a larger event will be detected post-injection than was present in the baseline.

9.4.6 Groundwater monitoring

The technical risk addressed by groundwater modelling concerned leakage up faults or wellbore into the Dilwyn Formation (depth 800 m) and into the near-surface Port Campbell Limestone. Water samples were obtained at 6-monthly intervals from three wells in the Dilwyn Formation and 21 in the Port Campbell Limestone (de Caritat et al. 2012). All of these were pre-existing wells and the wells in the Dilwyn were several kilometres upstream (in terms of aquifer flow) from the site (see Chapter 13).

This monitoring programme was actually driven by public concern about water security.

The arrival function for the Dilwyn measurements was essentially zero, so this aspect of the programme was largely to reassure uncertain or unconvinced members of the community. The contingent sensitivity was unknown, as the necessary work was not done to evaluate the possible reactions between rock matrix, water, and CO_2 to determine the likely size of a geochemical signal.

The arrival function in the Port Campbell Limestone was small but not zero, and was not quantified in terms of any definite mechanisms. If it is assumed that CO_2 could enter the aquifer in the area of surveillance within which the available wells lie, the probability of any one of these intersecting the resultant plume of acidified water was very small indeed, perhaps 1 in 200. This is because the plumes were inherently small and the aquifer flow rates also small (Jenkins et al. 2012). The contingent sensitivity was correspondingly very small: in short, if CO_2 were to enter the Port Campbell Limestone (which is very unlikely) there is small chance (perhaps of the order of 1:200 or less) of seeing it. This may well be a feature of *any* groundwater monitoring. The limitation is that the area of surveillance is likely to be large, the number of wells entering the aquifer limited, and the knowledge of where to look for CO_2 even more limited.

There was quite substantial variability in the measured properties of the Port Campbell Limestone samples, both in space and time. For example the bicarbonate concentrations showed a variability of ±150 in a mean of about 500 mg L^{-1}. It is not known what level of change in bicarbonate would be a cause for concern, but if it were small (say 100 mg L^{-1}) then the associated false alarm rate would be unacceptable, at around 1 in 3.

While there were clearly difficulties with the rigorous interpretation of the groundwater programme, this should not overshadow its effectiveness in answering more restricted, but still important questions. Showing that samples from a particular farmer's borehole were not changing, to within the typical variability in the area, was an important reassurance to the farmer. Any difficulties arose in attempting to extrapolate those results in space and time.

9.4.7 Soil gas monitoring

Soil gas monitoring addresses the risk that CO_2 may find its way up wellbores or faults and accumulate in the vadose zone, i.e. the region between the surface and the water table. The technique used collected small volumes of gas from around 2 m depth, and these were analysed (at the site) for CO_2, methane, oxygen, nitrogen, helium, and the ^{13}C carbon isotope (Chapter 14). The injected CO_2 at Otway has a different carbon isotopic signature to local environmental CO_2, although this may not always be the case in storage projects (Leuning et al. 2008). Data were collected once a year, at roughly 100 m spacings, across an area of about 4 km × 4 km.

It is important to be clear that this was not a flux measurement. Measurements of CO_2 flux from the surface are often made with small accumulation chambers, yielding the flow rate of CO_2 out of (or into) the ground, in molecules per square metre per second. The soil gas method used did not yield fluxes, forgoing them in the interests of more detailed compositional measurements (especially of the ^{13}C tracer). Fluxes cannot be deduced from concentration data without detailed knowledge of the permeability of the vadose zone, which in the case of the Otway Project was not possible because of the expense and the likely difficulties that would have been encountered in getting the agreement of landowners to undertake extensive drilling programmes.

The arrival function, contingent sensitivity, and false alarm rate for the Otway soil gas programme are therefore all unknown. Despite this, the technique was useful as a warning flag, although when to decide to raise the flag is a matter of definition. The technique can indicate what proportion of each of its samples appear to be unexplained CO_2, but not what the flux of CO_2 through the vadose zone might be; there is no information available on how long it took the unexplained CO_2 to accumulate in the region from which the sample was taken.

Since soil gas monitoring is widely used (or at least discussed), more detail is merited. As the CO_2 concentration in the Otway soil samples showed a huge variation (over several orders of magnitude) in space and time, determination of a "baseline" was not easy. However, the data did show two correlations. One was between the ^{13}C fraction and the CO_2 concentration, which relates to root growth: cells prefer to process the lighter ^{12}C rather than the heavier ^{13}C. This correlation, though not a very tight one, was similar, whether pre- or post-injection and suggested that there was little if any extra, unmetabolised CO_2 entering the samples.

There was also a tight correlation between oxygen and CO_2 concentrations in the soil gas, which reflected respiration: each molecule of oxygen metabolised by burning carbohydrate produces one molecule of CO_2. Deviations from this correlation (higher CO_2 than expected from the oxygen concentration) signal extra CO_2 not of respiratory origin. In the Otway data, the volume fraction of "unexplained" CO_2 was small, at most 2% and in most samples 1% or less.

The difficulty in interpretation, as emphasised, was that it was not known what to do with this number as it could not be correlated to a flux. The actual interpretation was more qualitative: the correlations were similar and explicable, whether pre- or post-injection; consequently no warning flags were raised.

The soil gas programme at Otway was run on a similar basis to the groundwater programme, with a single senior scientist working part-time. Casual help was enlisted during surveys, which typically took a week. Access to farmers' land was more of an issue than for water sampling, because of the density of the grid of points.

9.4.8 Atmospheric monitoring

Atmospheric monitoring at Otway was focused on leakage to atmosphere from spatially small areas (Chapter 15). In the context of the risk assessment, this meant wellbore leakage and possibly leakage up faults. The concentration of CO_2 was measured at a single location about 700 m downwind (in the prevailing wind) from the wells. In distinction to the case of soil gas, it was possible to infer fluxes from concentrations, because the physics of dispersion (and hence dilution) into the turbulent atmosphere is fairly well understood.

The sensor at Otway was sited some way from the wells so that plumes from leakage would be carried to it from a reasonably large area, given the usual range of wind directions. However, this meant signals were likely to be small, because of dilution, which was why the method was sensitive only to spatially small sources (such as wellbore leaks). Spatially extended sources would start out generating low concentrations of CO_2 and dispersion would then make this even lower.

The conditional sensitivity of the technique has been quantified at about 5000 t year^{-1} (Jenkins et al. 2012). The false alarm rate for these levels of emission is currently being better quantified with controlled release experiments during Stage 2B, but is probably around 1 in 10. The arrival function was not estimated.

As with other monitoring techniques, atmospheric monitoring required significant development of methods, in the hands of experts in the area. During Otway Stage 1 it was not an off-the-shelf method and it still is not, but sensors in particular have made remarkable progress in the past 5 years. It is now possible to contemplate affordable networks of precise atmospheric sensors, which could offer an economical and unobtrusive mode of surveillance of several square kilometres.

9.5   Conclusions

The various techniques deployed during Otway Stage 1 tell a qualitative story, consistent with "no leakage" and "no environmental impact". Some important quantification was achieved for key techniques, and most of the programme's limitations could be addressed with the benefit of experience. In any case, these limitations have not been an issue with regulators or other stakeholders, and have only been highlighted here because extrapolation to larger and perhaps more controversial projects requires care with the logic.

Obviously the main desideratum for a bigger project would be the systematic evaluation of the arrival function for the various domains of monitoring. This would involve approximations and uncertainties, but these could be made explicit, tracked, and improved with time. Developing these arrival functions would be an integral and foundational part of the preparation for risk assessment, although it would be necessary to be realistic about the time and expertise needed to do this.

The following general observations can be made regarding monitoring:

1. An M&V programme must be closely aligned to the risk assessment, and these risks should be as precisely stated as possible. A measurement must have the clearly-defined capacity to rule out a risk.

2. Risks may include societal risks, and the mitigation for these may include monitoring programmes; these should be designed on exactly the same logical and scientific principles as any other monitoring.

3. Some (probably not all) of the risks which a project mitigates by monitoring will be subject to regulatory scrutiny. It will be important to develop an agreed nexus between the words of regulations, and the numbers delivered by monitoring.

4. For each monitoring method, there should be at least some idea of the contingent sensitivity, the arrival function, and the false alarm rate. Establishing these properties even approximately may well be challenging, but almost any quantification is better than nothing.

5. A monitoring programme will need communication plans to convey results at appropriate levels of detail to various stakeholder groups. Particular attention should be paid to dealing with false alarms.

6. Defensible space should be created for exploratory monitoring. This has to be clearly separated from monitoring and *verification*. It should be possible for methods to be developed, inevitably producing odd or unexplained results, without these becoming the subject of regulatory controversy or alarm in the community.

Some of the more specific technical needs identified through the Otway Project include:

> redundancy in key downhole sensors, especially pressure sensors

> development of unobtrusive, permanently-installed seismic sensors

> networks of atmospheric sensors

> combining soil gas with soil flux measurements

> budgeting for at least one optimally-placed groundwater monitoring well

> better prior modelling of arrival function, sensitivity and false alarm rate.

More detailed discussions on the various monitoring techniques used in the Otway Project are provided in the following seven chapters.

9.6   References

Christensen, N.B., Sherlock, D. and Dodds, K. 2006. Monitoring CO_2 injection with cross-hole electrical resistivity tomography. Exploration Geophysics 37, 44, doi:10.1071/EG06044.

Cook, P.J., Causebrook, R., Michael, K. and Watson, M. 2013. Developing a small scale CO_2 test injection: experience to date and best practice. Report by CO2TECH for IEAHGG, Cheltenham.

de Caritat, P., Hortle, A., Raistrick, M., Stalvies, C. and Jenkins, C. 2012. Monitoring groundwater flow and chemical and isotopic composition at a demonstration site for carbon dioxide storage in a depleted natural gas reservoir. Applied Geochemistry (May), doi:10.1016/j.apgeochem.2012.05.005.

Dodds, K., Daley, T., Freifeld, B., Urosevic, M., Kepic, A. and Sharma, S. 2009. Developing a monitoring and verification plan with reference to the Australian Otway CO_2 pilot project. CO2CRC Publication No: JOU10-2154.

Enting, I., Etheridge, D. and Fielding, M. 2008. A perturbation analysis of the climate benefit from geosequestration of carbon dioxide. International Journal of Greenhouse Gas Control 2(3) (July), 289–296, doi:10.1016/j.ijggc.2008.02.005.

Haugan, P.M. and Joos, F. 2004. Geophysical Research Letters 31, L18202, doi:10.1029/2004GL020295.

Jenkins, C., Cook, P.J., Ennis-King, J., Undershultz, J., Boreham, C., Dance, T. and de Caritat, P., et al. 2012. Safe storage and effective monitoring of CO_2 in depleted gas fields. Proceedings of the National Academy of Sciences of the United States of America 109(2) (January 10): E35–41. doi:10.1073/pnas.1107255108.

Leuning, R., Etheridge, D., Luhar, A. and Dunse, B. 2008. Atmospheric monitoring and verification technologies for CO_2 geosequestration. CO2CRC Publication No. JOU08-1050.

Li, R., Dodds, K., Siggins, A.F . and Urosevic, M. 2006. A rock physics simulator and its application for CO_2 sequestration process, Exploration Geophysics 37, 67, doi:10.1071/EG06067.

Mathieson, A., Midgely, J., Wright, I., Saoula, N. and Ringrose, P. 2011. In: Salah CO_2 storage JIP: CO_2 sequestration monitoring and verification technologies applied at Krechba, Algeria. Energy Procedia 4 (January), 3596–3603. doi:10.1016/j.egypro.2011.02.289.

Shaffer, G. 2010. Long-term effectiveness and consequences of carbon dioxide sequestration. Nature Geoscience 3 (June (7)), 464–467.

Sharma, S., Cook, P.J., Berly, T. and Anderson, C. 2007. Australia's first geosequestration demonstration project— The CO2CRC Otway Basin Pilot Project. APPEA Journal 7, 259–268.

Shell Canada 2010. Quest Carbon Capture and Storage Project, Volume 1: Project Description, Appendix A: Measurement , Monitoring and Verification Plan, Carbon.

Sherlock, D. and Toomey, A. 2006, Gravity monitoring of CO_2 storage in a depleted gas field: a sensitivity study, Exploration Geophysics 59, 37–43.

Soroka, M.W. 2012. A review of existing best practice manuals for carbon dioxide storage and regulation. CO2CRC Report No. RPT12-3552 (September).

Urosevic, M., Pevzner, R., Shulakova, V., Kepic, A., Caspari, E. and Sharma, S. 2011. Seismic monitoring of CO_2 injection into a depleted gas reservoir—Otway Basin Pilot Project, Australia. Energy Procedia 4, 3550–3557, doi:10.1016/j.egypro.2011.02.283.

Watson, M., Bowden, A., Causebrook, R., Dance, T., Hennig, A., Kaldi, J. and Pershke, D. et al. 2007. The CO2CRC Otway Project quantitative risk assessment with newly acquired data. CO2CRC Report No. RPT07-0787 (August).

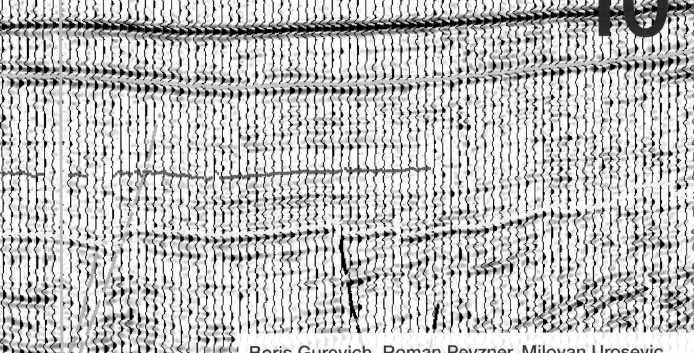

Boris Gurevich, Roman Pevzner, Milovan Urosevic,
Anton Kepic, Valeriya Shulakova, Eva Caspari

10. 2D AND 3D SEISMIC INVESTIGATIONS FOR STAGES 1 AND 2C

10.1 Introduction

Active time-lapse seismic monitoring was an important component of the Otway Project Stage 1. The main objective of a seismic monitoring programme is to provide assurance that the injected CO_2 remains confined to the storage reservoir. However, from the start, it was recognised that there were substantial challenges in monitoring the migration of the CO_2 plume within the Waarre C reservoir. These challenges arose from the fact that the reservoir (Waarre C Formation) was just a few metres thick, geologically complex and rather deep (approximately 2000 m). Furthermore, it was recognised that due to the presence of residual methane gas in the formation, the effect of the injection of a gas mixture (containing approximately 80% carbon dioxide and 20% methane) on elastic properties was likely to be very small, making the detection of the injected gas in the Waarre C reservoir even more difficult.

At the same time, it was likely the presence of even relatively small amounts of gas mixture in overlaying aquifers would cause significant changes in elastic properties, and hence a detectable time-lapse seismic signal. Thus high resolution time-lapse seismic information could potentially offer a means to provide assurance that no significant amount of gas had leaked into the overlaying aquifers—an indirect assurance that the CO_2/CH_4 mixture had remained in place in the storage reservoir.

With this in mind, the seismic programme for the Otway Project Stage 1 was designed to maximise the possibility of detecting very small time-lapse signals from the Waarre C reservoir, and at the same time to detect any time-lapse signals in overlaying strata. It was intended that a subsequent phase of the Otway Project (Otway Stage 2C) involve injecting and monitoring of up to 30,000 t of the CO_2/CH_4 mixture directly into a saline aquifer, accompanied by time-lapse seismic monitoring.

Stage 2C would then provide the opportunity to test the leakage modelling undertaken for Otway Stage 1. To date Project 2C has not been carried out, but what has been undertaken is extensive modelling of the acoustic signature of the proposed 30,000 t injection which is in any case relevant to the recognition of leakage if it were to occur during Stage 1. Therefore in this chapter, the discussion on the design of the seismic monitoring programme is preceded by a description of the detailed experimental, theoretical, and computational modelling of the time-lapse signal and analysis of seismic repeatability (time-lapse noise). This section outlines all the components of the research into the seismic response of the CO_2 injection into a Waarre C reservoir, followed by the modelling of the time-lapse response of possible leakage into an aquifer. Next, the field experiments aimed at assessing the repeatability of the 2D and 3D time-lapse seismic survey at the Otway site are described, followed by a description of the main 4D monitoring programme, which accompanied Stage 1 of the Otway project, including the survey design, acquisition, processing and analysis of the acquired 4D seismic data. Another important component of the monitoring programme, namely time-lapse borehole (VSP) surveys, is also considered in some detail. Finally, a laboratory study of acoustic responses on the injection of supercritical CO_2 into Otway sandstones is described.

As the investigations and particularly the seismic modelling for Otway Stages 1 (Waarre Formation) and 2C (Paaratte Formation) were undertaken in parallel, this chapter deals with work undertaken for both stages in an integrated manner.

10.2   Modelling seismic response of injected CO_2 in Stage 1

The applicability of seismic monitoring depends on the effect of CO_2 injection on the seismic response. Several factors impact on whether this effect is detectable, such as geological heterogeneity of the reservoir, rock and fluid properties at reservoir conditions as well as data quality and repeatability of land seismic surveys. Because of these variable factors, it is important to perform site-specific seismic modelling to estimate the detectability of changes in the seismic signature caused by CO_2 injection. Such modelling studies have been performed for a number of CO_2 sequestration sites such as Ketzin (Kazemeini et al. 2010) and the Aztbach–Schwanenstadt gas field (Rossi et al. 2008).

The seismic time-lapse signals from a CO_2/CH_4 injection into the depleted Naylor gas field were investigated and modelled as part of CO2CRC Otway Stage 1. The seismic monitoring programme for Otway Stage 1 included a baseline 3D surface seismic and 3D VSP survey in 2007–08, two monitoring surface seismic surveys in 2009 and 2010 as well as one 3D VSP monitoring survey in 2010.

10.2.1   Rock physics modelling

In order to model the change of the acoustic rock properties caused by CO_2/CH_4 injection, the Gassmann fluid substitution workflow was applied, using an approximate method for solving the Gassmann's equation suggested by Mavko et al. (1995), based on the P-wave modulus (M), without the use of shear-wave velocity. In this workflow, the P-wave modulus of a rock saturated with "fluid 2" was computed from the P-wave modulus of a rock saturated with "fluid 1", the P-wave moduli of the fluid mixtures the composition of the rock and the porosity.

In case of the Waarre C reservoir, fluid 1 was a mixture of formation brine and residual gas (mainly CH_4) and fluid 2 was a mixture of brine, CO_2 and CH_4. The in-situ brine properties were computed by the empirical formula of Batzle and Wang (1992), while the CO_2/CH_4 mixture properties were obtained from the flow simulation result (Ennis-King et al. 2010) by solving the equation of state of the GERG 2004 model (Kunz et al. 2004). To calculate the fluid bulk modulus of the brine/CO_2/CH_4 mixture, the Wood's mixing rule was applied. The use of Wood's equation assumes uniform saturation, a reasonable assumption for sandstones at seismic frequencies. The P-wave modulus of the solid grain material was computed by averaging the upper and lower Hashin-Shtrikman bounds of the clay and quartz volume fractions obtained from the well logs (Mavko et al. 1998).

To perform the fluid substitution modelling for the Waarre C reservoir, a 3D reservoir model populated with the described parameters was needed. For this purpose, information was used from an acoustic impedance (AI) inversion, the static geological model, flow simulation and the well log data.

10.2.2   Reservoir model

In order to construct a reservoir model, datasets in different domains, such as time and depth, and on different scales, had to be combined. First, a reservoir model in time was constructed, with seismic time horizons of the AI inversion using the RokDoc-ChronoSeis package (Ikon Science). Subsequently, a layer cake method was applied to convert the time model into a depth model, where the depth surfaces were fixed to the well tops in the CRC-1 and Naylor-1 wells. To fit the depth-based flow simulation grid into a RokDoc depth model, the top and bottom horizons of the Waarre C reservoir were replaced by the corresponding depth surfaces of the flow simulation grid.

Another issue was that the static geological model, which provided porosity estimates, was on a finer scale than the acoustic impedance inversion result. Therefore it was decided to build two models with different levels of detail in their reservoir properties (fine scale and coarse scale).

In Model 1 the porosity of the static geological model was used. The reservoir model was subsequently populated with calculated AI values and the P-wave moduli of the grain material, determined from the well data by applying co-located krieging, with the porosity model as a soft property. Since the logs in CRC-1 and Naylor-1 were measured at different times post- and pre-production respectively, fluid substitution was performed so that the AI values corresponded to 20% residual gas saturation. The fluid substitution was then applied to the 3D reservoir model to match the gas saturation pre-injection (2008), using the prediction of the flow simulations. Outside the reservoir, the AI model was built from the AI inversion volume of the 2008 seismic baseline data.

In Model 2, the AI volume from the inversion of the 2008 seismic baseline data was used. In this case the static porosity model was smoothed; by applying a low-pass filter; structures on a much smaller scale were smoothed out. The P-wave modulus of the grain material was set to the value for quartz.

Figures 10.1 and 10.2 show a comparison between the AI and porosity values of Model 1 and Model 2. Model 1 was much more detailed than Model 2, but the average values of AI and porosity in the two models were of the same order of magnitude. In summary, Model 1 honoured the static geological model, while Model 2 represented the inversion result of the seismic data.

10.2.3   Modelling results

Fluid substitution modelling was undertaken using the 2008 conditions of pre-injection, the injection of 35,000 t (in 2009) and the final total of 66,000 t (in 2010) for the CO_2/CH_4 mixture. For Model 1, the modelling was only carried out in areas where the porosity values were above 6%, since smaller porosity values produced unrealistic results. For Model 2 (smoothed porosity), the porosity values were above 6% for the entire model.

Figure 10.3 shows a map of the absolute difference in AI between 2010 and 2008 for Model 1 and Model 2. The difference is of the same order of magnitude for both models. Overall, the AI in the reservoir decreases from 2008 to 2010, since brine is replaced by the CO_2/CH_4 mixture. As expected, the difference in AI reflects the distribution of CO_2 mass fraction for 2010 predicted by the flow simulations. This prediction suggests that the CO_2 migrated below the gas cap and not into the gas cap located at the top of the reservoir near Naylor-1. This explains the smaller changes in this part of the reservoir. The largest change in AI occurs half-way between Naylor-1 and CRC-1.

Qualitatively, the same results were obtained for 2009 with a smaller decrease in AI compared to 2010. For comparison with the field data, the fine-scale model was used in the next section. The modelled scenarios correspond to the monitoring surface seismic data for 2009 and 2010 and to the 3D VSP data for 2010. In order to compare the modelling results to the seismic data, seismic forward modelling was performed by convolving the calculated AI volumes with a statistical wavelet, extracted from the surface seismic data for 2008.

Figure 10.4 shows a comparison between the synthetic and surface seismic data for 2008 and the time-lapse signal for 2008–10. The left column contains the synthetic section (a) compared to the field data (c) in 2008. The right column shows the predicted time-lapse response (b) and a slice of the time-lapse difference volume obtained

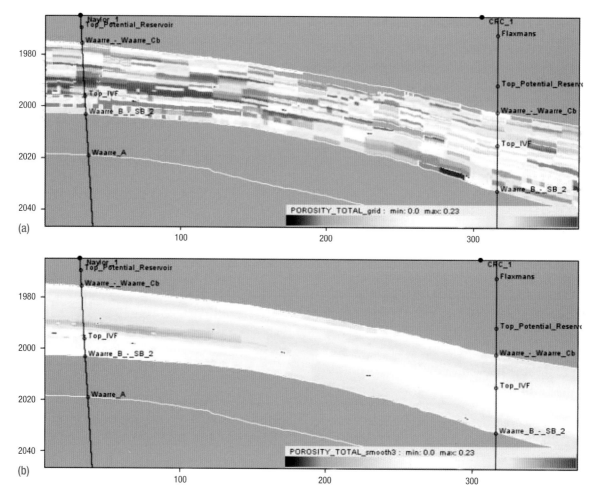

Figure 10.1: Porosity cross-sections through Model 1 (a) and Model 2 (b).

from the seismic data (d). Clearly, the modelled signal level is significantly smaller than the time-lapse noise in the surface seismic data.

In summary, both models (fine and coarse scale) lead to similar results in the absolute difference in acoustic impedance and resemble the main features of the seismic data. The coarse-scale model represents the amplitude distribution of the field data more adequately (not shown here), since it is directly based on the AI inversion result. However, simple averaging of the porosity might not represent a proper up-scaling to the resolution scale achieved in AI inversion. A better match between the datasets might be gained by using the AI inversion result

to build a porosity model. In general the seismic modelling study confirmed that the time-lapse signal was too small, or the time-lapse noise too high, to be able to detect a signal from injection of the Buttress gas using surface seismic data.

10.3 Modelling seismic response of CO_2 leakage for 2C

One of the objectives of CO2CRC Otway Project Stage 1 was to explore the ability of geophysical methods to detect and monitor leakage of CO_2 from the primary container into overlaying strata. The approach taken was

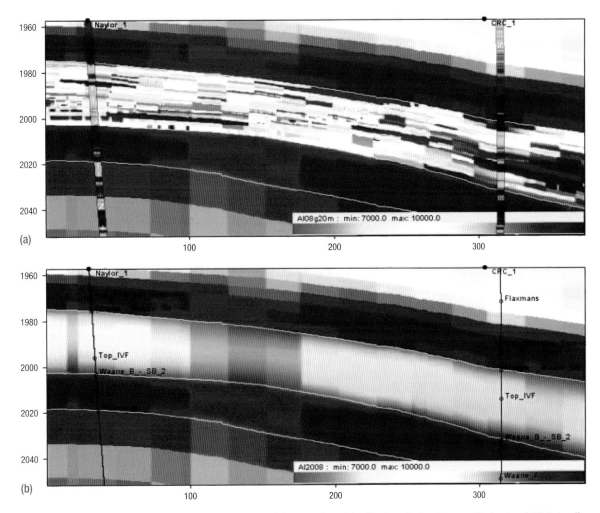

Figure 10.2: Acoustic impedance cross-sections through Model 1 (a) and Model 2 (b) along the line between Naylor-1 and CRC-1 wells.

to numerically model several scenarios of leakage of the gas into the overlying Paaratte Formation. The numerically modelled seismic signal (the so-called synthetic seismograms) was then compared to the time-lapse images acquired at the Otway site. This comparison (discussed later in this chapter) showed that no signal comparable to the synthetic signal was observed in the time-lapse seismic images obtained from field data, which in turn indicates that there was no detectable secondary accumulation of the CO_2 in the overlying sediments.

A second approach is planned as part of Otway Stage 2C, namely to experimentally simulate the leakage by injecting CO_2 into a saline aquifer, and performing time-lapse

geophysical measurements to determine the detectability of these simulated leakages in the subsurface. For this purpose, it is proposed to inject up to 30,000 t of gas mixture (80% CO_2 / 20% CH_4) into the Paaratte Formation, at a depth of about 1400 m. Before such an injection experiment is undertaken, the feasibility of geophysical monitoring was assessed using computer modelling. While this modelling study simulated the single-well injection scenario, it could also be considered a leakage scenario, corresponding to leakage of CO_2 via a well due to imperfect cementation.

To examine the detectability of the CO_2 plume using time-lapse seismic data, it is necessary to estimate the time-lapse signal and time-lapse noise. The time-lapse signal can be modelled using flow simulations, fluid substitution and seismic forward modelling. The level of time-lapse noise can be estimated from existing time-lapse Stage 1 3D data in the area. As mentioned earlier, from 2008 to 2010 three repeat 3D seismic surveys and a number of borehole seismic surveys were performed in the area as part of Stage I of the CO2CRC Otway Project (Urosevic et al. 2011; Pevzner et al. 2011). Of the three 3D surveys, those performed in 2009 and 2010 had the highest data quality, due to the use of a more optimal seismic source (IVI minivib) and high fold. Thus existing time-lapse data can provide direct information about the amplitude, level and character of the time-lapse noise.

10.3.1   Reservoir modelling

Modelling of the time-lapse signal is based on the results of reservoir simulations, which in turn requires a static geological model as input. The static model for 2C was built using the horizons extracted from the regional seismic data (CCG 2000). In this model, the injection interval was represented by a 7 m thick sand interval (known as Paaratte Zone 1) encased between two impermeable layers.

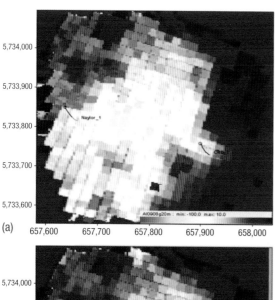

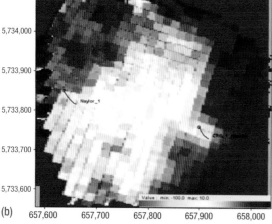

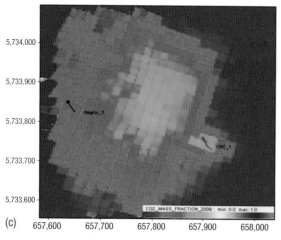

Figure 10.3: Absolute difference of AI (m/s*g/cm³) between 2009 and 2008 for Model-1 (a) and Model-2 (b) averaged over the depth interval of the reservoir zone. (c) CO₂ mass fractions for 2009 averaged over the depth interval of the reservoir zone.

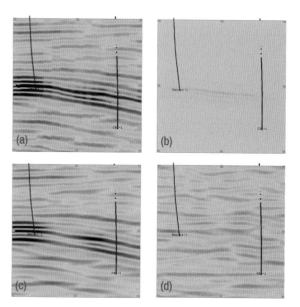

Figure 10.4: Synthetic (a) and field data sections (c) along in-line section 87 from 2008 survey and corresponding time-lapse differences between 2010 and 2008 surveys (b and d).

The flow simulations were performed using Eclipse300© software for two injection scenarios involving injection of 10,000 and 30,000 t of the CO_2/CH_4 gas mixture into Zone 1 through the CRC-2 well. The phase behaviour of the CO_2/CH_4 mixture was modelled using the Peng-Robinson (1976) equation. The two-phase (brine-gas) flow modelling utilised permeability functions adapted from the published core flood data for CO_2-brine flow (Bennion and Bachu 2008; Perrin and Benson 2009).

10.3.2   Model of elastic properties

Modelling the effect of the 2C plume on seismic data consisted of three main steps:

1. building an elastic model of the subsurface before injection

2. estimating the changes of elastic properties likely to occur due to the presence of the plume

3. computing the seismic response with and without the plume.

The first step was to build an elastic model of the subsurface without the plume. Information on elastic properties of the subsurface came from Stages 1 and 2A well log data and seismic data (VSP and surface seismic). The log data had the highest vertical resolution and therefore could be used to model the effects of fine layering. However, well log information was only available in a small number of well intervals. On the other hand, information from seismic data had broader spatial coverage but much lower vertical resolution. As the results of flow simulations, both the reservoir zone and the plume have very small thicknesses (7 m and 2–4 m, respectively). Thus the main challenge of the seismic programme was detection of a very thin plume. Therefore, it was essential to model the effects of thin layering as adequately as possible. To this end, an elastic model of the subsurface was built using well logs.

First, a one-dimensional (1D) model was built using well logs from the CRC-2 and CRC-1 wells and estimates were made of the changes in elastic properties for the different plume thicknesses. These models were later used to perform accurate and efficient full-waveform modelling of the seismic data. A 3D model of the subsurface was then built by laterally interpolating and extrapolating log

data from three wells along horizons mapped from 3D seismic data. To estimate changes of elastic properties away from the well, the flow simulation results were used, based on the porosity in the static geological model and the interpolated log data. These input parameters, particularly wet velocity and porosity, needed to be consistent. To achieve this, the velocities and densities within the 2C injection interval were adjusted while the porosity model was kept unchanged, since it determined the volume of the fluids in the flow simulations. The geological model was populated with P-wave and S-wave velocities based on empirical porosity-velocity relations obtained from the log data in the CRC-2 well. This model was later used for 2D seismic forward modelling along a line crossing the plume, which was designed to account for the finite lateral extent of the plume and lateral variations of its thickness.

10.3.3   Changes of rock properties due to CO_2/CH_4 injection

In order to predict the change in the elastic properties, the Gassmann fluid substitution workflow was used (Smith et al. 2003). In the case of the proposed 2C Paaratte injection interval, the elastic properties of the rock saturated with brine and the injected gas (mixture of CO_2 and CH_4) were calculated from the elastic properties of the rock saturated with formation brine, the properties of the fluid mixture, the solid grain material and the porosity. To assume realistic fluid saturations and properties the flow simulation results for 10,000 t and 30,000 t of injected CO_2/CH_4 were utilised.

The flow simulations predicted that 12% of the injected gas would be dissolved in the formation brine by the end of the injection. Since CO_2 is much more soluble in water than methane, the assumption was made that it is predominantly the CO_2 that dissolves, giving an average gas composition of 77% CO_2 and 23% CH_4 in the free gas. From this gas composition, the elastic properties of the free gas could be computed by an equation of state based on the GERG 2004 model (Kunz et al. 2004) implemented by J Ennis-King (personal communication).

The in-situ brine bulk modulus was computed from the empirical formula of Batzle and Wang (1992), while the brine density was obtained from the flow simulation results.

The fluid bulk modulus of the CO_2/CH_4/brine mixture was computed with the Wood's mixing rule. Since the Paaratte Zone 1 is dominated by relatively clean quartz-rich sandstone, the value of quartz was used for the bulk modulus of the rock. The matrix density was calculated directly from the density and porosity logs for the 1D model and kept constant (2.7 g/cm³) for the 3D model. The dry bulk modulus of the rock was determined from the sonic velocities of the baseline models (fully brine saturated conditions) by solving Gassmann's equation, which was in turn applied to calculate the elastic properties of the rock after injection.

The modelling was performed in 1D for different plume thicknesses ranging from 1 m to 6.9 m; an average gas saturation of 57%, was obtained from the flow simulations. For the 3D model, the change of P- and S-velocities, density and acoustic impedance (AI) was computed for each grid point of the reservoir zone. Then, for each lateral location, the effective thickness of the plume was calculated as the sum of the thicknesses of those cells which show over 5% change in AI. Figure 10.5 shows these thicknesses in map view for 10,000 and 30,000 t injections, respectively. The average relative change of AI within the plume for both scenarios was determined to be between 10 and 15%, and the thickness varies from 1 to 2.5 m for the 10,000 t injection and between 2 and 5 m for 30,000 t injection into the Paaratte Formation.

10.3.4 Seismic forward modelling of the time-lapse signal

Since the geometry of the overburden in the Otway site area is relatively flat, forward modelling was done using the 1.5D reflectivity algorithm implemented in OASES software (MIT). 1.5D modelling simulates 3D seismic wave propagation in locally 1D earth. To reduce the computation time, for all traces the simulations for the baseline case were performed for "flat geometry" elastic properties extracted from the CRC-2 well log data. Next, every trace from the actual Otway 3D seismic dataset (pre-stack) was substituted with the synthetic trace with the corresponding offset. The resulting data were processed to the final stacked volume using a standard processing flow. By doing so, precisely the same offset/angle distribution

was obtained as in the field data. This was important as, in the presence of gas, seismic amplitudes can significantly vary with offset. Figure 10.6 shows the comparison of the stacked synthetic "baseline" and field data along an in-line direction near the CRC-2 well location. For the monitoring case, the rock properties were altered in the gas saturated interval, with the thickness obtained from flow simulations and fluid substitution. A set of 3D volumes representing different plume thicknesses while being still 1D (e.g. laterally infinite) was then computed. To simulate the finite lateral extent of the plume, synthetic "monitor"

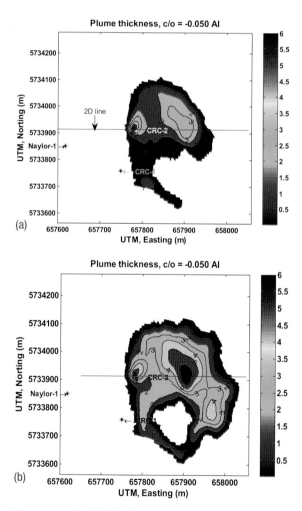

Figure 10.5: Modelled thickness of the 2C plume (m) for 10,000 t (a) and 30,000 t (b) of injected CO_2/CH_4, calculated for relative changes in AI that are greater than 5%. The red line indicates the 2D line for the forward modelling.

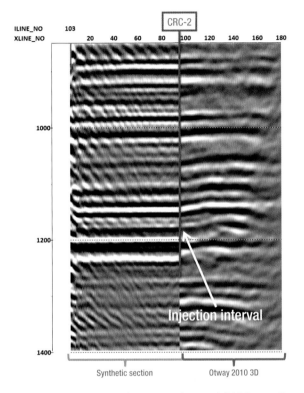

Figure 10.6: Comparison between synthetic and field Otway 3D baseline (2010) data along an in-line direction near the CRC-2 well location. The arrow indicates the injection interval.

survey volumes were computed by the interpolation between these volumes according to the plume thickness maps (Figure 10.5).

To verify the validity of this workflow, 2D finite difference modelling was performed along an in-line direction (baseline and monitor) using a fourth-order elastic 2D FDTD algorithm with a fine spatial grid (cell size of 0.5 × 0.5 m). The 2D modelling showed almost the same amplitude of the time-lapse signal as the 1.5D modelling.

10.3.5 Analysis of plume detectability

Following the repeatability analysis (described in the next section; see also Pevzner et al. 2011), it was assumed that random additive noise was the main contributor to the overall time-lapse noise pattern for the Otway site. To mitigate risks associated with underestimation of the noise level, several noise level scenarios were considered.

The noise modelling workflow comprised several steps. Two 3D volumes were created containing random noise uniformly distributed between –1 and +1. These volumes were filtered with a zero-phase filter such that the amplitude spectrum of the noise matched the amplitude spectra of the synthetic seismic data computed in the previous step; i.e. noise was modelled to guarantee a constant signal/noise ratio over the whole frequency band. Finally these noise volumes were migrated (using the Stolt F-K Migration) to produce non-zero spatial correlation for noise traces of the type which can be expected in field data (due to migration).

These two noise realisations were then mixed with the synthetic data in order to produce several signal/noise scenarios. To quantify the level of time-lapse noise, a normalised root-mean-square (NRMS) difference between the two different vintages a and b was used (Kragh and Christie 2002) for seismic data at the anticipated two-way travel time of injection (~1186 ms) within a time window of 60 ms. This is represented by the equation:

$$NRMS = 200\% \; RMS(a\text{-}b)/(RMS(a)+RMS(b)) \qquad (1)$$

To evaluate the time-lapse noise level, the following scenarios were considered:

> The L0 scenario has a smaller noise level than the field data (median NRMS value 19%).

> The L1 scenario with a median NRMS value of 28% is closest to the 2009/2010 4D seismic field data (median NRMS value is 30%).

> L2 and L3 scenarios exhibit stronger noise than the field data (median NRMS values 41 and 54% respectively).

Figure 10.7 shows a set of in-line sections from the resulting difference volumes at all four noise levels for a 10,000 t injection into the Paaratte Formation. From this figure it was concluded that the 3D seismic survey should be able to detect 10,000 t of injected CO_2/CH_4 mixture at the noise levels L0–L2. These include the actual noise level observed for field 4D data, as well as the second noise level (L2), which contains approximately 1.5 times more noise than L1. In the extreme case (L3, the level of noise twice that in the field data), detection of the injection could be problematic. A similar plot for 30,000 t injection into the Paaratte Formation (Figure 10.8) provides a greater level of confidence that the plume will be detectable at all the noise levels.

10.4 Time-lapse repeatability in Stage 1

In the last decade, time-lapse seismic reflection surveys have proved to be a powerful tool for sensing potential changes in a reservoir related to oil/gas production (Lumley 2001), steamflood for enhanced oil recovery (Waite and Sigit 1997; Isaac and Lawton 2006) or CO_2 storage (Eiken et al. 2000; Vandeweijer et al. 2009). Understanding the factors influencing repeatability and the possible limitations of the surface reflection seismic method is crucially important for the success of this type of survey

(Calvert 2005). There are many published case studies of offshore 4D seismic surveys and the factors affecting repeatability of marine seismic data are relatively well known. The biggest issues affecting repeatability of offshore seismic surveys are believed to be variations in acquisition geometry and positioning errors (Landrø 1999). The issue of spatial variations in the nature of the overburden and its influence on the sensitivity of time-lapse (TL) seismic reflection survey data is addressed by Malme, Landrø and Mittet (2005). In general, land seismic data repeatability is reported to be poor (Vandeweijer et al. 2009; Vedanti and Sen 2009), but there are no comprehensive reports or studies covering TL land seismic issues. A recently published paper dedicated to a feasibility study of land seismic monitoring of CO_2 sequestration (Rossi et al. 2008) does not evaluate the predicted time-lapse signal versus expected repeatability.

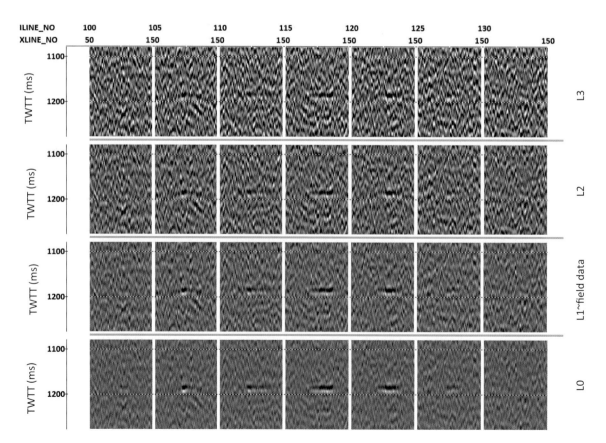

Figure 10.7: Set of in-lines from the difference volume, 10,000 t injection, L0–L4 noise level scenarios.

Given that factors controlling success in land seismic for imaging the subsurface (such as properties of signal and noise as well as limitations related to field operations) are different for land and marine surveys, the repeatability of a land seismic survey should also be very different from that in a marine survey. For example, land acquisition geometry can be repeated with a high degree of accuracy; hence, positioning errors are not significant for land time-lapse surveys. On the other hand, the pattern of source generated noise can rapidly change across the survey area. Also, the level of ambient noise related to wind, rain or local traffic is much higher in land data. Source-receiver coupling and scattering in the weathered zone may vary widely throughout the year, depending on the soil saturation and underground water level.

The main factors affecting land seismic repeatability are velocity variations in the shallow part of the subsurface,

variations in source parameters and geophone coupling (Bakulin et al. 2007; Ma et al. 2009). General repeatability issues of both impulse and vibrator source signatures have been analysed by Aritman (2001). Experimental observations of changes in the subsurface are addressed in Beaty and Schmitt (2003) and Meunier et al. (2001), while a modelling study on the subject of land data repeatability was treated in the paper by Al-Jabri et al. (2008).

In order to determine the most appropriate seismic methodology for Otway seismic work, a suite of land time-lapse seismic surveys were undertaken. In this chapter the focus is on three repeated 2D surveys obtained with the same geometry, but with different sources (weight drop and vibrator) and near-surface conditions varying from survey to survey, due to changes in soil saturation. The purpose of the repeated 2D acquisition test was to test 3D survey parameters, and ascertain the key issues

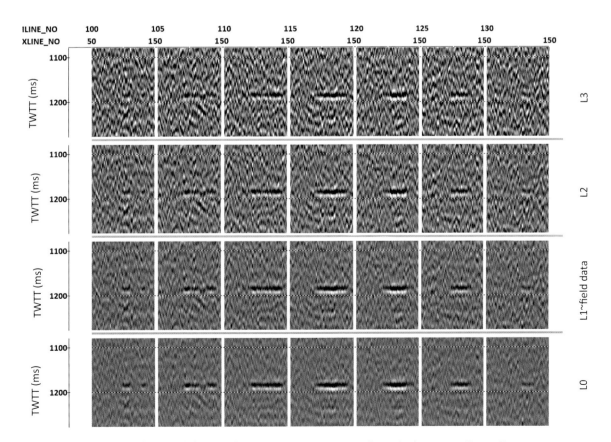

Figure 10.8: Set of in-lines from the difference volume, 30,000 t injection, L0–L4 noise levels scenarios (Stage 2C).

related to land data repeatability, while the 3D surveys were acquired in order to monitor CO_2 storage. A comparative study was possible due to the co-location of these tests and as a consequence it was possible to rigorously evaluate the influence of factors that affect TL land repeatability, notably source type, ambient non-repeatable noise, and changes in near-surface conditions related to variations in seasonal water saturation.

A comparison of 2D versus 3D repeatability was also undertaken, as it had been suggested that 2D time-lapse seismic surveys may be a cost-effective alternative to 3D time-lapse seismic surveys (Vandeweijer et al. 2009). Given the likelihood that there will be many onshore CO_2 storage projects in the future, the cost effectiveness of TL seismic surveys is likely to be an important issue.

10.4.1 2D pre-baseline seismic surveys

A unique set of repeated 2D trial surveys was collected over a period of 3 years within the Otway Project Stage 1 programme. This comprised seven repeated 2D seismic lines collected using the same geometry and acquisition system. Data were acquired at different times of the year (the water table and soil saturation are low in the summer and high in the winter) and with different sources (several different weight drop energies and a vibrator).

To determine the key limiting factors affecting repeatability of land seismic surveys, three different sets of seismic data were analysed:

1. 2008 line recorded in the dry season with a free-fall 5 kJ weight drop (WD) source
2. 2008 line also recorded in the dry season, but with an IVI minivib (MV) source
3. 2007 line recorded during the wet season with a free-fall 5 kJ weight drop source.

Several cases were analysed: (1) same near-surface conditions but data acquired with different sources (2008 data); (2) different near-surface conditions (wet versus dry), data acquired with similar sources (two weight drop surveys); and (3) different sources and near-surface conditions. The acquisition parameters are shown in Table 10.1. Receiver and source points were located at the same position (source

points between receiver points) along the same line for all the surveys. The length of the record for WD surveys was 3 s, with a 1 ms sampling interval, while the minivib (MV) survey was recorded with a 10–120 Hz liner 12 s sweep and a 3 s listen time. The average fold of this 2D data in the central part of the line is greater than 80.

Examples of repeated shots for three Otway Stage 1 2D lines are presented in Figure 10.9. The overall variation in signal-to-noise ratio across the survey records can be clearly seen. The differences in the source-generated noise pattern are particularly significant. Most of the differences can be related to the source type, and to some extent to the ground conditions (dry and wet periods). In general, the level of ambient noise and intensity of the ground roll is much higher for data acquired with the weight drop source. This can be attributed to variations in the distribution of energy versus the frequency; MV data do not have much energy at low frequencies (for Otway, the sweep signal started at 10 Hz); this decreases the overall level of the ground roll. Moreover, the relative proportion of surface waves generated by the WD varies more along the line than for the MV.

The weight drop data acquired in 2007 as part of Otway Stage 1 provided better signal-to-noise ratio than the weight drop data acquired in 2008. This can be explained by

Table 10.1: 2D test line acquisition parameters (Otway Stage 1). Both IVI minivib and the weight drop were used in 2008.

Source Type	Weight Drop	Weight Drop
Date	June 2007	November 2008
Weather condition	Wet	Dry
Total number of source positions	158	155
Total number of receivers	162	156
Source/receiver point spacing, m	10/10	10/10
Number of channels	162	156
Offset range, m	5–1605	5–1545
Reference to dataset	WD 2007	WD 2008
Source type	Weight drop	Weight drop
Date	June 2007	November 2008

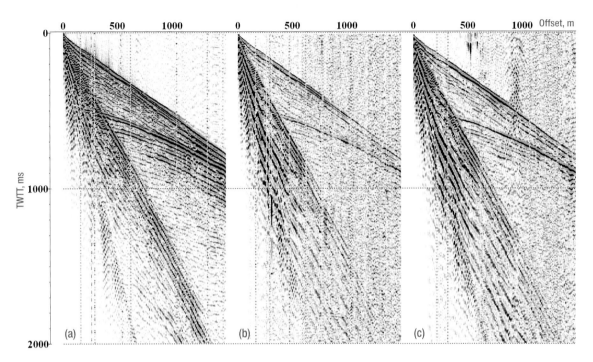

Figure 10.9: 2D pre-baseline test line, raw data example, common shot gathers; (a) IVI minivib data, 2008, (b) weight drop data, 2008, (c) weight drop data, 2007.

the difference in the saturation level of the near surface. The WD 2007 data were acquired during the wet season when soil was fully saturated with water, whereas the WD 2008 data were acquired during summer (dry season).

Processing of the data was performed with ProMax (Landmark, Halliburton) and RadExPro (DECO Geophysical) processing software systems. Similar amplitude-preserving processing flows were used for all three datasets, though it was impossible to use exactly the same flows. Variation in near-surface conditions (variable water table level) resulted in differences in the magnitudes of static corrections between the surveys. Similarly, variations in the near-surface ground saturation produced quite different coherent noise patterns. Nonetheless, the relative amplitudes between surveys are preserved through the creation and application of a single time-varying gain function.

Several routines were used to cross-equalise the Stage 1 data. Initially, post-stack static shifts were computed by cross-correlation between corresponding pairs of traces. Computed time shifts demonstrate a much higher consistency in the surveys acquired over 1 year (2008) than the surveys

acquired across 2 years. The time shifts for the 2008 surveys do not exceed 1 ms, whereas the datasets acquired over 2 years during wet and dry periods respectively exhibit static shifts as large as 3 ms. Moreover, the distribution of the shifts displays long-period anomalies. This is most likely related to changes in ground saturation, i.e., the depth of the water table. It should be noted that non-migrated data were analysed to avoid migration artefacts due to a short line length (approximately 1500 m).

For the next phase, a single shaping filter (Wiener) was designed for each pair of surveys by averaging across the filters computed for each pair of corresponding traces from the two surveys. The filter length was 100 samples with 0.1% of random noise being added. Data characterised by a wider spectral bandwidth was "degraded" to fit the bandwidth of the poorer quality set. The lowest data quality and narrowest band width were observed for weight drop data acquired in the 2008 summer dry season. The final energy normalisation was accomplished by deriving a single scalar computed in a 400–1100 ms window.

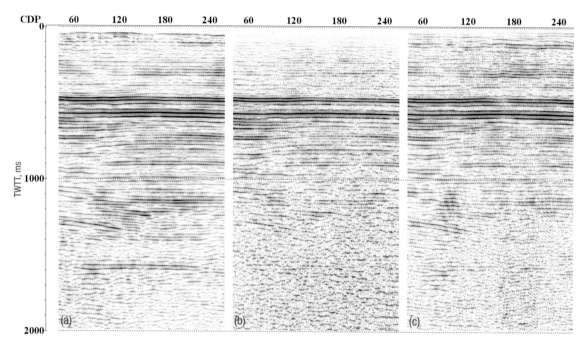

Figure 10.10: Processed and cross-equalised stacked sections, Otway Stage 1; (a) 2008 IVI minivib, (b) 2008 weight drop data and (c) 2007 weight drop data. Minivib section is of the highest quality, weight drop data recorded in dry period are of the worst quality.

10.4.2   Repeatability analysis of 2D data

Cross equalised sections are shown in Figure 10.10. Only the central parts of the lines, where the fold was maximal (CDP range of 40–260), were included in the analysis.

The 2008 weight drop data are characterised by a significant increase of random noise towards the lower part of the section. The reason for this is that the same amplitude decay function was used to compensate for the amplitude losses in all three surveys. The 2008 WD survey produced the weakest signal and, hence, the poorest signal-to-noise ratio, which further deteriorated over time. Consequently, amplitude compensation simply boosted the ambient noise that prevailed over the signal at later times.

To evaluate repeatability of these three lines, the NRMS difference sections were computed using equation 1. For each pair of traces, the trace NRMS attribute was produced, using a 60 ms sliding window. The results are displayed in Figure 10.11.

Analysis of NRMS sections show that, in all cases, relatively high repeatability coincides with strong reflectivity zones. This is to be expected, as high reflectivity means the best signal-to-noise (S/N) ratio, while no reflectivity or random noise scores the lowest S/N ratio and is not repeatable. Interestingly, there is no significant difference between non-repeatability computed for the different combinations of surveys. Moreover, the most repeatable appears to be a combination of the MV 2008 and WD 2007 Otway Stage 1 surveys.

10.4.3   Influence of ambient noise

In order to quantify the impact of random noise on the repeatability, each set was transformed into a section of S/N ratio attribute (Figure 10.12). The S/N ratio was estimated between pairs of consecutive traces on each survey in the same time windows as the NRMS value using the equation:

$$SN_i = \sqrt{\frac{[g_{i,i+1}]_{MAX}}{1 - [g_{i,i+1}]_{MAX}}} \qquad (2)$$

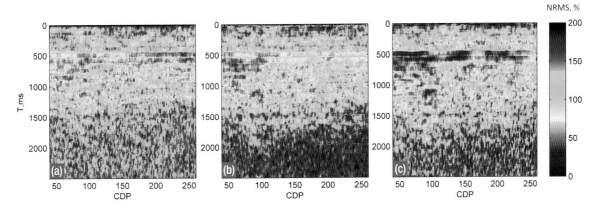

Figure 10.11: NRMS sections: (a) minivib 2008 versus weight drop 2007; (b) minivib 2008 and weight drop 2008; (c) weight drop 2008 and weight drop 2007.

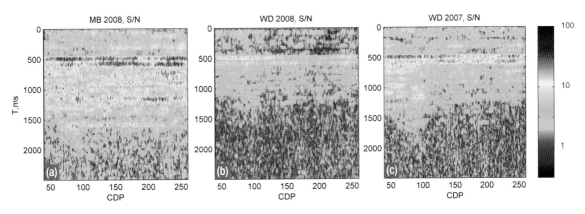

Figure 10.12: Signal-to-noise (S/N) ratio sections. Minivib sections displays, by far, the highest S/N attribute values.

where i is the trace number, $g_{i,i+1}$ is the normalised cross-correlation function between i and $i + 1$ traces and $[g_{i,i+1}]_{MAX}$ is its maximum value (Hatton, Makin and Worthington 1986). Equation 3 allows us to obtain a S/N ratio assuming that noise is additive, uncorrelated and has zero mean value.

The S/N attribute sections show that the MV was a more powerful source than the WD as it scores much greater S/N values for the same ground conditions (MV 2008 versus WD 2008). MV data quality was the best among the three different datasets.

Figures 10.11 and 10.12 show that the horizons with a greater S/N ratio correspond to those with higher repeatability. To analyse this relationship further, in Figure 10.13 time-averaged NMRS values are plotted

against noise-to-signal (N/S) values. The maximum N/S value is picked for a pair of surveys at each time sample; a linear relationship is apparent between the N/S and NRMS computed values.

Subsequently, simple mathematical modelling was used to determine the principle limitation of the NRMS measure due to the presence of non-repeatable random noise, using the following workflow to investigate the relationship between S/N and NRMS to:

> produce a synthetic seismic section containing only one horizontal event across 120 traces using an Ormsby wavelet, 10-20-60-80 Hz

> to simulate a pair of "baseline" and "repeated" surveys, produce an appropriate S/N ratio within the same bandwidth of the signal (10-20-60-80 Hz)

> compute S/N ratio and NRMS value using a 50 ms sliding time window, as applied to the real data case

> choose a different intensity of random noise and repeat the previous steps.

The results of the modelling are presented in Figure 10.14. For comparison with real measurements, the colour points represent pairs of NRMS and minimum S/N, meaning that the noisiest survey for each pair is controlling the seismic repeatability.

The results of this study shows strong agreement between modelled and measured non-repeatability. It should be noted that the generated seismic non-repeatability includes the ambient noise level effect. Most of the measured points corresponding to relatively high S/N values are located above the modelled curve. This is a consequence of the fact that besides ambient noise, there are other factors influencing repeatability in the real data case. For very low S/N ratios, measured values lie below the modelled curve. Such a small discrepancy can be expected for the steep part of the curve, where NRMS values are quite high. This plot demonstrates that it is impossible to achieve a NRMS value of less than 40% if the S/N ratio at the target horizons is lower than 5.

Repeatability of MV2008 to WD2007 (i.e. the best quality pair of surveys) is higher than the repeatability of WD2008 to WD2007 (i.e. the worst quality pair of surveys), demonstrating that the source type has less effect on repeatability, relative to the effect that S/N ratio or the power of the source.

10.4.4 Influence of near-surface conditions

Figure 10.9 shows raw data shots for two weight drop datasets acquired during wet and dry seasons. The corresponding results of processing are shown in Figure 10.10. These figures suggest that changes in near-surface conditions are very important for repeatability of time-lapse seismic surveys. Typically a much lower signal level was observed during dry periods (WD 2008) because of changes in soil hardness, particularly for impact sources such as the weight drop.

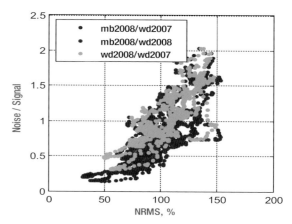

Figure 10.13: Cross-plot of N/S versus NRMS values measured from data.

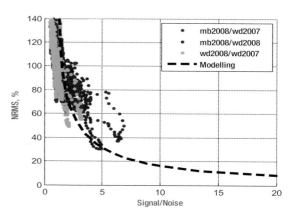

Figure 10.14: Influence of ambient noise to repeatability. The black line represents modelling results; colour dots obtained by measurements of S/N ratio and NRMS on processed and cross-equalised 2D sections.

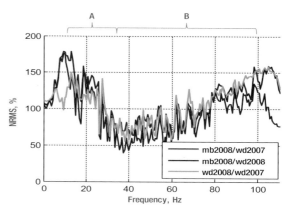

Figure 10.15: Repeatability of different spectral components of 2D test line. RMS difference computed in 0–1200 ms time window using processed and cross-equalised stacked sections.

The condition of the top soil (the agricultural layer that is overturned frequently by deep ploughing) combined with a rugged buried karstic limestone near-surface feature create scattering that will vary with changes in the underground water level. In Al-Jabri et al. (2008), stacked sections obtained by simulating different models of near-surface properties for this terrain showed that near-surface scattering could increase NRMS values from 20 to 40%.

Changes in near-surface conditions will generate variable source-generated noise patterns and the proportion of the energy lost to ground roll will vary, which in turn changes the ratio between the signal and noise. This leads to "frequency-dependent repeatability," as illustrated in Figure 10.15, where NRMS computed from cross-equalised stacked sections in the 400–1400 ms window is displayed as a function of frequency. All three pairs of surveys showed very similar behaviour. The frequency range of ~10–30 Hz (A), which is heavily influenced by ground roll, shows higher non-repeatability values in comparison to the central part of the spectrum, which is dominated by the primaries. For higher frequencies, the S/N ratio becomes poor, i.e. the NRMS becomes high, as the spectrum is affected by the ambient noise. For weight drop (a weak source), the repeatability decreases significantly in the high frequency range. Hence, weight drop is not a preferred source as its performance deteriorates with frequency.

These test line investigations made it possible to come to several important conclusions while planning 3D surveys. First, in a climate such as that in the region of the Otway Project, it is necessary to acquire data at the same time of the year to reduce the impact of the variability of the near-surface conditions. Second, the most powerful source available at the time of survey should be used, even if this means that it is necessary to change the source type of the monitoring survey. In the case of the reflector depths at the Otway site, neither impact nor vibrating sources would produce a dominant frequency higher than 50 Hz (Figure 10.15).

10.5 Time-lapse surface seismic monitoring for Stage 1

The objective of the seismic monitoring programme within Stage 1 of the Otway Project was two-fold: to attempt to detect changes in the Waarre C reservoir due to the injection of the CO_2/CH_4 mixture, and to ascertain if there was any detectable leakage of the gas into the overlying strata. Results of the modelling of the time-lapse seismic signal discussed in the previous sections indicated that the task of detection of the CO_2/CH_4 injected into the Waarre C depleted gas reservoir was extremely challenging and for there to be any possibility of detecting stored CO_2 in the reservoir, exceptionally high quality repeatable time-lapse seismic surveys would be necessary. The highest possible repeatability would also be required to maximise the prospect of small amounts of CO_2/CH_4 in the overburden strata being detectable. To this end, a comprehensive seismic monitoring programme was conducted, combining repeated 3D surface seismic surveys and repeated borehole seismic (VSP) surveys (discussed in the next section). This combination was designed to optimise broad spatial coverage of 3D seismic with higher repeatability and S/N ratio from the borehole seismic methods.

10.5.1 Data acquisition

In 2008, the first baseline 3D seismic data for Stage 1 was acquired. The survey area was 1.6 km × 1.9 km. The subsequent surveys for Otway Stage 1 were acquired at the beginning of 2009 and 2010 with the same acquisition geometry but including additional shooting sequences, with the intention of them being utilised to form the baseline surveys for Otway Stage 2. The parameters of all the four surveys are listed in Table 10.2.

The main goal of TL seismic at the Otway site was to determine whether it was possible or not to image CO_2 stored in the Waarre reservoir, using time-lapse effects between surveys. However, even small changes in near-surface conditions, such as changes in the water table, ground noise, and receivers and source positioning, can affect data differencing results and the repeatability of successive surveys. While positioning errors were practically

Table 10.2: Acquisition parameters for the surface seismic and 3D VSP surveys.

	Survey I	Survey II	Survey III
Date	VSP: December 2007 Surface seismic: December 2007 – January 2008	January 2009	January 2010
Source	Weight drop, Hurricane 10, 750 kg, free fall from 1.2 m, 4 stacks	IVI minivib, Linear sweep 12 s, 10–150 Hz, 4 sweeps per source point	
Number of source lines	VSP: 15 Surface seismic: 29	29	
Source line source spacing	200 m	100 m	
Source spacing	20 m	20 m	
Number of source points	VSP: 1139 Surface seismic: 2181	VSP: 0 Surface seismic: 2223	2223
Receiver parameters: VSP			
VSP tool	3C Schlumberger VSI, 10 level	N/A	3C Schlumberger VSI, 8 level
Acquisition interval	1485–1620 m		1500–1605 m
Receiver spacing along borehole	15 m		15 m
Receiver parameters: surface seismic			
Recorder	Seistronix, EX-6	Seistronix, EX-6	
Receiver type	10 Hz, single geophones	10 Hz, single geophones	
Number of active channels	440	873	
Number of receiver lines	10	10	
Receiver line spacing	100 m	100 m	
Receiver point spacing	10 m	10 m	
Recording pattern	Orthogonal cross-spread pattern, odd source lines were recorded by the first 5 receiver lines, even source lines by 6–10 receiver lines	Orthogonal cross-spread pattern, all	

non-existent for the last three surveys, the source type was changed from a low energy weight drop (WD) source in 2008 to the more powerful IVI minivib (MV) source used in 2009–10 (Figure 10.16). As described previously, variations in the S/N ratio as a function of the source strength relative to the background noise level are crucial. The source type is less important. Given that the time-lapse set for Stage 1 was also intended as a baseline for the Stage 2 of the Otway Project, a MV was chosen as the preferred source because it provided data with a S/N ratio that was twice that produced with WD.

The Otway 3D survey undertaken in 2008 was acquired with two swaths: I: receiver lines (RL) 1–5 (435 channels), and II: receiver lines 6–10 (437 channels). Odd-numbered source lines (SL) 1–27 and 28 were acquired with RL 1–5. Even-numbered source lines SL 2–24 and 29 were

acquired with RL 6–10. Receivers were fixed for the duration of the survey. The acquisition geometry was designed to yield 100 fold data with a maximum offset of 2150 m. For the subsequent 2 years, data fold was doubled by using all RL for each source point (SP). The acquisition parameters are given in Table 10.2. The acquisition geometry is shown in Figure 10.17.

10.5.2 Data processing

A critical aspect of the study was data processing and equalisation of the 2008 baseline survey and the 2009 and 2010 monitoring surveys. False time-lapse anomalies could easily be created by various factors such as source-generated noise or residual static shifts. The difference between impact and vibrating sources can be striking when

Figure 10.16: Seismic signal sources used in Otway project: IVI minivib.

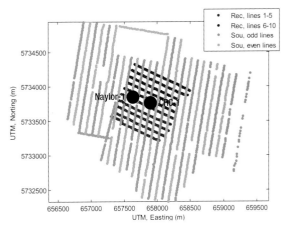

Figure 10.17: Surface seismic acquisition geometry.

it comes to the generation of coherent noise (Figure 10.18) and removal of such noise was an important task.

Despite the vastly different spectral contents and S/N ratios between the surveys, processing and cross-equalisation of the three vintage sets (2008, 2009, 2010) was attempted by adopting somewhat different approaches to time-lapse processing of seismic data.

Seismic data processing was done for the reduced geometry layout of year 2008. Both 2009 and 2010 volumes were reduced to match the baseline survey. Processing was then repeated for the full geometry layout for the years 2009 and 2010. All seismic volumes from 2008 to 2010 were processed simultaneously, using one processing flow which was focused on preserving reflection amplitudes. Processing of the data was performed with ProMAX (Landmark, Halliburton) and RadExPro (DECO Geophysical) processing software systems and also MatLab.

The result of seismic data processing was two sets of volumes:

› reduced geometry layout

   – three 3D migrated cubes in time domain (2008, 2009 and 2010) following standard processing

   – three cross-equalised 3D migrated cubes in time domain (2008, 2009 and 2010)

   – three 3D cubes of difference (2008–09, 2009–10, 2008–10)

   – three 3D cubes of NRMS parameters (2008–09, 2009–10, 2008–10)

   – elevation statics for sources and receivers

   – residual statics for sources and receivers

   – RMS velocities

   – interval velocities (single function) for migration

› full geometry layout

   – two 3D migrated cubes in time domain (2009 and 2010)

   – two cross-equalised 3D migrated cubes in time domain (2009 and 2010)

   – one 3D cube of difference (2009–10)

   – one 3D cube of NRMS parameters (2009–10)

   – elevation statics for sources and receivers (the same as in the "reduced" case)

   – residual statics for sources and receivers

   – RMS velocities (the same as in "reduced" case)

   – interval velocities (single function) for migration (the same as in the "reduced" case).

The processing sequence and parameters of the processing steps are summarised in Table 10.3.

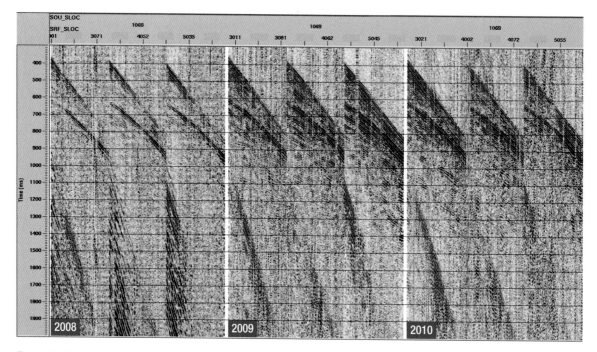

Figure 10.18: Raw data example, using the same common shot gather from the 2008, 2009 and 2010 surveys.

10.5.3   Data cross-equalisation

4D data analysis usually employs data subtraction in order to detect possible changes in seismic signatures. Before the subtraction, data need to be brought to the same amplitude, frequency and phase content. This is done by performing so-called cross-equalisation. The cross-equalisation post-processing sequence is given in Table 10.4.

The first step following the data input in RadExPro software was ensemble equalisation, which was performed in a 200–2000 ms time window. Different soil conditions (mostly relating to water saturation) and seismic data required time cross-equalisation between these datasets. Most of these time differences were in the long wavelength region. The post-stack static shifts were calculated and applied to match the reflector geometry of the different surveys. The 2009 MV dataset was selected as a reference dataset because of its relatively high quality. Two other surveys (2008 and 2010) were processed to match the reference dataset.

To calculate these time shifts, cross-correlation was used between corresponding traces from pairs of cubes (2008 – 2009, 2010 – 2009 and 2010 – 2008). Calculated static shifts were applied to the 2008 and 2010 volumes in order to bring them to the 2009 cube (reduced geometry) and to the 2010 volume to match 2009 data (full geometry). The cross-correlation function was calculated in the 430–1400 ms window and a correlation coefficient threshold of 0.1 was used; this meant that all traces that display lower correlation were excluded from further analysis. The average time shift between the two cubes was around 2 ms for the 2008 – 2009 pair and even less for the 2010 – 2009 pair.

After time equalisation, a matching filter needed to be calculated and applied in order to bring the data to a common phase and amplitude spectra. The filter was applied to 2009 and 2010 volumes, in order to match the 2008 volume. The shaping filters were calculated with the following parameters: filter type – Wiener with length of 100 samples, white noise level – 0.01, correlation coefficient threshold – 0.8. The filter design window was 400–1400 ms, which excluded the target horizon from the analysis.

Table 10.3: Surface seismic data processing sequence.

Procedure	Parameters
Data input	SEG2 data input
Geometry assignment	Applied from field generated SPS files
Correlation with sweep signal	Linear sweep 10–150 Hz
Binning	Bin size 10 m × 10 m
Trace dc removal	Type of central value: Mean
Trace editing	Kill bad traces
Elevation statics	Final datum elevation: 30 m; Replacement Velocity: 1800 m/s
Bandpass filtering	7-10-150-160 Hz
Trace muting	Top muting
Spike and noise burst edit	Spike detection threshold value: 10; operator length: 100
Automatic gain control	500 ms, applied before deconvolution and removed after
Spiking deconvolution	Minimum phase (2008), zero-phase (2009–10); Decon operator length: 70, operator "white noise" level: 0.1
Bandpass filter	7-10-140-150 Hz
Interactive velocity analysis	2 iterations; final grid: 100 m × 100 m, 30%–NMO muting
Residual static correction	2 iteration (max power autostatics)
Automatic gain control	500 ms, applied before radon filter and removed after
Radon filtering	Number of P-values: 500 modelled noise subtraction; applied in cone window
Automatic gain control and NMO	Applied before FX-deconvolution and removed afterwards; (AGC window: 500 ms, NMO muting: 90%),
FX-deconvolution	Wiener Levinson filter, 4–500 Hz
Pseudo AGC	
True amplitude recovery	Spherical divergence correction as 1/(time*vel**2)
CDP stacking	Method of summing: mean power scalar for stack normalisation 0.5
TV whitening	5-15-90-180 Hz
Pad 3D stack volume	In-lines 1–160, x-lines 1–195
FXY-deconvolution	Number of in-lines in filter: 7; number of x-lines in filter: 7; frequencies: 4–180 Hz
Migration	Explicit FD 3D time migration

Filters were computed across a large number of pairs of traces (threshold of 0.8). The final matching filter was then obtained by stacking all individually computed filters into a single filter, which was then used to cross-equalise phase and amplitude spectra between pairs of surveys.

After data filtering, a second iteration of time shifts was calculated and applied. Calculated static shifts were applied to the 2008 and 2010 volumes in order to bring them to the 2009 cube (reduced geometry) and to the 2010 volume to match the 2009 data (full geometry).

In order to balance seismic volumes in amplitudes, the time-variant gain (TVG) function was calculated for all cubes (amplitude characters are very similar for the 2009 and 2010 volumes but not for the 2008 volume). Finally the TVG function was applied only to the 2008 data to bring their amplitudes to the level of those from the 2009 and 2010 volumes.

10.5.4 Results

Figure 10.19 shows in-line cross-sections after cross-equalisation extracted from the three volumes obtained with the reduced geometry layout (2008, 2009 and 2010). Figure 10.20 shows 2009 and 2010 sections obtained with full geometry. Figure 10.19 demonstrates that data sets

Table 10.4: Data equalisation sequence.

Procedure	Parameters
Data input	SEGY data input in RadExPro software
Data balancing	RMS ensemble equalisation in 200–2000 ms time window. Applied to all volumes
Time shifts I	Applied to: 2008 and 2010 → 2009 (reduced geometry), 2010 → 2009 (full geometry). No smoothing
Shaping filter	Applied to: 2009 and 2010 → 2008 (reduced geometry), 2010 → 2009 (full geometry)
Time shifts II	Applied to: 2008 and 2010 → 2009 (reduced geometry), 2010 → 2009 (full geometry). No smoothing
Time variant gain function	Applied only to 2008

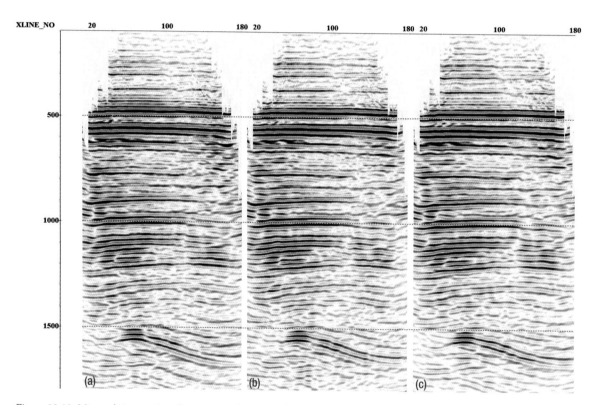

Figure 10.19: Migrated time section after cross-equalisation. In-line 87, reduced geometry: 2008 (a), 2009 (b) and 2010 (c).

with vastly different noise patterns (Figure 10.18) could be matched successfully. These cross-equalised volumes were used to investigate time-lapse effects for assurance monitoring during Otway Stage 1.

Figure 10.21 shows NRMS values computed in a 60 ms window between (a) the 2010 and 2008 surveys, reduced geometry; (b) between the 2010 and 2009 surveys, full geometry; and (c) between 2010 and 2008 3D VSP surveys. Scale 0–200% corresponds to the possible range of NRMS values, 0: signals match perfectly, 200%: signals anti-correlate. At the reservoir level (1.5 s), the NRMS is around 20% which is exceptionally good for land time-lapse seismic.

Figure 10.22 shows the results of preliminary processing performed after the completion of the first monitoring survey in 2009. This initial processing was interpreted to show a noticeable difference between the two surveys (Urosevic et al. 2011). However, acquisition and processing

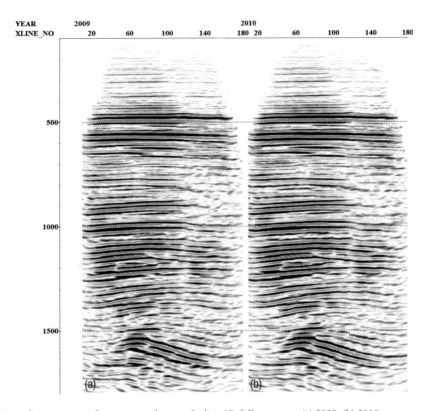

Figure 10.20: Migrated time section after cross-equalisation. In-line 87, full geometry: (a) 2009, (b) 2010.

of the second monitoring survey in 2010 allowed the simultaneous reprocessing of the three volumes: baseline (2008) and two monitoring surveys (2009 and 2010). The results of this final processing along in-line 87, reproduced in Figure 10.23, do not show a clear time-lapse signal.

The same is true for all horizons. Figure 10.24 shows an in-line section extracted from (a) the 3D migrated cube; (b) difference sections between 2009 and 2008; and (c) between 2010 and 2008 based on the final processing of the three volumes. For comparison, Figure 10.24(d) shows the modelling result for a simulated scenario of leakage of 7000 t of gas mixture. This comparison suggests that no significant amount of CO_2 has escaped from the Waarre C reservoir and migrated into overlying strata.

Comparison between the recorded and modelled time-lapse signal for the Waarre C reservoir level was shown in Figure 10.4. As discussed in the modelling section, even with the best possible processing, the time-lapse signal

from the injection of gas mixture into the depleted gas reservoir was below the level of time-lapse noise.

10.6 Downhole seismic methods for Stage 1

A downhole seismic monitoring programme was designed to complement the surface seismic programme (see also Chapter 11). The main advantage of downhole methods is the much higher S/N ratio due to the placement of the seismic receivers in the borehole (where the environment is extremely quiet) and shorter ray paths compared to ray paths of reflected waves in surface seismic data. The result is that in general, downhole seismic methods can be expected to provide a cleaner image and lower levels of noise.

The downhole seismic programme consisted of time-lapse 3D vertical seismic profiling (VSP) in the CRC-1 well

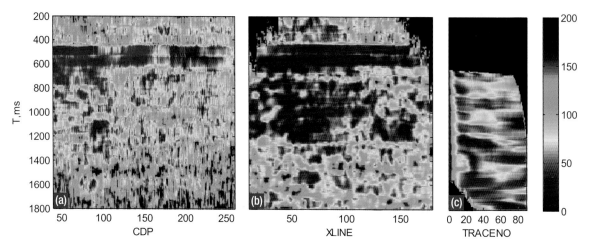

Figure 10.21: NRMS differences sections along an in-line direction between 2010 and 2008 surveys – reduced geometry (a); between 2010 and 2009 surveys – full geometry (b) and between 2010 and 2008 3D VSP surveys (c).

and several zero-offset and offset VSP surveys in both the Naylor-1 and CRC-1 wells.

3D VSP is a seismic method where a receiver array consisting of several seismic receivers (usually three-component geophones) is placed at a fixed position in a borehole above, but relatively close, to the target horizon, while the shot points are distributed on the free surface over a wide area around the borehole. A common objective of 3D VSP is to provide an image of the target horizon based on reflected waves travelling from various shot points and reflected from the target horizon and then detected by the borehole receiver array. Due to the large number of shot points required, it was convenient to acquire 3D VSP simultaneously with a 3D surface seismic survey, such that waves from the same shots were recorded simultaneously by a surface 3D array and a downhole receiver array.

Offset VSP is a downhole seismic method that employs multiple shooting of the seismic source at a fixed shot point on the surface, with the receiver array placed at various positions in the borehole. Offset refers to the distance of the shot point from the wellhead. Zero-offset VSP (ZVSP) is a variant of the offset VSP method where the offset is much smaller than the depth to the target horizon, and therefore wave propagation from the source to the receiver array can be considered to be along the borehole. The primary objectives of zero-offset VSP

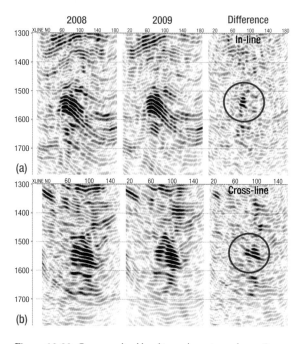

Figure 10.22: Cross-equalised baseline and monitor cubes at Otway and their difference, initial 2009 processing. (a) In-lines and their difference (b) Cross-lines and their difference. Anomalous difference at Waarre C is circled.

and offset VSP are to analyse the characteristics of the subsurface in the immediate vicinity of the borehole, using the transmitted (direct) waves propagating along the borehole and to provide information for seismic-to-well calibration.

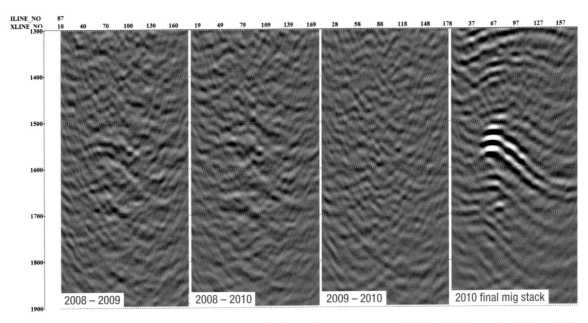

Figure 10.23: Difference sections along in-line 87 between baseline and monitoring surveys. Results of final processing is shown in the last column.

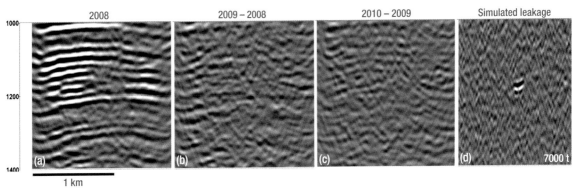

Figure 10.24: In-line sections extracted from 3D migrated cube (a), difference sections between 2009 and 2008 (b) and between 2010 and 2009 (c), and the modelling result for simulated Stage 1 scenario of leakage of 7000 t of gas mixture (d).

10.6.1 Time-lapse 3D VSP data acquisition

The time-lapse 3D VSP consisted of a pre-injection baseline survey shot simultaneously with the December 2007–January 2008 surface seismic survey and a monitoring survey, shot simultaneously with the January 2010 surface seismic survey. The baseline survey was acquired with a 10-level, three-component Schlumberger VSI downhole tool and a weight drop source (modified Hurricane F10 concrete breaker). The repeat 3D VSP was acquired in the same well with an eight-level VSI downhole tool and IVI minivib as the seismic energy source in January 2010. The objectives of the repeat survey were to provide a high resolution subsurface seismic image that could be used for comparison with potential subsequent monitoring surveys is to obtain an image that could be compared to a baseline survey for time-lapse analysis.

The baseline (2007) survey included 1148 source positions arranged along 14 source lines. The repeat (2010) survey consisted of 2224 source positions along 28 source lines

Table 10.5: Acquisition parameters of zero-offset VSP surveys in CRC-1 well.

	Survey I	Survey II
Date	December 2007	January 2010
VSP downhole tool	3C Schlumberger VSI tool	3C Schlumberger VSI tool
Source	Weight drop, concrete breakers Hurricane Force 9 and 10, operational weight 720 kg and 1425 kg respectively	Vibroseis, IVI minivib, at 9000 lb, 12.5 s sweep 10–150 Hz
Acquisition interval	457–2211 m	517–1900 m
Receiver spacing along borehole	15 m (517–1575 m) 7.5 m (1575–2211 m)	15 m (517–1455 m) 7.5 m (1455–1900 m)
Zero-offset VSP shot point location	Azimuth 105.5°, offset 89.7 m	Azimuth 105.5°, offset 89.7 m
Offset VSP shot point location	Azimuth 15.5°, offset 1024.0 m	Azimuth 15.5°, offset 1024.0 m

(Figure 10.25). Hence the repeat survey deployed almost twice as many sources as the baseline. Acquisition parameters of VSP and surface seismic surveys are given in Table 10.5.

As mentioned above, the 3D VSP surveys were acquired at the same time as the corresponding 3D surface seismic surveys. Each of these baseline and monitoring surveys took approximately 14 working days to acquire. The 3D surface seismic survey took an additional four or five days to deploy geophones and cables and retrieve them after the survey.

To operate the EX-6 seismic recording system, 1 hour of each working day one was spent distributing batteries before commencing the shooting, and 30 minutes to retrieve them. This meant that the surface seismic acquisition took 10–15% more time than the 3D VSP. Another important difference between the 3D surface seismic and 3D VSP acquisition was the influence of various types of surface noise. Borehole seismic acquisition is not very sensitive to surface noise sources (such as wind, rain, trucks and farming activities). Several delays in surface seismic operations were caused by strong wind or rain and local traffic. Even when weather conditions were less severe and allowed the survey to proceed, they still affected the data quality, whereas data quality for both 3D VSP surveys was generally good.

10.6.2   3D VSP data processing

The 3D VSP data processing (using the Z component only) was performed by Schlumberger (Campbell and Shujaat

2010). The same processing approach as for the 3D surface seismic was employed: the baseline and monitoring surveys were processed together, using only those source points that were present in both surveys (reduced geometry). The monitoring survey was also processed separately using all the source points (full geometry).

The processing was split into two phases. In the first phase, the data from baseline and repeat surveys were processed to obtain the best migrated images from the respective datasets. In the second phase, the data from both surveys were processed to obtain images that could be used for time-lapse analysis between the two surveys.

The processing flow is summarised in Figure 10.26. A 1D model was constructed from well logs. The velocities and anisotropy were calibrated with the 3D VSP times. By ray tracing through the model, it was possible to estimate tomographic residual statics, which were then applied to the deconvolved data prior to migration. The calibrated anisotropic 1D model was used in the migration to produce several image volumes.

10.6.3   3D VSP data cross-equalisation

As in the case of the surface seismic data, the time-lapse analysis of the baseline and monitoring 3D VSP surveys required the data to be cross-equalised between the two surveys. Despite the fact that the 3D VSP and surface seismic surveys used the same shot points, the 3D VSP surveys were cross-equalised independently from the surface seismic data for the following reasons:

> Different illumination of the Waarre C horizons using downhole and surface receivers and differences in the velocity model used for imaging resulted in noticeable differences in the topography of the reflectors.

> Incident angles of the reflected waves that contributed to the VSP images are different from those which contributed to the surface seismic image. This meant that differences could be expected between the two images due to dependence of the reflection coefficient on the incidence angle.

> Amplitude correction approaches utilised in VSP and surface seismic processing were also different.

The cross-equalisation workflow consisted of the following steps:

> Post-stack static shift analysis was performed to match the 2007 and 2010 data. Due to the limited vertical extent of the image, the entire trace length was selected to compute cross-correlations between corresponding pairs of traces. In principle, it was expected to see virtually no time shifts between the surveys after the model-based static correction had been applied during the processing. Indeed, the standard deviation of the shifts was ~0.55 ms.

> Single matching filters were derived and applied to match 2010 to 2007 data. For all pairs of traces with

a correlation coefficient of 0.8 and greater, a 200 ms matching Wiener filter with white noise factor of 0.01 was computed. These filters were averaged to achieve a single filter, which was applied to the 2010 datasets to match the amplitude and phase spectra of the wavelet of the 2007 data. The resulting filter turned out to be almost perfectly zero-phase, which meant that the deconvolution applied during the processing to both 2007 and 2010 data produced wavelets with very similar phase despite the use of different source types. Single scalar (one per volume) amplitude balancing was applied to all volumes using the RMS absolute amplitude computed in a 1450–1850 ms time window.

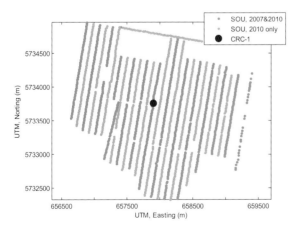

Figure 10.25: 4D VSP source point locations. Green points represent source locations for both the baseline and monitoring surveys, blue lines for the monitoring survey only.

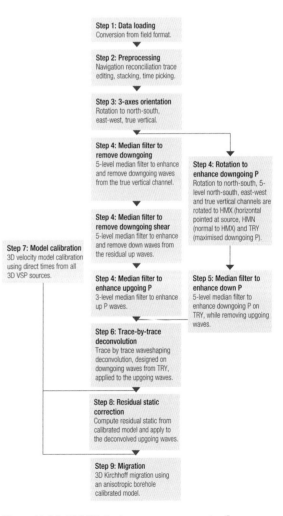

Figure 10.26: 3D VSP single component processing flow.

The results of the processing of the time-lapse 3D VSP datasets and their difference are shown in Figure 10.27(a). The corresponding results after the cross-equalisation are presented in Figure 10.27(b). It is clear that cross-equalisation has significantly improved the repeatability of the time-lapse VSP data.

Unfortunately the grid used for the migration of 3D VSP data was chosen without considering orientation of the surface seismic binning grid or orientation of the source lines. In order to use the same picked horizons or do trace-by-trace comparison with surface seismic, the 3D VSP data was interpolated onto the 3D surface seismic binning grid.

Different velocities used for the migration of surface seismic and VSP data, together with a different migration aperture, resulted in a small difference in the observed shape and the dip of the Waarre C horizon (Figure 10.28). This made further cross-equalisation between surface seismic and VSP data impossible. However, nearly all events identified in

the 3D VSP data can be traced to the 3D surface seismic and vice versa (Figure 10.29).

10.6.4 Time-lapse analysis of 3D VSP data

In order to analyse repeatability of 3D VSP data, an NRMS volume was computed using cross-equalised pairs of surveys. The same parameters for the computations (60 ms sliding window) were used as in surface seismic (Pevzner et al. 2010), so that the results could be compared across different types of surveys.

Figure 10.30 compares the achieved repeatability of both 4D surface seismic and 4D VSP surveys. Time-lapse VSP surveys demonstrated an outstanding level of repeatability with a median NRMS value of 16%.

Seismic amplitudes at the Waarre C horizon measured in 2007 and 2010 and the difference 3D VSP volume are

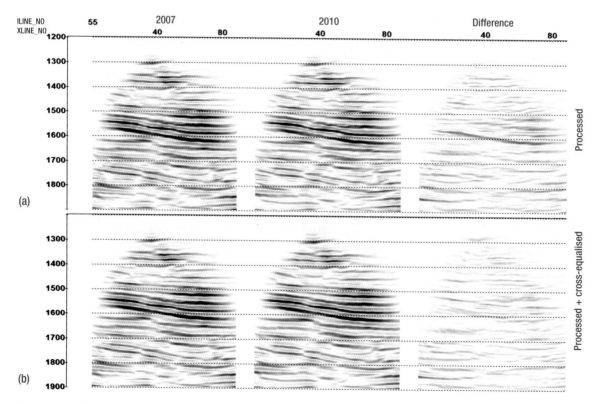

Figure 10.27: 4D VSP processing result before (a) and after (b) cross-equalisation.

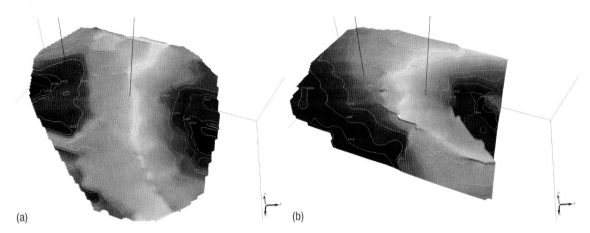

(a)   (b)

Figure 10.28: Waarre C horizon picked on 3D surface seismic (a) and 3D VSP (0° central dip) data (b).

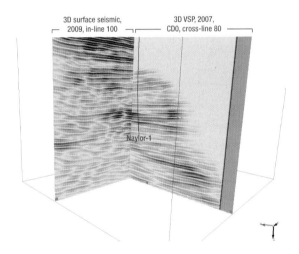

Figure 10.29: Comparison of 3D surface and 3D VSP data.

shown in Figure 10.31. It is evident that the raw amplitude has a very strong dependence on the distance from the CRC-1 well and is not sensitive to other factors. This can be explained by a range of factors including the influence of the migration aperture (varying with source offset) and features of the particular VSP migration software, directivity of the receiver (these volumes were produced from the Z component only), amplitude decay compensation and probably AVO effects (image points located further away from the well have disproportionate influence from bigger incidence angles). These factors were not compensated during the processing. Nevertheless, the amplitude of the difference volume does not show such a radial pattern

and all amplitude variations follow geological features of the reservoir. It is important to note that these amplitude variations do not exceed the background repeatability level, i.e. the amplitude differences observed elsewhere in the difference volume.

Finally, the results of the time-lapse analysis were compared with the numerically modelled time-lapse signal from the CO_2/CH_4 injection. Figure 10.32 shows the maps of NRMS differences at the Waarre C horizon level obtained from surface seismic data, from 3D VSP and from numerical modelling. As discussed earlier, the repeatability of the surface seismic data was too low to detect the time-lapse signal. For the time-lapse 3D VSP data, the achieved NRMS difference was also below the level required to confidently detect the signal, but at least of similar order of magnitude. Furthermore, the predicted time-lapse signal for Model 2 was somewhat higher and suggests that the VSP data may have the level of repeatability comparable to the level required to detect the signal. However, quantitative interpretation of the time-lapse 3D VSP data is very challenging due to variable fold and offset distribution along the reservoir.

10.6.5 Zero-offset and offset VSP

As part of the seismic programme of the Otway Project, a number of ZVSP and offset VSP surveys were acquired in the Naylor-1, CRC-1 and CRC-2 wells. Here the focus is on the ZVSP and offset VSP surveys acquired in the

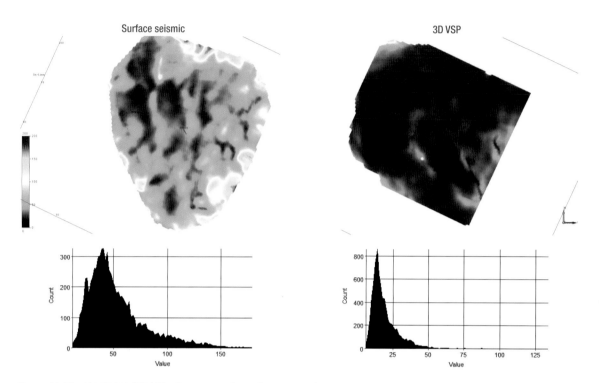

Figure 10.30: 2007/2010, NRMS value computed over the Waarre C horizon in 60 ms window for surface seismic (left) and 3D VSP (right) surveys, histograms of NRMS values are shown below maps.

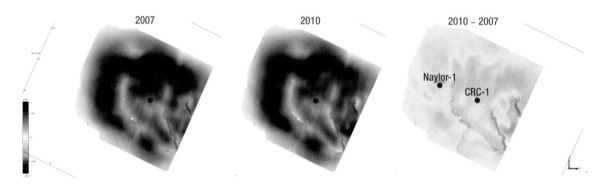

Figure 10.31: Amplitude on Waarre C horizon, central dip 0°.

CRC-1 well in 2007 (pre-injection) and 2010 (post-injection), as these repeated surveys make it possible to assess repeatability of the ZVSP data.

Acquisition parameters of both VSP surveys are presented in Table 10.5. As with the corresponding surface seismic and 3D VSP surveys, the 2007 ZVSP was acquired with a weight drop source, and the 2010 survey with a more powerful seismic vibrator source (IVI minivib).

Examples of Z component VSP seismograms are shown in the Figure 10.33, where spherical divergence correction and ensemble equalisation had been applied. Only repeated positions of receivers are presented. It can be seen that the IVI minivib (2010) generated a much wider spectrum than the weight drop (2007). The character of the wavefield generated by both sources was found to be very similar.

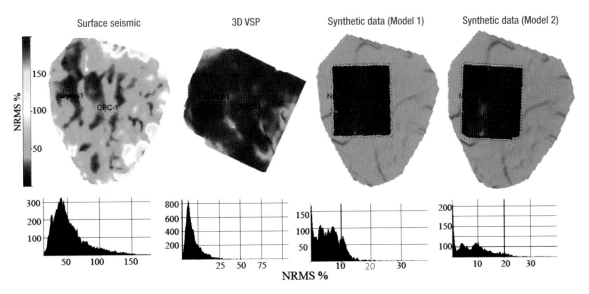

Figure 10.32: 2008/2010, NRMS value computed over the Waarre C horizon 60 ms window for surface seismic (left) and 3D VSP (middle) surveys and the predicted NRMS difference for 2008/2010 for Model 1 and 2 (right); histograms of NRMS values are shown below maps.

Transit times for the two VSP surveys show a reasonably good agreement. Depth drift between the two surveys was within 1 ms. This drift could probably be explained by the influence of attenuation on the phase of the wavelet obtained after correlation of viboseis data, given that transit times were picked along the principal extremum of the wavelet on MV data and along first sign change on weight drop data. However, this showed that if a precise transit time analysis is required, a change of the source type between baseline and monitoring survey can introduce unwanted discrepancies.

To analyse the repeatability of reflected wave amplitudes, it was necessary to equalise the wavelet on both vibroseis and weight drop data. An important feature of VSP data is that the emitted wavelet is known due to recording of both direct and reflected waves. For the purposes of time-lapse data cross-equalisation, a shaping filter was computed for each pair of corresponding traces using the descending wavefield to match 2007 data to 2010 data, and applied to the result of the wave field separation. This procedure automatically takes into account possible instability of source properties. A fragment of the seismogram containing the reflection from the Waarre C horizon after the cross-equalisation is presented in Figure 10.34. In each pair

of traces, the first trace represents 2007 data and the second, 2010 data. The same process was applied to the offset VSP data.

It is possible to conclude that repeatability of VSP data is quite high. The NRMS difference along strong horizons is below ~20–30% for both zero-offset and offset VSP data; this is a very good value for pre-stack data. Another important observation that can be made from repeated zero-offset VSP obtained in the CRC-1 borehole is that no significant changes in reflectivity caused by CO_2 injection could be observed at the location of the injection well. This is consistent with the predictions of the forward modelling and the analysis of 3D surface seismic data.

10.6.6   Seismic anisotropy from VSP

VSP data also provide an opportunity to analyse azimuthal anisotropy (variation of seismic velocity with azimuth of wave propagation). Changes in pore pressure or thickness of the reservoir could change stress field parameters in the overburden and observation of changes in shear wave anisotropy could be used to monitor it (Olofsson et al. 2003). Existence of strong shear wave anisotropy in the Otway Basin was first reported by Turner and Hearn

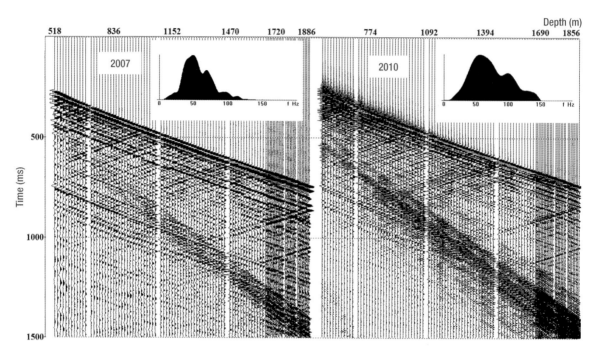

Figure 10.33: Zero-offset data example, Z component seismogram and amplitude spectrum of the direct wave.

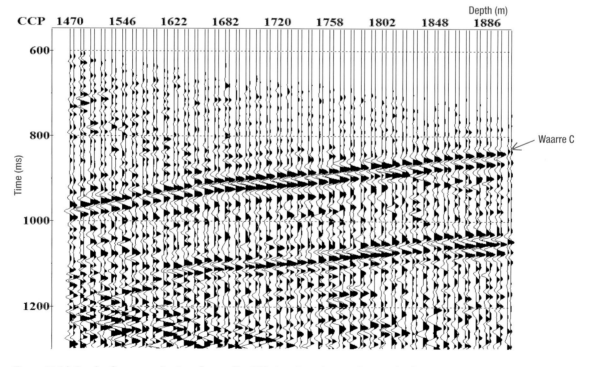

Figure 10.34: Result of cross-equalisation of zero-offset VSP data. In each pair of traces, the first trace represents 2007 data and the second, year 2010 data.

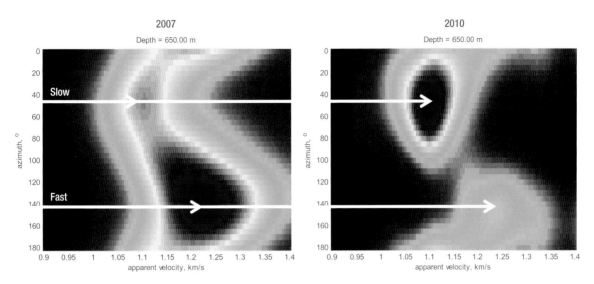

Figure 10.35: Results of multi-component velocity analysis of repeated zero-offset VSP data acquired in CRC-1 borehole for of depth level of 650 m.

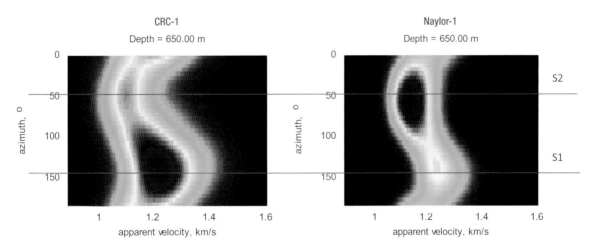

Figure 10.36: Examples of azimuthal velocity spectra for the CRC-1 and Naylor-1 boreholes.

(1995). Recently Pevzner et al. (2009) proposed an approach for multi-component velocity analysis of 3C zero-offset VSP data. The method was applied to the data from the Naylor-1 and CRC-1 wells. The results showed similar anisotropy magnitude for both boreholes, while the fast wave azimuth corresponded to the orientation of the regional stress field.

The results of multi-component velocity analysis of repeated zero-offset VSP data acquired in the CRC-1 borehole for one depth level are presented in Figure 10.35. It can be seen that the azimuth of polarisation of fast (~140°) and slow (~50°) shear waves as well as corresponding velocities remain unchanged. Figure 10.36 shows the results of the same analysis for the CRC-1 and Naylor-1 wells. Again, the anisotropy estimates are consistent between the neighbouring wells. Independently, azimuthal anisotropy can be estimated from 3D VSP data.

10.7 Laboratory studies of CO_2 acoustic response as an adjunct to field studies

10.7.1 The value of laboratory studies

Laboratory measurements on core samples provide petrophysical and geomechanical properties of a reservoir and can be used for the calibration of seismic data. Quantitative understanding of the acoustic response of sandstones from the injection site following complete or partial saturation with supercritical CO_2 ($scCO_2$) is important for time-lapse seismic imaging of CO_2 plume migration. As the injection of $scCO_2$ can potentially cause some alteration or damage of the constituent rocks, acoustic responses before and after the CO_2 injection were tested on Otway core samples.

Wang and Nur (1989) pioneered laboratory experiments involving CO_2 injection. They injected CO_2 in hexadecane-saturated sandstones and measured the acoustic response. Xue and Ohsumi (2004) studied CO_2 injection effects on the P-wave velocity and deformation of the water-saturated Tako sandstone. Shi et al. (2007) used acoustic P-wave tomography for monitoring and quantification of $scCO_2$ saturation during injection into the brine-saturated Tako sandstone. Lei and Xue (2009) injected gaseous and $scCO_2$ into a water-saturated sandstone under controlled pressure and temperature conditions and monitored the dependency of P-wave velocity and attenuation on saturation. Shi et al. (2011) conducted a comprehensive study of $scCO_2$ injection into brine-saturated sandstone using computer tomography methods. The long-term effect of brine/$scCO_2$ on the mechanical properties of sandstones was studied by Zemke et al. (2010). Recently, Siggins et al. (2010) performed the first measurements on samples from the CRC-1 well. The acoustic responses were measured during gaseous and liquid CO_2 injection into brine-saturated samples.

In the Otway Project, the effects of $scCO_2$ on the acoustic properties were investigated on five rock samples extracted from the CRC-2 well (drilled post-Stage 1). The study was designed to answer three main questions: (1) what are the elastic properties of the host sandstone and can the standard technique of Gassmann's fluid substitution be applied to calculate the elastic properties of saturated sandstone using standard laboratory measurements on the dry sample; (2) how will CO_2 injection change acoustic velocities; and (3) will it cause any damage in the host rock?

To answer these questions CO_2 was injected into either dry or brine-saturated samples, to obtain the acoustic velocities of the samples for a range of CO_2 saturations from 0 to 100%.

10.7.2 Sample description

Measurements were taken from a total of five sandstone core plug samples from the CRC-2 well. These core samples were extracted from the proposed injection interval (from 1442 to 1529 m) in the Paaratte Formation. The formation itself comprises a sequence of deltaic and shallow marine sediments which contain a succession of interbedded fine-to-coarse-grained quartz sandstones (potential reservoirs) and carbonaceous mudstones (baffles or seals). Five samples representing different facies were taken from the centre of the reservoir in order to investigate the effects of geological heterogeneity. One sample (1442.1 m) was cut parallel to the bedding plane; the other samples were cut normal to the bedding plane. The helium porosities and permeabilities of the samples measured by an automated porosimeter (Coretest Co.) and permeameter (AP-608) are shown in Table 10.6 along with other petrophysical properties. The porosities vary from 14 to 25% and the permeabilities are widely scattered, from 0.2 to 10,000 mD.

Table 10.6: Petrophysical properties of CRC-2 samples used for determining acoustic response.

Core Sample	Cutting Direction**	Density, kg/m³	Porosity, %	Permeability, mD
1442.1	H	1809	26*	10,000
1500.83	V	1806	24.69	5500
1509.6	V	1947	25.33	120
1513.82	V	2142	16.14	17
1526.9	V	2178	14.23	0.19

* Estimation from log data.

** H and V mean that the sample is cut parallel or normal to bedding, respectively.

Sample 1442.1 m is a homogeneous sandstone with only a few laminations parallel to the bedding plane. This porous and permeable sandstone contains well-sorted, fine to medium, rounded grains of predominately quartz with minor feldspar; the matrix comprises mica, with kaolinite and chlorite clay as a weak cement. Sample 1500.83 m is also a fine-grained and well-sorted quartz sandstone, but with the occasional mottled structure resulting from bioturbation (sand filled burrows) and some small quartz pebbles. It exhibits some gradational bedding and fine cross-bedded laminations. Sample 1509.6 m contains distinct wavy carbonaceous laminations within a medium to very coarse grained cross-stratified sandstone. Mica-rich clays and coaly flakes are common within the matrix. Sample 1513.82 m is a cemented sandstone; the original fabric is a very fine- to fine-grained clean quartz sandstone, but dolomite pervades throughout the pore space, coating grains and reducing the porosity. Sample 1526.9 m is from the Skull Creek Mudstone, which is below the Paaratte reservoir; it is poorly laminated to apparently structureless and contains abundant centimetre-scale, rounded to sub-rounded pyrite nodules and intense bioturbation.

10.7.3 Experimental protocol

The experimental procedure is described in detail below. The first two steps were the same for all the samples. During the third step, three samples (1509.6 m, 1513.82 m, 1526.9 m) were saturated with brine while the other two (1442.1, 1500.83 m) were saturated with CO_2. Subsequent steps were only applied to the CO_2-saturated samples.

1. Ultrasonic compressional and shear velocities of a dry sample were measured at room temperature at confining pressures of up to 60 MPa in steps of 4 MPa.

2. The experimental system was heated up to 45°C and the ultrasonic measurements of P- and S-wave velocities on the dry sample were repeated.

3. A confining pressure of 12 MPa was applied to the sample and it was flooded with brine with an injection pressure of 6 and 9 MPa. The ultrasonic velocities were measured again on the fully saturated sample at confining pressures from 10 to 70 MPa in steps of 4 MPa. The salinity of the brine used was 1500 ppm

of 50% NaCl and 50% KCl, which is typical of the field conditions. During the third step, two samples (1442.1 m, 1500.83 m) were flooded with CO_2 at 6 MPa instead of the brine. The ultrasonic velocities were then measured at the same conditions as for the samples saturated with brine.

4. CO_2 was continuously injected into the sample until pore pressure reached a critical point of 9.3 MPa, at which stage CO_2 became supercritical. The ultrasonic velocities were then measured again at confining pressures from 15 MPa to 70 MPa in steps of 4 MPa.

5. CO_2 was released from the sample by reducing pore pressure to zero but keeping the confining pressure at a constant level of 12 MPa. The sample was flooded with the brine at an injection pressure of 9 MPa. The salinity of the brine was the same as for the first three samples. The ultrasonic velocities were then measured again at confining pressures from 15 to 70 MPa in steps of 4 MPa.

6. The confining pressure was fixed as 30 MPa and $scCO_2$ (temperature 45°C, pressure 9.3 MPa) was injected into the brine-saturated sample with an injection rate of 1 mL/min. Ultrasonic velocity measurements were taken during $scCO_2$ flooding.

7. Brine with the same concentration of salts, was flooded into the sample at an injection rate of 1 mL/min and the ultrasonic velocities were measured during the process of the CO_2 replacement.

Residual $scCO_2$ saturation in the sample was estimated by measuring the amount of brine volume removed (and collected) from the sample. The total amount of CO_2 stored in the sample (the volume of $scCO_2$ and the amount of CO_2 dissolved in the brine) was estimated by weighing the sample after the CO_2 injection and at the end of the experiment, when the pore pressure was reduced and the CO_2 released from the sample.

10.7.4 Results

The obtained results are illustrated for the 1500.83 m sample (Figures 10.37 to 10.40) but description of the results acquired on some other samples are also considered. Figure

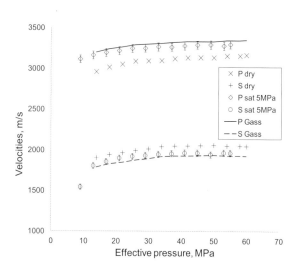

Figure 10.37: P-wave and S-wave velocities and Gassmann fluid substitution (brine) for CRC-2 reservoir sandstone excavated from the depth of 1500.83 m. $T = 45°C$.

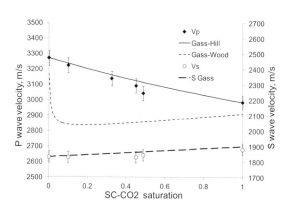

Figure 10.39: Ultrasonic velocities versus scCO$_2$ saturation for sample 1500.83 in comparison with Gassmann-Hill and Gassmann-Wood theoretical predictions.

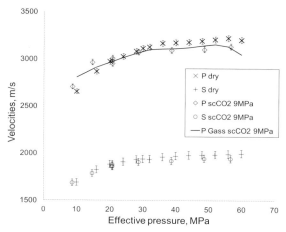

Figure 10.38: Conditions same as in Figure 10.39 but with the scCO$_2$ saturated sample.

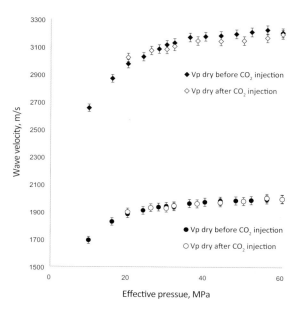

Figure 10.40: Ultrasonic velocities measured on dry sample 1500.83 before and after the experiments.

10.37 shows ultrasonic P- and S- velocities measured on the dry and brine-saturated 1500.83 m sample at different confining and pore pressures. The results are presented against the effective pressure, where the effective stress coefficient is assumed to be equal to 1. Velocities measured on dry sandstones at stresses of 10–20 MPa exponentially saturate to a linear trend that is observable at higher stresses. Such stress dependencies are typical for sandstones and are consistent with the empirical relationship reported by Eberhart-Phillips et al. (1989), explained theoretically by the dual-porosity model developed by Shapiro (2003)

and confirmed in a number of publications (e.g. Becker et al. 2007; Pervukhina et al. 2010).

Experimentally measured velocities are in good agreement with velocities calculated from dry velocities using Gassmann's relations for all the samples that are cut normal to bedding. It is worth noting that while for effective pressures above 20 MPa the differences between velocities measured on saturated samples and obtained with Gassmann's substitution are within the experimental error, the velocities measured for lower stresses substantially exceed Gassmann's predictions. This discrepancy may be caused by local flow (squirt) effects which cause an increase of rock elastic moduli at ultrasonic frequencies (Mavko and Jizba 1991). At the ultrasonic frequencies used in the experiment (0.5 MHz), local pressure has no time to equilibrate between equant pores with aspect ratios of ~1 and the thin grain contacts with aspect ratios of less than about 0.001, These grain contacts are very compliant and typically get closed at low effective stresses (~20 MPa for the samples studied) and consequently at the higher stresses local squirt does not affect elastic properties and the velocities of saturated rock can be calculated from the measurements on dry samples using Gassmann's equations.

Gassmann's predictions strongly overestimate P-wave velocities experimentally measured on the brine-saturated sample 1442.1 m, which was cut parallel to bedding. This sample was first saturated with gaseous CO_2 that became supercritical when the pore pressure was increased, reaching 9.3 MPa. The pressure was then released and eventually the sample was flooded with brine. The deviation from Gassmann's predictions of the velocities measured on the saturated sample can be explained by the fact that the sample was rich in weak clay cement which could be overdried and destroyed with CO_2 injection and the subsequent pore pressure increased. At the same time, S-wave velocity changes were in a good agreement with the predicted changes due to changes in density and did not exhibit any damage.

The velocities in fully $scCO_2$ saturated samples 1442.1 m and 1500.83 m were measured at step 3 of the experimental procedure. The results for the sample 1500.83 m at the pore pressure of 9 MPa are shown in Figure 10.38 in comparison with Gassmann's predictions for the $scCO_2$ properties. For both the 1442.1 m and 1500.83 m samples, the experimental and predicted results were in good agreement, that is, the discrepancies between experimental results and Gassmann's fluid substitution were within the experimental error.

Ultrasonic velocities measured during the injection of $scCO_2$ into a brine-saturated sample at confining stress of 30 MPa and pore pressure of 9.3 MPa are shown in Figure 10.39 for sample 1500.83 m in comparison with Gassmann-Wood's and Gassmann-Hill's theoretical predictions. Velocities were measured during $scCO_2$ injection into brine-saturated sandstone at a temperature of 45°C, with a confining pressure of 30 MPa and a pore pressure and $scCO_2$ injection pressure of 9.3 MPa. Velocities corresponding to 100% $scCO_2$ saturation are taken from Figure 10.38. For both samples, 1442.1 m and 1500.83 m, $scCO_2$ saturation of up to ~50% was reached; the velocities at 100% of $scCO_2$ saturation were taken from the previous experiment on the injection of $scCO_2$ into a dry sample. In the case of sample 1500.83 m, the compressional velocity followed the Gassmann-Hill's model up to ~40% of CO_2 saturation and decreased almost linearly with the increase of $scCO_2$ saturation. At the maximum saturation of ~50% reachable in this experiments, V_p still exceeded P-wave velocity at 100% $scCO_2$ saturated sample by about 50 m/s. The overall increase in the shear velocity was caused by density decrease and did not exceed 70 m/s or ~4% of the initial S-wave velocity. For sample 1442.1 m the compressional velocity remained constant up to an $scCO_2$ concentration of about 20% and then abruptly dropped. With the further increase of $scCO_2$ saturation from 30% to 100%, the velocity did not change (within the accuracy of experimental measurements). The overall drop in compressional velocities when the $scCO_2$ saturation increased from 0 to 50% was ~7% for both samples and with the further increase of the $scCO_2$ saturation, V_p dropped 0% and 3% for samples 1442.1 and 1500.83, respectively. The changes in ultrasonic velocities due to the $scCO_2$ saturation were noticeably different for these two samples. It is worth noting that the orientation of the samples was different: sample 1442.1 m had been cut parallel to the bedding plane while sample 1500.83 m was cut perpendicular to it.

Finally, compressional and shear velocities of the dry sample 1500.87 m before and after the CO_2 injection, are shown in Figure 10.40. It can be seen from this that the observed changes are small and within the experimental error. This fact shows that CO_2 injection in this sample caused no damage or alteration detectable by acoustic methods. Dry velocities after the CO_2 injection were measured for one sample only and further measurements will be necessary.

The CO_2 injection experiments carried out on the suite of sandstones from the CRC-2 well confirmed applicability of the Gassmann's fluid substitution method to brine (for four out of five samples) and to $scCO_2$ (for both tested samples). On injection of $scCO_2$ into brine-saturated samples, the samples exhibited observable perturbation of ~7% of compressional velocities with the increase of CO_2 saturation from 0% to the maximum (~50%). The acoustic velocities measured on the dry samples before and after the CO_2 injection showed no noticeable difference, which implies that no significant damage or alteration occurred due to the injection.

10.8   Conclusions

Exceptionally challenging conditions for monitoring CO_2 storage at the Otway site required creative thinking and implementation of a new and comprehensive seismic programme. High spatial data density coupled with high fold and high quality processing produced excellent repeatability at the target level of both 3D surface reflection and 3D VSP data. However, the very small time-lapse seismic signal presented a challenge for data processing and subsequent analysis of the differences between successive seismic surveys. A possible but very subtle seismic anomaly was identified after injection of 33 Kt and 66 Kt of CO_2-rich gas, respectively. This was encouraging, considering that weak seismic sources and two different source types were used for the time-lapse studies. However, after subsequently comparing the time-lapse seismic signals extracted from surface seismic and VSP, it appears that any changes in the reservoir are likely to be very subtle and that very minor changes in the processing sequence(s) could produce markedly different results. This is to be expected as differencing small numbers yields high instability in the results. It is therefore expected that another round (or more) of reprocessing will be necessary to converge towards unique difference cubes. This is also the case for 3D VSP processing.

Specific conclusions from the modelling are:

> Reservoir simulations show that a plume in the Paaratte Formation resulting from the proposed Stage 2C injection of 10,000 to 30,000 t of the

CO_2/CH_4 mixture, is likely to be quite thin (maximum of 7 m) with a large lateral extent (up to 400 m).

> The fluid substitution shows that expected changes of acoustic impedance in the plume will vary between 10–15%. The average absolute change of acoustic impedance is about 900 m/s*g/cm^3.

> The seismic 1.5D modelling shows that this change of acoustic impedance will result in a change of reflection amplitude of the target reflection by up to 100%.

> Analysis of these amplitude changes together with noise (estimated from the time-lapse field data in the area) suggests that a plume resulting from a 10,000 t injection is likely to be detectable by 3D seismic using the same seismic source and acquisition parameters employed in the 2009 and 2010 surveys.

> The same analysis provides a high level of confidence that the proposed Stage 2 experiment involving a 30,000 t plume would be detectable in the Paaratte Formation using a 3D seismic survey.

4D VSP data acquired for Otway Stage 1 in 2007 (baseline) and 2010 (monitor) were processed to migrated volumes and subsequently cross-equalised using exactly the same technique as developed for surface Otway 3D time-lapse surface seismic data. The main conclusions can be summarised as follows:

> The 4D VSP data demonstrate a superb level of repeatability in comparison to all other time-lapse datasets acquired for the survey area (with median NRMS value on the target horizon of ~16%).

> Amplitudes of the seismic events in the migrated volume show a strong variation with distance from the borehole; this can be explained by the influence of such factors as directivity of receivers and/or migration aperture. During quantitative interpretation of the 4D VSP survey, these factors must be taken into account and/or the processing flow improved.

> In order to compare the migrated volumes obtained from 3D surface seismic and 3D VSP surveys, the velocity models used for migration should be matched.

> 4D VSP provides a more sensitive tool for imaging reservoir changes in the vicinity of the well, while 4D surface seismic is essential for monitoring changes in the overburden (such as possible leakages).

> Permanent near-surface installations should be considered as a part of future Otway Project monitoring programme because they could combine many advantages of the borehole seismic with those of the surface seismic method.

> Repeated zero-offset and offset VSP data recorded in CRC-1 well show good levels of repeatability of direct wave arrivals. No changes in the characteristics of direct waves were detected.

It can be safely concluded that 3D surface seismic shows no evidence of any CO_2-rich gas leaking into units overlying the Waarre C Formation and that therefore no significant amount of injected CO_2 has escaped from the reservoir. Time-lapse VSP difference appears to show that a seismic anomaly is present between the two wells as suggested by reservoir simulation studies. The work demonstrated that, as expected, the stored CO_2 in the Naylor depleted gas field did not produce any detectable difference in the time-lapse seismic signal, a conclusion that is likely to have implications to other depleted gas fields with residual methane present.

10.9   References

Al-Jabri, Y.Y.M., Urosevic, M and Sherlock, D. 2008. The effect of corrugated limestone and the changing of the near surface conditions on CO_2 monitoring programme at Naylor-1, CO2CRC Otway Project, Victoria, Australia. First EAGE CO_2 Geological Storage Workshop, Extended Abstracts, A12.

Aritman, B.C. 2001. Repeatability study of seismic source signatures. Geophysics 66, 1811–1817.

Bakulin, A., Lopez, J., Herhold, I.S. and Mateeva, A. 2007. Onshore monitoring with virtual-source seismic in horizontal wells: Challenges and solutions. SEG Technical Program Expanded Abstracts 26, 2893–2897.

Batzle M. and Wang, Z. 1992. Seismic properties of pore fluids, Geophysics 57, 1396–1408.

Beaty, K.S. and Schmitt, D.R. 2003. Repeatability of multimode Rayleigh-wave dispersion studies. Geophysics 68, 782–790.

Becker, K., Shapiro, S.A., Stanchits, S., Dresen, G. and Vinciguerra, S. 2007. Stress induced elastic anisotropy of the Etnean basalt: Theoretical and laboratory examination, Geophys Res Lett 34, L11307.

Bennion D.B. and Bachu, S. 2008. Drainage and imbibition relative permeability relationships for supercritical CO_2/brine and H_2S/brine systems in intergranular sandstone, carbonate, shale, and anhydrite rocks. SPE Reservoir Evaluation and Engineering 11(3), 487–496.

Calvert, R. 2005. Insights and methods for 4D reservoir monitoring and characterization. Society of Exploration Geophysicists and European Association of Geoscientists and Engineers, ISBN 1-56080-128-X.

Campbell, A. and Shujaat, A. 2010. Borehole seismic processing report, 3D VSP Survey, CO2CRC, CRC-1 3D VSP, Baseline and first repeat survey, Otway Basin, Australia. CO2CRC, Schlumberger.

CGG 2000. Curdie Vale 3D Seismic data processing, 2000, Final Report, CGG Perth Processing Centre.

Eberhart-Phillips, D., Han, D.-H. and Zoback, M.D. 1989. Empirical relationships among seismic velocity, effective pressure, porosity and clay content in sandstone: Geophysics 54, 82–89.

Eiken, O., Brevik, I., Arts, R., Lindeberg, E. and Fagervik, K. 2000. Seismic monitoring of CO_2 injected into a marine aquifer. SEG Technical Program Expanded Abstracts 19, 1623–1626.

Ennis-King, J., Dance, T., Boreham, C., Xu, J., Freifeld, B., Jenkins, C., Paterson, L., Sharma, S., Stalker, L. and Underschultz, J. 2010. The role of heterogeneity in CO_2 storage in a depleted gas field: history matching of simulation models to field data for the CO2CRC Otway Project, Australia. 10th International Conference on Greenhouse Gas Control Technologies (GHGT-10), Amsterdam, The Netherlands, 19–23 September.

Hatton, L., Makin, J. and Worthington, M.H. 1986. Seismic data processing: theory and practice. Blackwell Scientific, ISBN 0632013745.

Isaac, J.H. and Lawton, D.C. 2006. A case history of time-lapse 3D seismic surveys at Cold Lake, Alberta, Canada. Geophysics 71, B93–B99.

Kazemeini, S.H., Juhlin, C. and Fomel, S. 2010. Monitoring CO_2 response on surface seismic data; a rock physics and seismic modeling feasibility study at the CO_2 sequestration site, Ketzin, Germany, Journal of Applied Geophysics 71, 109–124.

Kragh, E. and Christie, P. 2002. Seismic repeatability, normalized RMS, and predictability. The Leading Edge 21, 640–647.

Kunz, O., Klimeck, R., Wagner, W. and Jaeschke, M. 2004. The GERG-2004 wide-range equation of state for natural gases and other mixtures. GERG TM15-2007.

Landrø, M. 1999. Repeatability issues of 3-D VSP data. Geophysics 64, 1673–1679.

Lebedev, M., Pervukhina, M., Mikhaltsevitch, V., Dance, T., Bilenko, O., and Gurevich, B. 2013, An experimental study of acoustic responses on the injection of supercritical CO_2 into sandstones from the Otway Basin. Geophysics 78(4) D293-D306.

Lei, X. and Xue, Z. 2009. Ultrasonic velocity and attenuation during CO_2 injection into water-saturated porous sandstone: Measurements using difference seismic tomography: Physics of the Earth and Planetary Interiors 176, 224–234.

Lumley, D.E. 2001. Time-lapse seismic reservoir monitoring. Geophysics 66, 50–53.

Ma, J., Gao, L. and Morozov, I. 2009. Time-lapse repeatablity in 3C-3D dataset from Weyburn. CO_2 Sequestration Project. Frontiers + Innovation – 2009 CSPG CSEG CWLS Convention, Calgary, Canada, Expanded Abstracts, 255–258.

Malme, T.N., Landro, M. and Mittet, R. 2005. Overburden distortions – implications for seismic AVO analysis and time-lapse seismic. Journal of Geophysics and Engineering 2, 81–89.

Mavko, G. and Jizba, D. 1991. Estimating grain-scale fluid effects on velocity dispersion in rocks: Geophysics 56, 1940–1949.

Mavko, G., Chan, C. and Mukerji, T. 1995. Fluid substitution: estimating changes in Vp without knowing Vs. Geophysics 60, 1750–1755.

Mavko, G., Mukerji, T. and Dvorkin, J. 1998. The rock physics handbook: tools for seismic analysis of porous media: Cambridge University Press.

Meunier, J., Huguet, F. and Meynier, P. 2001. Reservoir monitoring using permanent sources and vertical receiver antennae: The Cere-la-Ronde case study. The Leading Edge 20, 622–629.

Olofsson, B., Probert, T., Kommendal, J. and Barkved, O. 2003. Azimuthal anisotropy from the Valhalla 4C 3D survey. The Leading Edge, Dec. 2003, 1228–1235.

Peng, D.Y. and Robinson, D.B. 1976. A new two-constant equation of state. Industrial and Engineering Chemistry: Fundamentals 15, 59–64. doi:10.1021/i160057a011.

Perrin, J.-C. and Benson, S.M. 2009. An experimental study on the influence of sub-core scale heterogeneities on CO_2 distribution in reservoir rocks., Transport in porous media, DOI 10.1007/s11242-009-9426-x, 2009 (published online: 30 June 2009).

Pervukhina, M., Gurevich, B., Dewhurst, D.N. and Siggins, A.F. 2010. Applicability of velocity–stress relationships based on the dual porosity concept to isotropic porous rocks. Geophysical Journal International 181, 1473–1479.

Pevzner, R., Gurevich, B. and Duncan, G. 2009. Estimation of azimuthal anisotropy from VSP data using multicomponent velocity analysis, 71st EAGE Conference and Exhibition incorporating SPE Europec 2009, Extended Abstracts, P-182.

Pevzner, R., Bona, A., Gurevich, B., Yavuz, I., Shaiban, A. and Urosevic, M. 2010. Seismic anisotropy estimation from VSP data: CO2CRC Otway Project case study. SEG Technical Program Expanded Abstracts 29, 353–357.

Pevzner, R., Shulakova, V., Kepic, A. and Urosevic, M. 2011. Repeatability analysis of land time-lapse seismic data: CO2CRC Otway Pilot Project case study. Geophysical Prospecting, doi: 10.1111/j.1365-2478.2010.00907.x.

Rossi, G., Gei, D., Picotti, S. and Carcione, J.M. 2008. CO_2 storage at the Aztbach–Schwanenstadt gas field: a seismic monitoring feasibility study. First Break 26, 45–51.

Shapiro, S.A. 2003. Elastic piezosensitivity of porous and fractured rocks: Geophysics 68, 482–486.

Shi, J.Q., Xue, Z. and Durucan, S. 2007. Seismic monitoring and modelling of supercritical CO_2 injection into a water-saturated sandstone: Interpretation of P-wave velocity data. International Journal of Greenhouse Gas Control 1(4), 473–480.

Shi, J. Q., Xue, Z. and Durucan, S. 2011. Supercritical CO_2 core flooding and imbibition in Tako sandstone— Influence of sub-core scale heterogeneity. International Journal of Greenhouse Gas Control 5(1), 75–87.

Siggins A.F., Lwin, M. and Wisman, P. 2010. Laboratory calibration of the seismo-acoustic response of CO_2 saturated sandstones: Int J Greenhouse Gas Control 4, 920–927.

Smith, T.M., Sondergeld, C.H. and Rai, C.S. 2003. Gassmann fluid substitutions: A tutorial. Geophysics 68, 430–440.

Turner, B. and Hearn, S. 1995. Shear-wave splitting analysis using a single-source, dynamite VSP in the Otway Basin. Exploration Geophysiscs 26, 519–526.

Urosevic, M., Pevzner, R., Kepic, A., Shulakova, V., Wisman, P. and Sharma, S. 2010. Time-lapse seismic monitoring of CO_2 injection into a depleted gas reservoir—Naylor Field, Australia. The Leading Edge 29, 936–941.

Urosevic, M., Pevzner, R., Shulakova, V., Kepic, A., Caspari, E. and Sharma, S. 2011. Seismic monitoring of CO_2 injection into a depleted gas reservoir-Otway Basin Pilot Project, Australia. Energy Procedia 4, 3550–3557.

Vandeweijer, V.P., Benedictus, T., Winthaegen, P.L.A. and Bergen, F. 2009. To a cost effective approach for monitoring CO_2 storage sites.71st EAGE Conference and Exhibition, Amsterdam, The Netherlands, Expanded Abstracts, P108.

Vedanti, N. and Sen, M.K. 2009. Seismic inversion tracks in situ combustion: A case study from Balol oil field, India. Geophysics 74, B103–B112.

Waite, M.W. and Sigit, R. 1997. Seismic monitoring of the Duri steamflood: Application to reservoir management. The Leading Edge 16, 1275–1278.

Wang, Z. and Nur, A. 1989. Effect of CO_2 flooding on wave velocities in rocks and hydrocarbons. Soc Petr Eng Res Eng 3, 429–439.

Xue, Z. and Ohsumi, T. 2004. Seismic wave monitoring of CO_2 migration in watersaturated porous sandstone. Explorations Geophysics 35, 25–32.

Zemke, K., Liebscher, A. and Wandrey, M. 2010. Petrophysical analysis to investigate the effects of carbon dioxide storage in a subsurface saline aquifer at Ketzin, Germany (CO2SINK). International Journal of Greenhouse Gas Control 4, 990–999.

Tom Daley, Barry Freifeld, Tony Siggins

11. SEISMIC AND MICROSEISMIC MONITORING[1]

11.1 Introduction

The Naylor-1 monitoring completion, a unique and innovative instrumentation package, was designed and fabricated at Lawrence Berkeley National Laboratory (LBNL). The intent was to deploy a suite of instrumentation for monitoring subsurface reservoir conditions. The instruments included pressure-temperature gauges, fluid sampling equipment at multiple depths, multi-use seismic sensors, and a packer for zonal isolation. All the instruments needed to fit into a small diameter mono-bore with an additional restriction due to a casing patch above the reservoir zone. The deployment method chosen was semi-permanent (retrievable), using 0.75″ diameter "sucker" rods typically used for oil field pump jacks. Instrumentation was installed between

26 September and 14 October, 2007 in the pre-existing borehole Naylor-1. Figure 11.1(a) shows a schematic of the Naylor-1 instrumentation layout, and Figure 11.1(b) shows deployment operations. Three U-tube geochemical samplers were deployed, with one located near the top of the residual CH_4 gas cap and two located beneath the gas-water contact. The 21 geophones (Figure 11.1(c)) were used to perform three distinct seismic measurements-high resolution travel time (HRTT), walkaway vertical seismic profiling (WVSP), and microseismic monitoring.

11.2 High resolution travel time (HRTT) monitoring and offset VSP

The active seismic programme using the Naylor-1 sensors included high resolution travel time (HRTT) monitoring using sensors located at reservoir level and walkaway vertical seismic profiling (WVSP) using a string of nine vertical geophones located above the Waarre reservoir horizon. The surveys conducted are summarised in Table 11.1. The six locations used for conducting HRTT measurements and the 22 shot point locations used in the WVSP surveys conducted in April 2008 and May 2008 are shown in Figure 11.2.

[1] This work was partially supported by the GEOSEQ project undertaken for the Assistant Secretary for Fossil Energy, Office of Coal and Power Systems through the National Energy Technology Laboratory, of the U.S. Department of Energy, under contract No. DE-AC02-05CH11231.

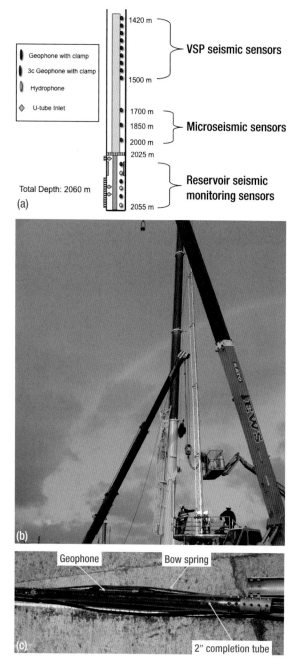

Figure 11.1: (a) Monitoring instrumentation deployed in the Naylor-1 borehole. (b) Deployment of the Naylor-1 bottom hole assembly. September 2007. (c) Three-component geophone with bow-spring mounted on completion tubing used in the Naylor-1 monitoring well.

11.2.1   Pre-injection testing

The initial seismic data collected in October 2007, using a small weight drop source, indicated that greater source strength would be required, given the depth of the reservoir. The December 2007 data collection, using a Rocktec Hurricane Force 9 source, was able to provide a signal to the shallowest vertical component geophones but the data had a low signal-to-noise ratio. To get a better signal-to-noise ratio, a larger Hurricane concrete breaker (www.rocktec.co.nz) was employed, with rated energy of 2720 Nm, versus 1125 Nm for the smaller Force 9. The third dataset, collected in January 2008 with the larger source and more source stacking, had good quality on nearly all sensors above the reservoir. Using the Rocktec Hurricane Force 10 seismic source, baseline data was collected at the six HRTT sites shown in Figure 11.2. The three geophones in the Waarre reservoir, however, were still showing a lack of signal. In the microseismic monitoring, a signal on these sensors could be seen, and it is possible that attenuation within the reservoir was part of the problem, with poor sensor coupling also likely.

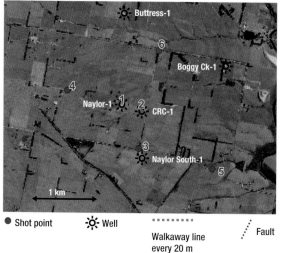

Figure 11.2: Location photo/map of HRTT sites (blue dots) and wells. HRTT sites 1 and 2 are on the well pads of Naylor-1 and CRC-1, respectively. The dashed blue line shows the 22 WVSP shot points. The indicated fault is a schematic representation of the projected surface location of the reservoir bounding fault.

Figure 11.3: Hurricane impact seismic source, with previous shot locations (holes).

The results of the January 2008 data collection were summarised by Daley and Sherlock (2008). Because of the large number of blows needed to increase signal-to-noise ratio using the Hurricane Force 10, considerable surface disruption occurred in the dairy paddocks (20 to 40 holes of about 1 m^2), as shown in Figure 11.3. It was determined that explosive shots (400 g at 3 m depth) would provide the best data quality with the lowest impact on the land surface for conducting subsequent HRTT and WVSP measurements. Despite the large number of source blows, geophones within the reservoir still had a poor signal-to-noise ratio, limiting useful data from the reservoir interval using surface sources.

The pre-injection testing of seismic sources for active source experiments (October 2007 to April 2008, Table 11.1) produced two main conclusions: first, explosive sources were the best sound source; and, second, the reservoir level sensors had too low a sensitivity to the surface source to allow analysis. Therefore the active

source analysis focused on reflected energy from below the Waarre C reservoir zone. This geometry is shown in schematic form in Figure 11.4, where the seismic energy travels through the reservoir twice, i.e. going down and coming back up. This reflected energy was recorded by the sensors in the Paaratte Formation, as well as by the microseismic monitoring sensors just above the Waarre C Formation. Potential changes in reservoir reflections needed to have a larger energy or amplitude change than the residual change follow the differencing of first arrivals (which did not pass through the reservoir).

11.2.2 Expected HRTT response

It was difficult to carry out satisfactory seismic monitoring at the Otway site for two main reasons: first, the near surface has large seasonal changes in the groundwater table; and, second, the expected seismic response from CO_2 displacing CH_4 was very small. An additional complication was the presence of a prominent buried karst surface at the top of the Port Campbell Limestone which interfered with the seismic ray path. Using a petrophysical model based on White's patchy saturation model with the NIST-14 equation of state for CO_2/methane, the change in P-wave velocity is very small change—of the order of 10 m/s, or 0.3% (Figure 11.5(a)). This assumes the residual CH_4 gas saturation of 25% would be displaced by an 80:20 CO_2/CH_4 mix. A more detectable seismic response can be expected if the gas cap in the Waarre were pushed down so that gas displaced brine. In this case the P-wave velocity change was calculated as about 100 m/s, or 3% (Figure 11.5(b)). This implies about 0.4 ms change in travel time for a 25 m thick plume. Greater change would be expected if pre-injection saturation was less than 25%.

Table 11.1: Otway Project seismic surveys.

Date	Source Type	Number of Locations	Walkaway Shots	Report Date
October 2007	Surface acoustic (Hartley Source)	1	No	None
October 2007	Small weight drop	3	No	None
December 2007	Rocktec Hurricane Force 9	3	No	5 January 2008
January 2008	Rocktec Hurricane Force 10	6	No	16 March 2008
April 2008	Explosive (400 g @ 3 m)	6	Yes, 22 shots	9 April 2008
May 2008	Explosive (400 g @ 3 m)	6	Yes, 22 shots	20 May 2008
November 2008	Explosive (400 g @ 3 m)	6	Yes, 22 shots	N/A

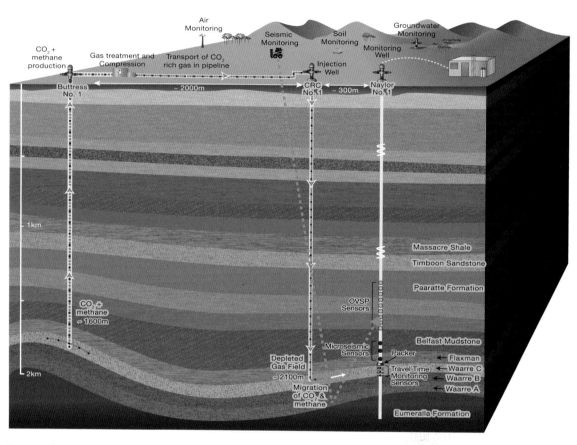

Figure 11.4: Otway site, including seismic sensors in Naylor-1 borehole, with a schematic example active source ray path shown in blue.

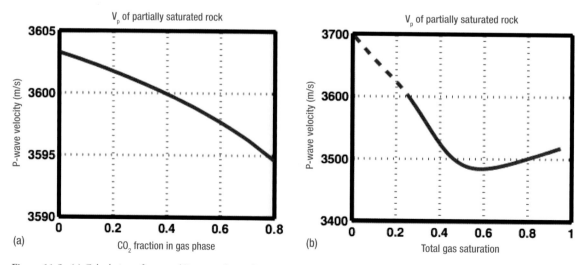

Figure 11.5: (a) Calculation of expected P-wave velocity for Waarre C reservoir for varying CO_2 saturation. (b) Calculation of expected P-wave velocity for CO_2 displacing brine. The solid line indicates the potential change at Otway, allowing for the pre-existing gas saturation (dashed line). Calculation by Dr Jonathan Ajo-Franklin.

11.2.3   Post-injection analysis

Explosive source HRTT data were collected in April 2008 (at the beginning of injection; Figure 11.6) and May 2008 (1 month after injection). A processing flow was developed to minimise the near-surface changes and search for travel-time changes in the reservoir. This analysis included minimising travel-time and amplitude changes in the first arriving waves. These first arrivals passed through the near surface, but not through the reservoir. Therefore they could be used to minimise effects related to shallow material. Figure 11.7 shows a comparison of data from April and May 2008.

The time-lapse analysis developed using the April/May data included the following processes:

1.   Remove/minimise time change from near surface (including shot variation) using cross-correlation of first arrival energy (150 ms time window.)

2.   Remove/minimise near surface amplitude change with normalisation based on first arrival RMS amplitude (50 ms time window).

3.   Calculate change in later arrivals (coda) using moving window correlation/variance analysis (25 ms time window, 0.125 ms interval).

At the time of the May 2008 seismic acquisition, about 5200 t of CO_2 had been injected. This was too small an amount to be seismically "visible", but the data could be used to assess detectability limits. The first arrival times showed a good repeatability (about 2 ms), but there had been little seasonal groundwater change between these two datasets. Figure 11.8(a) shows the amplitude variance for each of the six source sites. Residual change is still evident in the first arrivals and in borehole tube-wave energy. Figure 11.8(b) shows the moving window time shift data, which, as expected, do not have an observable (i.e. consistent) time shift associated with the Waarre C reservoir.

In addition to the six travel-time monitoring sites, a "walkaway" survey was run with source points about every 10 m between the Naylor-1 and CRC-1 boreholes (Figure 11.9). These data were analysed for amplitude and travel-time changes. Again as expected, the results showed

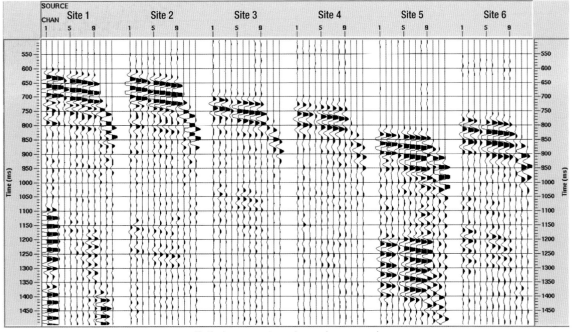

Vertical component geophones above reservoir

Figure 11.6: Data collected during the April 2008 seismic survey at the six HRTT locations showing good signal-to-noise ratios.

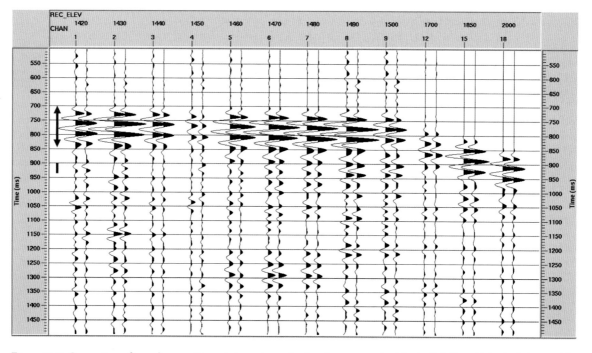

Figure 11.7: Comparison of recordings on 12 vertical geophones for data from source site 4 in April (left) and May (right) 2008. The red arrow indicates the time window used for first arrival normalisation (slightly shifted for each pair), and the blue line indicates the length of the time window used for "moving window" analysis.

no detectable change after only 1 month of injection, but again the resolution or detectability could be assessed with these data.

11.2.4 Estimate of HRTT precision and comparison to field data

The time-lapse data from April and May 2008 were analysed for the RMS time shift in three windows representing "background noise". The levels measured in the field data could be compared to the expected change due to CO_2, and compared to an estimate of best expected precision. The best expected precision was calculated based on the Cramer-Rao bound of delay time (Silver et al. 2007). For given acquisition parameters the Cramer-Rao lower bound (Weinstein and Weiss 1984) can be calculated using the following parameters:

   f = centre frequency
   rho = correlation coefficient

T = window length
B = fractional bandwidth
SNR = signal-to-noise ratio

Of these parameters, SNR was the key parameter that could be controlled, hence the need for high SNR and the use of explosive sources. Using Otway field results for the parameters, it was found that the Cramer-Rao estimate for Otway HRTT was about 0.07 ms to 0.7 ms. This defined the maximum expected precision for the HRTT experiment and indicated that the predicted change of 0.04–0.4 ms may have been attainable.

Figure 11.8 shows a comparison of the calculated HRTT field data precision, the Cramer-Rao bound and the delay time change estimated by petrophysical modelling. About half of the data points exceed the estimated change, meaning these data are too noisy to allow detection. The data points below this level indicate detection is possible, but will require repeatability as good as or better than that of the April and May datasets.

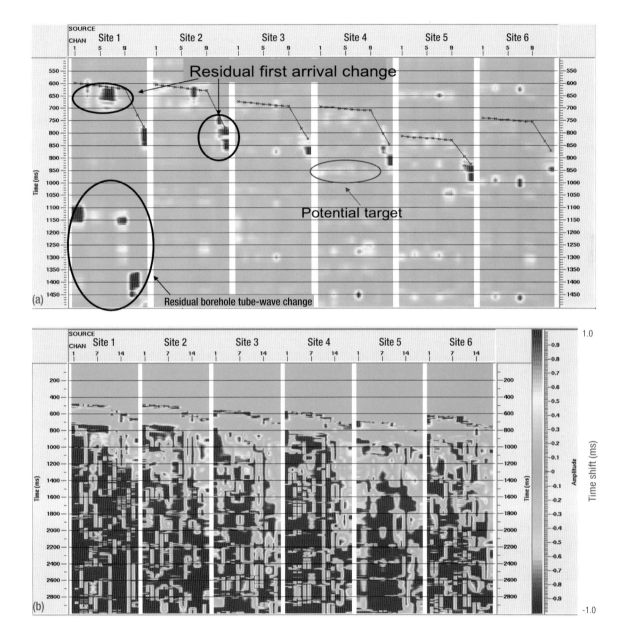

Figure 11.8: (a) Time-lapse change amplitude, between the April 2008 and May 2008 HRTT measurements. Red indicates more variance between datasets. Large variance at first arrival times indicates that better normalisation of first arrival energy is required for evaluating the small changes expected from reflections in the reservoir horizon (such as the indicated potential target). (b) Time-lapse time shift in a moving window for each of the six source sites. The target time window is 1400–1900 ms. The time shift is inconsistent in this window. While no effect from the injected CO_2 was expected to be seen, a decrease in travel time (green to blue colours) was predicted for CO_2 displacing CH_4.

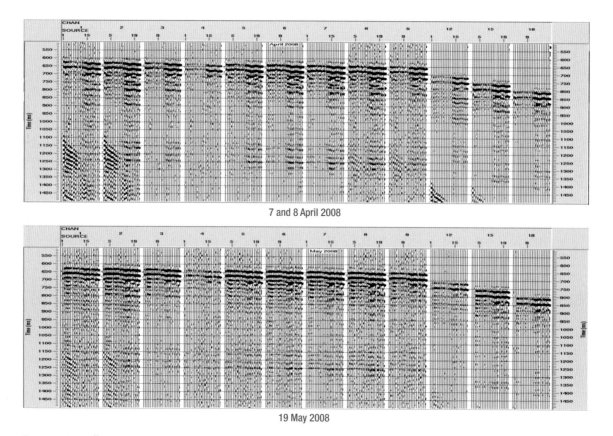

Figure 11.9: Walkaway type seismic survey data for the 12 vertical geophones in Naylor-1 well.

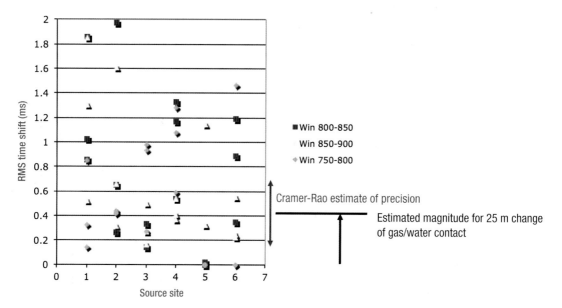

Figure 11.10: The RMS time shift between April and May 2008 datasets for three time windows representing typical "noise" levels and for the six monitoring sites and for three sensors at each site. Also shown is the estimated magnitude of time shift for a 25 m change in the gas/water contact (the maximum expected at Naylor-1) and the estimated theoretical precision of the Cramer-Rao bound.

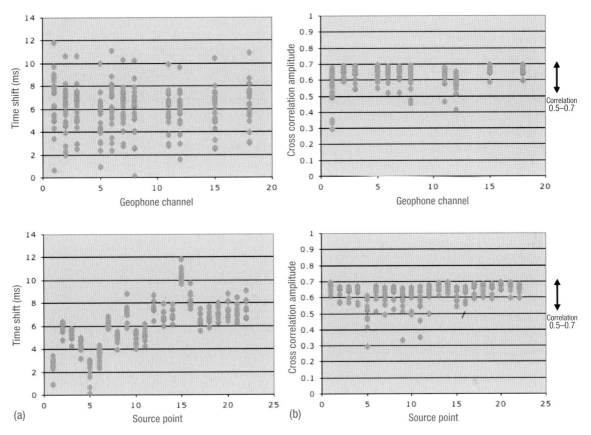

Figure 11.11: Time shift and amplitude correlation coefficient for all source points and geophone channels. The data shown are differences between April and November 2008 surveys; (a) time shift, (b) correlation coefficient.

In November 2008 the first true "post-injection" seismic data were collected (the previous survey from May 2008 was only a month after commencing injection). The analysis indicated that the data repeatability was not sufficient to allow detection of a CO_2 induced response. The previous analysis indicated the repeatability from April 2008 to May 2008 was only marginally sufficient, whereas the November 2008 data indicated a much larger variation of about 6 ms in April-to-Nov. time shift compared to 1–2 ms shift for April-to-May (Figure 11.11). The calculated residual time shifts in a moving window for the HRTT source points (Figure 11.12(a)) had scatter in the data (from 2 to –2 ms shift) which overwhelmed an expected change of about 0.4 ms. Similarly, Figure 11.12(b) shows the amplitude change, also in a moving window. For amplitude, the residual change in the direct first arrival was again larger than any later arrival which travelled through the CO_2 reservoir zone.

This meant that changes in the seismic "coda" could not be attributed to the CO_2 injection. This analysis was consistent for the HRTT source points and the walkaway data. To understand the possible cause of the lack of results, time shifts and amplitude correlation coefficients were compared for all geophone channels and all source points (Figure 11.11). The results showed that no single geophone or source point had unusual variation, suggesting that the problem with detection limits was consistent over the entire experiment.

It was therefore concluded that the limitations of the HRTT experimental work were both instrumentation (electrical noise and sensor clamping) and geological (the pre-existing gas-in-formation which limited the seismic response).

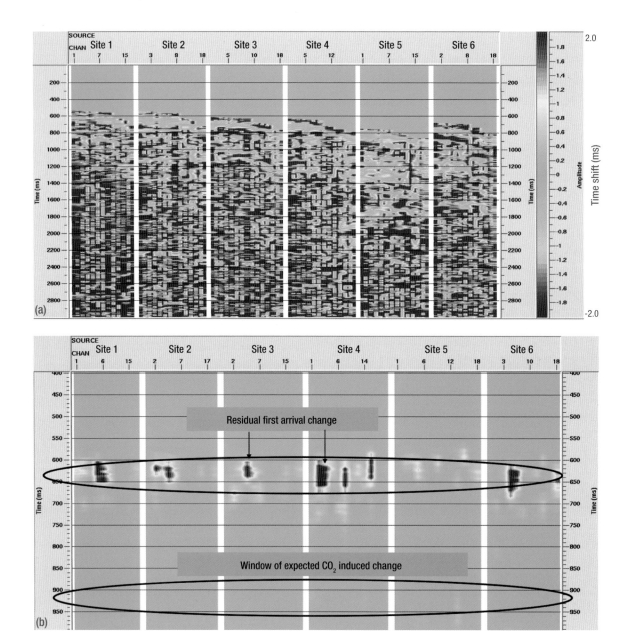

Figure 11.12: (a) Calculated time shift between the April and November 2008 HRTT surveys. The scatter in the data is larger than the expected CO_2 induced change, and therefore no detection is possible using HRRT in this instance. (b) Calculated residual amplitude change for the November 2008 survey. The residual change in the first arrival window (approximately indicated) is larger than later changes including the indicated window of expected change, meaning that data repeatability was insufficient to allow detection of CO_2 induced changes in later arrivals.

11.3   Passive seismic monitoring

11.3.1   Microseismic events

Laboratory experiments have shown that, as rock samples approach failure, the acoustic emission rate (AE) increases dramatically. In an analogous fashion, activation of pre-existing faults in the field, due to effective stress changes, particularly pore pressure changes due to the injection of fluids, may result in the generation of microseismic (MS) events, which can be monitored with geophysical instrumentation such as accelerometer, hydrophone or geophone arrays. In sedimentary rocks, which are usually the focus of CO_2 injection, there have been a number reports in recent years of MS activity associated with fluid injection, particularly water flooding, including by Peterson at al. (1996), Warpinski et al. (1996), Branagan et al. (1997), Nolen-Hoeksema and Ruff (2001), Zoback and Zinke (2002) and Oye and Roth (2003). Similar MS activity, following CO_2 injection in the liquid phase and most probably in the critical state at elevated pore pressures, may also occur, depending on the in-situ stress state and the structural geology of the target horizon.

MS monitoring has the potential to reveal faults via clustering of MS events and to aid the investigation of the progress of CO_2 gas or fluid fronts during injection. Since sedimentary rocks are softer than the crystalline rocks encountered in many hot dry rock (HDR) geothermal fields, any event magnitudes will be lower and there will be problems in detecting them because of the low signal-to-noise ratio. In general, the Richter magnitude or local magnitude, M_L, of these induced events is typically between +1 and +2 in crystalline rock. In sedimentary rock, event magnitudes can be as low as –2. The lower magnitudes are approaching the noise floor of conventional geophones used in wells. Therefore the emphasis on monitoring has to be on establishing accurate source locations and refocusing these "event clouds" in order to delineate fracture geometries, reveal fracture activation and define fluid flow paths. MS monitoring may provide sufficient warning, in the case of fault reactivation, to allow remedial measures to be taken, such as injection pressure control, to ensure that there is minimal chance of leakage of CO_2 into potable aquifers or to the surface.

MS activity follows injection of reservoir pore fluids at pressures approaching the critical state for mobilisation of fractures. These events most probably arise from faults and fractures that are optimally oriented for slip. Consequently, the events will be dominated by shear arrivals. It should be noted that MS activity has been shown to be associated with the advancement of fluid pressure fronts (Shapiro 2000), and these fluids may not necessarily be injected CO_2 but in-situ pore fluids such as formation water or hydrocarbons. Geomechanical modelling of reservoirs, which includes pore pressure, poroelastic effects and stress coupling (Angus et al. 2010), indicates that the pattern of displacement of faults and fractures around injection or production wells follows the geometry of observed microseismic clouds. In order to further quantify the fracture displacements in the field, it is desirable to calculate the fault-plane solutions from the first motions of the recorded seismograms. This requires a well-calibrated seismic array using test firings, to determine velocity models, together with borehole imaging at reservoir depths. If there are a sufficient number of MS events and a high degree of precision in the recording system, then fault plane solutions, which will give the moment tensor acting at the source, can be determined. The spectral content of the MS events can yield parameters such as event radius (according to the Brune, 1970 model) and stress change at the event source. Furthermore, if the event source radii are known, estimates of the area over which fluid may flow in the fractured zone can be made, since the fractures become areas of increased permeability.

11.3.2   Microseismic monitoring using Naylor-1 deep sensors

After initial installation, the microseismic monitoring was activated at the CO2CRC Otway site in order to collect pre-injection background data. The seismic recording system was remotely accessible via the internet for data transfer and changing of recording parameters. Each microseismic event was recorded with an 8 s length at 2000 samples per second per channel, and a 3 s pre-trigger length (leaving 5 s of post trigger recording). The trigger parameters were changed multiple times during the initial recording period to try and reduce the number of "false" triggers, mainly from electrical noise bursts.

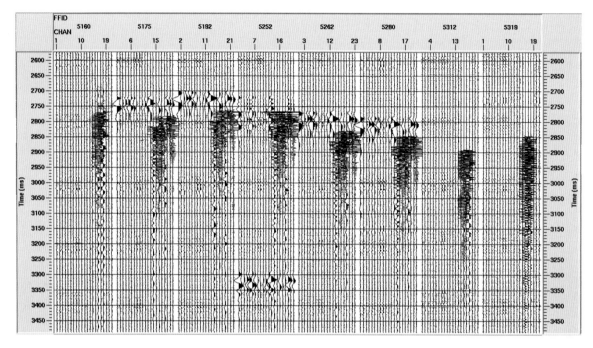

Figure 11.13: Eight microseismic events from the time period of the 2008 CRC-1 perforation shots. The second through sixth events (from the left) are each recorded within 25 s of a perforation shot (timing was not synchronised). The other three events are separated by hours and are "natural" events. Event 5160 (far left) is shown in detail in Figure 11.16.

Several thousand triggers were recorded, with almost all of the events inspected being electrical noise triggers. Many of the other events had signals on only one or two sensors and were most likely due to near-borehole noise. Modification of the trigger parameters during the course of injection lowered the "false" trigger rate to about five events per week. Individual inspection of events found only a few events of interest. A good system functional test was obtained when five perforation shots were performed in the CRC-1 well during perforation of injection casing in the injection zone. These shots all generated events with similar characteristics, demonstrating the performance of the seismic string. The move-out of low-frequency energy observed on sensors within the Waarre reservoir implies functional coupling of these three sensors, which had not "seen" any arrivals in the active source HRTT surveys. This discrepancy in observed signals remains unexplained.

11.3.3 CRC-1 perforation shots

In early February 2008, the CO_2 injection well, CRC-1, was perforated with a wire-line perforation gun. This activity provided the potential opportunity to detect seismic events with a known source location, with about a 300 m lateral offset, within the reservoir zone. Figure 11.13 shows all of the events within 2 days of the perforation shots which appear to be true seismic events. Most notably, every perforation shot was accompanied by a triggered event—events 5175, 5192, 5252, 5262 and 5280, which correspond to shots at depths of 2062.5, 2061, 2057, 2055 and 2053 m, respectively, in CRC-1. The timing between perforation and recording system was unfortunately not synchronised, but all the shots had just one event (ranging from 7 to 25 s before the recorded shot time), and no other events were triggered within minutes. These five perforation events can be compared to the other three events in Figure 11.13 (5160, 5312 and 5319). The non-perforation events had arrivals only on channels 16, 17, and 18—the 3-C sensor at 2000 m depth. The perforation events had arrivals on sensors at

1850, 2000, 2030, 2040 and 2050 m depth. Therefore, it was concluded that these five events were all caused by the perforation shots in CRC-1.

The perforation shot events had a low frequency energy arrival about 50 ms before the high frequency arrival. The low frequency arrival at first appeared to be electrical cross-talk noise because of the lack of move-out, except on the deeper sensors below the packer where move-out was seen. It is possible this low frequency energy was a combination of seismic energy and spurious electrical noise. Event 5252 was the only event which had an appearance of low frequency energy, at 3300 ms, separate from the high frequency arrivals. Figure 11.14 shows a true amplitude plot of three perforation events, all of which show the large low-frequency arrival on channel 19 (the sensor at the top of the Waarre-C reservoir), in the gas zone and in the casing patch of Naylor-1. It is possible that the large amplitude on channel 19 was responsible for electrical cross-talk noise on other channels. Nonetheless the consistent delay between the low and high frequency components indicated seismic propagation, as did the move-out seen in the low-pass filtered data plot (Figure 11.15).

An important conclusion from Figure 11.15 is that the three geophones in the Waarre reservoir (channels 19, 21 and 23) were sufficiently well coupled to be able to see the energy of the perforation shots. This was important because the controlled source effort (high resolution travel time monitoring, HRTT) did not obtain sufficient signal-to-noise ratio for these three geophones.

The move-out of the perforation events was difficult to interpret since the propagation was largely horizontal. Nonetheless, the low-frequency move-out between the 2030 and 2050 m sensors (channels 19 and 23) was about 7 ms, giving an apparent velocity of 2850 m/s. While the high frequency component had little move-out from 2000 to 2050 m, the move-out was approximately 35 ms between the 2000 m and 1850 m sensors, giving an apparent velocity of 4300 m/s. This high velocity could indicate propagation in the steel (casing or sucker rod), or simply horizontal propagation with high apparent vertical velocity.

The spectral content of a perforation shot event is shown in Figure 11.16. The signal is broadband with a peak at 20–30 Hz (the low frequency arrival) and energy within 20 dB to 300 Hz. In addition to the power line notch filters, a 15–500 Hz band pass filter was applied.

11.3.4   Natural microseismic events

The three non-perforation-shot events shown in Figure 11.13 are representative of the few non-electrical noise events recorded in the initial monitoring. The signal is impulsive and broadband with about 350 ms of coda. Some events were seen in the upper section of the string (1420–1500 m) but never on more than one or two sensors. Of these three events, one appeared to have a separate P and S phase arrival, allowing estimation of source distance (Figure 11.17). The identification of the S-wave is supported by the increased energy on the vertical component, because the vertical component is orthogonal to a horizontally propagating wave. The 90 ms of P-to-S time delay in event 5160 implies a distance of about 488 m (for V_p = 3.7 km/s and V_s = 2.2 km/s). This distance is approximate, since the actual velocity depends on the propagation path.

During the beginning of injection in April 2008, data were recorded continuously for about 3 weeks. These data were searched for events, with few observed. A typical event is shown in Figure 11.18. To look for trends in seismicity, the time of detected events was plotted against injection wellhead pressure, as shown in Figure 11.19. No clear pattern emerged and there was no significant seismic activity associated with the injection. It is notable that the injected CO_2 had by this time reached the Naylor bounding fault. This is the first known case of a fault seal being tested by a sequestration pilot and the lack of induced seismicity is therefore an important observation. A notable gap in seismicity was observed during November and December 2008, with a few events in January and February 2009.

11.4   Microseismic monitoring using surface stations

Microseismic monitoring at the Otway site, using a near-surface seismometer, commenced in September 2007, prior to the injection of CO_2. The initial installation consisted

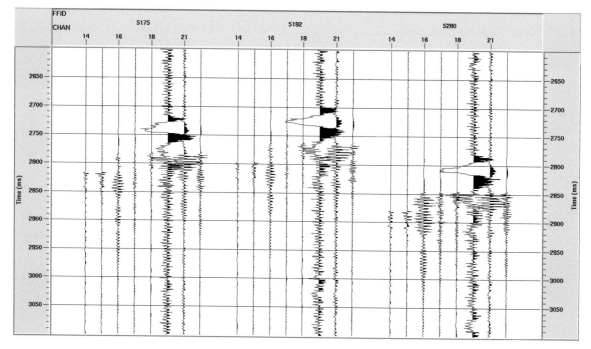

Figure 11.14: Three perforation events are shown at true relative amplitude. The dominant signal is a low frequency arrival on channel 19, with a later high frequency arrival on the other channels. Channels 1–13 had no signal and 20, 22, 24 (hydrophones) are not shown.

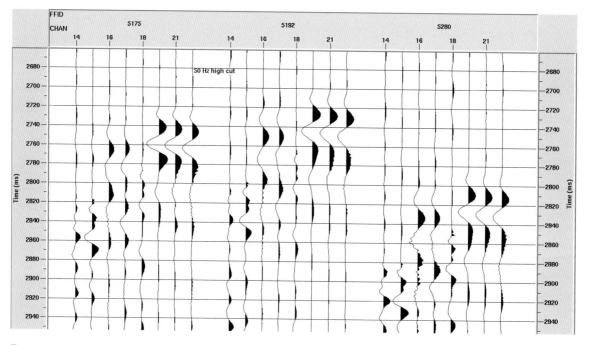

Figure 11.15: This figure shows the same three events as in Figure 11.14, with a 50 Hz low pass filter applied and trace amplitude equalisation. The move-out of the event on the three sensors within the reservoir (channels 19, 21 and 23) indicates the event was a seismic arrival (and not electrical noise or cross-talk).

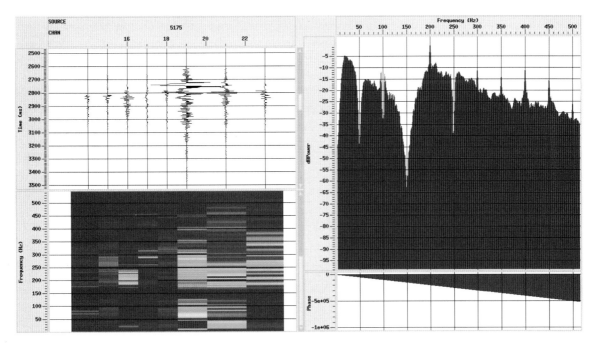

Figure 11.16: Spectral analysis of event 5280. The total average spectra is in red on the right, while the individual trace spectra are on the left; with the seismograms showing amplitudes (red is high amplitude, blue is low amplitude). The low frequency peak is at 20–30 Hz, while the high frequency is peaked at 200–300 Hz.

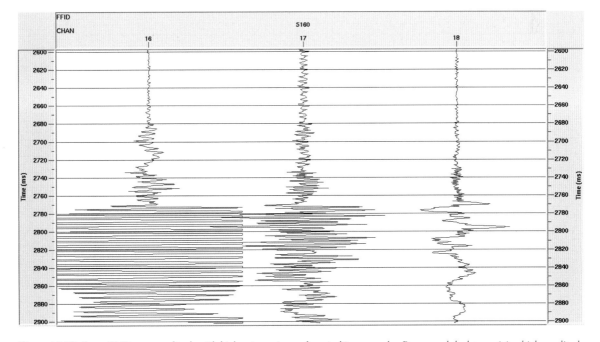

Figure 11.17: Event 5160, true amplitude with high gain to view early arrival interpreted as P-wave and the later arriving high amplitude S-wave. Channels 16, 17 and 18 are from the three-component geophone at 2000 m with channel 18 being vertical. Note that for this horizontally propagating event, the P-wave is on the horizontal component, while the vertical component has a large amplitude for the S-wave. Other channels have no arrivals associated with this event.

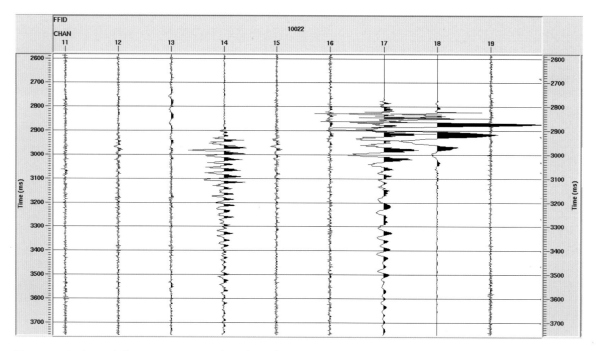

Figure 11.18: A record of a microseismic event from the Naylor-1 geophone array using three 3-component sensors.

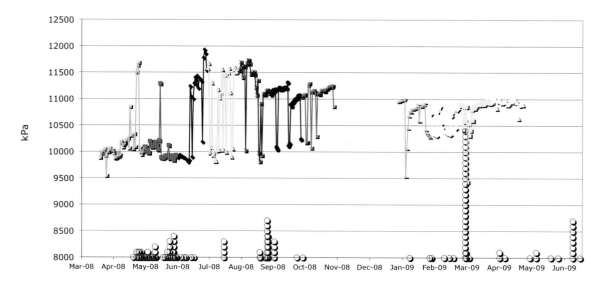

Figure 11.19: Occurrence time of microseismic events detected by the Naylor-1 seismic sensors (blue circles) plotted with CRC-1 wellhead injection pressure. No consistent relationship is evident between injection pressure and seismic activity. Figure courtesy of Dr John Peterson.

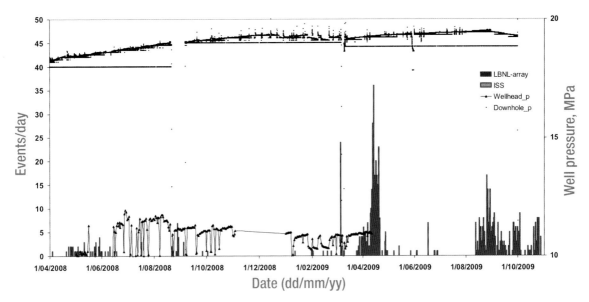

Figure 11.20: Downhole well pressure and "microseismic" activity during Stage 1 CO_2 injection.

Figure 11.21: One of the three microseismic monitoring installations at the Otway site in the vicinity of the Naylor-1 monitoring well. The seismometer was solar powered as was the radio. The microwave dish for the 5.8 GHz digital LAN communications network is visible. Triaxial geophones were installed at shallow depths (10 m and 40 m) down the borehole beside the central mast. Data were transmitted to a seismic computer approximately 2 km distant.

of a commercially available, two-channel seismometer (International Seismic Systems, MS unit) connected to a pair of triaxial geophones installed downhole in an abandoned water well at 10 m and 40 m depths, in the vicinity of the Naylor-1 monitoring well (see Figure 11.21). This system supplemented the downhole array of geophones and hydrophones installed at reservoir depth by Lawrence Berkley National Labs (LBNL) during the installation of a range of instruments in Naylor-1, in October 2007 (discussed previously). However, the subsurface array became increasingly unreliable and eventually failed after some months of operation, leaving only the surface station as the primary microseismic monitoring tool.

The shallow geophone installation remained in operation despite some down time due to power failures in the region. This equipment recorded events at 16-bit resolution, based on triggers derived from a short-term to long-term (STA/LTA) averaging algorithm. Only events which triggered both geophones simultaneously were accepted and transferred via a digital radio link to a remote computer incorporating the Otway MS database. "Simultaneous" in this context means triggers which occurred within a time window defined by the passage of a compressional wave over the distance between geophones, which in this case was 30 m. The intention with this first installation was to determine the level of natural background of microseismic activity at the Otway site prior to CO_2 injection.

11.4.1   Stage 1 results

During the injection of more than 60,000 t of CO_2 over the period from 18 March 2008 to 29 August 2009 the MS installations recorded episodes of low-level MS activity, again with M_l values around 0.1. However, these events could not readily be associated to the injection pressure history (Figure 11.20). In addition, the lack of an extended geophone array made event locations error prone. Mains power to the region was highly variable and there were many periods when the monitoring system computer was off-line.

Nevertheless, there were a small number of low amplitude events recorded each week at apparently random depths of local magnitude, M_l, near zero, which could be considered as natural background micro seismicity (Figure 11.22). However, the distinction between seismic and other sources of transient noise remains unresolved.

11.4.2   Stage 2 results

In March 2011 the shallow MS installation was upgraded with the addition of two more wells, each containing two triaxial geophones at 10 m and 40 m depth respectively. The complete array comprised six geophone stations arranged in a triangular pattern around the new injection well, CRC-2. The upgrade was made in anticipation of the Stage 2B phase of experiments at Otway during which a small volume of CO_2 (100 t) was to be injected in CRC 2 along with a series of water injections and productions. This experiment was designed to provide an estimate of residual trapping in a saline aquifer at Otway and is described more fully by Zhang et al. (2011) and in Chapter 17. Figure 11.22 shows the location and Figure 11.23 the sensitivity of the expanded microseismic array at Otway. A significant number (> 100) of small events, Ml –1 to –2 were recorded by the monitoring system over the period of the Stage 2 experiments undertaken in the period May 2011 to September 2011. However, only one event triggered all six geophones in the array

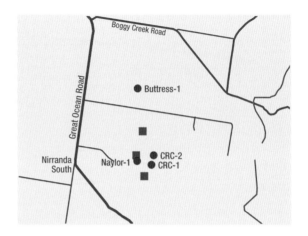

Figure 11.22: The microseismic array at the Otway site. The geophone well locations are shown in blue while production, injection and monitoring wells are brown. Buttress-1 is the CO_2 source well.

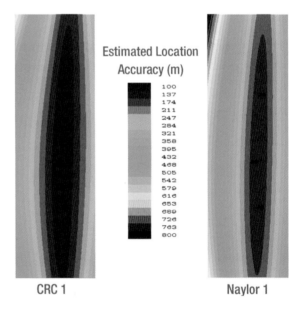

Figure 11.23: Calculated sensitivity of the Otway MS surface array. Uncertainty in metres in depth (z) is shown for events near the CRC-1 and Naylor wells locations. The vertical scale is 0 m to 1600 m.

and this was associated with a CO_2/water injection. The rest of the recorded "events" appear to be of cultural origin or transient electrical noise, although there may be some very low-level natural seismicity.

11.4.3 Induced seismicity

In general there has only been a low level of induced microseismicity reported at CCS sites worldwide, including at experimental sites. There have been suggestions made that injection of critical state CO_2 is less likely to generate microseismic activity than, for example, water injection. In terms of compressibility there may be some basis to this argument, as critical state CO_2 at reservoir temperatures and pressures has a low bulk modulus (~70 MPa at a pore pressure of 20 MPa and 80°C) and therefore high compressibility compared to formation water (bulk modulus of 2.4 GPa at 20 MPa and 80°C). In other words, critical state CO_2 can be considered as a "soft" fluid. In addition, the low density and high relative permeability of the formation to critical state CO_2 differ considerably from water and may be less likely to activate small shear faults. However, it is generally accepted that elevated pore pressures and injection rates are the major factors in fault reactivation and associated seismic activity, rather than fluid compressibility (Chapter 7). Verdon (2010) concludes that "despite differences in compressibility, viscosity, density and relative permeability between the fluids, CO_2 and water have similar induced patterns of seismicity". It may be that, by avoiding known fault structures at CCS injection sites, the risk of induced seismic activity can be greatly reduced.

11.5 Conclusions

A multipurpose monitoring instrumentation package was installed in the Naylor-1 well. As part of this package, seismic sensors were installed with two planned uses—microseismic monitoring and active source monitoring. Initial testing showed that the seismic sensors had some electrical noise problem (with some leakage to ground observed) and also poor coupling to the borehole. This was probably a problem with the bow spring clamps which were a unique design for this sucker rod deployment. Additionally, the sensors below the packer, in the reservoir, did not provide useful response to active source signals despite observation of perforation shots during perforation of the injection well, CRC-1.

Microseismic event monitoring began on 29 January 2008, with very few events observed; most triggers were due to electrical noise. Continuous (24/7) recording was undertaken for approximately 1 month during initial injection, with no evidence of missed events and therefore the system was returned to triggered recording. Adjusting trigger parameters to minimise "false" triggers was part of the first few months of monitoring. Background observation levels showed about two triggers a day and approximately two true events a week. This lack of activity and the very small size of the events indicate there was no fault activity associated with injection of CO_2 at the Otway site. In 2010, the Naylor-1 geophone string had an electrical short and stopped operation. The microseismic monitoring programme has continued up to the present day using shallow borehole surface stations.

The active source monitoring involved choosing (at an early stage) an appropriate source, given that there was poor sensor coupling (requiring high source signal). The decision was made to use an explosive source. The problems with sensors in the reservoir required the use of reflection/coda analysis which further limited the signal-to-noise ratio. The data indicated that the sensitivity of the seismic response (both travel time and amplitude) to CO_2 injected into the Waarre Formation was insufficient for detection of microseismic events. This problem was compounded by the presence of residual methane, and the expected very small seismic response for CO_2 displacing CH_4 in the reservoir.

A major technical challenge in microseismic monitoring, particularly at sites such as the Otway site where CO_2 injection is into permeable sand formations and saline aquifers, is the very low amplitude level of recorded events. Peak amplitudes may be in the order of 10^{-7} m/s and are close to the noise floor of geophones deployed in wells. This results in an associated problem, namely that of distinguishing between recorded "events" and "cultural" noise such as transient events due to noises in pipe work and electrical pulses. For example, during the Otway Stage 2 experiments, more than 100 events were recorded but only one event triggered all six triaxial geophones in the shallow array. At the Otway site there was little or no low-level induced seismicity associated with CO_2 injection. Finally, it is important to note that there was no induced seismicity despite the fact that the CO_2 reached a significant fault zone.

11.6   References

Angus, D.A., Kendal, J.M., Faber, Q.J., Segura, J.M., Skachkov, S., Crook, AJI. and Dutko, M. 2010. Modelling microseismicity of a producing reservoir from coupled fluid-flow and geomechanical simulation. Geophysical Prospecting 58, 901–914.

Branagan P., Peterson R., Warpinski N. and Wright T. 1997. Results of multi-site experimentation in the B-sand interval: Fracture diagnostics and hydraulic fracture intersection. Topical Report, Gas Research Institute, GRI-96/0225, DOE/MC/30070-5218, May 1997.

Brune, J.N. 1970. Tectonic stress and the spectra of seismic shear waves. J. Geophys Res 75, 4997–5009.

Daley, T.M. and Sherlock, D. 2008. Otway Project: Naylor-1 high resolution travel time monitoring initial results of January 2008 Acquisition. CO2CRC Report, 16 March 2008.

Nolen-Hoeksema, R.C. and Ruff, L.J. 2001. Moment tensor inversion of microseisms from the B-sand propped hydrofracture, M-Site, Colorado. Tectonophysics 336, 163–181.

Oye, V. and Roth, M. 2003. Automated seismic event location for hydrocarbon reservoirs. Computer and Geosciences 29, 851–863.

Peterson, R.E., Wolhart, S.L., Frohne, K.H., Branagan, P.T., Warpainski, N.R. and Wright, T.B. 1996. Fracture diagnostics research at the GRI/DOE Multi-site Project: Overview of the concepts and results. SPE Paper 36449, presented at the 1996 Society of Petroleum Engineers annual technical conference, Denver, Colorado.

Shapiro, S.A. 2000. An inversion for fluid transport properties of three-dimensionally heterogeneous rocks using induced microseismicity. Geophys J. Int. 143, 931–936.

Silver, P.G., Daley, T.M., Niu, F. and Majer, E.L. 2007. Active source monitoring of crosswell seismic travel time for stress induced changes. Bulletin of Seismological Society of America 97(1B), 281–293.

Verdon J.P., Kendall J-M and Maxwell S. 2010. A comparison of passive seismic monitoring of fracture stimulation from water and CO_2 injection. Geophysics 75(3), MA1–MA7.

Warpinski, N.R., Wright, T.B., Uhl, J.E., Engler, B.P., Drozda, P.M, Peterson, R.E. and Branagan, P.T. 1996. Microseismic monitoring of the B-Sand hydraulic fracture experiment at the DOE/GRI Multi-Site Project. Paper SPE 36450 presented at the 1996 SPE Annual Tech. Conf. and Exhibition, Denver, October 6–9.

Weinstein E. and Weiss, A. 1984. Fundamental limitations in passive time delay estimation—Part II: Wide-band systems. IEEE Trans. Acoustical Speech Signal Processing: ASSP-31:1064-1068.

Zhang Y., Freifeld B., Finisterle S., Leahy M., Ennis-King J., Paterson L. and Dance T. 2011. Single-well experimental design for studying residual trapping of supercritical carbon dioxide. Int. J. of Greenhouse Gas Control 5 (2011), 88–98.

Zoback, M.D. and Zinke J.C. 2002. Production—induced normal faulting in the Valhalla and Ekofisk Oil fields. Pure and Appl Geophys 159, 403–420.

Chris Boreham, Barry Freifeld,
Dirk Kirste, Linda Stalker

12. MONITORING THE GEOCHEMISTRY OF RESERVOIR FLUIDS

12.1 Introduction

Geochemistry was a critical component of the overall monitoring and verification (M&V) strategy implemented for the CO2CRC Otway Project Stage 1.

Reservoir fluids (gases and formation waters) were sampled and analysed from the Buttress-1 well both pre- and post-injection of the CO_2-rich fluid (injected CO_2) and during the drilling of the CRC-1 injection well. However, the main geochemical focus of the Otway investigations was on sampling of the Naylor-1 monitoring well at the wellhead, and especially the reservoir fluids (pre- and post-CO_2-injection) via U-tubes (Figure 12.1) placed in the bottom hole assembly (BHA). The U-tubes were able to sample the change in fluids at a series of levels (Figures 12.1 and 12.2) where initially formation water was present but displaced over time by the injected CO_2. Tracers were added at CRC-1 and these compounds were also recovered at Naylor-1, quantified and compared with the evolution of CO_2 and CH_4 contents of the gas. Sampling by U-tubes at Naylor-1 first commenced soon after the BHA installation on October 2007 and continues now (2014).

During the course of Otway Stage 1, various operational issues relating to the system design and reservoir fluids were encountered and some delays were experienced until novel solutions were developed. Particular difficulties were encountered during the deployment of the BHA containing the reservoir level tools, due to the utilisation of a production well that was both narrow (3½″ completion) and contained a casing patch at a critical depth close to the injection formation (Figure 12.2).

12.2 Sampling the Buttress-1 well

Throughout the course of the gas injection period, Buttress-1 gases were collected both before the separator at the Buttress production plant and at the CRC-1 injection well before the gases were deployed subsurface. Buttress gas was sampled at the wellhead during a pre-injection flow test in June 2006 and again during the CO_2 injection phase of the Otway Stage 1 test in 2008–09 (see Chapter 4). Between the wellheads at Buttress-1 and CRC-1, evacuated sample cylinders were attached at surface access points using a SS NPT ¼″–½″ connector. Following a short gas flush to expel residual air, the connector was tightened, the gas cylinder valve opened and the gas pressure allowed to equilibrate for 1 minute. Solid wax was taken from the Buttress-1 CO_2 processing plant (see Chapter 4) where it accumulated in the first stage of the separator at reservoir pressure (approximately 9.4 MPa) but at a temperature that had dropped to approximately 35°C. Wax was also sampled in late 2007 at the Boggy Creek CO_2 production facility (Boreham et al. 2008).

12.3 Sampling the CRC-1 injection well

Fluid samples were collected from the CRC-1 injection well using a Schlumberger MDT modular formation dynamics tester. The MDT is a device that enables the collection of fluid samples at selected depths after a well has been

drilled. These fluid samples were then analysed and used to determine the composition of the formation fluid at that depth interval. MDT samples were collected from six different intervals in the CRC-1 well, to characterise the composition of the formation water occurring within the Dilwyn, Paaratte, Flaxman, Waarre C and Waarre A Formations.

The post-production gas cap in the Naylor Field at the CRC-1 well placed the gas-water contact (GWC) within the Waarre C interval at the end of gas production in 2004. Subsequently, the GWC fell as the gas cap partially refilled, while residual gas remained, saturated to approximately 20%. The composition of this pre-production gas was likely to be reflected in the recovered free gas from the MDT test (taken during the drilling of the CRC-1 well in March 2007) in the overlying Flaxman Formation. Here the free gas was 87.7% methane and 0.86% CO_2 with a wet gas content (% C_2-C_5/C_1-C_5) of 9.5% (Boreham et al. 2008).

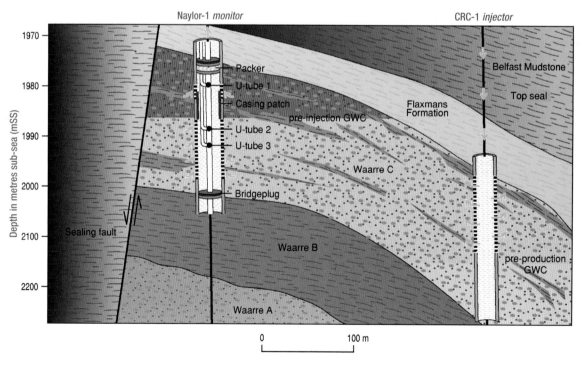

Figure 12.1: Schematic cross-section of the sampling system in the Naylor-1 well showing the relationship between the positioning of the U-tubes, the reservoir and the injection well.

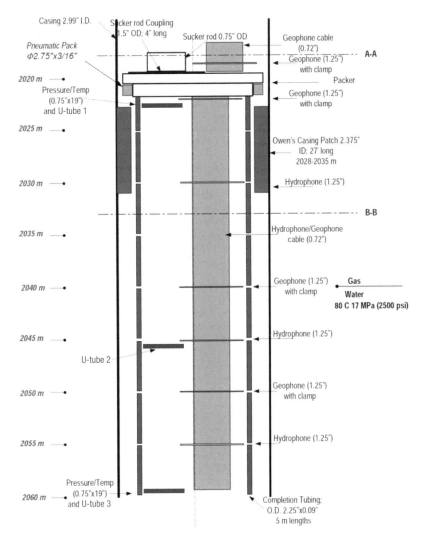

Figure 12.2: Schematic of the bottom hole assembly in the Naylor-1 observation well at the CO2CRC Otway Stage 1 site showing the distribution of equipment below the single packer in the Waarre C Formation interval being tested.

During the drilling of the CRC-1 well, mud gas (representing initially free and dissolved gas) was continuously extracted from the top of the "shaker" and analysed by gas chromatography for methane and individual C_2-C_5 wet gases, as part of the mud logging operation undertaken by Baker Hughes. Analyses were reported on the mud gas composition at 0.5 m intervals from 32 m to total depth at 2249 m. Additionally, mud gas was collected in Isotubes™ at 100 m intervals to 900 m, then at 20 m intervals to the top of the Flaxman Formation at around 2030 m, followed by 1 to 3 m intervals over the Flaxman Formation and Waarre Formation Unit C to 2083 m and finally at 10 m intervals throughout the lower Waarre and Eumeralla Formations to total depth (Boreham et al. 2008).

12.4 Sampling the Naylor-1 monitoring well

12.4.1 Downhole conditions

A tri-level U-tube installation allowed sampling of the residual gas cap in the shallowest U-tube (U1) and two progressively deeper intervals (U2 and U3) below the pre-injection gas-water contact (Figures 12.1 and 12.2). The injection of the CO_2 began on 18 March 2008. During the course of the CO_2 injection, there was lowering of the GWC as the Naylor structure filled with the CO_2-rich fluid until eventually all U-tubes produced gas only. Integration of the compositional results of the U-tube-derived fluids provided information on fluid flow and the temporal subsurface geochemical interactions of the CO_2, methane, formation water and rock (Underschultz et al. 2008). An integral part of the geochemistry M&V programme involved conservative tracers (CD_4, Kr and SF_6) being added to the Buttress-1 feed gas at the CRC-1 wellhead. The tracers separately tagged the injected CO_2 and methane components (Section 12.5). The first positive detection of the tracers was used to confirm breakthrough of the injected fluid at the monitoring well, initially as part of a dissolved CO_2 front and eventually the arrival of a supercritical CO_2 phase (Boreham et al. 2011). Once all U-tubes in Naylor-1 were delivering only gas at the surface, a second pulse of tracers (SF_6 and R-134a) was added to the injected CO_2 to monitor in-reservoir mixing

and diffusive processes within the expanding gas cap with significant residual water saturation. Furthermore, the tracers provided assurance of safe geological storage of CO_2 (Stalker et al. 2009a), with tracers being analysed as part of environmental monitoring systems in the near-surface and atmosphere (see Chapters 14 and 15).

Prior to commencement of monitoring at Naylor, a natural gas sample was taken at the wellhead in late 2006, approximately 2 years after cessation of commercial gas production, by which time the depleted natural gas reservoir had repressurised to its near pre-production level. By 2008 in the Naylor-1 monitoring well at the Waarre C Formation level (Figure 12.1) the following conditions applied:

> depth of monitoring in well: 2028–2065 m

> bottomhole pressure: 19,500 kPa

> bottomhole temperature: 85°C

> pressure accuracy: 0.02% of full scale (pressure at Naylor-1 was 5000 psi or 34.5 MPa)

> temperature accuracy: ± 0.5°C with a resolution of 0.005°C.

> cabling: TEC (tubing encapsulated conductor) Baker Hughes temperature rating to 177°C: collapse pressure rating 20,000 psi.

> three single-component geophones for HRTT monitoring, three 3-component geophones above reservoir for microseismics and nine single-component geophones for reflection surveys.

> three U-tubes (LBNL).

12.4.2 Sampling system

Subsurface fluid samples were collected from the reservoir level in the Naylor-1 monitoring well, through a specifically designed triple U-tube sampling array (Freifeld et al. 2005; Underschultz et al. 2008; Freifeld et al. 2009). Each of the three U-tubes consisted of two ¼″ stainless steel tubing lines (0.152″ i.d) connected by a tee with a check valve and a cylindrical 40 mm stainless steel inlet filter of 0.6 m length (Figures 12.2 and 12.3). The BHA was deployed

A-A

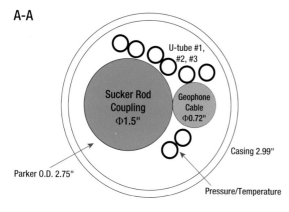

Figure 12.3: Cross-section of tools and communications systems from injection interval to surface in the Naylor-1 monitoring well.

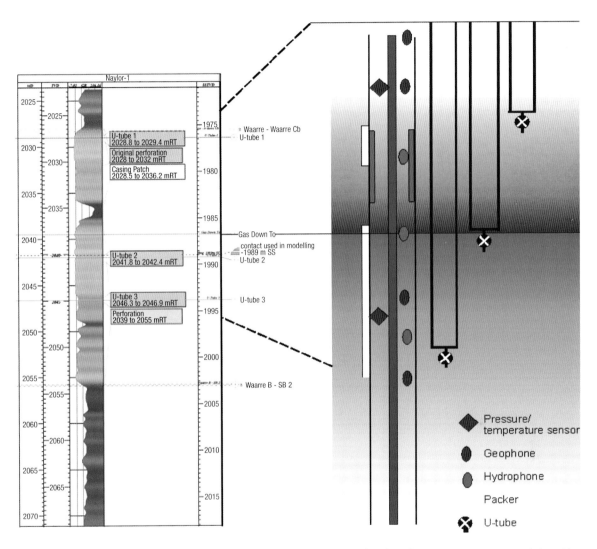

Figure 12.4: Location of U-tubes in the Naylor-1 monitoring well. Well logs and geological interpretation are presented alongside a schematic figure of the bottom hole assembly. This illustrates the location of the three U-tubes: U-tube 1 is in the gas cap, U-tube 2 just below the gas-water contact and U-tube 3 deeper in the residual gas column.

from surface to approximately 2 km down the Naylor-1 well, with each U-tube volume of approximately 58 L. Formation fluid entered the wellbore through perforations in the casing at 2028–2032 and 2039–2055 m drill depth from the rotary table (mRT), which extended above and below the U-tube filter intakes. The upper U-tube (U1) at 2028.8–2029.4 mRT had access to the residual gas cap, while the two lower U-tubes (U2 and U3) were below the post-production gas-water contact (GWC) at 2039.5 mRT; U2 located at 2041.8–2042.4 mRT; and

U3 at 2046.3–2046.9 mRT (Figure 12.4). The U-tubes were isolated from the remainder of the wellbore using an inflatable packer set at 2022 mRT (Figure 12.2). The U-tube sampling locations were selected so that any changes in the chemistry of the methane gas cap could be monitored, the initial breakthrough of CO_2 at Naylor-1 could be observed close to the GWC, and constraints on storage capacity and filling of the Naylor structure could be timed with eventual sampling of the injected CO_2 from the lowermost U-Tube.

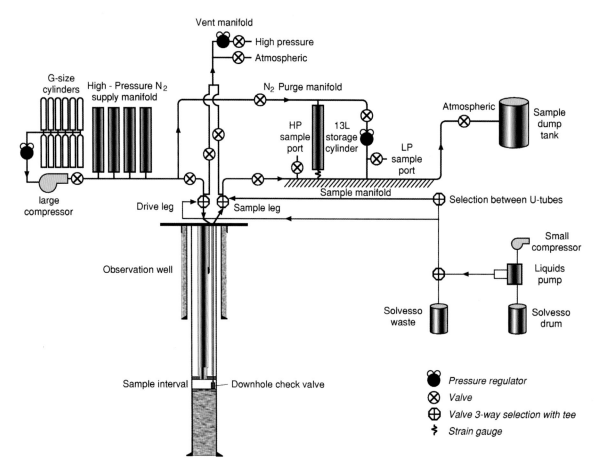

Figure 12.5: Simplified schematic of the Naylor surface monitoring facility incorporating the U-tube sampling and the solvent delivery/retrieval systems. Note only a single U-tube is shown for simplicity. Modified from Freifeld and Trautz (2006).

The Naylor surface facility consisted of a sample collection system for high pressure fluid and high and low pressure gas samples (Figures 12.5–12.7). U-tubes were sampled by opening the venting lines that terminated at the surface facility, where the wellhead pressure rapidly dropped and consequently formation pressure exceeded the pressure inside the U-tube, resulting in the check valve opening and formation fluid filling the U-tube.

Gas was obtained directly from U1, as this had access to the residual methane gas cap, producing self-lifted gas at surface at reservoir pressure (wellhead pressure of U1 is ~15.7 MPa). Gases were collected in both Swagelok SS cylinders (150 mL capacity and 34.5 MPa-rated with either one or two 34.5 MPa-rated stem valves with ¼″ NPT

female end fittings) and Isotubes (110 mL capacity, rated to 690 KPa and coupled to an Isotech wellhead sampler) connected in series (Figure 12.5). The SS cylinder was first filled to formation pressure, then isolated and the gas pressure slowly released to flow gas through the Isotube. The procedure was repeated and the Isotube removed once the pressure had fallen to 345 KPa (this sample was considered as a field backup sample). The SS cylinder was then repressurised to reservoir pressure and disconnected from the sampling line. This venting process resulted in variable cooling of the SS cylinder and Isotube and was a probable source of compositional fractionation that was evident when comparing the results from both types of samples (Boreham et al. 2011). Following analyses in the laboratory, an Isotube subsample of the high pressure

Figure 12.6: Naylor surface monitoring installation.

gas in the SS cylinders was taken using a fill-purge cycle (repeated five times). The SS cylinder was then vented, washed with dichloromethane, evacuated and recycled back to the Otway Project site. Since different laboratories were used for different types of analyses, the same Isotube sample was not necessarily used for all the analyses.

The two deeper U-tubes, positioned below the GWC, initially accessed the formation water, which did not flow to surface. Therefore, a N_2-assisted lift (N_2 down the drive leg or upstream lines, as shown in Figure 12.5) was required to push the formation water to surface for sampling. The U-tube lines were initially flushed with high pressure N_2 at 24.1 MPa, and then the N_2 pressure was released to atmosphere, allowing fresh formation fluid to fill the U-tube through the opened downhole check valve. High pressure nitrogen (24.1 MPa) was reintroduced to the upstream ¼″ lines of the U-tube being sampled. This closed the downhole check valve and mobilised the fluid sample up the ¼″ downstream line (sample leg in Figure 12.5) to surface at formation pressure (~13.8 MPa) for water and dissolved gas sampling.

Formation water samples for chemical and isotopic analyses were collected from U2 and U3 prior to self-lift. To collect the formation water samples, high pressure N_2 was used to drive 10 L of formation water to waste in a 120 L dump tank. The water was allowed to flow through high pressure and temperature in-line probes at a pressure of 1.38 MPa. In-line probe readings were

Figure 12.7: Inside the Naylor surface facility showing (top right and clockwise): high pressure gas sample and isotube in series; gas-liquid separator for low pressure exsolved gas sampling into isotube; wide view of high and low pressure sampling showing in-line 10 micron filters (rde, yellow and green casing) and strain gauge readout signifying weight of fluid filling the outside 13 L storage cylinders.

recorded after at least 8 L had passed through the system. Fluid flow was then diverted through a small SS cylinder (150 mL) and a further 3 L of water were discarded, after which the cylinder fluid pressure was allowed to build-up to 13.8 MPa. The formation water sample was then isolated and retained for later off-site subsampling and laboratory analyses.

Dissolved gas samples were collected from the 13 L holding cylinder immediately following the formation water sampling. After the initial 13 L of water were discarded through the water sampling procedure, fluid flow was diverted to the 13 L SS sample-holding cylinder, to formation pressure (Figure 12.5). Subsamples were then taken in two Swagelok SS cylinders connected in series (HP sample port in Figure 12.5). Slowly releasing the pressure from the top of the 13 L sample-holding cylinder allowed the evolved solution gas to flow through a low pressure perspex water trap (1 L) and then through an Isotube (LP sample port in Figures 12.6 and 12.7). When the pressure in the outside 13 L container had dropped to 345 KPa, a low pressure Isotube sample was taken as representative of the dissolved gas.

12.5   Injecting tracers at the CRC-1 injection well

As a major concern of stakeholders in any carbon dioxide storage project is the ability to show that the site is secure, the addition of suitable tracer components can be valuable to:

> verify that injected CO_2 is present

> determine the onset of CO_2 breakthrough at monitoring sites

> monitor for leakage (breach of seal, well, via faults, etc.)

> demonstrate that CO_2 at the surface is not sequestered CO_2

> aid in the understanding of the change in reservoir characteristics with CO_2 saturation increase (e.g. effect of change of wettability, heterogeneity or other geological uncertainties on storage)

> model risk and uncertainty

> demonstrate to the public that everything possible is being done to make sure that in the unlikely event of CO_2 leakage, it can be detected.

Without the addition of a tracer in the Otway Project Stage 1 study, it may have been difficult to determine whether any CO_2 present was from (1) the reservoir itself, (2) migration from high CO_2 accumulations in the region, which may be potentially similar in isotopic composition to that being stored or (3) atmospheric CO_2. The Otway area is a known area of CO_2 accumulation and therefore some accounting of these sources might well have been necessary. A review was undertaken to evaluate the site and potential tracers to be used and to

determine the best methods for their injection, collection and analysis for deployment in Stage 1 (Stalker et al. 2006; 2009a; 2009b).

Tracers are required to be chemically inert, environmentally safe, non-toxic, persistent and stable for purposes of long-term monitoring. Low-volume usage (for cost, availability and ease of handling and injection) and sensitivity to detection by analytical methods is also of importance. It is important to identify the characteristics of the site as well as other major constraints such as analysis of background conditions. The factors taken into consideration when deciding to use tracers are quite site specific.

One potential tracer for the Otway site was the carbon dioxide itself. In order to evaluate whether it could act as a chemical tracer in its own right, an inventory of likely carbon dioxide sources was prepared with bulk compositions, and where possible, carbon isotopic compositions measured or inferred.

The source of the injected CO_2 for Otway Stage 1 was from the Buttress-1 gas field. During injection, it was sampled regularly to determine any variation in bulk composition or carbon isotopes over the 66,000 t injection in Stage 1. Results (Section 12.7.1) show minor variation in the composition over time. The CO_2 at Buttress-1 is suspected to be from an igneous source (Figure 12.8), confirmed by its carbon isotopic composition of $\delta^{13}C$ –6.7‰.

The target reservoir, the Waarre Formation in the nearby Naylor-1 field, contains around 1% CO_2 as measured in U-tube samples taken before CO_2 injection commenced. While the Naylor-1 carbon dioxide may be isotopically distinct from the injected CO_2 (i.e. it has a more organic signature of $\delta^{13}C$ –11.0‰), other sources of CO_2 could have had more similar values to the injected gas from Buttress. For example, CO_2 from the processing or burning of fossil fuels is more likely to be close to that of CO_2 from oil/gas emplacement or organic processes such as respiration in soils, making it more difficult to differentiate between the signatures of the different gas sources. Conversely, the benefit of such an isotopically enriched CO_2 gas source with a signature –4 to –7‰ is that the CO_2 will be isotopically distinct from soil gas CO_2 resulting from photosynthesis and biological processes. Thus, the CO_2 gas itself may act as a tracer at the soil surface for purposes of monitoring and detection, so long as biological and respiratory processes have not affected the CO_2.

There are many small gas fields in the Otway Basin, and these natural gas accumulations often have quite varied ratios of methane to carbon dioxide. In fields with low CO_2 contents (< 5%) the carbon isotopic compositions may vary from those typical of igneous sources (Watson et al. 2004; Stalker et al. 2009a).

To identify the source of CO_2 in some of these natural gas fields from the Otway Basin, 11 samples underwent helium isotope analysis (Watson et al. 2004). Helium concentrations were between 153 and 5940 ppm, while isotopes measured, R_c/R_a ($R_c = {}^3He/{}^4He$ of sample and

$R_a = {}^3He/{}^4He$ in air of 1.4×10^{-6}) were between 0.03 and 3.03. The range at one end is typical of crustal sources while the latter is dominated by a mantle-derived source of helium. However, the use of the helium data as a potential tracer becomes difficult due to sample handling. The conclusion was that any "in-situ" materials that could potentially act as tracers were potentially subject to far too many variables, ranging from the varied CO_2 sources in the region to difficulties in performing specialised analyses, and so alternative and additive tracers were then evaluated. Furthermore, adding tracers to the injected-CO_2 provided an opportunity to develop skills and expertise in this area for future CCS projects.

12.5.1   Which tracers will be used?

Nimz and Hudson (2005) provided a framework to evaluate potential tracer chemicals. They primarily focused on noble gas tracers (helium, neon, argon and xenon) as well as sulphur hexafluoride and the use of ^{14}C-labelled CO_2 due to their inert nature and long-term stability.

Other gases and liquids were considered (Stalker et al. 2006; 2009a) that included chlorofluorocarbons (CFCs), perfluorocarbons (PFCs), hydroflurocarbons (HFCs) and hydrochlorofluorocarbons (HCFCs). Most of these have been used in tracer applications ranging from medical research, atmospheric monitoring, and groundwater pollution (Stalker et al. 2006, 2009a).

Each chemical type has different strengths and weaknessess; e.g. radiogenic $^{14}CO_2$ would be expected to behave very like CO_2 in its various phases and travel with the injected-CO_2. However, approvals to use such a substance, the social perspective of using a radiogenic product, sourcing the product and finding a laboratory that could measure it routinely would have been more difficult. Furthermore, as an absence of $^{14}CO_2$ in subsurface fluids relative to the atmosphere also acts as an indirect measurement of CO_2 source, this benefit would be lost. This is of importance if the results reported by Cenovus Energy (Cenovus Energy Report 2011) in relation to the potential leak of CO_2 at the Weyburn-Midale CO_2 monitoring and storage project site in Canada are considered. At Weyburn, the investigators used the abundance of "young" ^{14}C carbon dioxide versus "old" CO_2 depleted in ^{14}C as a major marker for an absence

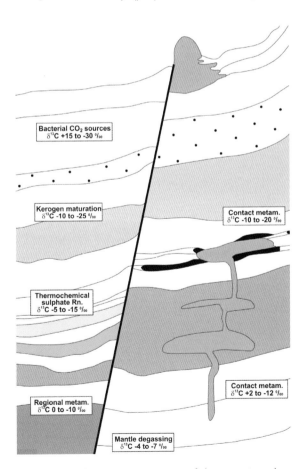

Figure 12.8: Schematic representation of the range in carbon isotopic composition of different CO_2 sources in the subsurface. after Thrasher and Fleet 1995.

of "old" carbon dioxide either leaking to the surface or being reworked from the deeply buried and ancient fossil fuels containing no ^{14}C. It was therefore judged unwise to adopt $^{14}CO_2$ as a tracer at the Otway site.

A further consideration for the test was the abundance of methane, both in the injected gas stream and in the natural gas field at Naylor-1. Little was known about the behaviour of methane and carbon dioxide or how it would interact in the subsurface in terms of mixing. It was clear from laboratory experiments (Sackett and Chung 1979) that there is no exchange of carbon between them at the storage reservoir conditions in Naylor-1. Perdeuterated ("heavy") methane (CD_4) was proposed as a marker for understanding the behaviour of methane in the system as well as for the carbon dioxide and this influenced the decision to use a novel tracer.

Thus, keeping in mind some of the key factors of tracer behaviour, availability and analytical complexity, the following tracers were chosen for the Stage 1 interwell test:

> Batch 1:

   – Sulphur hexafluoride (SF_6): 312 kg used

   – Perdeuterated methane (CD_4): 2000 L used

   – Krypton (Kr): 20,000 L used

> Batch 2:

   – 1,1,1,2-tetrafluoroethane (R-134a): 62 kg used

   – Sulphur hexafluoride (SF_6): 50 kg used.

The second tracer injection occurred later in the test. While it had been planned to add a pulse of xenon (Xe) with SF_6 to the test, Xe had become prohibitively expensive and it was difficult to source. An alternative, R-134a, was proposed. It was readily available, low cost and could be analysed using the same methods and equipment as other tracers.

There was discussion on the benefits of SF_6 and R-134a, given that they were already being monitored for baseline purposes for the Otway atmospheric programme. A large excess of SF_6 was deemed necessary for it to be detected above atmospheric background levels in the event of a point source surface leakage. There was agreement that, during the design of the tracer experiments, there had to be good communication between the different monitoring teams to ensure none of the different monitoring methods was compromised.

12.5.2 How much tracer to use and when?

The tracer amounts were based on a 30% residual methane gas saturation, transit distance of 300 m from injector to monitoring well, a need to saturate a 15 m radius around the injection borehole, reservoir porosity of 15% and a downhole pressures of around 20.7 MPa (3000 psi). This corresponded to approximately 1 month of injection of the proposed 100,000 t of CO_2 over a 24-month period. Also, while there are partitioning coefficient data for all of

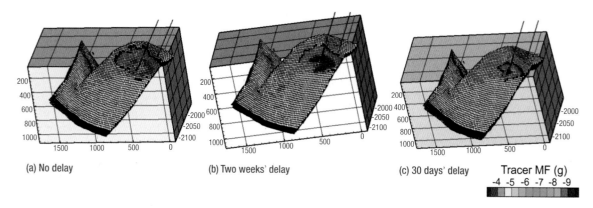

(a) No delay   (b) Two weeks' delay   (c) 30 days' delay   Tracer MF (g)
                                                                                                   -4 -5 -6 -7 -8 -9

Figure 12.9: Degree of saturation of tracer near the injection wellbore, modelled for (a) no delay where the tracer was injected prior to gas injection, (b) where approximately 2 weeks or 1000 t of CO_2 injection occurred and (c) where 30 days of injection occurred before the tracer was added. Figures provided by Jonathan Ennis-King.

these tracers in *n*-octanol, there are no data for those in supercritical CO_2. Therefore there could be unanticipated changes in behaviour during the injection and migration of these fluids.

Furthermore, it was also considered that there could be significant losses of tracer around the wellbore as injection commenced. Therefore various tracer injection scenarios were simulated to determine the best time for the introduction of the tracers. Too soon and the tracers may well have been the saturating solution that remained around the injection borehole. Too late, and the CO_2/CH_4 stream could have diluted the tracers to the extent that they would fall below detection limits (Figure 12.9). Therefore, the tracers were introduced after 1000 t of CO_2 had been injected and approximately 2 weeks into the start of injection.

The tracers were injected in two discrete pulses. The main pulse containing SF_6, Kr and CD_4 was injected over the 4 and 5 April 2008. The second pulse was added on 21 January 2009 and contained SF_6 and R-134a.

12.5.3 How were tracers injected?

Consideration was given to whether the tracers should have been injected as a pulsed injection (single injection, undiluted) or fed into the storage gas in a long-term controlled manner. The choice to use a single pulsed injection was made for a number of reasons, ranging from the likelihood that the operational design would be far simpler and less expensive, that the volume of tracer used would be lower, and that it would provide a better understanding of arrival time and mixing processes. The pulsed injection system was also easier to integrate with the surface facilities.

A slip-stream injector system was designed and built for Stage 1 to facilitate the insertion of a volume of gaseous tracers into the wellbore being swept by Buttress gas on the way down the CRC-1 well (Figure 12.10). G-size cylinders were attached to a section of the slip-stream and the connection opened to allow the gas pressure to equilibrate. The connection was then sealed and approximately 10 volumes of Buttress gas swept through the slip-stream to take the tracer down the well to the

Waarre Formation. The section was vented to atmosphere and sealed before reopening the G-size cylinder again to pressure equilibrate. These steps were repeated until each G-size cylinder reached < 350 kPa (< 50 psi) pressure to be effectively emptied (Stalker et al. 2009b).

CD_4 injection commenced on 4 April 2008 with the first cycle injected over 6 minutes. After six cycles of equilibration, flushing and venting as described above, all CD_4 was injected from that first cylinder after approximately 30 minutes. During this time, the injected-CO_2 gas flow rate averaged 121.7 t/day. A G-size cylinder of SF_6 (52 kg SF_6) followed the CD_4, with four cycles of injection taking place over a 1 hour 11 minute period, with a flow averaging 126.5 t/day. Three G-size cylinders followed containing Kr, then SF_6 and a final Kr cylinder. The final cylinder of Kr and four more cylinders of SF_6 were added the following day over a 5-hour period to complete the tracer injection. The consequence of employing such a

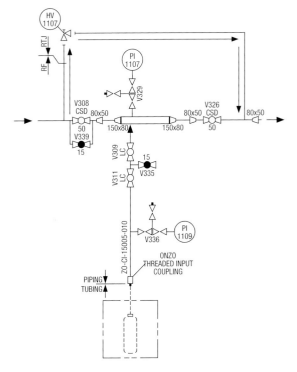

Figure 12.10: Slip-stream design for injecting gaseous tracers at CRC-1. The arrows show how the Buttress gas flushes through the system taking an aliquot of tracer down the well each time the valves are opened.

system was that it was possible to use virtually all the tracer gas contained within each G-size cylinder, maximising efficiency in tracer gas delivery. At the same time, it was possible to minimise risk of site contamination.

A port was fitted to allow subsampling of gas from the slip-stream. Pre-evacuated cylinders were attached to enable sampling of the supercritical mix prior to its delivery downhole at CRC-1. This allowed the sampling of injected CO_2 just prior to its injection into the CRC-1 well and could be compared with samples taken at Buttress-1 and via U-tubes in Naylor-1. However, this comparison proved impossible, as the former samples contained excessively high tracer contents due to some unswept dead space in the slip-stream sampling line.

The second tracer injection pulse occurred over a 7-hour period on 21 January 2009, a matter of a few days after U-tube 3 reached self-lifting. A total of five E-size cylinder of SF_6 (10 kg SF_6 each) and one G-size cylinder of R-134a was added using the same sample loop and procedure as for the first injection.

There was no obvious evidence of contamination around the CRC-1 site. This is always a potential concern during the deployment of tracers. These tracers were gaseous at STP and so readily dispersed if a spill were to occur. However, more recent experience with liquid tracers shows that the general handling of liquids relative to gases appears easier.

There was no capacity to test the soils in the vicinity of the CRC-1 well for the presence of tracers as part of the soil gas monitoring programme due to the fact that the Enviroprobe had to be pushed some distance into the soil. The infrastructure surrounding the slip-stream prevented the deployment of the soil gas sampling equipment for safety reasons.

Sampling procedures are described for the Naylor-1 monitoring well in Section 12.4. The samples recovered by U-tube for formation fluids, CO_2 and hydrocarbon gases had sufficient pressures to also be analysed for tracer contents. Sampling still continues (2014) and samples are batched and analysed for tracers, composition and isotopes of CO_2. In addition, atmospheric samples, soil gas, groundwater headspace and groundwater samples continue to be tested for presence of tracers.

12.6 Analytical methods

12.6.1 Gas chromatography of gaseous components

In the laboratory, a pressure regulator was attached to the SS cylinder containing the high pressure gas with the outlet pressure set to 100 kPa and connected to the gas chromatograph. A manual valve (Isotech Instruments) was screwed onto the Isotube and the dead space evacuated via a needle inserted through the septa and attached to a vacuum pump. The manual valve was opened, a small volume of gas extracted using a gas tight syringe, the gas expelled from the syringe, the syringe then filled, the pressure relieved to atmosphere and the desired contents of the syringe immediately injected into the gas chromatograph.

The molecular composition of the gas was determined using an Agilent 6890 gas chromatograph (GC) fitted with a series of $1/8''$ packed columns and with the GC oven held isothermal at 100°C (Boreham and Edwards 2008). The gas sample flowed at 30 mL/min for 1 minute through two gas valves (0.5 mL and 2 mL sample loops) connected in series. Injection through the 0.5 mL loop in a helium carrier gas resulted in the analysis of C_1-C_5, C_{6+}, O_2, N_2 and CO_2 with thermal conductivity detection. The mol % was determined against an external synthetic natural gas standard (mol %) with 62.95 C_1 or CH_4, 10.0 C_2, 10.1 C_3, 0.46 i-C_4, 2.99 n-C_4, 0.50 i-C_5, 0.51 n-C_5, 0.32 C_{6+}, 6.96 N_2, 5.21 CO_2. The experimental error for CO_2 (one standard deviation) was ± 2% of the reported value and the detection limit for CO_2 was 0.02 mol % (five times the signal-to-noise). Samples with high air content (> 10 mol %), due to leaking cylinders or ineffective flushing of Isotubes and those with elevated wet gas contents due to condensate build-up, were excluded from further consideration.

12.6.2 Gas chromatography of liquid hydrocarbons

Gas chromatography on the liquid and solids was conducted using an Agilent 6890 gas chromatograph fitted with a cross-linked methyl silicone capillary column (HP-1 PONA 50 m 0.2 mm i.d. × 0.5 μm film thickness). The wax (in a saturated solution of dichloromethane) and neat condensate were separately injected (0.5 μL) using a split/splitless injector operated in the split mode (500:1) at a temperature of 300°C. The oven was programmed from 30°C (isothermal 20 minutes) to 310°C at 8°C min^{-1} increments, with a final isothermal period of 10 minutes. The carrier gas was hydrogen with a linear flow velocity of 35 cm s^{-1}. Hydrocarbons were detected using a flame ionisation detector (FID).

12.6.3 Carbon isotopic analysis of individual gas compounds

Gas chromatography-combustion-isotope ratio mass spectrometry (GC-C-IRMS) was used to determine the carbon isotopic composition (all results reported in per mil VPDB) of CO_2 and the C_1-C_5 hydrocarbons. The gas components were resolved using a Poraplot Q fused silica column (25 m × 0.32 mm i.d.; Chromapak) with He carrier at 7.6 psi. The injector temperature was held at 150°C. For methane analysis, 25 μL of natural gas in a gas-tight syringe at atmospheric pressure was injected with a split flow of 200 mL/min and the GC held isothermal at 30°C. An independent injection was used to measure the wet gases (C_2-C_5). Following the elution of ethane, the GC oven temperature was increased to 230°C at 20°C min^{-1} and then held isothermal for 10 minutes. Splitter flow and injection volume were varied in order to keep peak intensities between 0.5 and 4V. As a minimum, all analyses were run in triplicates and the experimental error was ± 0.3‰ (one standard deviation).

12.6.4 Formation water chemical and isotopic analyses

The MDT sample chambers held approximately 450 mL of water. The high pressure formation water sample containers held approximately 150 mL of fluid at 13.8 MPa.

The amount of water decreased as the proportion of supercritical fluid increased in sampling prior to self-lift from U2 and U3. The MDT and U-tube fluid samples were sub-sampled for the various analyses. Upon opening the pressurised vessel pH, EC, temperature and Eh were measured immediately. A Jenway 3540 dual channel pH/conductivity meter with a Jenway 924 005 pH probe and Jenway 027 013 conductivity cell was used for the determination of pH and EC and an ATC probe for temperature. The probes were calibrated using two point standard techniques. ORP/Eh measurement was carried out using a Jenway 3510 pH/ORP meter and Jenway 924 003 ORP probe reported in mV relative to the platinum electrode.

The water samples were analysed for total alkalinity by titration of 0.5–1 mL water with hydrochloric acid (~0.02 N HCl). The equivalence point of the titration was determined colourimetrically, using a mixture of methylene blue and methyl red as an indicator. Subsequently, total iron (Fe), silica (SiO_2), phosphate (PO_4) and nitrate (NO_3) were determined using a HACH D 2800 spectrophotometer and according to the methods HACH 8008, 8185, 8190 or 8192.

The remaining water was filtered through a 0.45 μm cellulose acetate filter and subsampled for analysis of major and minor anions by ion chromatography (IC), stable isotopes of H, O and C by isotope ratio mass spectrometry (IRMS) and major, minor and trace elements by inductively coupled plasma—optical emission spectroscopy (ICP-OES) and—mass spectroscopy (ICP-MS). Subsamples for ICP-OES and ICP-MS analyses were acidified to pH < 3 with 10 N HNO_3 immediately after filtering (to prevent precipitation) and stored in plastic vials.

IC and ICP-OES was used for Br$^-$, Cl$^-$, F$^-$, SO_4^{2-} which were determined using a Dionex ion chromatograph series 4500i. ICP-OES samples were analysed for major, minor and trace elements (B, Ba, Ca, Fe, K, Mg, Mn, Na and Si) using a Varian Vista AX CCD Simultaneous ICP-OES. Accuracies of analytical data were generally better than 2%. IC and ICP-OES analyses were run with internal quality control samples and accepted only if precision and bias were within ± 5%. About 10% of the samples were spiked and accepted only if recovery was within ± 10% of the expected values.

ICP-MS analysis (an Agilent 7500cs with an Octopole reaction system) was used for elements including Ag, Al, As, B, Ba, Be, Cd, Ce, Co, Cr, Cs, Cu, Dy, Er, Eu, Fe, Ga, Gd, Ge, Hf, Ho, K, La, Li, Lu, Mg, Mn, Mo, Nb, Nd, Ni, Pb, Pr, Rb, Re, Sc, Se, Sm, Sn, Sr, Ta, Tb, Th, Ti, Tl, Tm, U, V, W, Y, Yb, Zn and Zr. The ICP-MS samples were run with internal quality control samples and standard reference materials, and accepted only if precision and bias were within ± 5%.

^{18}O and ^2H (H_2O) and ^{13}C (DIC) contents were determined using a Finnigan MAT 252 mass spectrometer. $\delta^{18}O$ values of water were measured via equilibration with He-CO_2 at 32°C for 24–48 hours in a Finnigan MAT gas bench and analysed using continuous flow. The δ^2H values of water were measured via reaction with Cr at 850°C using an automated Finnigan MAT H/Device. $\delta^{18}O$ and δ^2H values were measured relative to internal standards that were calibrated using IAEA SMOW, GISP, and SLAP standards. Data were normalised following Coplen (1988) and expressed relative to VSMOW where $\delta^{18}O$ and δ^2H values of SLAP are –55.5‰ and –428‰, respectively. CO_2 from dissolved inorganic carbon (DIC) was liberated by acidification using H_3PO_4 in a He atmosphere and analysed by continuous flow. $\delta^{13}C$ values are expressed relative to VPDB. Many samples were analysed twice and the precision (1σ) based on replicate analyses is: $\delta^{18}O$ = ± 0.1‰; δ^2H = ± 1‰; $\delta^{13}C$ = ± 0.1‰.

12.6.5   Carbon isotope analysis of CO_2

Gas samples were analysed for the $\delta^{18}O$ of CO_2 using approximately 0.05 to 1 cc of gas from the Isotubes sampled using a gas syringe and then injected, via a proprietary valve designed to minimise air contamination, into a He stream and a GC attached to a Finnigan 252 mass spectrometer using a Conflo III. The ^{13}C and ^{18}O isotope analyses were measured using CF-IRMS methods. CSIRO standard CO_2 gas and laboratory carbonate standards were used to correct the analyses onto the VPDB scale. The gas (CPR CO_2 $\delta^{13}C$ = –6.84; $\delta^{18}O$ = 16.63) and carbonate standards (CSIRO carbonate $\delta^{13}C$ = –13.46; $\delta^{18}O$ = –5.26 and PRM-2 $\delta^{13}C$ = 1.15; $\delta^{18}O$ = –17.63) have been calibrated using solid international carbonate standards NBS19 ($\delta^{13}C$ = +1.95; $\delta^{18}O$ = –2.20; Coplen

et al. 1995) and LSVEC ($\delta^{13}C$ = –46.60; $\delta^{18}O$ = –26.50; Coplen et al. 2006) that were reacted with 104% phosphoric acid in vacuutainers at 25°C overnight and injected into the same GC system.

12.6.6   Analysis of tracer compounds from U-tube samples

Each of the tracers could be measured by more than one method. Methods varied depending on the medium analysed—groundwaters, air, pressurised gases, reservoir fluids and soil gas samples. Each had to be handled differently because of the medium the tracers might be in, and the concentration levels relating to each type of sample.

As the Project initially relied on third parties to analyse the tracer samples, it was not always possible to use the same instrumentation at the same institution, thus delaying analysis. This meant that there would only be periodic releases of data back to the research staff to observe changes to the concentration of tracers (including the identification of breakthrough). Significant time was spent on method development to enable equipment that might not have been the best analytical tool for some of the tracers to be reconfigured; this added to the time and cost of analysis. Subsequently Geoscience Australia took over most of the analyses. Having equipment for analysis on-site was found to be much more successful during the Stage 2B operation, having learned the lessons of Stage 1 sample analysis protocols.

As samples were analysed by different providers at different times it was important that the same standards, as well as blank samples and duplicates, were used to check for consistency in the measurements. Gas samples were carefully equilibrated and subsampled so that multiple institutions had the same set of gas standards to compare results.

During repeat analysis of some older samples it was found that over time the tracer responses had declined, giving false low values. The conclusion here was that the tracer compounds (present in "trace" quantities) could adhere to the vessel walls, which were Isotubes, and this limited the shelf life of the samples. This problem was compounded by the batching of the samples for analysis in the early period of sampling. In spite of these experiences, the dataset was found to be sufficiently consistent over the sampling period to provide useful data.

Gas samples collected from all three U-tubes were analysed for molecular and carbon isotopic composition at Geoscience Australia prior to being sent on for specialised tracer analysis. As it was anticipated that future tracer deployments might include other noble gases, the method development included analysis of xenon (Xe) even though it was not injected as a tracer for Stage 1.

The main instrumentation used to analyse for tracers from the pre-injection to about 600 days of sampling by U-tube was a Hewlett Packard 5890 II gas chromatograph (GC) interfaced to a VG AutoSpecQ Ultima mass spectrometer (MS) using a GCMS-SIR method as described by Boreham et al. (2007) and Stalker et al. (2009a). A gas-tight syringe was used to inject the gas sample (250 µL) into the heated injector (250°C). The head pressure of the column was set up to 172 kPa (25 psi) with a split flow of 23 mL/min. The GC column used was a PLOT fused silica column (50 m × 0.32 mm i.d.) coated with a 5A molecular sieve. The oven temperature programmes were as follows: CD_4 isothermal at 40°C (one run), SF_6 isothermal at 40°C (one run) while Kr and Xe were analysed for simultaneously on an isothermal 250°C run. Xenon had been identified as a potential future tracer in the Stage II Otway Project, and also gave an indication of the levels of air contamination in the samples.

Different single ion monitoring (SIM) programmes were written to optimise the analysis and quantitation of the three tracers and xenon. The diagnostic mass to charge (m/z) ratios used were as follows: CD_4 m/z = 20.0564, SF_6 m/z = 126.9641, Kr m/z = 83.9115 and Xe m/z = 131.9041. The mass spectrometer was tuned to 1300 resolution (electron energy 70 eV; electron multiplier 250 V; filament current 200 µA; source temperature 250°C). The overall sensitivity allowed for the GCMS in SIM mode to detect sub ppb levels of CD_4, and similar for the krypton if the sample was air-free (Boreham et al. 2007). This particular GCMS method required a specialised sample injection port to minimise atmospheric contamination. Frequent use of blanks, column burn-offs, calibrations and replicates were carried out to constrain analytical

error. He-only injections gave average (daily over 10 days of analyses) "blank" responses of 0.2 ppb, 30 ppb and 5 ppb for CD_4, Kr and SF_6, respectively.

GC-ECD (electron capture detector) can improve SF_6 detectability by a factor of 100. In order to obtain some preliminary results to screen for the arrival of injected-CO_2 and first evidence of tracer in the subsurface, some samples were analysed on a 5890 Agilent gas chromatograph with ECD interfaced to a custom-built sample inlet. The equipment is a CSIRO custom-built piece of equipment designed to measure the presence of low concentrations of SF_6 in ground waters. The equipment allows different volumes to be injected by way of different sample loop sizes and dilution steps. A pre-column was used to clean and dry the samples using magnesium perchlorate before the sample was sent to a 5A molecular sieve column. Compounds that were not of interest were backflushed off the column and vented, leaving SF_6 to be measured. Detection limits are believed to be comparable with similar equipment described in Clark et al. (2004) of as low as 0.03 pmol^{-1}. An isothermal oven temperature of 60°C was used, with the detector set at 250°C.

SF_6 was also analysed at Geoscience Australia by GCMS-SIR (Agilent 5973) by monitoring mass-to-charge 127. A thick-film methylsilicone fused silica capillary column (BP-1, 50 m × 0.25 mm, 1 µm film thickness) was used under a constant flow of helium carrier flow of 1 mL/min. The column oven was sub-ambient at –20°C and under these conditions CD_4, SF_6 and Kr were partially separated with SF_6 eluting within 2 minutes. The analysis was performed in triplicate with repeated injections during a single data collection. A gravimetrically prepared gas mixture of 10 ppm (v/v) SF_6, Kr and CD_4 in helium was supplied by CoreGas together with a 1:10 dilution of the gas standard in He were used as calibration for determining the tracer concentrations in the gas samples. Reproducibility of the tracer concentrations was ± 10% of the reported value.

For R-134a (1,1,1,2-tetrafluoroethane), the ANU Micromass AutoSpec was used for GCMS-SIR analysis and operated at a resolution of 1500 and monitoring masses 83.011 and 102.009 with a dwell time of 200 ms. A gas-tight syringe was used to inject the gas sample (200 µL; triplicate injection at 1 minute intervals) into the heated (250°C) split injector (25 mL/min split flow) and onto a thick-film methylsilicone fused silica capillary column (BP-1, 50 m × 0.25 mm, 1 μm film thickness). Helium carrier flow was held constant at 1.1 mL/min. Helium was used as carrier gas under a constant pressure of 25 psi and the column oven temperature was held isothermal at 40°C where the first R-134a peak eluted around 4 minutes. Gravimetrically-prepared gas mixtures of 11 ppm (v/v) and 1.1 ppm (v/v) (CoreGas) were serially diluted into He-filled stoppered glass bottles (sealed with a pre-drilled hole in the cap with a 5 mm thick silicone septa) to give 110, 11 and 1.1 ppb R-134a working standards for calibration. Reproducibility of the tracer concentrations was ± 10% of the reported value.

Clearly the inability to analyse gases from a single supplier created difficulties for the Project, with the main issue being result delivery time. Dedicated tools are an important feature to provide timely and compatible datasets over time particularly for monitoring and verification purposes.

The detailed calibrations and testing conducted on each piece of equipment, while very costly, provided a robust framework for data comparison.

12.7 Composition of hydrocarbons

12.7.1 Buttress-1 supply well

Buttress-1 gas, which varied slightly in composition over the course of injection ranges from 72–78 mol % CO_2 (average 75.4 mol %), 3–6% C_2-C_5 wet gases and 2–3% N_2 with the balance methane. The variation in the C_{6+} liquids content shows a seasonality control, with generally higher molecular weight liquids during the warmer months. The carbon isotopic value of the Buttress-1 supply gas, which averaged –6.7‰ (Boreham et al. 2011) had its origin in the geologically recent influx of volcanogenic CO_2 into the Otway Basin (Watson et al. 2004). The occurrence of waxes in the Buttress gas is discussed later.

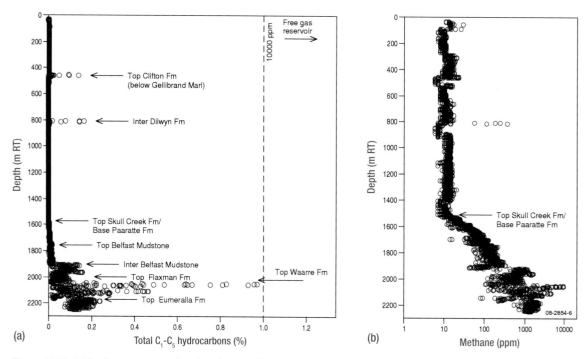

Figure 12.11: Molecular composition of mud gas from mud log of CRC-1 versus depth (m) showing (a) total C_1-C_5 gaseous hydrocarbons (%), and (b) methane (ppm; note logarithmic scale).

12.7.2 CRC-1 injection well

The pre-production gas in the overlying Flaxman Formation included methane (87.7%), CO_2 (0.86%) and a wet gas content (% C_2-C_5/C_1-C_5) of 9.5% (Boreham et al. 2008). From the mud gas analysis, a maximum total hydrocarbon concentration of 9730 ppm occurred at the top of the Waarre Formation Unit C reservoir (Figure 12.11(a)). Surprisingly, a free gas zone as identified from MDT in the overlying Flaxman Formation (see above) had a lower maximum concentration (1742 ppm total hydrocarbons; Figure 12.11(a)). Total hydrocarbon and methane concentrations followed background levels (around 10 ppm methane; Figure 12.11(b)) from near surface down to the base of the Paaratte Formation at around 1500 m, suggesting the Skull Creek Formation formed an effective seal to upward migration of hydrocarbons. The gas kick (up to ~0.2% total hydrocarbons at 458 m and 812 m, Figure 12.11(a)) resulted from carbide addition (to estimate drilling mud circulation time) and its rapid decomposition to acetylene. At greater depths, methane concentrations show an accelerated increase (Figure 12.11(b)), possibly due to upward diffusion of

gas and/or early gas generation (Mango and Jarvie 2009). However, gas contents showed an abrupt increase below the intra-Belfast Formation shales (Figure 12.11(a)), which acted as the major seal of the gas column below (Boreham et al. 2008).

The CO_2 concentration, determined using the Isotube gas samples, showed the highest concentration of 3413 ppm within the Flaxman Formation. No obvious relationship between methane and CO_2 contents was found (Figure 12.12 and Boreham et al. 2008).

The average $\delta^{13}C$ methane of the mud gases from the Waarre Formation was –32.9‰. The consistency in the $\delta^{13}C$ of methane in the Waarre Formation over a 10-fold range in hydrocarbon concentration, suggests that there were no major baffles or impervious barriers that would have resulted in reservoir compartmentalisation. However, on closer inspection, the total hydrocarbon content did show "offsets" coinciding closely with the less permeable facies within the Waarre Formation Unit C (Figure 12.13). This was likely to be associated with the influx of gas as the Waarre Formation repressurised, slowly returning

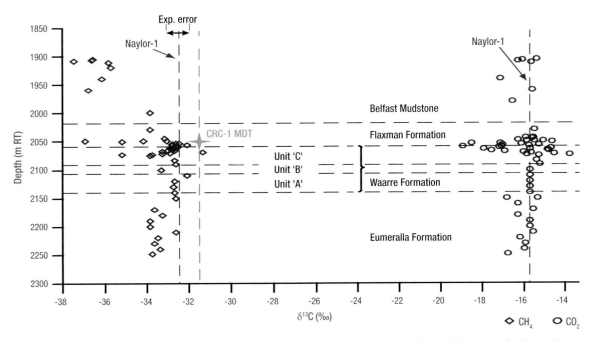

Figure 12.12: Carbon isotopic composition (‰) of methane and CO_2 of mud gas in Isotubes from CRC-1 versus depth (m). Otway air = δ^{13} C CO_2 –8‰; Buttress 1 = δ^{13} C CO_2 –6.7‰.

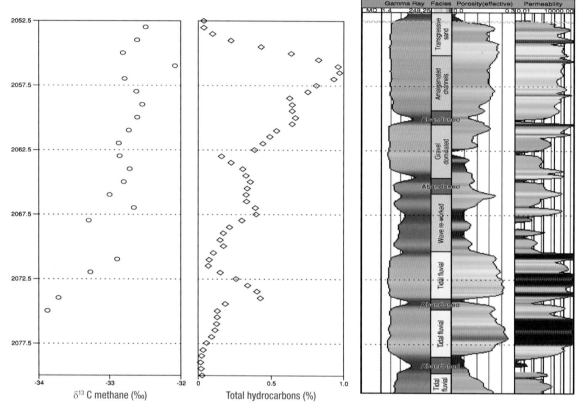

Figure 12.13: Carbon isotopic composition of methane, mud gas concentration (% total hydrocarbons), gamma ray, interpreted facies, porosity and permeability throughout the Waarre Formation.

to equilibrium conditions following production from the Naylor structure. However, there was no difference in the $\delta^{13}C$ methane over the main Waarre Formation Unit C reservoir (Figure 12.12), suggesting no carbon isotopic fractionation across these baffles. Methane in the Flaxman MDT gas was slightly heavier (enriched in ^{13}C) by 0.9‰ compared to the Naylor-1 methane (Figure 12.12), but still within experimental error. Similarly, methane from the deeper Eumeralla Formation ($\delta^{13}C$ –33.5‰), although, on average, slightly depleted in ^{13}C compared to Naylor-1 gas, was still within experimental error. On the other hand, the $\delta^{13}C$ of methane in the Belfast Mudstone showed a general trend of increasing depletion in ^{13}C with distance from the gas saturated zone underlying the Flaxman/Waarre successions (Figure 12.11). This trend is consistent with isotopic fractionation whereby the lighter isotope ($^{12}CH_4$) preferentially migrates fastest through the lower permeability seal rock.

12.7.3 Formation water

The primary difficulty encountered in the sampling of formation fluids from the recently drilled wells was contamination of the fluids by drill mud. During the Otway drilling, the mud entered the porous and permeable formations and mixed with the formation fluid, contaminating the fluid. At CRC-1 a tracer was added to the drill mud to determine if the extent of mixing between drill mud and formation fluid could be quantified. The tracer used was fluorescein, an organic chemical that fluoresces at a characteristic wavelength on UV excitation, enabling its measurement at low concentrations. The fluorescein content of the drill mud, as well as the chemical composition of the mud, was measured during drilling and quantities of tracer were added twice daily to the drill mud over a period of 18 days (from 745 m to total depth) in order to establish a fixed fluorescein baseline value of 5 ± 1 ppm w/v. This baseline value could then be used to quantify the drill mud contamination of the MDT samples. Some of the main additives used in the drill mud at CRC-1 included KCl, NaOH and Na_2CO_3 resulting in elevated pH, K, Cl and Na content. The MDT fluid composition needed to be "corrected" for the amount of mud contamination to enable the true formation fluid composition to be determined.

The amount of drill mud contamination determined from the fluorescein content varied from 4 to 20%. Examination of the formation water composition after correcting for the drill fluid contamination based on the fluorescein content indicated that the results were inconsistent and the formation water composition determined through this method were unreliable. A second method based on regressing the MDT data using the mud chemical composition was applied and proved to be quite successful. Calculated drill mud contamination using this method ranged from 5 to 44%. Table 12.1 summarises the MDT sample information including the calculated drill mud

contamination determined using the fluorescein tracer, and the calculated formation water composition using the fluorescein and regression methods. The results of the regression method were consistent with DST formation water sample analyses from similar horizons in the region (Table 12.1). However it should be pointed out that the DST data was highly unreliable, as some degree of drill fluid contamination was common.

As shown by Table 12.1, the formation waters in the Waarre C are moderately saline and have an Na-Ca-Cl composition. The pH is circum-neutral and the sulphate content low. The calculated formation water compositions for the Dilwyn and Paaratte Formations have low salinity and are of the Na-Ca-Cl type for the Dilwyn and Na-Cl for the Paaratte. However, the mud composition changed rapidly during drilling, resulting in a large potential error associated with corrections using these values. Nonetheless, the data suggest that these formations do have less saline water than the deeper Waarre, are easily distinguished from the target reservoir formation water, and clearly indicate

Table 12.1: Calculated pH and major ion chemistry (mg/L) of the MDT samples after correction for drill mud contamination using the fluorescein tracer and regression methods. The lower part of the table shows the DST sample composition for two regional wells.

Depth	Formation	Contamination (%)	pH	Ca	K	Mg	Na	Cl	SO$_4$
Fluoresceine tracer corrected									
700	Dilwyn	19.9	7.15	234	1522	132	1616	4966	231
1278.6	Paaratte	10.3	8.88	10	188	6	160	577	4
2049.2	Flaxman	4.4	7.27	123	10501	16	4543	19632	898
2054	Waarre C	11.4	6.67	520	−1516	74	5588	10045	−27
2075.5	Waarre C	11.4	7.04	502	120	72	5814	12548	109
2138.3	Waarre A	16.1	7.29	2120	1368	16	3779	14731	411
Regression corrected									
700	Dilwyn	33.7	7.13	271	24	158	1742	4021	252
1278.6	Paaratte	12.8	8.87	10	1	6	94	331	-17
2049.2	Flaxman	43.7	7.06	197	61	22	4497	13821	520
2054	Waarre C	5.4	6.70	487	79	70	5521	11047	60
2075.5	Waarre C	11.6	7.04	502	81	72	5816	12525	107
2138.3	Waarre A	20.7	7.27	2243	53	17	3736	14039	353
Regional	**Well**	**Formation**							
1820	Howmains 1	Waarre	6.10	900	135	100	6550	13575	46
1494	North Paaratte 1	Waarre	7.61	1035	91	84	5981	10700	120

there is no hydraulic continuity between the formations. The MDT sample from the Flaxman Formation was predominantly gas and the results of the water composition calculations should be considered to have a very large error associated with them. Any future requirements of the Flaxman water chemistry will need careful consideration prior to inclusion of the results of the MDT.

12.7.4   Pre-breakthrough dissolved gas

The CO_2 (N_2-free basis) of the exsolved gas averaged 7.5 mol % (Figure 12.14), the CO_2/methane ratio was very low at 0.08 while the CO_2/ethane and CO_2/propane ratios were 2.8 and 13, respectively (Figure 12.15). The N_2/CH_4 averaged 0.84, reflecting the mixing of 13 L of N_2 in the 13 L capacity surface storage cylinder as part of the sampling procedure. The average $\delta^{13}C$ CO_2 was –13.0‰ for the CO_2 gas released from both U2 and

U3 (Figure 12.16). Isotopic equilibrium between the dissolved and gas phase CO_2 at 20°C (surface facility separation temperature) was calculated to be –1.1‰, suggesting that further depletion in ^{13}C for the initial exsolved CO_2 from U2 and U3 compared to U1 free CO_2 was likely to involve some other process, such as a kinetic Rayleigh-type distillation or mixing (Boreham et al. 2011).

12.7.5   Pre-breakthrough free gas

Naylor-1 natural gas from the Waarre Formation gas cap was sampled in late 2006 at the wellhead during a pre-injection flow test and before the installation of the BHA. The composition of this gas was 86.6% methane, with a wet gas content (% C_2-C_5/C_1-C_5) of 9.4% and 0.12% for C_{6+} liquids; the CO_2 content was slightly lower than the Flaxman MDT gas at 0.09% (Boreham et al. 2008).

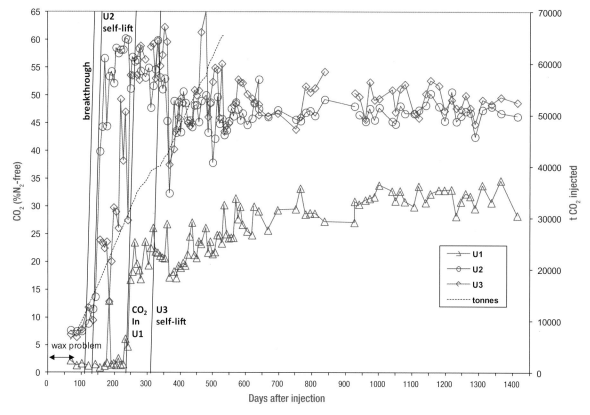

Figure 12.14: Time series (to 1450 days), of U-tubes 1, 2 and 3 showing CO_2 content (mol % as N_2-free basis) of samples collected at Naylor-1 well, and cumulative tonnes of mixed supercritical CO_2-hydrocarbon gas injected into the CRC-1 well.

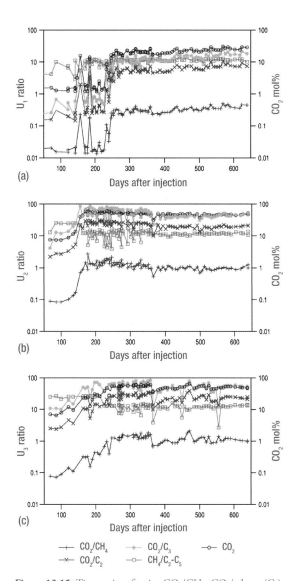

Figure 12.15: Time series of ratios CO_2/CH_4, CO_2/ethane (C_2), CO_2/propane (C_3), CH_4/wet gases (C_2-C_5) and CO_2 mol % for (a) U-tube 1, (b) U-tube 2 and (c) U-tube 3. Note the log scale on primary and secondary y-axes.

Based on the carbon isotopic composition of the CO_2 at –15.8‰, its origin was considered to be mixed with an input of isotopically lighter (depleted in ^{13}C) CO_2 from an organic source (Boreham et al. 2008). Once the U-tube installation was deployed, gas samples from the upper U-tube resulted in a slightly different composition. The average of the U1 gas from pre-injection to just before breakthrough of CO_2 gave a more refined gas composition

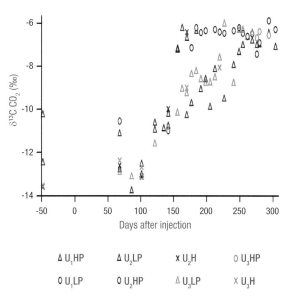

Figure 12.16: Time series of U-tubes 1, 2 and 3 showing the carbon isotopic composition (^{13}C ‰) of CO_2. Note: HP = high pressure gas sample collected either in a high-pressure SS cylinder or a low pressure Isotube; LP = low pressure gas from depressurising formation water to 345 kPa and the released gas collected in an Isotube.

of 85.8% methane, a wet gas content (% C_2-C_5/C_1-C_5) of 9.5% and 0.84% in C_{6+} liquids, with an elevated CO_2 content of 1.2% (Figure 12.14) and CO_2/CH_4, CO_2/ethane and CO_2/propane ratios averaging 0.02, 0.39 and 0.82, respectively (Figure 12.15). However, the largest variance between the wellhead sample and that collected via the U-tube was the enrichment in ^{13}C at –11.0‰ for the latter (Figure 12.16). The reason for this difference remains unclear, although the higher concentration of the CO_2 in the U-tube-derived gases may minimise isotopic fractionation effects during sampling. Similarly depleted CO_2 (average –16.0‰; Figure 12.12) was found in the CRC-1 mud gases where CO_2 contents were very low (below 0.34%; Boreham et al. 2008). Nevertheless, the carbon isotopic composition of CO_2 formed a natural tracer with at least a 4.3‰ difference between the carbon isotopic composition of the "heavy" injected and "light" indigenous CO_2. On the other hand, the carbon isotopic composition of the gaseous hydrocarbons were the same, with methane from Naylor-1 wellhead and U1 gases, having $\delta^{13}C_{methane}$ of –32.5 and –31.9‰, respectively, and identical (within experimental error) to methane from Buttress-1 and CRC-1 (Boreham et al. 2008; 2011).

12.7.6   Post-breakthrough dissolved gas

Arrival (breakthrough) of the injected CO_2 occurred at U2 between 100 and 121 days after injection with a measurable increase in CO_2 content above a baseline of 7.5 mol % CO_2 (N_2-free) at day 121 (Figures 12.14 and 12.17). This timing corresponded to 10,000–12,700 t of CO_2 having been emplaced in the Naylor structure via the CRC-1 injection well. After this time, U2 displayed a consistent rise in CO_2 mol % with an abrupt increase in CO_2 mol % between 142 and 156 days. By day 177, U2 transition from N_2-assisted lift of formation water to a self-lifting gas to surface, was complete after the injection of 21,100 t of CO_2-rich fluid. The transition to self-lifting in U2 corresponded to the downward movement of the GWC to within the vicinity of the U2 inlet; a movement of at least 2.3 m (to top of U2 inlet filter) over the pre-injection GWC level.

Over the same time period, for U2 the CO_2/methane, CO_2/ethane and CO_2/propane ratios increased rapidly from their low pre-breakthrough ratios to a maximum ratio of ~1.4, 31 and 86, respectively (Figure 12.15). With ever increasing higher exsolved gas contents from breakthrough and over the transition period, the N_2/methane ratio decreased accordingly while the C_1/C_2-C_5 ratio remained constant at around 25. This latter ratio was approximately a two fold increase over the pre-breakthrough ratio and that of the injected CO_2-rich fluid. The higher C_1/C_2-C_5 ratio during this transition period was a consequence of the preferential dissolution of methane compared to wet gases from the critical fluid phase to the aqueous phase. They provided compelling evidence for a dissolved CO_2 "front" immediately preceding the arrival of the supercritical CO_2 phase.

The interpretation of the U3 compositional results was further complicated by varying degrees of wellbore mixing (Boreham et al. 2011), although compositional changes

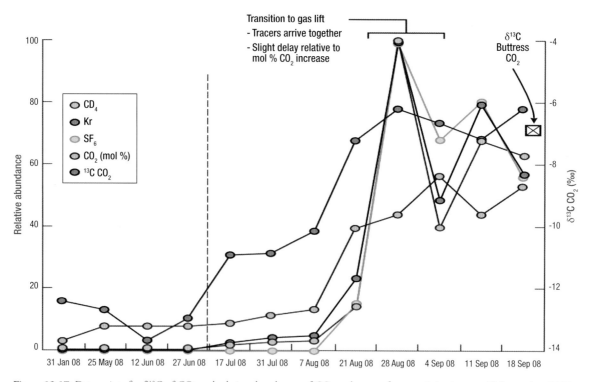

Figure 12.17: Data points for $\delta^{13}C$ of CO_2 and relative abundances of CO_2 and tracers from pre-injection to mid-September, 2008, in U-tube 2 to around the time that it became self-lifting.

during the transition of self-lifting generally followed similar trends to U2. However, transition of self-lifting for U3 occurred over a much longer time frame commencing between day 226 and 247 and completed on day 303 with the cumulative injection of ~39,000 of CO_2-rich fluid (Figure 12.14).

By comparison, the (always) self-lifting U1 gas did not indicate the arrival of injected CO_2-rich fluid until day 247, with a consistent increase up to approximately 20 mol % CO_2 by day 261. The time of the initial rise in CO_2 content at U1 post-dates the final rapid rise in CO_2 content in U3 (Figure 12.14).

The carbon isotopic composition of CO_2 formed a natural tracer, with U1 $\delta^{13}C$ CO_2 at –11.0‰ average for the free gas. For U2, there was an abrupt increase in $\delta^{13}C$ of approximately 2‰ on day 121 (Figure 12.16), coincident with the initial rise in mol % CO_2 (Figure 12.14). A larger enrichment in ^{13}C was observed concomitant with the major change in CO_2 concentration between days 142 and 156; the transition from N_2-assisted gas lift of formation water to self-lifting gas was complete by day 177. There, the carbon isotopic signal of the allochthonous CO_2 completely overwhelmed the indigenous CO_2. After this time and up until the final isotopic measurement (day 303), the CO_2 carbon isotopic composition remained constant, with an average of –6.5‰ ($\sigma = 0.4$‰, $n = 13$), which was the same isotopic value found in Buttress-1 gas (average –6.7‰). The $\delta^{13}C$ CO_2 from U3 also showed the enrichment in ^{13}C on day 121 concomitant with the changes at U2, suggesting limited wellbore access to gas for the deeper U3 level at the time of breakthrough at U2. By comparison, the change in $\delta^{13}C$ CO_2 for U1 remained relatively small throughout the time that U2 and U3 had taken to stabilise at the $\delta^{13}C$ value of the injected CO_2 (Figure 12.16). However, there was a trend towards enrichment in ^{13}C before the major increase in CO_2 mol %, indicating very minor access to CO_2-rich fluid from below. Wellbore mixing at U1 was confirmed with the two transient spikes in CO_2 content in U1 at days 156 and 184 (Figure 12.14). A consistent increase in $\delta^{13}C$ of CO_2 in U1 only occurred with the increase in CO_2 mol % and rapidly stabilised within a couple of weeks to the same isotopic value as the injected CO_2.

12.7.7 Post-breakthrough free gas

From day 177 for U2 and after day 303 for U3, no further water was produced and the lower two U-tubes thereafter delivered only free gas to the surface. This represented the arrival of the supercritical fluid phase and the downward movement of the GWC to the level of the respective U-tube intakes. CO_2 contents in the free gas consistently remained lower than around 60 mol %, which was well below the Buttress-1 gas which averaged 75.4 mol % CO_2. Following an initial high value of 51–60 mol % CO_2 between days 170 to 345, the CO_2 content remained fairly constant at 47 ± 2 mol % from day 380 to day 1413 (January 2012) meaning that it remained compositionally stable for 34 months. Over the same extended time period U3 averaged 49 ± 4 mol % CO_2 (Figure 12.14).

The CO_2/CH_4 ratio for U2 increased rapidly to an average of approximately 1.4 between days 177 to 345 and then remained constant at 1.0 ± 0.1 from day 380 onwards (Figure 12.15(b)), which was the same as U3 (1.1 ± 0.2; Figure 12.15(c)). This compared to a CO_2/CH_4 ratio of 3.5 for the injected gas at CRC-1. The C_1/C_2-C_5 ratio at U2 of the free gas dropped from a high of 25 during the transient dissolved CO_2 front, down to an average of approximately 12 after day 163, which was back to the level of the pre-breakthrough ratio (Figure 12.15(b)). U3 showed a similar transition from a drier to a wetter gas, although over a longer transition time (Figure 12.15(c)).

By comparison, the always-self-lifting U1 gas did not indicate the arrival of injected gas until day 247 with a consistent increase up to approximately 20 mol % CO_2 on day 261 followed by a gradual rise to an average of over 30 mol % CO_2 (moving average of 5 = 31.7 ± 0.6 mol %) between days 972 and 1413 (Figure 12.14). The time of the initial rise in CO_2 content at U1 post-dated the final rapid rise in CO_2 content in U3. A similar overall trend was seen in the CO_2/CH_4, CO_2/ethane and CO_2/propane ratios at U1, resulting from the mixing of low-CO_2 Naylor-1 gas with the CO_2-rich injected gas (Figure 12.15(a)). Before mixing (up to day 226) CO_2/CH_4, CO_2/ethane and CO_2/propane ratios averaged 0.02, 0.39 and 0.82, respectively. After the arrival of the CO_2-rich gas and after the mixed gas had initially stabilised at CO_2

approximately 20 mol %, the ratios were 0.29, 5.5 and 11.9 (average of days 261 to 450) increasing to a stable average value of 0.51, 11.0 and 25.0 after day 972, respectively. The pre-injection Naylor-1 gas was slightly wetter (CH_4/C_2-C_5 = 8.6) compared to the mixed gas signal after the arrival of the CO_2-rich gas with CH_4-C_2-C_5 = 10.5 average of days 261 to 450 (Figure 12.15(a)) and CH_4/C_2-C_5 = 12.9 average after day 972. All U1 gases were much wetter than the injected gas at CRC-1 (CH_4/C_2-C_5 = 16.9). Significantly, there were three early but transient rises in CO_2 mol % in U1 on days 157, 184 and 233 and all were attributed to wellbore mixing.

12.8   Formation water composition and behaviour

12.8.1   U-tube wellbore sample mixing

Due to the engineering complexity of installing a series of packers beneath the narrow diameter casing patch, multi-level sampling was performed in the slim Naylor-1 borehole without proper zonal isolation for the single inflatable packer installed at 2022 mRT (Figure 12.2). This led to some uncertainty as to the source of the fluids being sampled given that the volumes extracted (approximately 30 L before the U-tube sample was taken) were comparable to the wellbore volumes (Figures 12.2 and 12.3). While the original gas production perforations near U1 were patched by the previous field operator, log evidence indicated that this patch was installed too deeply, leaving at least a metre of open perforations. The installation of U1 (sampling aperture between 2028.8 and 2029.4 mRT; Figure 12.2) at this depth took advantage of these open perforations to sample near the top of the gas cap. During the Naylor-1 recompletion effort, an additional length of casing (2039 to 2055 mRT) was perforated to permit the lower two U-tubes, sampling aperture between 2041.8–2042.4 mRT (U2) and 2046.3–2046.9 mRT (U3), to sample deeper in the reservoir and beneath the pre-injection GWC at 2039.5 mRT (Figure 12.4).

A series of U-tube extended flow tests, with the aim of investigating potential cross-contamination processes, was undertaken over 2 days in December 2009 at a time when all U-tubes were self-lifting gas. The time series geochemical data collected using a field quadrapole mass spectrometer (Freifeld and Trautz 2006) showed that U1 fluids were distinct, both in composition and tracers, from U2 and U3. This result substantiated the flow modelling where the very small pressure drawdown produced on flowing the U tubes was insufficient to overcome gravitation forces associated with rising CO_2-rich gas approximately 10 m from the level of U2, up to and mixing with the CO_2-poor gas at U1 (Boreham et al. 2011). Nevertheless, immediately after breakthrough the resultant U2 lighter CO_2-poor gas was expected to communicate with U1, and this could be clearly observed on days 157 and 184, which was attributed to transient wellbore mixing (Figure 12.14). On the other hand, there was greater potential for mixing of fluids collected from U2 and U3, which were placed closer together, with an average 4.5 m depth difference. Given the volumes sampled, it was expected that the fluid would mix between the lower two U-tubes when both were producing water. This was borne out by the similar water chemistry, prior to self-lift (Kirste et al. 2009). Once U2 went to self-lift, which happened abruptly over 2 weeks (Figure 12.14), the U-tubes sampled different fluids over a much more restricted catchment, because the injected gas and formation water had very different densities; U2 sampled an upper zone producing gas, whereas U3 sampled a mixture of gas from this zone and water from a lower zone; the larger density difference allowed for mostly isolation of the two U-tubes during this phase. Over a protracted transition to self-lift for U3, which took approximately 6 weeks, the water fraction declined and eventually self-lift ensued. Compositional measurements at U3 over this period reflected a varying combination of dissolved gas, free gas and variable mixing. Once both U2 and U3 were self-lifting, the wellbore between them contained mostly gas. With little density difference between the lower two U-tubes, mixing was expected to occur and it is evident from Figure 12.14 that the compositions at this time were very similar.

12.8.2   Chemical contamination

A total of 11 water samples from U-tube 2 and 26 water samples from U-tube 3 were collected covering a period from pre-injection to self-lift. The operational problems with the U-tube system in the early stages of the project introduced concerns about the ability to achieve baseline water chemistry conditions. All the pre-injection water samples showed evidence of KCl (kill fluid) contamination (~15%) introduced during the recompletion of Naylor-1 to reperforate and install the U-tube bottom hole assembly. The kill fluid content decreased to under 0.5% in the samples collected after injection began and remained low (predominantly less than 3%) for the remainder of the sampling period. Baseline conditions were considered to have been achieved by day 53.

12.9   Constraining CO_2 breakthrough

12.9.1   Behaviour of CO_2

The design of the BHA resulted in varying degrees of wellbore mixing that could be explained by a flow model involving strong gravitation control whereby mixing was greatest at times of minimal density contrasts between different fluids in the wellbore (see above). Notwithstanding these limitations, the fluid chemistry results led to well constrained estimates on the timing of breakthrough of the injected CO_2-rich fluid at the monitoring well and in-reservoir processes within the developing gas cap, and ultimately allowed predictive models to be robustly accessed.

Breakthrough, defined here as the first instance of the positive detection of added tracers, was unequivocally supported by the timing of results (Figure 12.14, see also Figure 12.18) at U2 between sampling on day 100, which had no indication of allochthonous fluids, and day 121, by the simultaneous detection of an increase in mol % CO_2 and enrichment in ^{13}C CO_2. Reservoir simulation and flow modelling in the Waarre-C reservoir unit, predicted breakthrough between 4 and 8 months after injection (Underschultz et al. 2011; Jenkins et al. 2012). The observed breakthrough in the Naylor Field was at the earlier end of the range forecast by dynamic modelling. Furthermore, the dynamic models did not have the resolution to predict very small changes in CO_2 saturation and only considered breakthrough where the cell saturation increased by 20%. Therefore, the breakthrough

event was not that significant in the terms of ability to model the system, which was largely governed by the free gas compositions during the transition to self-lift and thereafter.

In U1, the pre-injection high pressure free gas averaged 1.5 mol % CO_2 ($\sigma \pm 0.3$ mol %, $n = 26$, air and N_2-free basis to day 226), while the low pressure sample gases, representing the dissolved gas below the GWC, from U-tube 2, averaged 7.5 mol % CO_2 ($\sigma \pm 0.1$ mol %, $n = 3$, air and N_2-free basis to day 100) before breakthrough and self-lift. This difference reflected the preferential solubility of CO_2 in the aqueous phase relative to the other gases present (Boreham et al. 2011).

The observed free-to-dissolved gas relationship could be compared with modelled compositions and was used to ascertain the performance of the U-tube sampling methodology. In general, the model results matched the measured values fairly well, except for CO_2 which showed the largest variance and was outside experimental error (Boreham et al. 2011). The relatively high solubility of CO_2, even at the lower pressure and temperature of sampling, meant that a greater proportion of CO_2 stayed in the aqueous phase and thus the exsolved concentration was less than that modelled. Although facile CO_2/water/rock reactions could have reduced the amount of exsolved CO_2, the design of the surface sampling also caused compositional fractionation whereby gas with a higher CO_2 content was exsolved later in the depressurisation procedure.

The relative success of the predictive model enabled the generation of a mixing model to evaluate the evolution of the aqueous phase gas content in terms of end-member gas phase contribution (Boreham et al. 2011; Underschultz et al. 2011). The initial composition was represented by U1 day 170 (selected because it appeared to have little impact from wax) and the CO_2-rich end-member by U2 day 184. The mixing model showed that the CO_2 content rose rapidly to greater than 50% CO_2 occurring when only 15% of the CO_2-rich end-member was present. The data for U2 and U3 after breakthrough and prior to gas lift, did not exceed approximately 50% CO_2 and the U3 data with CO_2 contents around 20–30% from day 150 to day 212 would have only needed a contribution of 2–5% of the CO_2-rich end-member to reach those values.

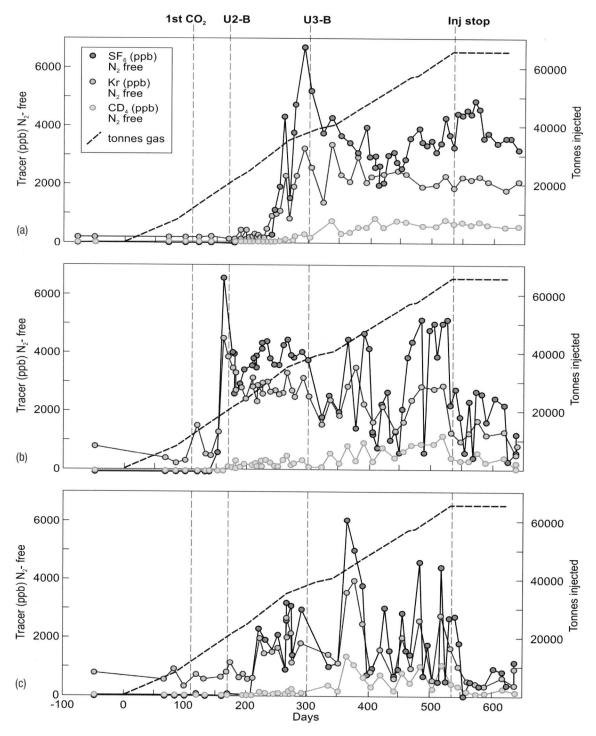

Figure 12.18: Tracer data to post-CO$_2$ injection. (a) is U-tube 1, (b) is U-tube 2 and (c) is U-tube 3. "1st CO$_2$" is the arrival (101–121 days after injection) of the CO$_2$ and tracers at U-tube 2. "U2-B" is self-lift of U-tube 2 (177 days) and then "U3-B" is the self-lift of U-tube 3 (303 days). Injection ceased at the end of August 2009 after 530 days.

This suggested that the sampled water had to be a mixture of water interacting with the CO_2-rich phase and water essentially at the initial reservoir conditions; only a very small proportion of the water sampled was from proximal to the GWC. The configuration of the U-tube sampling system resulted in each water sample being a mixture of water sourced from the length of the perforated interval. The mixing ratios derived from the model indicated that additional constraints on flow modelling and geochemical modelling could be applied to better understand the process of reservoir filling.

One of the difficulties encountered during sampling was determining whether purely aqueous phase samples were collected, or if a mixture of aqueous phase and gas phase was drawn into the U-tube. The CO_2 content itself did not provide a definitive measure; however, the relative solubilities of the hydrocarbons and the tracers resulted in significant differences between the exsolved and the gas phase content. It was clear that the CH_4/C_2 data gave a good indication of when samples were purely exsolved ($CH_4/C_2 > 30$) and when the gas phase was included in the sample ($CH_4/C_2 < 25$) (Boreham et al. 2011). Comparing the predicted exsolved composition indicated that U2 was dominated by the gas phase by day 163 and in U3 by day 226.

The carbon isotope data (Figure 12.16) for CO_2 could also be used to determine the mass proportion (mixing ratio) of injected CO_2-rich gas that combined with the initial in-place CO_2-poor dissolved gas. For a two-component system ($\delta^{13}C$ CO_2 initial gas for U1, U2 and U3 from day 68 and $\delta^{13}C$ CO_2 initial free gas equal to –6.7‰ average for Buttress-1 gas, with no isotope exchange), allowing for an experimental error of ± 0.3‰ in measured values, there was reasonable agreement between the fraction of injected CO_2 calculated using the two independent methods (compositional and isotopic). This supported the concept of injected CO_2 initially arriving in a dissolved state (Boreham et al. 2011). The observed carbon isotopic fractionation ($\Delta\delta^{13}C$) between the gaseous CO_2 (at U1) and DIC (at U2) of approximately 10.5‰ was slightly greater than the expected fractionation of 8–9‰ at the surface temperature. Pre-breakthrough at U1, $\delta^{13}C$ $CO_2(g)$ averaged –11.0‰ ($CO_2(g)$ – HCO_3^- isotopic fractionation (α) equal to 1.0085 at 20°C (Mook et al. 1974); and U2 $\delta^{13}C$ DIC averaged 0.55‰, suggesting a Rayleigh-type

process may have taken place during the depressuring that occurred when collecting the low pressure Isotube gas sample. During depressuring of the holding cylinder, the majority of the gas was released to the atmosphere prior to collection of the low pressure Isotube exsolved gas sample, making such a process likely.

12.9.2   In-reservoir mixing and equilibration

The CO_2 content in U2 and U3 after gas lift (~60 mol %) was appreciably lower than the average CO_2 content of the injected fluid (75.4 mol %). This most likely represented local mixing of the injected fluid with the methane-dominant residual gas in the pore space along the migration pathway and within the vicinity of the Naylor-1 wellbore. The injected gas most likely travelled under strong buoyant forces until it reached the gas cap. Within the gas cap the injected gas was denser than the methane, leading to the injected gas spreading laterally and mixing with the native methane. At U1, the arrival of CO_2-rich fluid resulted in the CO_2 content being considerably lower, at approximately 20 mol %. The Waarre-C reservoir was very heterogeneous with high permeability regions inter-dispersed with zones of low permeable shaly baffles (Dance et al. 2009; Underschultz et al. 2011). Reservoir models showed CO_2-rich fluid in permeable zones within the residual gas cap between U1 and U2 at the wellbore (Underschutz et al. 2011). It was apparent from the U1 results that CO_2-rich fluid could be emplaced even higher up in the residual gas column under the right geological conditions. Given the limitation of only one observation borehole in a complex geological system it is unlikely that a single plausible scenario could have been identified. However, the bulk behaviour of the system was important because the compositions of the sampled gas informed any estimate for storage volume available within the Waarre-C reservoir (Underschultz et al. 2011).

The decrease in CO_2 from 55–60 mol % immediately after gas lift of U2 and U3 to an average of approximately 48 mol % after 1413 days of sampling was significant (Figure 12.14). This was also accompanied by a gradual increase of CO_2 from 20 mol % up to approximately 30 mol % in U1. These shifts in the CO_2 mol % for the three U-tubes were most likely in response to the gas cap moving towards equilibrium between the free gas and gas dissolved in the residual formation water. The existence of a horizontally continuous GWC was unlikely (Underschultz et al. 2011) due to the presence of many permeability baffles and porosity/permeability heterogeneity within the Waarre-C reservoir (Dance et al. 2009).

As the gas cap expanded, dynamic fluid flow modelling predicted the retention of an estimated 40–50% residual water in the pore space and based on CO_2 core flooding experiments on sandstone from CRC-1 (Underschultz et al. 2011). The attenuation to lower CO_2 content with time for U2 and U3 was likely to be due to the mixing of the initial (residual) CO_2-poor gas with the (arriving) CO_2-rich injected gas. Mass balance calculations of 20% residual methane gas with 1.5 mol % CO_2 and 30% CO_2 injected gas, which was similar to the gas composition at U2 and U3 from day 972 (Figure 12.14). Dissolution was also another mechanism for CO_2 attenuation. Although there was 52 mol % CO_2 in the free gas, there was 92 mol % CO_2 in the dissolved gas at equilibrium—dissolution being a relatively rapid process (Shevalier et al. 2004). Since different volumes of residual water and gas occupy the available pore space, the amount of CO_2 that can be dissolved in the proximal residual water is significant (approximately 47 kg CO_2 per tonne of formation water (Zhenhao et al. 1992) (http://www.geochem-model.org/models.htm) and this would have impacted on the overall composition of the free gas, especially in the case of U2 and U3. The effects of mixing and dissolution were competing processes, with the quantitative contributions attenuating the CO_2 content. The 7-month intervening period from the initial gas lift at U2 (after day 170) to stabilisation at a CO_2 content (after day 972) suggested rapid attainment of equilibrium. On the other hand, the gradual increase of CO_2 mol % at U1 to a constant composition 2 years later, indicated slightly slower processes within an existing gas column.

12.10   In-reservoir behaviour of tracers

Sampling of fluids in-reservoir by U-tube has occurred since the bottom hole assembly (BHA; Figure 12.2) was installed in October 2007 in the Naylor-1 well. Initially baseline samples were obtained prior to injection and demonstrated low CO_2 concentrations in the U-tube samples from the three different levels and an absence of tracers. Sampling was then undertaken initially every 2 to 3 weeks and then on a weekly basis from day 163. From day 604 sampling continued on a fortnightly basis changing on day 1413 to 6-weekly intervals. Sampling has now extended to over 2000 days or approaching 6 years with samples analysed through to January 2014.

The initial frequency of sampling may not have been sufficient to see the tracers arriving at a different time to the CO_2. There was some evidence to suggest that this could be the case based on information from the first Frio Brine test (Freifeld et al. 2005). In the case of Otway Stage 1, the project was not designed with this degree of detail in mind to sample at such short intervals, which in retrospect was a mistake.

Defining and announcing the date of breakthrough with the confirmation of the tracer data, to support the observations with the CO_2 concentration increases was not straightforward. This is illustrated by the results from U-tube 2 presented in Figure 12.17. Clearly, taken in isolation the plot shows how difficult it is to observe breakthrough from a sample taken at the point where no subsequent data exists. After many more samples were taken and analysed (Figure 12.17), the breakthrough time became much clearer. The $\delta^{13}C$ of CO_2 was found to be a very sensitive isotopic marker for breakthrough, while the tracers and the abundance of CO_2 show only very small relative changes at the breakthrough point. Only by collecting all the different types of data and reviewing them together post-breakthrough could a definitive date be applied to breakthrough at U-tube 2.

Defining breakthrough in U-tube 3 was more complex, with sampling moving back and forth from gas-lift to water-rich samples on a weekly basis for over 6 weeks, before finally remaining in the supercritical gas-lift state between day 212 and 226 (23 and 30 October 2008).

The arrival of tracers and CO_2 in U-tube 1 was somewhat earlier than anticipated. This was in part due to modelling scenarios where the only driving mechanism to allow CO_2 or tracer to reach U-tube 1 in the gas cap was expected to be by diffusion. However, the presence of increased levels of CO_2 and consistent presence of tracers (they were somewhat variable prior to this time) were observed by day 233. By slightly adjusting the extent of permeability baffles between CRC-1 and Naylor-1, the observations of CO_2 and tracers in U-tube 1 in the gas cap could be satisfactorily modelled. Subsequent to this work, partition coefficients between supercritical CO_2 and water have been measured for all tracers used by the CO2CRC over the years and models are currently being re-run with this new information (Myers et al. 2013).

As mentioned previously, R-134a and SF_6 were injected in a second tracer pulse within a few weeks of U-tube 3 becoming fully self-lifting (i.e. injected CO_2 had filled past the deepest sampling point in the well). Monitoring for R-134a commenced and there was a level of uncertainty as to how the tracer would interact with the existing CO_2 and CH_4, and if the new tracer would be seen in the U-tubes at all.

Preliminary results showed that R-134a arrived in U-tube 3 after 11 June and before 16 July 2009 taking 141–176 days transit time. U-tube 2 showed some presence of R-134a by the 3 September, some 231 days after the injection of this second pulse of tracers. The second SF_6 pulse was not differentiated from the background ppm levels of SF_6 established after the injection of larger quantities in batch 1.

The tracer concentrations (Figure 12.18) do not appear to be typical bell shaped breakthrough curves (e.g. like those seen at the Frio Brine 1 test; Freifeld et al. 2005). This was due to the fact that the Otway site was a tilted fault block with complete closure—thus no fluids were migrating away from the area, but rather were conserved within the accumulated fluids. Thus the tracers and gases were added cumulatively resulting in the need for more complex modelling of the results over time.

The tracers used for Stage 1 were chosen for a variety of reasons (Section 12.5.1). However, in the case of these and other tracers used to date in carbon storage applications, there is a dearth of information on the partitioning between supercritical CO_2 and water (or formation fluids with higher salinities). Modelling software tends to use the Henry's coefficients (i.e. the partition coefficient between

water and n-octanol or water and air). Clearly there is a likelihood that the values may bear no relation with that in the subsurface (higher temperatures and pressures). This means for example that modelling may have limited value for understanding residual saturation. The lack of relevant partitioning data for the tracers used at Otway has now been resolved with a project near completion on the measurement of the tracers in supercritical CO_2 and water at a range of temperatures and pressures. Furthermore, the experiments have been repeated in a mixture of supercritical CO_2 and CH_4 with concentrations representative of the Buttress gas injected. This has enabled new modelling activities to take place (currently underway) and will provide new procedures for using tracer data in models.

12.11 Liquid hydrocarbons

While the focus of this chapter is on CO_2, CH_4 and their various states in the Waarre reservoir, it should be noted that small amounts of liquid (condensate) were recovered from the Naylor-1 natural gas, from the CRC-1 MDT gas upon pressure reduction from reservoir to atmospheric pressure and inside the flow lines of the Naylor-1 surface sampling facility. The gas chromatogram of the neat condensate from CRC-1 showed the light hydrocarbons ($< C_8$) dominated by n-alkanes and cyclic hydrocarbons but with relatively low contents of benzene and toluene (Boreham et al. 2008). The higher molecular weight compounds were dominated by a homologous series of n-alkanes with progressive decrease in relative abundance from n-C_9 to n-C_{30}. However, the minor content of waxy hydrocarbons ($> n$-C_{22}) showed a maxima at n-C_{27} in CRC-1 condensate. This wax input was even more pronounced in the gas chromatogram of the Naylor-1 condensate (Naylor-1 well completion report).

12.12 Solid hydrocarbons

In addition to liquid hydrocarbons, solid hydrocarbons occurred in the liquid separator lines at Buttress-1, causing initial disruptions to production and requiring continuing maintenance (Figure 12.19). A similar solid was periodically scraped from the inside of the surface piping at the nearby Boggy Creek CO_2 production plant

Figure 12.19: Buttress-1 CO$_2$ production facility showing scrubber valve (lh green cylinder) and flash pot (rh silver cylinder), where supercritical CO$_2$ is at 31–34°C and 9.5 Mpa. These components are most susceptible to wax build-up, blocking the internal piping (top right) and with limited prevention via daily wax discharge (bottom right).

operated by BOC, where it was stored in drums ready for periodic disposal. Strategies were developed for the Naylor-1 operation using Solvesso solvent flushing of the ¼″ SS tubing to control such a precipitated solid causing problems within the U-tube assembly.

The GC trace of the Boggy Creek solid was dominated by a homologous series of > C$_{22}$ waxy n-alkanes (Figure 12.20), which maximised at n-C$_{27}$ followed by a progressive decrease in relative abundance to > n-C$_{32}$ (Boreham et al. 2008). The Boggy Creek n-alkane profile was strikingly reminiscent of the small maxima envelope found in the condensates from CRC-1 and Naylor-1. This strong genetic relationship between the solid wax and the condensates was confirmed by the very similar carbon and hydrogen isotopes of their respective waxy n-alkanes (Boreham et al. 2008).

At Naylor-1, subsurface build-up of wax in the U-tube ¼″ SS lines prevented the delivery of U-tube fluids to surface, particularly in the upper U-tube sampling the gas cap.

At Buttress-1, precipitation of wax occurred in the production plant at the initial gas scrubber-flash pot (Figure 12.19). Here, the wellhead pressure during production was typically between 9.55 and 9.95 MPa but the temperature had fallen from a reservoir temperature of 85°C to between 31°C and 34°C at the surface. This temperature drop allowed wax to separate from the gas.

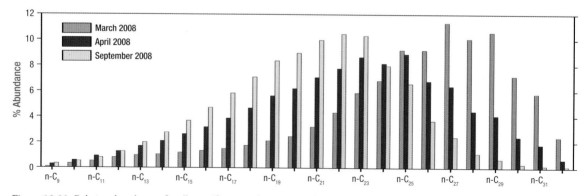

Figure 12.20: Relative abundance of n-alkanes (from gas chromatography) in wax from Buttress-1 collected over time (after Boreham et al. 2011).

At Naylor-1, the temperature gradient was much more severe due to lack of continuous production, resulting in wax accumulation in the narrow ¼″ U-tube lines in the near subsurface.

This wax build-up caused U1 to stop flowing gas in early April 2008 as baseline sampling was being established and after removal of only a relatively small volume of gas from the Naylor-1 well. Although a range of remedies were suggested (e.g. flowing hot water into the well via the ⅜″ line used for N_2-lifting of the initial kill fluid after BHA installation to threading a wire down the inside of the U tube ¼″ SS tubing), it was decided to dissolve the wax. After laboratory tests with common organic solvents, Solvesso-100™ (an industrial solvent composed dominantly of trimethylbenzene isomers, ExxonMobil, Houston, TX, USA) was chosen, based on its solvation properties, low volatility and low toxicity. With a solvent delivery and retrieval system (using a piston pump capable of 34.5 MPa outlet pressure), Solvesso-100 was pumped into either the drive or sample legs (Figure 12.5). Generally, it was found that if only a limited volume (10 L) of solvent was introduced into the sample leg's ¼″ SS tubing and allowed to soak for a period, there was sufficient reservoir pressure to self-lift the solvent back out of the line and into a waste drum. After all the U-tubes began to access only gas, preventative maintenance required Solvesso-100 to be introduced into the U-tubes on a monthly basis. It was important that all solvent was removed from the U-tube line before sampling formation gases. Since the solvent had a higher density than the formation gas, the presence of Solvesso-100 in the U-tube lines could be identified by a reduction in the surface pressure of that line using wellhead-mounted pressure gauges. Both lines were allowed to flow gas until the pressures on the two lines equilibrated. This was an indication that all the solvent was removed from the lines and that no blockages remained.

12.13   Conclusions

Sampling of gases and liquids in the Naylor-1 monitoring well presented a number of challenges that were exacerbated by the small diameter of the borehole and the presence of wax in the gas. There were some instrumental problems but the LBNL-designed U-tube system, performed very well and yielded critically important geochemical information. Gas geochemistry of samples taken at the Naylor-1 observation well provided a direct measure of breakthrough of the mixed CO_2/CH_4 fluid injected at the CRC-1 well. This was achieved in the context of a depleted natural gas reservoir. Both the molecular and carbon isotopic compositions of CO_2 and the added tracers showed positive responses at breakthrough, which occurred 100 to 121 days after injection began and after the addition of 10,000 to 12,600 t of mixed CO_2/CH_4. Since there was sufficient carbon isotopic differentiation between baseline and injected CO_2, both the chemical and carbon isotope data were useful in tracing the fate of the injected CO_2.

The tracers, which were deliberately chosen to be conservative, were successfully and efficiently injected by a slip-stream method. They were not "lost" near the well bore at CRC-1 but were injected after enough CO_2 had been injected so that the tracers did not saturate the near well bore area. Breakthrough curves for both the tracers and the injected CO_2 generally compare well although there are subtle differences in the timing of maximum abundances.

Following breakthrough, the CO_2 content in the free gas rose from 1.5 to approximately 60 mol %, which was well below the 75.4 mol % CO_2 in the slightly modified Buttress-1 injected gas. The difference was the result of mixing of the injected gas (as fluid) with methane-rich/CO_2-poor residual gas (approximately 20 % methane-saturated formation water) encountered along the 300 m migration distance between the injection and observation wells and with the CH_4-rich/CO_2-poor residual gas column. The composition of the produced U-tube gases evolved over many months. The GWC moved below the lowest sample point, down to approximately 48 mol % CO_2 for U2 and U3 and for the upper U-tube (initially within the residual gas column) up to a stable value of approximately 31 mol % CO_2 after 2 years. The reduction in mol % CO_2 of the free gas phase for U2 and U3 was likely to have been in response to partitioning between dissolved and free gas phases. The efficiency of mixing of the residual CO_2-poor residual gas, the introduced CO_2-rich injected gas and the residual formation water, involved a redistribution process over a short residence time. There was a less than optimum gas sweep of formation water as the GWC moved down with progressive filling of the Naylor structure. Importantly, the storage capacity of supercritical CO_2 within the Naylor closure required taking into account both the dissolved and residual free CH_4 already in the system, as well as the residual water saturation (Underschultz et al. 2011). Ignoring the impact of native free and dissolved CH_4 and not accounting for the pore space remaining occupied by residual formation water would lead to an erroneously large estimate of storage capacity.

The addition of a second tracer pulse after the injected CO_2 passed below U-tube 3 was still observed after a slightly longer transit time than the first tracer travel time. This tracer was also observed in the slightly shallower (5 m) U-tube 2, implying some sort of mobility and mixing. New R&D on determining relevant partitioning coefficients for the tracers used at Otway may go some way towards understanding the behaviour of the tracers and may provide more information on residual saturation and storage capacity.

Finally, assuming ongoing sampling and analysis, the U-tube results from each interval will continue to provide novel data on subsurface behaviour in a storage project many years after the end of CO_2 injection. It will also provide information on the stability of the plume over extended periods of time and on how CO_2 has impacted on the reservoir fluids over 6 years or more.

12.14   References

Boreham, C.J. and Edwards, D.S. 2008. Abundance and carbon isotopic composition of neo-pentane in Australian natural gases. Organic Geochemistry 39, 550–566.

Boreham, C.J., Underschultz, J., Stalker, L., Freifeld, B.,Volk, H. and Perkins, E. 2007. Perdeuterated methane as a novel tracer in CO_2 geosequestration. In: Farrimond, P. et al. (Eds.), The 23rd International Meeting on Organic Geochemistry, Torquay, England, 9–14 September 2007, Book of Abstracts, 713–714.

Boreham, C.J., Chen, J. and Hong, Z. 2008. Baseline study on sub-surface petroleum occurrences at the CO2CRC Otway Project, western Victoria, PESA Eastern Australasian Basins Symposium III, Sydney, 14–17 September 2008, 489–499.

Boreham, C., Underschultz, J., Stalker, L., Kirste, D., Freifeld B., Jenkins, C. and Ennis-King, J. 2011. Monitoring of CO_2 storage in a depleted natural gas reservoir: Gas geochemistry from the CO2CRC Otway Project, Australia. International Journal of Greenhouse Gas Control 5, 1039–1054.

Cenovus Energy Report. 2011. http://www.cenovus.com/operations/oil/docs/Cenovus-summary-of-investigation.pdf, November 2011.

Clark, J. F., Hudson, G. B., Davisson, M. L., Woodside, G. and Herndon, R. 2004. Geochemical imaging of flow near an artificial recharge facility, Orange County, CA. Ground Water 42, 167–174.

Coplen, T. B. 1995. Reporting of stable carbon, hydrogen, and oxygen isotopic abundances in Reference and intercomparison materials for stable isotopes of light elements: Vienna. International Atomic Energy Agency, TECDOC-825, p. 31–34.

Coplen, T.B. 1988. Normalization of oxygen and hydrogen isotope data: Chemical Geology (Isotope Geosciences Section) 72, 293–297.

Coplen, T.B., Brand, W.A., Gehre, M., Gröning, M., Meijer, H.A.J., Toman, B. and Verkouteren, R.M. 2006. New guidelines for $\delta^{13}C$ measurements. Analytical Chemistry 78, 2439–2441.

Dance, T., Spencer, L. and Xu, J.Q. 2009. Geological characterization of the Otway pilot site: what a difference a well makes. Energy Procedia 1, 2871–2878.

Freifeld, B.M., Trautz, R.C., Kharaka, Y.K., Phelps, T.J., Myer, L.R., Hovorka, S.D. and Collins, D.J. 2005. The U-tube: A novel system for acquiring borehole fluid samples from a deep geologic CO_2 sequestration experiment. Journal of Geophysical Research: Solid Earth 110(10), 1978–2012.

Freifeld, B.M. and Trautz, R.C. 2006. Real-time quadrupole mass spectrometer analysis of gas in borehole fluid samples acquired using the U-tube sampling methodology, Geofluids 6, 217–224. doi: 10.1111/j.1468-8123.2006.00138.x.

Freifeld, B.M., Perkins, E., Underschultz, J. and Boreham, C. 2009. The U-tube sampling methodology and real-time analysis of geofluids. 24th International Applied Geochemistry Symposium, 1–4 June 2009, Fredericton, N.B., Canada.

Jenkins, C.J., Cook, P.J., Ennis-King, J., Undershultz, J., Boreham, C.J., Dance, T., de Caritat, P., Etheridge, D.M., Freifeld, B.M., Hortle, A., Kirste, D., Paterson, L., Pevznera, R., Schacht, U., Sharma, S., Stalker, L. and Urosevic, M. 2012. Safe storage and effective monitoring of CO_2 in depleted gas fields. PNAS 109(2), E35-E41.

Kirste, D., Perkins, E., Boreham, C., Freifeld, B., Stalker, L., Schacht, U. and Underschultz, J. 2009. Geochemical modelling and formation water monitoring at the CO2CRC Otway Project, Victoria, Australia. 24th International Applied Geochemistry Symposium, 1–4 June 2009, Fredericton, N.B., Canada.

Mango, F.D. and Jarvie, D.M. 2009. Low-temperature gas from marine shales. Geochemical Transactions 10, 3 doi:10.1186/1467-4866-10-3.

Mook, W.G., Bommerson, J.C. and Stavermen, W.H. 1974. Carbon isotope fractionation between dissolved bicarbonate and gaseous carbon dioxide, Earth and Planetary Science Letters 22, 169–176.

Myers, M., White, C., Pejcic, B., Stalker, L. and Ross, A. 2013. Chemical Tracer partition coefficients for CCS. CSIRO Report #EP 133018. ANLECR&D Project 3-1110-012 Fundamentals of tracer applications for CO_2 storage. http://www.anlecrd.com.au/projects/fundamentals-of-tracer-applications-for-co-sub-2-sub-storage.

Nimz G.J. and Hudson G.B. 2005. The use of noble gas isotopes for monitoring leakage of geologically stored CO_2. In Thomas, D. and Benson, S. (Eds), Carbon dioxide capture for storage in deep geologic formations, Vol. 2, Amsterdam, Elsevier Press, 1113–1130.

Sackett, W. and Chung, M. 1979. Experimental confirmation of the lack of carbon isotope exchange between methane and carbon oxides at high temperatures. Geochimica et Cosmochimica Acta 43(2), 273–276.

Shevalier, M., Durocher, K., Perez, R., Hutcheon, I., Mayer, B., Perkins, E. and Gunter, W. 2004. Geochemical monitoring of the gas-water-rock interaction at the IEA Weyburn CO_2 monitoring and storage project, Saskatchewan, Canada. In: Wilson, M., Morris, T., Gale, J. and Thambimuthu, K. (Eds.), Proceedings of the 7th International Conference on Greenhouse Gas Control Technologies, Vol. II, Part 2, Elsevier, Amsterdam, 2135–2139.

Stalker, L., Boreham, C. and Perkins, E. 2006. A review of tracers in monitoring CO_2 breakthrough. CO2CRC Report RPT06-0070.

Stalker, L., Boreham, C. and Perkins, E. 2009a. A review of tracers in monitoring CO_2 breakthrough: Properties, uses, case studies, and novel tracers. In: Grobe, M., Pashin, J.C. and Dodge, R.L. (Eds.), Carbon dioxide sequestration in geological media—State of the science, AAPG Studies in Geology 59, 595–608.

Stalker, L., Boreham, C., Underschultz, J., Freifeld, B., Perkins, E., Schacht, U. and Sharma, S. 2009b. Geochemical monitoring at the CO2CRC Otway Project: tracer injection and reservoir fluid acquisition. Energy Procedia 1(1), 2119–2125.

Thrasher, J. and Fleet, A.J. 1995. Predicting the risk of carbon dioxide "pollution" in petroleum reservoirs. In: Grimalt, J. and Dorronsoro, C. (Eds.), Organic geochemistry: developments and applications to energy, climate, environment and human history, Selected papers from the 17th International Meeting on Organic Geochemistry, San Sebastian, AIGOA, 1086–1088.

Underschultz, J., Freifeld, B., Boreham, C., Stalker, L., Schacht, U., Perkins, E., Kirste, D. and Sharma, S. 2008. Geochemistry monitoring of CO_2 storage at the CO2CRC Otway Project, Victoria. The APPEA Journal, Volume 48 and Conference Proceedings (DVD), Extended Abstract.

Underschultz, J., Boreham, C., Stalker, L., Freifeld, B., Xu, J., Kirste, D. and Dance, T. 2011. Geochemical and hydrogeological monitoring and verification of carbon storage in a depleted gas reservoir: examples from the Otway Project, Australia, International Journal of Greenhouse Gas Control 5, 922–932.

Watson, M.N., Boreham C.J. and Tingate, P.R. 2004. Carbon dioxide and carbonate cements in the Otway Basin: Implications for geological storage of carbon dioxide. Australian Production and Petroleum Exploration Association Journal 44, 703–720.

Zhenhao, D., Nancy, M. and Weare, J. 1992. An equation of state for CH_4, CO_2 and H_2O II, Mixtures from 0 to 1000°C and from 0 to 1000 bar. Geochimica et Cosmochimica Acta 56, 2619–2631.

Patrice de Caritat, Allison Hortle, Dirk Kirste

13. MONITORING GROUNDWATERS

13.1 Introduction

The Otway Project is located within the Port Campbell hydrogeological sub-basin, which in turn corresponds to the structural feature recognised as the Port Campbell Embayment (Duran 1986), an onshore basement low, infilled with thick Late Cretaceous and Tertiary sediments (SKM 1999).

Groundwater is used extensively for irrigation, dairy and domestic purposes in the Otway Basin and is sourced from the unconfined to semi-confined Port Campbell Limestone. A second, deeper and confined aquifer, the Dilwyn Formation (Figure 1.5), also contains potable water, and has been used in the past for urban water supply. Both units are located well above the Waarre Formation, the target injection reservoir, and are separated from it by at least 1100 m of alternating aquitards and aquifers. Rigorous characterisation of the site (Dance et al. 2009; Chapter 5) indicated that the likelihood of injected CO_2 moving from the target reservoir into either of these

aquifers was remote. Nevertheless, the presence of these water resources meant that there was strong community interest in demonstrating the ongoing integrity of these resources in parallel with CO_2 storage.

A comprehensive assurance monitoring programme was initiated in 2006 to document the natural state of the shallow subsurface and define a baseline prior to the injection of CO_2 (Jenkins et al. 2012). As part of this programme, groundwater from pre-existing bores was sampled biannually from the shallow, unconfined Port Campbell Limestone Aquifer (21 bores) and from the deeper, confined Dilwyn Aquifer (3 bores) within a radius of ~10 km around the injection well CRC-1 (Figure 13.1). Standing water levels (SWLs) were monitored continuously in both aquifers where possible, using pressure and temperature data loggers to determine the transient aquifer flow rates and directions.

The aims of this chapter are to (1) describe in some detail the groundwater monitoring programme developed at the Otway site, both in terms of groundwater levels and groundwater composition; (2) review general methods, results and quality assessment and quality control (QA/QC) procedures; and (3) outline technological applications and innovations that may be useful at other sites.

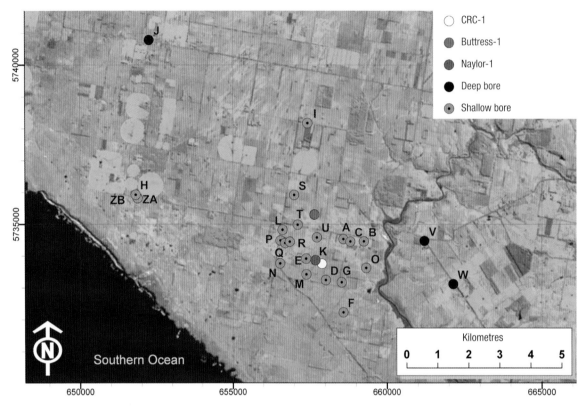

Figure 13.1: Location of the deep and shallow groundwater monitoring bores used in the Otway Project, and of the Buttress-1, Naylor-1 and CRC-1 wells discussed in the text.

13.2 Monitoring groundwater level

Permanent water level loggers, Solinst Levelogger Gold units, were installed in six open bores (piezometers) to measure and record water levels and temperature hourly. Four of these bores were owned and maintained by the Victorian government's water monitoring body (Department of Sustainability and Environment, DSE—now DEPI), one was owned by Wannon Water and one was privately owned (Figure 13.1). The data loggers were downloaded biannually in accordance with the conditions of access defined by DSE (Hennig et al. 2007b). One of the six bores also had a barometric pressure logger installed to allow correction for atmospheric variations. During each groundwater sampling trip, water levels were measured manually using a dip meter. In addition, the water level data were correlated with periodic measurements taken

manually by contractors for the DSE. The installation, downloading and reporting procedures are documented by Hennig et al. (2007a; b) and Hennig (2007).

Recorded data were compensated for barometric pressure variations and water density, using the software supplied with the data loggers and corrected to Australian Height Datum (AHD) for plotting. Barometric efficiency (BE) was estimated using methods described in Inkenbrandt et al. (2005) and Spane (2009). Data from the shallow aquifer, the Port Campbell Limestone, were corrected for barometric pressure using a BE of 1.0 and a density of 1.0. This was a freshwater, unconfined aquifer and any pressure change was transmitted directly to the aquifer. For the deeper, confined Dilwyn Aquifer, BE was determined to be 0.4, and water density 1.0. The corrected hydrographs do not produce smooth curves and the "noise" is due to the influence of earth tides. These small effects do not impact on the hydraulic head values.

Figure 13.2: Types of bores used to access groundwater for the Otway Project monitoring programme: (a) open bore (b) bore equipped with a submersible pump or (c) with a windmill, and (d) open bore pumped with a 4WD-mounted compressed air pump.

13.3 Monitoring groundwater composition

13.3.1 Field measurements

Access to groundwater in the Otway Project area was provided through the existence of a network of bores. These were either open bores (Figure 13.2(a)) or bores equipped with pumps, such as submersible or surface electric pumps (in many cases, Figure 13.2(b)) or windmills (more rarely, Figure 13.2(c)); no groundwater bore was drilled for the Project. The open bores were pumped with a 4WD-mounted compressed air environmental sampling (low flow) Bennett pump (Figure 13.2(d)).

Two aquifers were targeted for the monitoring exercise: (1) the shallow Port Campbell Limestone aquifer, which is intensively used by the local community and extends from the surface down to some 150 m (Nahm 1985), and (2) the high-water quality, deeper Dilwyn Aquifer, which lies between about 600 and 900 m depth (SKM 1999). The shallow aquifer is intercepted in the monitoring area by 21 bores (16 equipped and five open), the deeper aquifer by only four open bores (Table 13.1). Of these, a few bores became inaccessible after the first sampling trips, leaving 17 shallow and three deep bores the subject of regular monitoring.

The schedule of groundwater monitoring trips is shown in Table 13.2. Five sampling trips occurred before injection (18 Mar 2008 – 29 Aug 2009; Underschultz et al. 2011),

Table 13.1: Bore location and depth details, Otway Project.

Station	Bore Name	Landowner Code**	Bore/Pump Type	X# (m)	Y# (m)	Z# (m)	TD^ (m)
Shallow bores							
A	House Bore	CC	Electric	658586	5734533	49	50
B	BCGP* Firefighting Bore	BC	Electric	659250	5734461	30	80
C	Dairy Bore	CC	Electric	658817	5734455	41	175
D	House Bore	JB	Electric	658020	5733248	57	46
E	Back Paddock Bore	PD	Windmill	657393	5733919	47	9
F	House Bore	WJ	Electric	658588	5732230	48	61
G	House Bore	BF	Electric	658531	5733178	51	9
H	141240	DSE	Monitoring bore	651872	5735823	43.36	76.20
I	141241	DSE	Monitoring bore	657422	5738208	59.77	30.41
K	Naylor-1 Water Bore	PD	Open casing	657690	5733870	47	46
L	House Bore	MB	Electric	656621	5734826	56	9
M	Dairy Bore	GC	Electric	657399	5733423	55	67
N	House Bore	GC	Electric	656546	5733768	57	18
O	Paddock Bore	MC	Windmill	659328	5733628	27	15
P	House Bore	RB	Electric	656552	5734487	55	6
Q	Dairy Bore	RB	Electric	656706	5734408	53	9
R	Irrigation Bore	RB	Open casing	656857	5734450	61	36
S	North Paddock Bore	PP	Windmill	657004	5735918	62	Unknown
T	Dairy Bore	PP	Electric	657117	5734985	46	18
U	Dairy Windmill	PP	Windmill	657735	5734586	47	Unknown
ZA	141239	DSE	Monitoring bore	652000	5735993	48	34.82
Deep bores							
J	85937	DSE	Monitoring bore	652222	5740778	49.35	1685.54
V	84290	WW	Monitoring bore	661100	5734300	37.17	825.74
W	84291	DSE	Monitoring bore	662122	5733178	46.76	907.20
ZB	85942	DSE	Monitoring bore	651937	5736102	37	1250

* Boggy Creek Gas Plant

** DSE: Department of Sustainability and Environment;
WW: Wannon Water; others: private owners

Map Grid of Australia (Geocentric Datum of Australia 94) Zone 54

^ Total depth, as measured or reported by landowner (may be a best estimate)

Grey shading: bore sampled only once or a few times during the monitoring programme

two during injection, and a further five took place after completion of injection. The frequency of sampling was biannual until about 2 years after the completion of injection, at which point it was changed to annual. Sampling is ongoing as part of a long-term programme at the Otway Project site.

Five field parameters were recorded at regular intervals: acidity (pH), oxido-reduction (redox) state (*Eh*), electrical conductivity (EC), dissolved oxygen (DO) and temperature (*T*). These were measured using Thermo Scientific Orion electrodes in a flow cell closed to the atmosphere and Thermo Scientific Orion meters (Figure 13.3). Monitoring and recording of these parameters occurred at regular intervals (ranging from hourly at the beginning of pumping deep bores, to every 5 minutes when close to sampling) until three consecutive, stable set of values were recorded for all parameters.

Table 13.2: Groundwater monitoring sampling trips, Otway Project.

Trip #	Dates	Timing WRT Injection
1	6–13 Jun 2006	Pre-injection
2	8–12 Nov 2006	Pre-injection
3	5–12 Mar 2007	Pre-injection
4	13–18 Oct 2007	Pre-injection
5	2–6 Mar 2008	Pre-injection
6	27 Oct–1 Nov 2008	Syn-injection
7	16–21 Mar 2009	Syn-injection
8	24–29 Oct 2009	Post-injection
9	20–26 Mar 2010	Post-injection
10	18–23 Oct 2010	Post-injection
11	21–25 Mar 2011	Post-injection
12	19–24 Mar 2012	Post-injection
13	4–9 March 2013	Post-injection
14	13–18 March 2013	Post-injection

Grey shading: groundwater sampling undertaken during injection

Once the pumped water reached a stable composition—believed to be representative of the aquifer conditions—the decision was taken to sample for chemistry. The following samples were collected:

> one sample (100 mL high density polyethylene or HDPE bottle) for cation analysis (filtered to 0.45 μm and acidified with HNO_3)

> one sample (100 mL HDPE bottle) for anion analysis (filtered to 0.45 μm; not acidified)

> one sample (100 mL amber glass bottle) for iodide (I^-) analysis (filtered to 0.45 μm; not acidified)

> one sample (1000 mL HDPE bottle) for carbon isotopic composition of dissolved inorganic carbon (alkalinised with NaOH and precipitated using $SrCl_2$)

> one sample (1000 mL HDPE bottle or more, for low sulphate waters) for sulphur and oxygen isotopic composition of dissolved sulphate (acidified with HNO_3 and precipitated using $BaCl_2$)

Figure 13.3: Measurement of (a) alkalinity, (b) Fe^{2+} concentration.

> two samples (15 mL Vacutainer each) for oxygen and hydrogen isotopic composition of water (not filtered; not acidified)

> two samples (100 mL Grace Davison clear, crimp-top, glass serum bottle each) for dissolved gas/tracers analysis (sealed under water then crimped; not filtered; not acidified)

> field alkalinity titration (two at 10 mL each) using a Hach digital titrator and methyl orange indicator (not filtered; not acidified)

> field determination of reduced iron (Fe^{2+}) (25 mL) using a Hach DR/2800 spectrophotometer and 1,10 phenanthroline method.

Table 13.3: Analytical precision calculated from laboratory duplicates. The two columns on the right are in order of worsening precision.

Parameter/Element	Unit	Standard Deviation	Precision (%)	Parameter/Element	Precision, Ordered (%)
Al	mg/L	0.000121786	12	Ca	1
As	mg/L	6.12313E-05	3	Cl	1
B	mg/L	0.007801705	12	NO$_3$	1
Ba	mg/L	0.000822571	10	SiO$_2$	1
Br	mg/L	0.024985454	3	SO$_4$	1
Ca	mg/L	1.778365027	1	Mg	2
Cl	mg/L	2.923551588	1	Na	2
Cu	mg/L	6.89089E-05	3	As	3
F	mg/L	0.021752154	8	Br	3
Fe	mg/L	0.058072214	4	Cu	3
Hf	mg/L	0.00017732	13	Sr	3
K	mg/L	0.307256858	5	Fe	4
Li	mg/L	0.000225248	5	Mn	4
Mg	mg/L	0.347411098	2	Ti	4
Mn	mg/L	0.000799402	4	K	5
Na	mg/L	2.423406278	2	Li	5
Ni	mg/L	0.00045877	8	Zn	5
NO$_3$	mg/L	0.040357611	1	U	6
Rb	mg/L	0.000168458	7	V	6
Sc	mg/L	0.001373564	50	Rb	7
SiO$_2$	mg/L	0.201445615	1	F	8
SO$_4$	mg/L	0.614873938	1	Ni	8
Sr	mg/L	0.02203693	3	Ba	10
Th	mg/L	0.004133564	108	Al	12
Ti	mg/L	0.002947771	4	B	12
U	mg/L	0.00037362	6	Hf	13
V	mg/L	0.000407261	6	Sc	50
Zn	mg/L	0.000250416	5	Th	108

Field alkalinity and Fe^{2+} concentration were determined in the field by titration and spectrophotometry (Figure 13.3), respectively. Sample containers for chemical analysis were stored in a coolbox with ice bricks (except the C and S isotope bottles, in which the $SrCO_3$ and $BaSO_4$ precipitates instantly and remain stable). At the end of each field day, the samples were stored in a refrigerator (except the C and S isotope bottles) and kept cool until returned to a refrigerator or a cool room in the laboratory.

13.3.2 Laboratory analyses

The samples for cation analysis were split into two aliquots in the laboratory, applying the new gloves, new containers and triple rinse quality control measures described later. One aliquot analysed by inductively coupled plasma-atomic emission spectrometry (ICP-AES) using a Varian Vista AX CCD simultaneous inductively coupled plasma-atomic emission spectrometer for selected elements (of which Ba, Ca, K, Mg, Na and Si were used for data analysis). The other aliquot analysed by inductively coupled plasma-mass spectrometry (ICP-MS) using a PerkinElmer SCIEX ELAN 6000 for trace elements of which Ag, Al, As, Au, B, Be, Bi, Cd, Ce, Co, Cr, Cs, Cu, Dy, Er, Eu, Fe, Ga, Gd, Ge, Hf, Hg, Ho, In, La, Li, Lu, Mn, Mo, Nb, Nd, Ni, Pb, Pd, Pr, Rb, Rh, Ru, Sb, Sc, Se, Sm, Sn, Sr, Ta, Tb, Te, Th, Ti, Tl, U, V, W, Y, Yb, Zn and Zr were used for data analysis. The ICP-AES and -MS samples were run with internal quality control samples and standard reference materials, and accepted only if precision and bias were within ± 5%. About 10% of the samples were spiked and accepted only if recovery was within ± 10% of the expected values.

Anion analysis was by ion chromatography (IC) using a Dionex ion chromatograph series 4500i for Br^-, Cl^-, F^-, NO_3^-, PO_4^{3-}, SO_4^{2-} and by Ion Specific Electrode (ISE) using a Thermo Scientific Orion ISE Model 9653 for I^-. IC analyses were run with internal quality control samples and accepted only if precision and bias were within ± 5%. About 10% of the samples were spiked and accepted only if recovery was within ± 10% of the expected values. About 10% of the samples were rerun (laboratory duplicates) and accepted only if precision was within ± 10%. ISE analyses were performed only after the electrode's behaviour had been checked with several QC samples of different concentrations and all results were within ± 5% of the expected values.

The rinsed, filtered and air-dried $SrCO_3$ and $BaCO_2$ precipitate samples were analysed for C isotope composition of dissolved inorganic carbon ($\delta^{13}C(DIC)$) and for S isotope composition of dissolved sulphate ($\delta^{34}S(SO_4^{2-})$), respectively. Measurement of $\delta^{13}C$ followed the classic method in which isotopically representative CO_2 from carbonates was prepared by reaction of the sample with concentrated H_3PO_4 into a Y-tube. Parameters of H_3PO_4 preparation, reaction temperature and reaction time were carefully controlled during the phosphorolysis. Ultimately, the evolved CO_2 gas was introduced into the ion source of a VG 903 stable isotope ratio mass spectrometer and analysed for the $^{13}C/^{12}C$ ratios. Precision and bias using this technique were typically better than ± 0.2‰ V-PDB for $\delta^{13}C$ ($n = 10$ internal lab standards). Sulphur isotope ratios ($^{34}S/^{32}S$) of pure $BaSO_4$ were analysed using continuous flow-elemental analyser-isotope ratio mass spectrometry (CF-EA-IRMS). The instrumentation comprised a Carlo

Erba NA 1500 elemental analyser interfaced to a magnetic-sector mass analyser. Internal laboratory standards were characterised against International Standards and were rechecked periodically. Precision and bias of $\delta^{34}S$ ($BaSO_4$) using this technique were typically better than ± 0.7‰ V-CDT ($n = 10$ internal lab standards).

The samples for H, or deuterium (D), and O isotope analysis of water (δ^2H, δD, and $\delta^{18}O$) were analysed using a Finnigan MAT 252 mass spectrometer. δ^2H values of water were measured via reaction with Cr at 850°C using an automated Finnigan MAT H/Device and analysed via the dual inlet. $\delta^{18}O$ values of water were measured via equilibration with He-CO_2 at 32°C for 24–48 hours in a Finnigan MAT gas bench and analysed by continuous flow. δ^2H and $\delta^{18}O$ values were measured relative to internal standards that were calibrated using IAEA SMOW, GISP, and SLAP standards. Typical precision (1 s) of the analyses based on replicate samples was ± 1‰ V-SMOW for δ^2H and ± 0.15‰ V-SMOW for $\delta^{18}O$.

13.4 Interpreting groundwater results

The interpretation of the groundwater level data for flow rates and directions is a standard technique and does not require any specific modifications for CO_2 monitoring. Various parameters must be taken into account, including rainfall, temperature, instrument drift and any quantifiable extractive activities. The results for the two aquifers are discussed below.

13.4.1 Port Campbell Limestone

The baseline reduced water levels (RWL), i.e. relative to the Australian Height Datum, recorded prior to injection are shown for the Port Campbell Limestone in Figures 13.4–13.6. The spikes recorded in these data (as for the Dilwyn Aquifer data) represent times when the logger was removed from the borehole. Average monthly rainfall is included for comparison. It can be seen (Figures 13.5 and 13.6) that two piezometers (B141241 Station I and B124659 Station R) respond to rainfall changes. B141240 (Station H) shows only a slight, delayed response, suggesting the aquifer is more confined at this location (Figure 13.4).

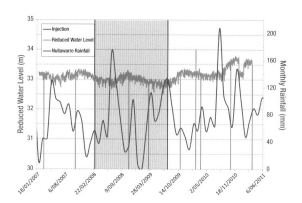

Figure 13.4: Reduced water level (RWL) and average monthly rainfall data for Bore 141240, Port Campbell Limestone aquifer (Station H).

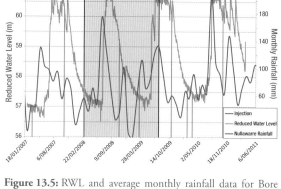

Figure 13.5: RWL and average monthly rainfall data for Bore 141241, Port Campbell Limestone Aquifer (Station I).

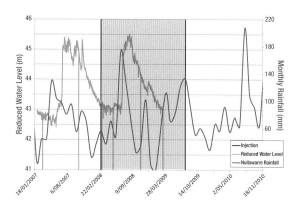

Figure 13.6: RWL and monthly rainfall data for Bore 124659 (Station R), Port Campbell Limestone Aquifer.

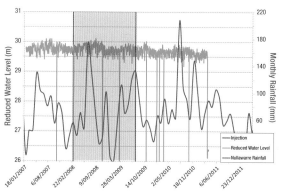

Figure 13.7: RWL for Bore B85937, Dilwyn Aquifer (Station J).

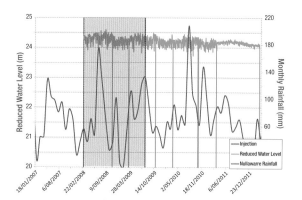

Figure 13.8: RWL for Bore 84290, Dilwyn Aquifer (Station V).

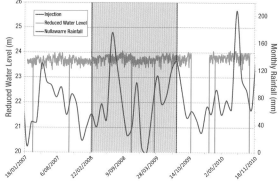

Figure 13.9: RWL and water column temperature for Bore 84291, Dilwyn Aquifer (Station W).

The fluctuation in hydraulic head due to rainfall is minor relative to the regional hydraulic gradient, which varies from 100 to 20 m in the vicinity of the Otway Project (Figure 13.10).

The temperature measured in these three shallow bores was constant over time (~15°C) at the point of measurement. This does not reflect the formation temperature due the difference between the measuring point and the screened interval (up to 50 m in B141240, Station H).

In spite of the response to rainfall, the unregulated groundwater use and the karstic nature of the Port Campbell Limestone, the overall potentiometric surface for the aquifer in this area (Figure 13.10) was consistent over time and independent of seasonal variations. The main pathway for regional flow was through the primary intergranular pore space (Bush 2009). Using the effective porosity and hydraulic conductivity values determined by Bush

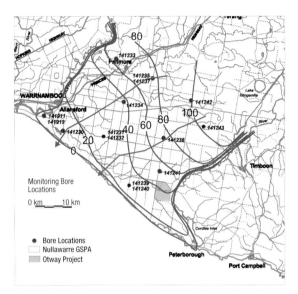

Figure 13.10: Potentiometric surface map for the Port Campbell Limestone (Otway Project area in green shading), modified after SRW (2005).

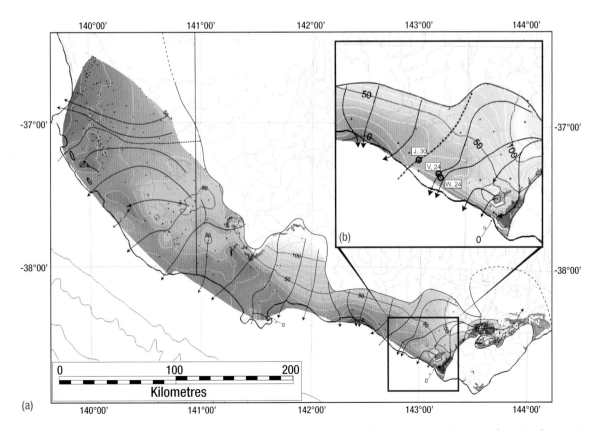

Figure 13.11: (a) Potentiometric surface map for the Lower Tertiary Sandy Aquifer (includes the Dilwyn Aquifer, taken from Bush (2009). Red dashed lines are groundwater divides. (b) CO2CRC monitoring stations and average RWL recorded.

(2009), the Darcy flow velocity through this aquifer in this area was between 4 and 15 cm/day, directed south-west, towards the coast. The mechanisms controlling flow were a combination of the topography of the water table and the permeability.

13.4.2   Dilwyn Aquifer

The baseline RWLs for the Dilwyn Aquifer are shown in Figures 13.7–13.9 and include the temperature recorded by the logger in Bore 84291 (Station W). A temperature decline observed in bore 84291 (Station W; not shown) is considered to result from a faulty temperature sensor as the data from all other loggers consistently recorded a single constant temperature at around 15°C and there is no climatic or geological reason for the decline.

Data from Stations J (B85937) and W (B84291) showed similar water level patterns, while the average RWL for Station J (29.7 m) was slightly higher than Station W (23.8 m). There were limited data from Station V due to installation problems and there were no trends visible in this data. The average RWL for this piezometer was 24.2 m. Stations J and W both appeared to have some limited seasonal fluctuation (Figures 13.7 and 13.9).

Figure 13.11 shows the regional density-corrected potentiometric surface map for the Lower Tertiary Sandy Aquifer, which includes the Dilwyn Aquifer, taken from Bush (2009). Regional flow within the aquifer is directed to the south-west. Bush (2009) describes a local- or intermediate-scale recharge zone to the south-east, which appears as a groundwater mound on Figure 13.11. This feature produces local flow lines that are not aligned with the regional flow direction. CO2CRC monitored wells and the average RWL for each have been added to Figure 13.11(b) and fit well within this regional model. Using the effective porosity and hydraulic conductivity values determined by Bush (2009), the Darcy flow velocity through this aquifer in this area is between 4 and 10 cm/day.

13.5   Groundwater composition

Only general examples of methods used to interpret groundwater compositions are given here, as detailed interpretations are presented elsewhere (de Caritat et al. 2013).

Examining the evolution of groundwater composition at a particular location as a function of time can be useful, whether the trend is linear, cyclic or otherwise. An example of this is shown as a time series in Figure 13.12.

In contrast, the particular behaviour of a population, or subpopulation of data, can be statistically and graphically represented using plots such as the boxplot (Figure 13.13), where the pre-injection data can be readily compared to the syn- and post-injection data. The median values (horizontal lines dividing the boxes) can also be compared. In the particular representation of the boxplot used here, derived from Tukey (1977) and Velleman and Hoaglin (1981), the median of a population is significantly different from another median if it is outside the notches (square brackets) that surround that other median value.

In Figure 13.14, maps of pre-injection and syn-/post-injection distributions of the $\delta^{13}C$ (DIC) are compared. The median values recorded at every site over the appropriate time period are represented by a symbol. The data were also spatially interpolated here using ordinary kriging. The same colour scale and classification is used for both maps, allowing the temporal differences to be seen at a glance. de Caritat et al. (2013) also show a similar map for the HCO_3^- concentrations.

The conclusions of the groundwater chemistry monitoring project are as follows (de Caritat et al. 2013):

›   The Port Campbell Limestone groundwater is fresh (EC = 801–3900 µS/cm), cool (T = 12.9–22.5°C), with near-neutral acidity (pH 6.62–7.45).

›   The deeper Dilwyn Aquifer groundwater is fresher (EC 505–1473 µS/cm), warmer (T = 42.5–48.5°C), and more alkaline (pH 7.43–9.35); it is also compositionally more heterogeneous than the Port Campbell Limestone groundwater.

›   Major processes controlling the composition of groundwater are evapotranspiration carbonate dissolution and cation exchange.

›   Comparing the composition, including the oxygen and hydrogen isotopic composition of H_2O and the C isotopic composition of DIC. Groundwater collected before, during and after injection shows

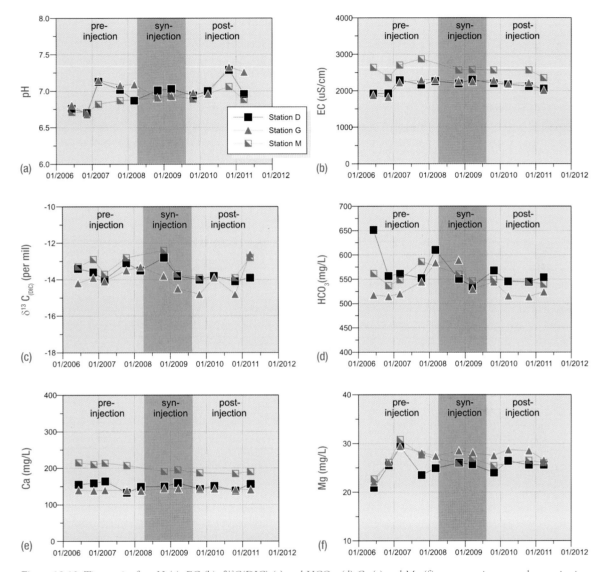

Figure 13.12: Time-series for pH (a), EC (b), $\delta^{13}C(DIC)$ (c) and HCO_3^- (d) Ca (e) and Mg (f) concentrations over the monitoring period at Stations D, G and M located down-gradient from the CO_2 injection well and accessing groundwater from the Port Campbell Limestone.

either no evidence of statistically significant changes or, where they are statistically significant, they are changes that are generally not consistent with CO_2 addition to the groundwater.

> As such, the groundwater monitoring programme detected no change in the quality of groundwater resources in the area that could be attributable to CO_2 leakage.

13.6 Operational issues relating to groundwater monitoring

Monitoring of groundwater levels over the long term is necessary to observe and account for variations due to climate. For example, in the Otway dataset, it was observed that there was an aquifer response to drought-breaking rains. Concomitant with this was the need to

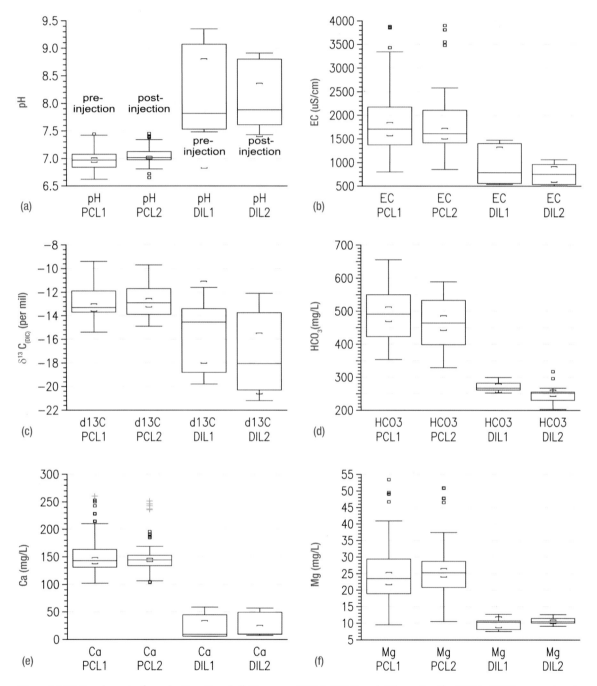

Figure 13.13: Boxplot pairs from the Port Campbell Limestone (PCL1, PCL2) and Dilwyn Aquifer (DIL1, DIL2) comparing the pre-injection data (left boxplot: PCL1, DIL1) to the syn- and post-injection data (right boxplot: PCL2, DIL2) for pH (a), EC (b), δ^{13}C(DIC) (c) values, and HCO$_3^-$ (d), Ca (e) and Mg (f) concentrations. Whiskers (T), notches ([ ]), outliers (□) and extreme outliers (+) as defined by Tukey (1977), Velleman and Hoaglin (1981) and Reimann et al. (2008).

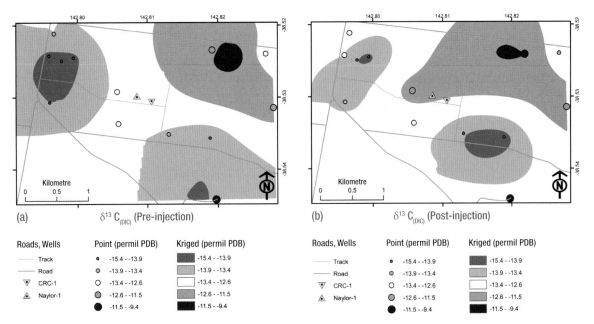

Figure 13.14: Kriged maps of the distribution of $\delta^{13}C(DIC)$ values in the Port Campbell Limestone groundwater representing pre-injection (a) compared to the syn- and post-injection (b) data.

also monitor those other parameters that cause natural variations in the aquifer. These included rainfall, water extractive activities and such things as mining. It was also necessary to maintain and calibrate the data loggers. Obtaining a reliable groundwater sample was a key aspect of the programme; some of the ways that this was done are summarised below.

13.6.1 Pumping times

Because the value of a groundwater analysis is contingent upon obtaining a water sample that is representative of the groundwater in the aquifer, emphasis was placed on obtaining a high quality, fresh groundwater sample. This was operationally managed by (1) allowing sufficient time for pumping prior to sampling (up to 5 hours of pumping for the deep bores), and (2) monitoring critical water quality parameters of the pumped water at regular intervals between the commencement of pumping and sampling for chemical and isotopic analysis.

For most shallow bores, a stable water composition was attained within an hour of pumping and monitoring, especially for dairy bores, which are used twice a day by

farmers to wash the dairy clean after milking. As mentioned above, pumping times were far more extended for the deep bores, which were pumped very slowly (1–2 L/min) using the configuration described below. Experience gained during this project indicated that pumping the deep Dilwyn Aquifer bores for 5 hours was sufficient to obtain a representative groundwater sample.

13.6.2 Sampling the deep bores

Sampling the deep Dilwyn Aquifer bores required an innovation to get the intake screen level with the perforated intervals in the bores' casing, which were well beyond the 120 m of reel available on the "normal" pump setup.

The Bennett pump is designed so that a low density polyethylene (LDPE) extension tubing can be attached. Besides the pump itself, the required length of tubing, weights and an inlet screen were necessary to sample deep bores.

For deep groundwater sampling, a rod-like intake screen unit and weights were attached at the bottom part of a length of LDPE tubing and lowered down the bore. The weights were necessary to counter the buoyancy of the LDPE tubing. At the other (upper) end, the tubing was connected to the bottom of the Bennett pump, which was lowered down the bore as deep as necessary (Sundaram et al. 2009).

Groundwater was lifted inside the tubing mainly by the pressure differential between the deeper end (higher hydrostatic pressure) and the shallower end (lower hydrostatic pressure) of the tubing under only a minimal drive from the pump. Once the water was inside the pump, it was pushed up the pump hose by the piston action driven by the compressed air. Pumping rates were quite slow, and even with the intake screen positioned right at the casing's slotted interval, it took some time to ensure that fresh groundwater was being pumped up. Continued monitoring of field parameters, as discussed elsewhere, was of paramount importance here.

In the Otway Project, success was attained in sampling groundwater from depths down to 850 m below ground level using the Bennett double piston, compressed air pump connected to lengths of either 700 or 800 m of $\frac{3}{8} \times 0.250$ Esdan food grade LDPE tubing rated for up to 120 psi pressure (Figure 13.15(a)). The tubing was made up from two lengths of 300 m and one or two lengths of 100 m, as required, joined together with Swagelok stainless steel connectors (Figure 13.15(b)). Two deep bores required 700 m of tubing, while the third required 800 m of tubing to reach the perforated intervals within the Dilwyn Aquifer. The intake screen unit was a QED stainless steel unit (Figure 13.15(c)). Three weights made of cut lengths of stainless steel rod, approximately 34 cm in length and 2.2 kg in weight each (Figure 13.15(c)), were used successfully for either 700 or 800 m of tubing. The tubing was connected to the intake of the Bennett pump (Figure 13.15(d)), which was then lowered into the bore (Figure 13.15(e)). At about 800–850 m depth (within the slotted interval), pumping rates of 1–2 L/min were achieved with this configuration.

It should be noted that some water and environmental agencies ban rapid pumping (purging) to remove the theoretical three bore volumes of water, especially for old bores whose casing may collapse if the contained water is removed. This leaves low-flow pumping, as discussed here, within (or as near as possible to) the slotted interval as the only viable alternative for obtaining a proper groundwater sample.

13.6.3 Groundwater temperature

Whereas water temperature measured in the flow cell is quite representative of the conditions in the shallow aquifer, the groundwater pumped up from the three deep bores cools down considerably as it slowly moves up the narrow tubing, which is itself surrounded by a large volume of progressively cooler bore water toward the surface. Thus, a further innovation was necessary to obtain a more realistic groundwater temperature for the Dilwyn Aquifer (this was implemented from Trip #10). The temperature of the deep groundwater was measured using an iButton temperature sensor sealed in a custom-made stainless steel housing attached to the bottom of the QED screen (Figure 13.15). The temperature obtained in this way was representative of in-situ conditions and was found to be much higher than previously reported temperature measured in the flow cell at the surface.

13.7   Quality control

To further ensure the quality of the analytical results obtained, a series of quality control (QC) measures were adhered to in the field and in the laboratory. These were as follows:

› Bores were pumped for an extended period of time to ensure a representative fresh groundwater sample (recently emerged from the aquifer formation) was collected, rather than stale, possibly oxidised and/or contaminated bore water.

› All sampling containers used were new.

› All containers, except Vacutainers, were rinsed three times with the water to be sampled, before the sample was collected.

› Operators filtering samples were required to wear new non-powder latex gloves for each sample.

› From Trip #8, tracer samples were collected under water in a bucket, as per the USGS method <http://water.usgs.gov/lab/sf6/sampling>.

Figure 13.15: Deep water sampling equipment: (a) reel of LDPE tubing; (b) connection between lengths of tubing; (c) intake screen with weights in background; (d) connecting the tubing to the pump; (e) lowering the Bennett pump in the bore after the tubing and before the pump hose; and (f) tripod and swivelling block used to lower and retrieve tubing.

> From Trip #3, single-use 0.45 µm WATERRA filter capsules were used instead of membrane filters in a filter holder.

> High quality chemicals were used (Milli-Q distilled, deionised water; distilled HNO_3).

> Care was taken to avoid contamination on site and in the lab.

> Samples were labelled in such a way that sample mix-up or sample label loss was virtually impossible.

> No smoking was allowed during sampling.

> Field duplicates were collected at the approximate rate of one for every 10 sites.

> In the preparation laboratory, approximately 10% of samples were split and submitted unidentified to the analytical laboratory (lab duplicates).

> Various blanks were prepared: travel blank (raw Milli-Q water that was taken in the field), rinse blank (field Milli-Q water put through the field filtering equipment in the field), lab blank (fresh, lab Milli-Q water).

At the conclusion of the baseline monitoring, a full quality assessment (QA) of the groundwater chemistry results was carried out (de Caritat et al. 2009), the main findings of which are given below.

13.7.1   Laboratory blanks

Lab blanks are aliquots of distilled, deionised water produced by the Millipore Academic Milli-Q apparatus in the hydrogeochemistry laboratory, Geoscience Australia. This water has a resistivity of 18.2 MΩ·cm at 25°C, monitored through a display on the apparatus. The aliquots are collected directly at the dispenser's outlet into 10 mL Sarstedt vials, previously rinsed three times also with Milli-Q water. Powder-free latex gloves are worn throughout this operation. The samples are not filtered. The aliquots for cation analysis (by inductively coupled plasma, or ICP, methods) are supplemented with a few drops of distilled HNO_3 as used in the field.

Lab blanks show the purity, at source, of the water used to rinse the filtering apparatus (Trips #1 and 2) or pressure

flask (Trip #3 and later) except for cation (ICP) analyses, where the results also reflect the purity of the added distilled HNO_3.

Lab blanks indicate that the Milli-Q water used was very pure, at the source, with nearly all analyses below detection level (< DL). Only in rare cases were > DL results noted (maximum recorded Mg 0.020 mg/L; F 0.031 mg/L; Al 0.002 mg/L; B 0.020 mg/L; Cu 0.001 mg/L; Fe 0.094 mg/L; Ni 0.002 mg/L; Ti 0.020 mg/L; and Zn 0.008 mg/L).

13.7.2   Travel blanks

Travel blanks are also aliquots of Milli-Q water, but instead of taking the samples straight from the dispenser in a laboratory environment, the aliquots were collected from the field container holding the Milli-Q water. The samples were not filtered. The aliquots for cation analysis were supplemented with a few drops of distilled HNO_3 in the field.

Travel blanks show how clean the water used to rinse the filtering apparatus (Trips #1 and 2) or pressure flask (Trip #3) is under field conditions (field container, dust, etc.), except that for cation (ICP) analyses the results also reflect the purity of the distilled HNO_3 added.

Travel blanks indicated that the Milli-Q water used in the field contained measurable (but comparatively low) concentrations of some major cations and anions (maximum recorded Na 0.462 mg/L; K 0.067 mg/L; Mg 0.041 mg/L; and Cl 3.178 mg/L) and small but detectable quantities of some trace elements (maximum recorded Al 0.009 mg/L; Fe 0.020 mg/L; Ti 0.020 mg/L; and Zn 0.002 mg/L). All other analyses were below the detection limit.

The Milli-Q water was used for rinsing equipment in the field (e.g. filtering apparatus, pressure flask) and then discarded, commonly leaving only a few drops of Milli-Q water on the sides of the pressure flask. Therefore, minor amounts of measurable species/elements were not a cause for concern because of the large dilution caused by adding much larger quantities of groundwater used for a triple rinse compared to the few drops of Milli-Q water.

13.7.3 Filter blanks

Filter blanks are aliquots of field Milli-Q water that have been through the filtering apparatus (Trips #1 and 2) or pressure flask and a new, disposable 0.45 μm filter capsule (Trip #3 and later), in the field.

Filter blanks show how using a filtering apparatus and filter membrane or, alternatively, a pressure flask and filter capsule, can contaminate a groundwater sample. The procedure potentially emulates what would happen to a real groundwater sample. Therefore filter blanks can give a realistic worst-case-scenario indication of field contamination, particularly when compared to the travel blank. For Trips # 1 and 2, samples were filtered using the filtering apparatus and filter membrane. From Trip # 3 onward, only a few of the real groundwater samples were pushed through the pressure flask and filter capsule; most samples did not require the use of the filter flask because the filter capsule could be connected directly to the pump outlet hose. Commencing with and including Trip #12, field procedures were modified so that the pressure flask was no longer required. For cation (ICP) analyses, the results also reflect the purity of the added distilled HNO_3.

Filter blanks gave the following worst-case-scenario indications of contamination caused by the pressure flask and filter capsule: maximum recorded Na 0.205 mg/L; Mg 0.015 mg/L; Al 0.002 mg/L; B 0.006 mg/L; Fe 0.023 mg/L; Ti 0.019 mg/L; W 0.005 mg/L; and Zn 0.008 mg/L.

It was possible to gain an indication of the maximum recorded contamination due to the pressure-filtering process without the effect of "contamination" of the field Milli-Q water, by subtracting the maximum recorded concentrations in the travel blank from the maximum recorded concentrations in the filter blank. This provided a more realistic indication of contamination due to pressure-filtering of groundwater in the field. The results of this were Fe 0.003 mg/L; W 0.005 mg/L; and Zn 0.006 mg/L (negative values omitted). Based on where these elements were detected in the various blanks, it is possible that the minor Fe and Zn contaminations were from the Milli-Q water or the HNO_3 acid, given that they were also detected

in the unfiltered lab and travel blanks. Conversely, the minor W contamination appear to have originated from the pressure flask or filter capsules (the measurable W value was in a Filter Blank from Trip #4), given that it was not detected in the unfiltered lab or travel blanks.

This analysis of blank QA/QC data showed that contamination from field procedures relative to concentrations measured in the groundwaters was negligible.

13.7.4 Laboratory duplicates

Lab duplicates are a split of selected samples prepared under clean conditions in the lab and submitted to the analytical lab without revealing that it is a duplicate. One lab duplicate was prepared approximately every tenth sample. The lab duplicate allowed quantification of the ability of the analytical laboratory to reproduce results (precision).

The results of the analysis of nine pairs of lab duplicates collected on Trips #1–5 are given in Table 13.3, which shows the analytical precision calculated as a standard deviation and precision. The standard deviation (s) is calculated by dividing the average absolute difference between pairs (R) by 1.128; the precision is expressed as a coefficient of variation (or relative standard deviation, in %) calculated by dividing s by the average values of the duplicate analyses (Greenberg et al. 1992). Precision can only be calculated for parameters/elements that have at least one pair of duplicates with both values greater than the detection limit (DL).

According to Greenberg et al. (1992), acceptable precision levels for metals, anions and other inorganic species are 10% at high concentrations/levels and 25% at low concentrations/levels, where the cut-off between low and high concentrations/levels is 20 times the method detection limit. The worst precision results were obtained for elements that are mostly found at low concentration in these duplicates (av Th = 4 × DL, av Sc = 3 × DL, av Hf = 1 × DL, av B = 67 DL, av Al = < 1 × DL, av Ba = 4 × DL) and, therefore, not overly concerning. Of these, Th, Sc and B are not satisfactory in terms of precision, given the Greenberg et al. (1992) criteria.

13.7.5   Field duplicates

Field duplicates consist of a second sample collected in the field, in exactly the same manner as the first sample, but some time after it (about half an hour, generally). This means that field duplicate data encapsulate not only the ability of the analytical laboratory to reproduce results but also temporal variability in the pumped water composition. This temporal variability within relatively short time frames can for instance be due to (1) a front of slightly different groundwater moving through the aquifer or mixing in the bore casing; (2) variable mixing between two or more aquifers feeding the same borehole or between fresh groundwater and stagnant bore water; or (3) changes in the pumping regime (e.g. changes in wind strength in the case of a windmill). Therefore, field duplicates can capture a range of poorly constrained variables and are notoriously poor (or remain unreported). Nevertheless, they are useful in the sense that they place boundaries around the natural and analytical variability of the data by addressing the question, "What if the sample had been collected at a slightly different time?"

The results of the analysis of eight pairs of field duplicates collected on Trips #1–5 are given in Table 13.4 and show the combined sampling and analytical precision calculated as a standard deviation and precision (see above).

As mentioned previously, according to Greenberg et al. (1992), acceptable precision levels for metals, anions and other inorganic species is 10% at high concentrations/levels and 25% at low concentrations/levels, where the cut-off between low and high concentrations/levels is 20 times the method detection limit. The worst precision results were obtained for elements that are found at low concentration in these duplicates (av Al = $2 \times$ DL, av I = $2 \times$ DL, av Cu = $4 \times$ DL, av Ba = $6 \times$ DL, av DO = $18 \times$ DL, av F = $14 \times$ DL) and are therefore not of concern.

Of the elements noted above as having poor precision based on the lab duplicates, Th and Sc were not detected above DL in the Otway field duplicates and therefore cannot be further commented on, except to say that very few samples have detectable Th and Sc (maximum concentrations recorded are 0.006 and 0.008 mg/L, respectively). The poor precision for B noted previously for lab duplicates

(12%) is mitigated by the fact that the more comprehensive field duplicates showed an acceptable precision for that element (9%; av B = $56 \times$ DL).

13.7.6   Other quality control checks

Two other ways of quickly checking a dataset for errors or anomalous samples are by plotting charge balance (CB) versus a salinity indicator, and total dissolved solids (TDS) against electrical conductivity (EC).

CBs are calculated by the software AquaChem (v. 5.1). Positive values indicate an excess of cations (in meq/L) relative to anions, whereas negative values indicate an excess of anions (in meq/L) relative to cations. In practice, CB values between –5% and +5% are considered excellent, and values between –10% and +10% acceptable. Beyond this range, water analyses are considered as requiring a check or reanalysis. Reasons for "poor" CB values can include poor sample preparation/preservation, poor quality of analysis or simply that a rare ion is present in the sample and was simply not analysed for. In the present study, only two samples exceed the –10 to +10% range, and by a minimal margin (CB = +10.07 and +10.09% respectively). This was not considered sufficiently important to reject these analyses.

TDS values were also calculated using the software AquaChem (v. 5.1), whereas EC values were measured in the field. The relationship between TDS and EC is dictated by the abundance (concentration in mg/L) and nature of ions (namely their charge, e.g., Na^+ vs Mg^{2+}, or Cl^- vs SO_4^{2-}) present in the water, because different ions have different masses (affecting TDS) and different charges (affecting EC). Normally groundwaters from a given area/aquifer have a similar TDS to EC relationship and plot along a well-defined line in the TDS-EC space. If a water sample plots well outside the general trend, this may be an indication of some error; alternatively there may well be a real explanation for this behaviour. Regardless, such a sample is worth looking into further. In the Otway Project, all the groundwater samples plot along a single line with a regression coefficient (r) of 0.98 (and slope of 0.61). No sample stands out as requiring further inspection.

Table 13.4: Combined sampling and analytical precision calculated from field duplicates. The values in the two right columns are in order of worsening precision.

Parameter/Element	Unit	Standard Deviation	Precision (%)	Parameter/Element	Precision, Ordered (%)
T	°C	0.166223404	1	EC	0
EC	μS/cm	6.427304965	0	pH	0
DO	mg/L	0.263741135	15	As	0
pH	pH units	0.019946809	0	$\delta^{18}O(SO_4)$	0
Eh	mV (SHE)	6.017287234	3	T	1
Al	mg/L	0.00177305	20	Ca	1
As	mg/L	6.92225E-06	0	Cl	1
B	mg/L	0.004779437	9	Na	1
Ba	mg/L	0.001966282	16	SO_4	1
Br	mg/L	0.030388313	4	$\delta^{13}C(DIC)$	1
Ca	mg/L	1.126904437	1	HCO_3	2
Ce	mg/L	2.3871E-05	3	K	2
Cl	mg/L	2.018437881	1	Mg	2
Co	mg/L	3.91175E-05	5	Ni	2
Cu	mg/L	0.000648743	17	Rb	2
F	mg/L	0.033736747	12	SiO_2	2
Fe	mg/L	0.014763103	4	U	2
Fe^{2+}	mg/L	0.044326241	9	$\delta^{34}S(SO_4)$	2
HCO_3	mg/L	10.89071284	2	Eh	3
Hf	mg/L	4.35957E-05	4	Ce	3
I	mg/L	0.007092199	17	Li	3
K	mg/L	0.131254117	2	Mn	3
Li	mg/L	0.000159873	3	Sr	3
Mg	mg/L	0.403328145	2	$\delta^{18}O(H_2O)$	3
Mn	mg/L	0.001002337	3	Br	4
Na	mg/L	1.660480567	1	Fe	4
Ni	mg/L	8.78935E-05	2	Hf	4
NO_3	mg/L	0.608876831	6	Se	4
Rb	mg/L	8.98324E-05	2	$\delta D(H_2O)$	4
Se	mg/L	0.000133769	4	Co	5
SiO_2	mg/L	0.334775398	2	V	5
SO_4	mg/L	0.455977501	1	NO_3	6
Sr	mg/L	0.021459318	3	Zn	7
Ti	mg/L	0.006471736	8	Ti	8
U	mg/L	7.58685E-05	2	B	9
V	mg/L	0.000110297	5	Fe^{2+}	9
Zn	mg/L	0.000384885	7	F	12
$\delta D(H_2O)$	‰ V-SMOW	4.236497944	4	DO	15
$\delta^{18}O(H_2O)$	‰ V-SMOW	0.633764507	3	Ba	16
$\delta^{13}C(DIC)$	‰ V-PDB	1.626787533	1	Cu	17
$\delta^{34}S(SO_4)$	‰ V-CDT	0.319764253	2	I	17
$\delta^{18}O(SO_4)$	‰ V-SMOW	0.026595745	0	Al	20

In summary, various blank and duplicate samples, as well as other tests, show that the data obtained for analysis for the establishment of the groundwater baseline in the Otway Project did not suffer from contamination issues: they were precise, internally consistent and of acceptable quality.

13.8 Conclusions

The present review of methodologies and activities relating to groundwater monitoring at the Otway site presents details on how we endeavoured to obtain the best quality data that are fit for purpose in a CCS context.

Monitoring groundwater levels is fairly straightforward and was done using a combination of depth-temperature (DT) or conductivity-depth-temperature (CDT) water level loggers permanently installed in selected bores. This was complemented by biannual to annual manual water level measurements using a dip meter in all open bores. State authorities conduct their own water level checks, and their data were also used.

Monitoring groundwater composition is more complex and requires time, understanding of potential pitfalls and appropriate equipment, to be carried out properly. The questions the field operator has to ask include: Where does the groundwater come from? How can a sample representative of the aquifer conditions be ensured? Answering the first question requires knowledge of the depth of the borehole and, more importantly, of the screened interval(s) from which groundwater is pumped. To answer the second question, the operator has to be alert to the fact that water that has been standing in a bore for some time is likely to have been modified from its original composition representative of aquifer conditions by various processes. These can include mixing with rainwater or with groundwater leaking from other intervals, interaction with the atmosphere, chemical reaction with the casing, action of microorganisms living in and around the bore and interaction with the objects/animals that have been dropped or have fallen into the bore.

The operator has to pump for a sufficiently long time to ensure that fresh groundwater is obtained. In a deep and wide bore, this can take several hours, especially when pumping with a low-flow ("environmental sampling") pump. One way of having confidence in the quality of the groundwater being pumped is to monitor basic water quality parameters, such as pH, temperature, electrical conductivity, and in a flow cell closed to the atmosphere. Often, these parameters will evolve during pumping through several transient stages that can represent water stratification in the bore, before reaching a stable composition when a steady flow has been established from the aquifer to the intake screen.

Groundwater compositional data of sufficient quality are obtained through a well-designed quality control process. This has to include "blind", field and lab duplicates, and certified reference materials in order to demonstrate precision and bias. These properties are especially important in long-term monitoring programmes, where samples are analysed in numerous batches over an extended period of time, and sometimes using different equipment or laboratories.

Finally, the groundwater collected before, during and after injection shows either no evidence of statistically significant changes or, where they are statistically significant, they are changes that are generally not consistent with CO_2 addition to the groundwater. As such, the groundwater monitoring programme detected no change in the quality of groundwater resources in the area that could be attributable to CO_2 leakage.

13.9 References

Bush, A. 2009. Physical and chemical hydrogeology of the Otway Basin, southeast Australia. PhD Thesis, School of Earth Sciences, University of Melbourne.

Dance, T., Spencer, L., Xu, J.-Q. 2009. Geological characterisation of the Otway Project pilot site: What a difference a well makes. Energy Procedia 1, 2871–2878.

de Caritat, P., Hortle, A., Raistrick, M., Stalvies C. and Jenkins, C. 2013. Monitoring groundwater flow and chemical and isotopic composition at a demonstration site for carbon dioxide storage in a depleted natural gas reservoir. Applied Geochemistry 30, 16–32.

de Caritat, P., Hortle, A., Stalvies, C. and Kirste, D.M. 2009. CO2CRC Otway Project groundwater baseline: Results of pre-injection monitoring of groundwater levels and chemistry. CO2CRC Report, 09-1557.

Duran, J. 1986. Geology, hydrogeology and groundwater modelling of the Port Campbell hydrogeological sub-basin, Otway Basin, SW Victoria. Geological Survey of Victoria, Unpublished report, 1986/24.

Greenberg, A.E., Clesceri, L.S. and Eaton, A.D. (Eds) 1992. Standard methods for the examination of water and wastewater. American Public Health Association, 18th edition.

Hennig, A. 2007. Field guide for installing, programming and downloading water level loggers CO2CRC Otway Project. Cooperative Research Centre for Greenhouse Gas Technologies, Canberra, Australia. CO2CRC Publication Number RPT07-0763.

Hennig, A., Hogan, G. and Kirste, D. 2007a. CO2CRC Otway Basin Pilot Project groundwater sampling and monitoring program: health, safety and environment plan. Cooperative Research Centre for Greenhouse Gas Technologies, Canberra, Australia. CO2CRC Publication Number RPT06-0332.

Hennig, A., Perkins, E. and Kirste, D. 2007b. Groundwater monitoring procedures for DSE owned SOBN Bores, Otway Basin Pilot Project. Cooperative Research Centre for Greenhouse Gas Technologies, Canberra, Australia. CO2CRC Publication Number RPT06-0333.

Inkenbrandt, P.C., Doss, P.K. and Pickett, T.J. 2005. Barometric and earth-tide induced water-level changes in the Inglefield Sandstone in southwestern Indiana. Proceedings of the Indiana Academy of Science 114, 1–8.

Jenkins, C.R., Cook, P.J., Ennis-King, J., Underschultz, J., Boreham, C., Dance, T., de Caritat, P., Etheridge, D.M., Freifeld, B.M., Hortle, A., Kirste, D., Paterson, L., Pevzner, R., Schacht, U., Sharma, S., Stalker, L. and Urosevic, M. 2012. Safe storage and effective monitoring of CO$_2$ in depleted gas fields. Proceedings of the National Academy of Sciences 109, E35–E41.

Nahm, G.Y. 1985. Groundwater resources of Victoria. Department of Industry, Technology and Resources, Victoria.

Reimann, C., Filzmoser, P., Garrett, R. and Dutter, R. 2008. Statistical data analysis explained: applied environmental statistics with R. John Wiley & Sons.

SKM [Sinclair Knight Mertz] 1999. Gippsland declining levels, groundwater trends in deep systems elsewhere in Victoria, Final Report, June 1999. Department of Natural Resources and Environment, Victoria, Monograph Series ISSN 1328-4495.

Spane, F.A. 1999. Effects of barometric fluctuations on well water-level measurements and aquifer test data. U.S. Department of Energy, Contract DE-AC06-76RLO 1830, Pacific Northwest Laboratory, Richland, Washington 99352.

SRW (Southern Rural Water) 2005. Annual Report Nullawarre Water Supply Protection Area management plan, year ending 30 June 2005. Report submitted to the Minister for Water, the Glenelg Hopkins and Corangamite Catchment Management Authorities.

Sundaram, B., Feitz, A.J., de Caritat, P., Plazinska, A., Brodie, R.S., Coram, J. and Ransley, T. 2009. Groundwater sampling and analysis—A field guide. Geoscience Australia Record, 2009/27: 95 pp.

Tukey, J.W. 1977. Exploratory data analysis. Addison-Wesley, Reading, Massachusetts.

Underschultz, J., Boreham, C., Dance, T., Stalker, L., Freifeld, B., Kirste, D. and Ennis-King, J. 2011. CO$_2$ storage in a depleted gas field: An overview of the CO2CRC Otway Project and initial results. International Journal of Greenhouse Gas Control 5, 922–932.

Velleman, P.F. and Hoaglin, D.C. 1981. Applications, basics, and computing of exploratory data analysis. Duxbury Press, Boston, MA.

Ulrike Schacht

14. SOIL GAS MONITORING

14.1 Introduction

Soil gas monitoring has been widely accepted as one option to detect CO_2 that has migrated from a storage site to the near surface, before leaking into the atmosphere. It has been applied to a number of CO_2 storage projects, such as Cranfield, USA (Romanak et al. 2012), Gorgon, Australia (Flett et al. 2009), In Salah, Algeria (Mathieson et al. 2011) and Weyburn, Canada (White et al. 2004).

The concentration of CO_2 and other gases in the soil is highly variable as a result of the geological and seasonal variability in the vadose zone as well as the variable level of biological activity (Amundson and Davidson 1990). A strength of soil gas monitoring is that it is relatively easy to measure the composition of the soil gas. It can also be used to collect samples for later analysis of tracers that are characteristic of the stored CO_2. These may have been artificially added (Jenkins et al. 2012; Martens et al. 2012), or may be distinctive natural isotopes such as

$^{13}CO_2$ (see below). A disadvantage of measuring only composition is that it is not possible to infer fluxes from concentrations, without having extra information about transport mechanisms through the vadose zone. Since a leak of CO_2 to the surface represents a flux of CO_2 soil gas measurements are difficult to interpret in terms of leakage, or limits on leakage, in a quantitative way.

It is standard practice to undertake soil gas surveys a number of times before injection of CO_2 commences (Schloemer et al. 2013); these baseline surveys are then a reference for post-injection data. If the post-injection data are similar to the baseline, this is taken to be evidence that leakage has not occurred. However, the meaning of "similar" is hard to define. For example, datasets may be the same to within the error bars, but if the error bars are large, then the method may have little statistical power. Ideally there would be some definite idea of what size of difference would be produced by a leak. But as mentioned previously, distinguishing changes of concentrations from changes in fluxes does require detailed knowledge of transport properties such as soil permeabilities, and these can be difficult to acquire.

14.2   Surficial geology

The surface and near-surface geology of any region has a significant impact on soil type and therefore on soil gas. The surface geology in and around the Otway site is dominated by the Hanson Plain Sands. This overlies a karstic surface at the top of the Port Campbell Limestone. At approximately 1 m sample depth either the Port Campbell Limestone or a transition zone between the Hanson Plain Sands and the Port Campbell Limestone, was encountered in the sampling programme. This transition zone was typically a clay-rich mixed sand and limey soil and was probably part of the karstic weathering profile. As the Hanson Plain Sands are not vertically extensive, it is likely that the Port Campbell Limestone is interacting with the soil gas system to a higher degree than might be expected from the geological map of the study area.

Occasionally the geology at the sample site comprised coastal dune deposits, which typically extended to greater than 1 m depth and it is assumed that a transition zone to the Port Campbell Limestone would occur at 1–3 m depth (Watson et al. 2006).

14.3   Soil gas sampling at Otway

An aim of monitoring at the Otway Project site was to determine if soil gas surveys could be used to detect leakage of any stored CO_2 that migrated to the near surface before leaking into the atmosphere. Pre-injection characterisation surveys to study the distribution and source of the existing soil gas were carried out in March 2007 and February 2008, followed by post-injection assurance monitoring surveys conducted in February 2009 and March 2010.

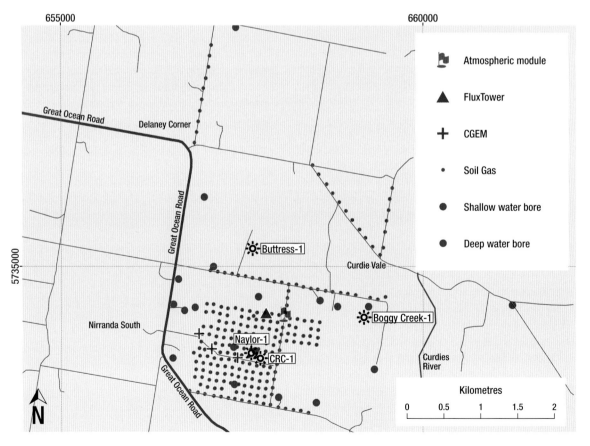

Figure 14.1: Otway Project site area and major monitoring locations; CGEM denotes the CanSyd monitoring stations; GW denotes "groundwater".

A direct-push soil gas sampling methodology was adopted for the study, as it minimised the risk of atmospheric contamination, had only minor impact on the surface and soil environment and had a fast, effective soil gas recovery rate (Watson et al. 2006). A Post Run Tubing system (PRT) was used to access the soil vadose zone and to acquire a sample. O-ring connections ensured that the PRT system had a vacuum-tight seal.

The PRT probes were pushed approximately 1 m into the soil and then jacked up approximately 0.2 m, leaving a 0.2 m, 2.54 cm diameter void in which soil gas could collect. During the jacking, the PRT retractable drive point dropped approximately 5 cm down the hole, at which point it caught onto the end of the drive point holder, allowing soil gas to enter the probe. A 1.5 m length of polytubing with the PRT adaptor attached to the end was inserted down the inside of the probe and screwed into the retractable drive point holder at the base of the rod. Approximately 150 mL or three times the volume of the polytube, was purged from the system via a syringe, to remove any atmospheric contaminants. The system was then sealed. Each probe was left in place for 1 hour prior to gas sampling. This allowed re-equilibration in the vadose zone environment following any disturbance caused by the rod insertion (Watson et al. 2006). After sampling, the rods were jacked out and cleaned before proceeding to the next hole. A maximum of half a litre of gas was sampled at each sample point. At each sampling point (Figure 14.1) the soil type, surface geology and other environmental conditions were recorded.

Each survey sampling programme targeted the same point, including vertically above the predicted CO_2 plume migration path within the injected (Waarre C) reservoir, points near each well, and vertically above any known natural CO_2 or hydrocarbon accumulations in the study area. Roadside reservations and government owned land was preferentially sampled for ease of access. However, privately owned land was accessed for many of the sample points. Some areas were inaccessible during sample surveys, because landowner permission was not granted. Figure 14.1 shows the sample point distribution for the four soil gas sampling surveys conducted between 2007 and 2010. A summary of total sample points for each soil gas survey is provided in Table 14.1.

Table 14.1: Sampling totals for the four soil gas surveys 2007–10.

Survey	Number of Samples	Successful Samples
March 2007	212	206
February 2008	153	142
February 2009	235	228
March 2010	156	139

Some samples were unsatisfactory due to contamination by atmospheric gases, insufficient gas recovery or groundwater entering the system. Probe insertion was sometimes hindered by wet sticky clays or very compacted soil (e.g. along road sites), and several attempts at placing the probe were required. Water-logged soils typically did not allow gas flow. Due to regular and normally heavy rainfall in winter time, surveying of soil gas could be challenging in terms of soil gas recovery (Watson et al. 2006). Surveying was thus limited to the summer months, when soil moisture levels were low and soil gas retrieval was more successful.

14.4 Analysis of soil gas

Molecular compositional analysis for CO_2 and CH_4 was performed at the end of each day on site. Advanced gas analysis was typically performed on a subset of samples at the end of each soil gas survey. These analyses included helium concentration, $\delta^{13}C_{CO_2}$ and sulphur hexafluoride (SF_6) compositional analysis. Details of the analytical techniques are given by Schacht et al. (2010).

During the 4 years of soil gas sampling, molecular composition analyses were performed successfully on 715 samples. Numerous analyses were repeated to check for accuracy; two or three analyses were run on sample point duplicates (i.e. a second sample at same location) and standards and blanks were run after every fifth to tenth sample. All samples analysed had sufficient concentrations of CO_2 to gain accurate analytical data. CH_4 concentrations were commonly below the detection limit of the GC. Table 14.2 details the average gas concentrations, ranges and other statistical data for each of the surveys reported here.

Table 14.2: Statistics for the dataset on CO_2 gas concentration (ppm) molecular compositional analysis (bd: below detection).

Statistic	Mar '07	Feb '08	Feb '09	Mar '10
Number*	206	142	228	139
Mean	6876	6616	4970	5578
StDev	± 6977	± 8328	± 13,327	± 11,025
Max.	42,982	53,450	101,531	87,241
Min.	371	292	bd	200

*Number of successful samples

Table 14.3: March 2007 survey subset for ^{14}C analysis.

Sample No.	CO_2 (ppm)	$\delta 13C_{CO_2}$	Radiocarbon age (BP years)	He (ppm)
487	9572	–16.0	55	7.83
54	21,219	–18.3	210	32.4
584	14,781	–21.7	Modern	10.6
613	4719	–15.8	Modern	10.0
666	20,036	–19.2	Modern	12.4
696	12,920	–13.5	Modern	8.97

14.5   Soil gas results

Figures 14.2 and 14.3 show changes in the concentration of soil CO_2 over the various surveys. The CO_2 concentration in the soil gas was generally higher in the north-east region of the study area. There, CO_2 levels were an order of magnitude greater than the average for the area. CO_2 levels greater than one standard deviation above the mean were occasionally associated with the petroleum wells in the region. CH_4 concentrations were also considered in this study. Typically the CH_4 concentrations were at the lower limit of detection for the gas chromatograph. It was concluded that effective evaluation of the CH_4 baseline was not possible and that there was no correlation of CH_4 concentration with any other parameter in the study area.

The $\delta^{13}C_{CO_2}$ isotopic analyses were performed on a subset of samples, in an attempt to characterise the source of the CO_2 in the soil gas. The sample subset was selected after molecular composition analysis so that a range of gas compositions could be analysed that were representative of the whole study area. These samples were compared to the $\delta^{13}C_{CO_2}$ values for natural gas compositions in the Otway area which range from –12‰ to –5‰ (Boreham et al. 2001). Isotopic analytical results are given in Figure 14.5.

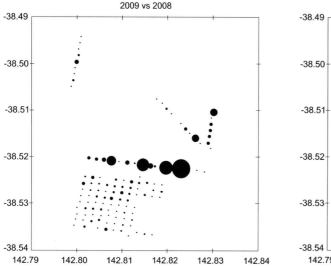
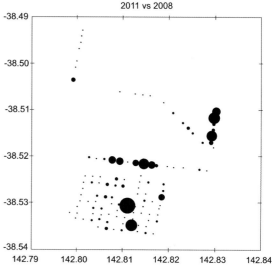

Figure 14.2: Changes in CO_2 concentration on soil gas samples, referred to pre-injection levels, shown schematically on the soil gas grid. Larger circles indicate larger changes, with little consistency from year to year (refer to Figure 14.1 for more accurate sample locations).

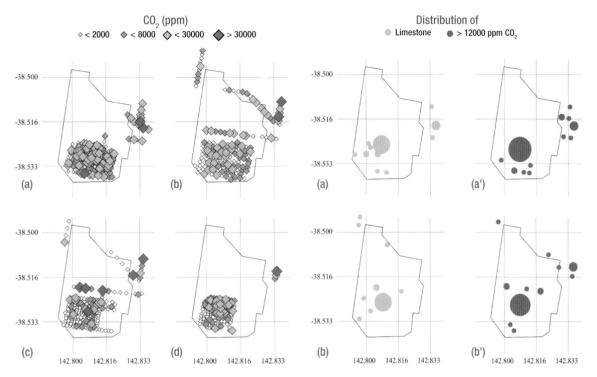

Figure 14.3: Variation in CO_2 soil gas concentrations across the study area for the four surveys: (a) March 2007, (b) February 2008, (c) February 2009, and (d) March 2010.

Figure 14.4: Distribution of limestone within the study area at 1 m depth and of elevated CO_2 concentration as determined for the two surveys: (a-a') March 2007, (b-b') February 2008. Note how the occurrence of limestone matches elevated CO_2 concentrations in soil gas samples of this area.

A subset of samples for ^{14}C isotopic analysis was selected from the 2007 survey, but sample numbers were limited due to the very high cost of this analysis. Six samples were selected to best represent the geochemical variability and distribution across the study area (Table 14.3).

While leakage to the surface was considered unlikely, tracers were added to the injected CO_2 at CRC-1, at the start of injection in April 2008, to be able to more confidently distinguished "natural" soil CO_2 from injected CO_2. To collect baseline data on the tracer SF_6, a subset of samples from the February 2008 survey was analysed (1 month prior to commencement of CRC-1 injection) followed by analysis of samples from the March 2010 survey. Samples were selected to best represent the geochemical variability and distribution across the study area. The SF_6 concentrations were below detection limit for all the soil gas samples analysed.

14.6 Interpretation of soil gas results

Near-surface geology and associated weathering define and modify soil gas composition. At the Otway site, this process could have acted for example through dissolution of the Port Campbell Limestone resulting in heightened CO_2 soil gas levels. It could have also acted indirectly for example through a heightened CO_2 concentration (from the Port Campbell Limestone) enhancing the micro-bacterial activity in the soil and further concentrating the CO_2. Areas where the occurrence of the Port Campbell Limestone is less than 1 m below the surface were likely to have elevated CO_2 concentrations, as seen in the northern part of the main sampling grid (Figure 14.4). However, the sampling method used could not accurately determine the depth of the limestone and therefore the correlation between limestone depth and soil gas chemistry could only be qualitative.

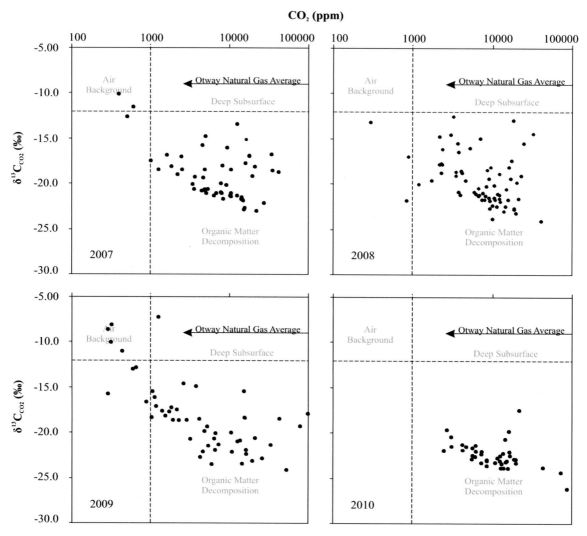

Figure 14.5: CO_2 concentrations vs $\delta^{13}C_{CO_2}$ values for the two baseline (2007–08) and two monitoring surveys (2009–10) presented here (Schacht et al. 2011). Otway average natural gas composition according to Boreham et al. (2001).

The CO_2 concentrations and $\delta^{13}C_{CO_2}$ values were compared to help determine the source of the CO_2 in the soil gas (Figure 14.5).

The majority of samples from all four surveys fell outside the range of composition described for CO_2 accumulations in the study area, with a few samples approaching air background composition.

It is thus interpreted that the variation in CO_2 concentration and isotopic value of the carbon in the CO_2 mainly is

a product of climatic factors affecting organic matter decomposition, and the permeability of the soil to allow atmosphere-soil gas interaction. A deep subsurface source (which would indicate deep gas migration to surface) is not apparent in the data.

While atmospheric contamination is likely to have occurred for a few samples, it is believed that the soil gas geochemistry of the one "outlying" sample in 2009 was due to preferential fractionation of the lighter $^{12}C_{CO_2}$, probably through a damaged sample bag (Schacht et al. 2011).

Six samples from the March 2007 survey were selected as representing the geochemical variability and distribution of CO_2 across the study area. As can be seen in Table 14.3, the radiocarbon ages for the subset were very recent, with the oldest sample having an age of 210 years BP conventional radiocarbon age (see Steiver and Polach 1977 for definition of terminology). CO_2 from deep sources would be much older than could be determined using ^{14}C. This suggested an overwhelmingly modern surface source for the soil CO_2, most likely from organic matter decomposition.

14.7   Conclusions

A four-phase study evaluated existing vadose zone soil gases at the Otway Project site. Carbon dioxide and methane were the main gases analysed and most CH_4 concentrations were below detection levels. Carbon dioxide concentrations ranged from atmospheric background levels of 0.033% up to about 10% of the total soil gas. The $\delta^{13}C_{CO_2}$ values were consistent with organic matter decomposition being the main source of the CO_2 in the soil gas (Figure 14.5). Modern radiocarbon ages obtained for the CO_2 from ^{14}C analysis, and tracer (SF_6) concentrations below the detection limit supported the conclusion that the CO_2 in the soil gas was from modern organic sources. There was a relative increase in CO_2 concentration measured in locations where the Port Campbell Limestone was near the ground surface, which implied additional CO_2 had been produced from the weathering of limestone (Figure 14.4). The soil gas surveys at the Otway site showed no evidence of leakage of any stored CO_2 into the soil.

14.8   References

Amundson, R.G. and Davidson, E.A. 1990. Carbon dioxide and nitrogenous gases in the soil atmosphere. Journal of Geochemical Exploration 38, 13–41.

Boreham, C.J., Hope, J.M. and Hartung-Kagi, B. 2001. Understanding source, distribution and preservation of Australian natural gas: a geochemical perspective. The APPEA Journal 41T(1), 523–548.

Flett, M., Brantjes, J., Gurton, R., McKenna, J., Tankersley, T. and Trupp, M. 2009. Subsurface development of CO_2 disposal for the Gorgon Project. Energy Procedia 1, 3031–3038.

Jenkins, C., Cook, P.J., Ennis-King, J., Underschultz, J., Boreham, C., Dance, T., de Caritat, P., Etheridge, D.M., Freifeld, B.M., Hortle, A., Kirste, D., Paterson, L., Pevzner, R., Schacht, U., Sharma, S., Stalker, L. and Urosevic, M. 2012. Safe storage and effective monitoring of CO_2 in depleted gas fields. PNAS 109(2), 35–41.

Martens, S., Kempka, T., Liebscher, A., Lüth, S., Möller, F., Myrttinen, A., Norden, B., Schmidt-Hattenberger, C., Zimmer, M., Kühn, M. and The Ketzin Group 2012. Europe's longest-operating on-shore CO_2 storage site at Ketzin, Germany: a progress report after three years of injection. Environmental Earth Sciences 67, 323–334.

Mathieson, A., Midgely, J., Wright, I., Saoula, N. and Ringrose, P. 2011. In Salah CO_2 storage JIP: CO_2 sequestration monitoring and verification technologies applied at Krechba, Algeria. Energy Procedia 4, 3596–3603.

Romanak, K.D., Bennett, P.C., Yang, C. and Hovorka, S.D. 2012. Process-based approach to CO_2 leakage detection by vadose zone gas monitoring at geologic CO_2 storage sites. Geophysical Research Letters 39, L15405.

Schacht, U., Boreham, C.J. and Watson, M.N. 2010. Soil gas baseline characterisation study—methodology and summary. CO2CRC RPT09-1714, 49.

Schacht, U., Regan, M., Boreham, C. and Sharma, S. 2011. CO2CRC Otway Project, Soil gas baseline and assurance monitoring 2007–2010. Energy Procedia 4, 3346–3353.

Schloemer, S., Furche, M., Dumke, I., Poggenburg, J., Bahr, A., Seeger, C., Vidal, A. and Faber, E. 2013. A review of continuous soil gas monitoring related to CCS—Technical advances and lessons learned. Applied Geochemistry 30, 148–160.

Steiver, M. and Polach, H. 1977. Reporting of ^{14}C data. Radiocarbon 19(3), 355–363.

Watson, M.N., Boreham, C.J. and Rogers, C. 2006. Soil gas baseline characterisation study—Winter 2005 Survey, Otway Basin Pilot Project. CO2CRC RPT06-0181, 29.

White, D.J., Burrowes, G., Davis, T., Hajnal, Z., Hirsche, K., Hutcheon, I., Majer, E., Rostron, B. and Whittaker, S. 2004. Greenhouse gas sequestration in abandoned oil reservoirs: The International Energy Agency Weyburn pilot project. GSA Today 14(7), 4–10.

David Etheridge, Ray Leuning, Ashok Luhar, Zoe Loh, Darren Spencer, Colin Allison, Paul Steele, Steve Zegelin, Charles Jenkins, Paul Krummel, Paul Fraser

15. ATMOSPHERIC MONITORING

15.1 Introduction

The CO2CRC Otway Project was one of the first geological storage projects to include comprehensive atmospheric monitoring as part of the assurance monitoring programme. This programme was designed to detect if any change had occurred in nearby soil, aquifers or atmosphere as a result of leakage of stored carbon dioxide. The development of methodologies able to attribute and quantify emissions was a major additional goal of the atmospheric monitoring research at the Otway site.

While leaks can potentially occur at various stages of the capture, transport, injection and storage process, the main focus for atmospheric monitoring is the detection of leakage of CO_2 from the geological reservoir. This is one of the main concerns of regulators, project operators and the public. Leakage associated with carbon capture and storage (CCS) in the atmospheric context is defined as the emission of gases to the atmosphere, particularly injected CO_2, but also gases that might accompany the injected CO_2

and gases such as methane (CH_4) that might be displaced from a storage reservoir. A number of potential leakage pathways have been proposed, such as via wells, faults, permeable cap rock, or a combination of these (Benson et al. 2005). Storage sites can be extensive, with subsurface reservoirs extending for several kilometres. Leakage to the atmosphere could be through point or diffuse sources and with locations that have varying degrees of uncertainty.

Although atmospheric techniques have only recently been applied to monitoring potential emissions from geological storage, gas fluxes from land surfaces to the atmosphere have been successfully determined for landfills, crop canopies, volcanoes and coal mines. These applications present some similar challenges to monitoring geological storage. A review of atmospheric monitoring techniques for measuring emissions from geological storage across a range of scales and their relevance to the Otway Project is given by Leuning et al. (2008).

Most atmospheric measurements provide composition (gas concentrations and isotopic ratios) over time from point locations. Ideally, however, atmospheric monitoring would measure land-air fluxes continuously across the storage site. The techniques that do measure flux directly, namely flux chambers and eddy covariance flux stations, have relatively small "footprints" compared to the dimensions of a geological storage site and would require intense and possibly impractical networks and frequent campaigns to ensure representative site coverage. Soil fluxes in particular are highly variable over time and location. Alternatively, the emission rates and locations of sources can be derived by using an inverse approach, which combines concentration measurements with atmospheric transport modelling to take dispersion into account. A number of inverse analyses were tested over the course of the Otway Project, the most successful being during Otway Stage 2, which used continuous concentration measurements at two points to derive the location and the emission rate of controlled releases of gas.

The requirements for detecting and quantifying leaks at rates that would compromise the effectiveness of long-term storage as a climate mitigation option are very different from the monitoring requirements for health and safety. The latter usually requires that the concentration thresholds of some gases in the air are not exceeded, set for example, at about 1% mole fraction for CO_2. Such situations could result only from extremely high leakage rates (that would totally compromise the effectiveness of the storage site unless immediate remedial action was taken) and would require a much simpler monitoring regime than for monitoring effective long-term storage at a geologically suitable and well-characterised site, which is the focus of this chapter.

15.2   Sensitivity

Several studies have concluded that for emissions reduction by CCS to be climatically-effective well into the future, leakage should be below about 0.01% per year, averaged across all CCS projects (Haughan and Joos 2004; Enting et al. 2008; Shaffer 2010). Geological considerations suggest that this level of retention is very likely for suitably selected and monitored sites (Benson et al. 2005). For a future commercial-scale storage project of 10 Mt CO_2, 0.01% per year corresponds to 1000 t CO_2 per year, which is the amount adopted here as a target leak rate for monitoring. The Otway Project, although one of the largest pilot projects in terms of amount of CO_2 stored (Figure 18.1), injected only 66,000 t of CO_2-rich gas

into a secure reservoir. However, it provided a useful platform to develop techniques for monitoring larger sites and emission rates and made use of emissions from activities such as the controlled release of gas as targets for detection.

Assurance monitoring is intended to assess a key requirement of CO_2 storage: that no injected CO_2 is detected outside of the storage reservoir. This requirement can be difficult to prove, compared for example to seismic or geochemical monitoring of the reservoir, where unambiguous observation of the injected fluid can be a conclusive result. The challenge for assurance monitoring is to demonstrate sufficiently high sensitivity, such that emissions would be detected and be quantifiable if they occurred at or above the target leak rate.

Confirmation that atmospheric and other assurance monitoring has achieved this sensitivity could involve detecting actual leaks, caused, for example, by over-pressuring the reservoir or injecting into unsuitable formations. However, intentionally causing leakage would obviously be impractical and undesirable. The alternative is to undertake controlled releases of gases at research facilities such as Ginninderra, near Canberra, Australia (Loh et al. 2009; Humphries et al. 2012), and ZERT, Montana, USA (Krevor et al. 2010), which have been used to simulate leakage in a field environment. At Otway, occasional emissions from activities at the surface of the site, such as generator exhausts and venting of gas storage tanks and, in particular, the controlled releases of gas at predetermined rates and times during Stage 2, were used to calibrate the sensitivity of the atmospheric strategy.

15.3   Simulated emissions and monitoring design

The atmospheric monitoring design at Otway was based on the simulated atmospheric signals resulting from a hypothetical leak (Etheridge et al. 2005; Leuning et al. 2008). Emissions of CO_2 and tracer gases were "released" in the CSIRO's prognostic meteorological and air pollution model TAPM (Hurley et al. 2005), configured for the Otway region. The perturbations in the concentrations of these gases around the area were simulated over 1-month periods. Emissions were assumed to be from a point source (such as a well) or a diffuse source (representing seepage through an area of ground) at a constant nominal rate of 1000 t CO_2 per year and with CH_4 and introduced tracers at the ratio of their concentrations in the injected fluid. The time series of the simulated concentrations showed small (less than 5 ppm) and infrequent CO_2 signals (due to variation in winds) within about 1 km of the source. It was clear from these simulations that precise and continuous measurements would be necessary and that tracers would help to reveal an emission above the background variability. Further input for the development of the monitoring came from the Ginninderra controlled release experiment, which showed that concentration perturbations of 1% or more above background are required to accurately determine the emission rate from concentration and atmospheric turbulence measurements (Loh et al. 2009). This also highlighted the need for precise and well-calibrated measurements of concentration at a number of point locations, rather than line of sight (open path) measurements, and the use of gas tracers as well as measurements of CO_2 itself.

The resulting atmospheric monitoring configuration at Otway consisted of measurements of composition and fluxes across several time and space scales. Composition was measured in-situ and using flasks samples from a 10 m high inlet at the atmospheric module (Figure 1.3). The location of the atmospheric module was guided by the model simulations, with the aim of achieving a balance between signal size and likelihood of receiving the emissions emanating from a range of potential source locations (such as from wells, or diffuse leakage from the subsurface plume) in the area. The module was sited to the north of the storage site to take advantage of the clean winds coming off the Southern Ocean to the south.

Continuous measurements of CO_2 concentration began in January 2007 with a CSIRO LoFlo analyser and from November 2009 with a cavity ring down spectrometer instrument (CRDS) (Envirosense 3000i, Picarro, California, USA) which also measured CH_4 and water vapour concentrations. Concentrations of CO_2, CH_4, carbon monoxide (CO), hydrogen (H_2) and nitrous oxide (N_2O) and the stable isotopes of CO_2 were measured in flask samples at the CSIRO Marine and Atmospheric Research GASLAB (Francey et al. 2003).

Concentrations of the introduced tracers sulphur hexafluoride (SF_6) and HFC-134a (CF_3CH_2F) were measured by gas chromatography-mass spectrometry (GC-MS) (Miller et al. 2008). The flask samples were collected typically every few months during "baseline" conditions (daytime, winds from the south-east to south-west sector). Flasks were also collected during specific campaigns such as during controlled release events or over a diurnal cycle.

The composition of the headspace air in three deep water wells was monitored using flask air sample measurements as part of the regulatory requirements of the Otway Project.

Continuous meteorological data and the fluxes of CO_2, H_2O and heat were measured using an eddy covariance flux station (Foken et al. 2012). Wind and atmospheric stability data were used in dispersion modelling and the CO_2 flux data were used to determine the contribution of the ecosystem flux to the measured CO_2 concentrations.

Soil CO_2 fluxes were also measured during several campaigns, each lasting 1 to 2 days, using a portable flux chamber (LI-8100, LiCor, Lincoln, Nebraska, USA).

Details of the measurements, instruments, calibrations, data management and archiving, as well as links to national and international networks, are described by Loh et al. (2011).

15.4   Background CO_2

A major difficulty for the atmospheric monitoring of CO_2 sources is that the fluxes of CO_2 from terrestrial ecosystems are high, causing significant variations in the background CO_2 concentration that could mask potential leakage emissions. This was especially so at the Otway site where the ecosystem is highly active for much of the year, with maximum fluxes (night time emissions due to respiration) of typically 10 μmol CO_2 m^{-2} s^{-1}. This flux is equivalent to the target leak rate (1000 t CO_2 year^{-1}) emanating from an area of less than 0.1 km^2. The atmospheric CO_2 perturbations for a hypothetical leak at this rate at Otway, simulated by transport modelling, were 2–3 ppm at 700 m from a point source and even less if the emissions were from a diffuse area (Leuning et al. 2008). These relatively small signals would need to be

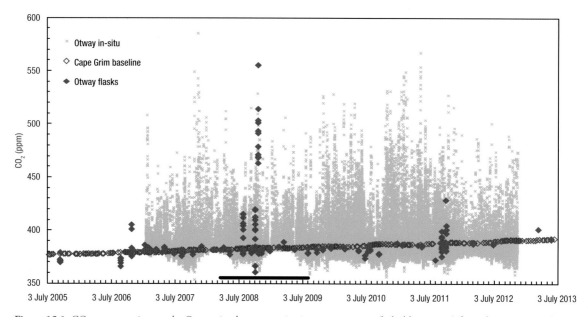

Figure 15.1: CO_2 concentrations at the Otway site: hour-mean in-situ measurements (light blue crosses) from the CSIRO LoFlo and the Picarro instruments and flask sample measurements (blue diamonds), from the 10 m inlet at the atmospheric module. All data are shown, including controlled release events and night-time flask sampling. The period of injection into CRC-1 is shown by the black line. Measurements of CO_2 at the Cape Grim Baseline Air Pollution Station, Tasmania, selected for baseline conditions (strong winds off the ocean), are shown in brown.

detected against background ecosystem CO_2 variations of tens of ppm, depending on the season, the time of day and atmospheric dispersion conditions at the time (Figure 15.1).

Stable atmospheric conditions allow the build up of emitted gases and cause larger composition changes, but these signals are useful only if the leaked CO_2 could be distinguished from other (mainly ecosystem) sources using tracers such as CO_2 isotopes or the injected tracers.

Several approaches were developed for the Otway Project to discriminate between ecosystem-derived CO_2 and potential leakage. They included establishing a baseline of CO_2 variations, measurement of the ecosystem flux with an eddy covariance flux station, estimating the effect of ecosystem fluxes on CO_2 concentrations with atmospheric dispersion models, the use of tracers to discriminate ecosystem CO_2 from other sources, and the measurement of the differences in CO_2 across the storage site.

15.4.1   Establishing a baseline

Establishing a baseline for a CCS project involves starting the measurement programme before CCS activities begin; the baseline can be used to identify any existing sources of CO_2, CH_4 and other (tracer) gases and to quantify their emission rates and variations in time. The main sources of CO_2 and CH_4 at the Otway site are the ecosystem and cattle respectively. Monitoring over at least an annual cycle was necessary to characterise the variations in flux due to temperature, moisture, sunlight, cropping and herd sizes. Small and infrequent signals were also expected from emissions from Melbourne and the Latrobe Valley power stations hundreds of kilometres to the east (Luhar et al. 2009). The variations in CO_2 concentration and fluxes at the site are shown in Figures 15.1 and 15.2. The maximum range in concentrations occurs during late winter and spring and the minimum range is during the dry months of February and March.

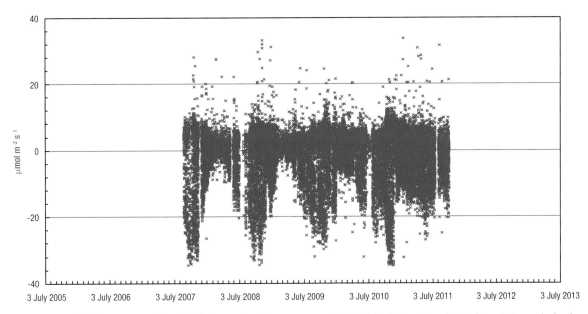

Figure 15.2: CO_2 fluxes measured by the EC flux station (hour means at a height of 4 m). Negative values indicate CO_2 uptake by the land surface.

15.4.2   Quantifying ecosystem CO_2

If the ecosystem CO_2 flux is accurately known and accounted for, then additional sources such as leaks from storage (if sufficiently large) would be seen as anomalies in the measured CO_2 fluxes or concentrations. Ecosystem fluxes were measured using the eddy covariance flux station and also modelled by a land surface scheme (CABLE) coupled to the meteorological module of TAPM (Luhar et al. 2009). The modelled fluxes compared reasonably well with the data, and were used by TAPM's numerical dispersion module to estimate the CO_2 concentrations expected at the atmospheric module. Additionally, a steady-state analytical dispersion model for the surface layer (Horst and Weil 1992) was extended to area sources and used for concentration calculations, in order to better represent the surface-layer meteorology and to incorporate the finer vertical resolution that can have significant influence on the modelled concentrations, especially during stable conditions at night. The model-simulated concentrations reproduced the main observed features (Figure 15.3), such as the seasonal and diurnal variations and their timing, but not well enough to accurately reveal the concentration increases that might indicate a small leak (at the target rate), especially during periods of stable conditions and strong ecosystem respiration.

15.4.3   The use of tracers

Tracers can significantly enhance the sensitivity of detection and quantification of monitoring in the atmosphere where CO_2 is naturally abundant and variable. Ideal tracers for atmospheric monitoring would have low and stable natural levels, be unique to the stored gas, be able to be precisely and continuously measured at low concentrations in the field environment and accompany the stored gas at all stages of the process (injection, migration in the reservoir and through to the surface if leakage occurs). From a practical perspective, tracers must be economical to use, readily injected and non-toxic. These requirements pose significant constraints on the selection of tracers for monitoring the atmosphere and the other domains of CCS.

The Otway Project provided a real world situation to test a number of tracers. These included tracers naturally present in the stored gas and those added during injection. The source gas at Otway was from the Buttress reservoir which contained a significant amount (about 20% by mole) of CH_4 and there was considerable residual CH_4 in the storage reservoir. Detection and quantification of leakage by monitoring CH_4 is more effective than CO_2 in most environments because of its lower and more stable concentration in the background atmosphere (Loh et al.

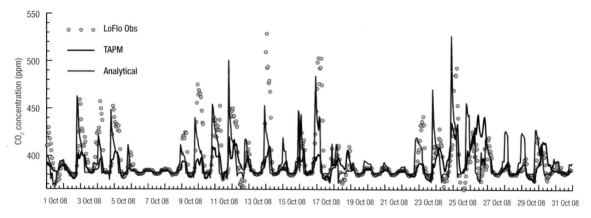

Figure 15.3: Observed CO_2 concentrations compared to simulations with an ecosystem model coupled to two atmospheric dispersion models (TAPM and analytical). The selected period (October 2008) was when the ecosystem was at its most active time in the seasonal cycle, causing large CO_2 variations.

2009; Etheridge et al. 2011; Luhar et al. 2013). Methane could be an important tracer gas when CO_2 is stored in depleted natural gas reservoirs or when the injected CO_2 contains residual CH_4, such as when it is separated from natural gas. Methane was a powerful tracer at Otway and contributed to the unambiguous detection of emissions from several controlled release events. Monitoring CH_4 in the atmosphere has benefited greatly from the advent of new continuous monitoring technology that is precise, stable and field deployable. One such instrument, a Picarro CRDS, was installed at the Otway atmospheric module in November 2009. A second instrument was installed at the visitor centre during the Stage 2 controlled release period in 2011. Together, these provided continuous CO_2 and CH_4 concentrations upwind and downwind of the Otway site during SSW and NNE winds. These instruments had accurate and stable calibrations (better than 0.1%) allowing precise differences in concentrations to be determined across the site.

The carbon isotopic composition of the injected CO_2 was also used as a tracer at the Otway site. The $\delta^{13}CO_2$ of the injected gas was –6.8 per mil, reflecting its magmatic origin, which was significantly different to the ecosystem $\delta^{13}CO_2$ of –28 ± 1 per mil measured over periods of several months. Furthermore, being of "fossil" age the injected CO_2 contained no ^{14}C, unlike CO_2 from the ecosystem. Both carbon isotopes were measured at various times during monitoring campaigns at Otway. Importantly,

monitoring with CO_2 isotopes uses the natural differences in their abundance between background and injected CO_2 and does not require the isotopes to be added. The main limitation for the use of CO_2 isotopes in atmospheric monitoring of the Otway Project was measurement technology, which relied on analysis of flask air samples using stable isotope mass spectrometry for $\delta^{13}CO_2$ and accelerator mass spectrometry for $^{14}CO_2$. The difference in atmospheric $\delta^{13}CO_2$ between Buttress and the local ecosystem was about 0.05 per mil per ppm of CO_2 signal. Continuous in-situ measurement capability for $\delta^{13}CO_2$ by CRDS was installed in late 2009 at Otway but several years later the technology is still being refined to quantify the small possible signals. Detection of the isotopic change associated with 10 ppm CO_2 increase from Buttress gas may soon be within the realm of the CRDS instruments, but this is ongoing work. Measurement of $^{14}CO_2$ on the other hand is still reliant on the relatively costly analysis by accelerator mass spectrometry of the CO_2 in flask air samples.

Several tracer gases were introduced into the fluid injected at Otway. Of these, SF_6 and HFC-134a were used for atmospheric monitoring. Both are synthetic gases with low and stable atmospheric background concentrations and have few if any sources local to the Otway site. The use of SF_6 as an atmospheric tracer was modelled on a mean molar concentration of 1 ppm in the injected fluid, which was assumed to stay constant as the fluid migrated.

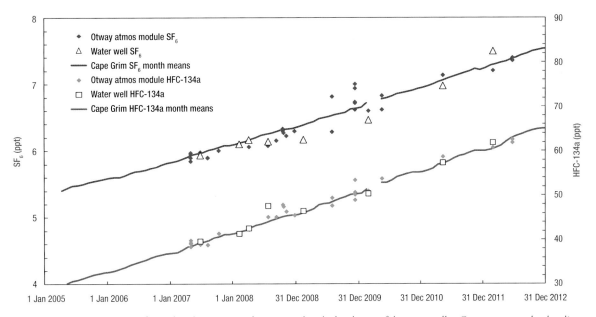

Figure 15.4: Concentrations of introduced tracers in ambient air and in the headspace of the water well at Otway, compared to baseline conditions at Cape Grim. Note the elevated values in December 2009 when gas was intentionally released from the Naylor observation well.

In practice, SF_6 was introduced in several pulses and moved at different rates to the bulk fluid in the reservoir (Chapter 12), leading to high concentration zones within the fluid. Measurements of SF_6 and HFC-134a involved flask sampling followed by specialised laboratory analysis (GC-MS) and consequently monitoring was limited to routine samples during periods of suitable winds, campaign samples collected during controlled releases and air sampled from the water wells. The results are shown in Figure 15.4. These synthetic tracers are sensitive indicators of the reservoir fluid and no unexplained anomalies were observed. The use of SF_6 helped detect and quantify the rate of emissions during a controlled release from the reservoir in December 2009 (Etheridge et al. 2011). However, while introduced tracers such as SF_6 and HFC-134a can be beneficial from a monitoring perspective, the use of synthetic tracer gases in the future might be limited to small-scale injections by practicality, cost and environmental concerns.

An automated sampling system was deployed at Otway for tracers that were not measured in the field, such as CO_2 isotopes and the introduced tracers. Samples were collected at predetermined times, or when wind directions

or CO_2 and CH_4 concentrations measured at the site indicated suitable conditions.

Tracers can also be used to identify emissions which are not due to CCS sources. For example, episodes of high CO_2 that are accompanied by elevated CO concentrations indicate a combustion origin, such as the diesel engines of the well drilling rig which produced an anomaly in March 2007 (Etheridge et al. 2011). Another example was a period of extreme CO_2 concentrations (up to 600 ppm) in spring 2008 which were traced to the ecosystem using the $\delta^{13}CO_2$ signal ($\delta^{13}CO_2 = -29$ per mil) during a period of intense respiration and a very stable nocturnal boundary layer, rather than emissions from the Otway facility ($\delta^{13}CO_2 = -6.8$ per mil) or from the nearby industrial CO_2 plant ($\delta^{13}CO_2 \sim 0$ per mil).

15.4.4   Upwind-downwind measurement

Air flowing across a source, such as CO_2 from a well, will have higher concentrations of CO_2 compared to upwind levels. The degree of enrichment depends on the source strength, atmospheric stability and the positions of the upwind and downwind detectors relative to the source. With this information it is possible to locate the position of the source and to estimate its strength (Loh et al. 2009; Humphries et al. 2012; Luhar et al. 2013). Precise and continuous measurements of the concentration difference are essential, requiring instruments located upwind and downwind that are intercalibrated to an uncertainty that is much smaller than the expected leakage signal, simulated at the Otway site to be 2–3 ppm. Only recently have instruments become available that are capable of maintaining this precision over long periods, that are suitable for field deployment (with infrequent operator intervention, rudimentary temperature control and intermittent power) and are cost effective.

Most of the monitoring period at the Otway site had only one measurement location (the atmospheric module) and used measurements at the Cape Grim station in Tasmania as a proxy for the upwind concentration during southerly winds. This needed an allowance for the sources and sinks between Cape Grim and the site, which, during winds from the south-east to south-west, are almost entirely due to the 4 km fetch over the land surface. Ecosystem fluxes were measured and simulated, tracers were used and data analysis techniques were developed to select conditions when ecosystem fluxes were minimal and meteorological conditions were best suited for emissions from potential leaks to be detected and quantified.

15.5   Data filtering

Using the atmospheric baseline measurements as a guide, optimal conditions of background wind directions, unstable atmospheric conditions and low ecosystem CO_2 fluxes were selected to reduce the concentration data to a record where anomalies would be more easily detected. Downward short wave radiation (essentially sunlight) measured at the flux station was found to be a suitable proxy for the combined stability and flux filter conditions. The CO_2 and CH_4 records filtered in this way are shown in Figure 15.5. Periods of high values, corresponding to stable periods (and dominant ecosystem respiration in the case of CO_2), are almost entirely removed by filtering.

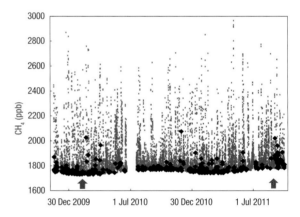

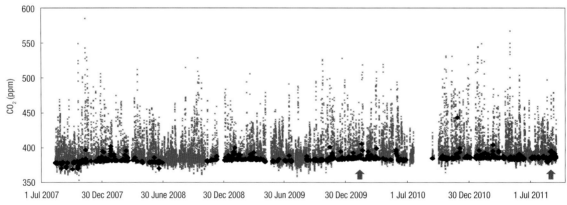

Figure 15.5: The Otway CO_2 (bottom) and CH_4 (top) records from continuous in-situ measurements (hourly mean concentrations from the LoFlo CO_2 and Picarro CH_4 monitors at the atmospheric module), before (blue) and after filtering based on downward shortwave radiation and wind direction (black). Arrows indicate periods of emissions during well drilling and controlled gas releases.

The filtered record was then used in a number of ways:

1. to reveal anomalies that might be due to emissions events, which could subsequently be attributed to sources by using tracers and wind trajectories (Jenkins et al. 2012; Etheridge et al. 2011)

2. to calculate statistics of hour to hour variations as an indication of background variability and thus the sensitivity of detection

3. to look for evidence of leakage by comparing CO_2 and CH_4 concentrations before and after injection or in wind sectors coming from the site compared to the adjacent background.

Elevated CO_2 and CH_4 concentrations were revealed in the filtered records during the drilling of the CRC-1 well in February and March 2007 and of CRC-2 in January and February 2010 and the controlled release of gases during Stage 2 in August and September 2011. The CO_2 emission rates during these events were between about one and three times the target leak rate of 1000 t per year. The remaining elevated values were likely to be caused by sources such as vehicles, burning off and livestock herds. The seasonal variation in background atmospheric CH_4 was also revealed.

Analysing the distributions of concentration differences is one way to deal with the biogenic fluxes and corresponding concentrations. Filtering first reduced the natural variability caused by stable atmospheric conditions and ecosystem emissions. The probability distribution of hour to hour CO_2 and CH_4 concentration differences during the selected periods then revealed the residual variability in the background wind sector. Probability distributions of hour to hour concentration differences, as the wind direction alternated between the background sector and the sector of interest (e.g. containing the injection and monitoring wells), then revealed if CO_2 or CH_4 concentrations in that sector were elevated above the background. An example, showing no apparent CO_2 source in the region of the CRC-1 well, is provided in Figure 15.6. Model simulations showed that a point source of 1800 t CO_2 per year and 13 t CH_4 per year would be easily detectable using this technique.

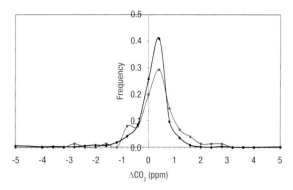

Figure 15.6: Histogram of hour to hour differences in CO_2 (ppm), showing the variability of CO_2 in the background (black) and the difference between the background and the potential source sector, centred on the CRC-1 well (blue). A source would be indicated if there was a shift in the blue curve to positive values above the range of background variability.

15.6 Bayesian inverse modelling

An inverse Bayesian probabilistic modelling framework was developed to combine concentration measurements, model and observational errors, and available prior source information to determine source emission rates and locations (Luhar et al. 2013). The required source-receptor relationship was calculated via a backward Lagrangian particle dispersion model driven by meteorological observations.

The controlled release activity at Otway during August–September 2011 provided an opportunity to develop and evaluate the inverse modelling technique. The two-point monitoring configuration using the Picarro CRDS instruments gave simultaneous upwind and downwind CO_2 and CH_4 concentration measurements and also provided the background concentration and its spatial variability/uncertainty, which were fed into the modelling. The inverse modelling system predicted both location (x_s, y_s, z_s) and emission rate (q) of the source, using data from a series of hourly periods given that the emission rate was constant with time. The predicted emission values compared favourably with the measured values (Table 15.1). As expected the results were better for CH_4 given its lower background concentration and variability and thus a higher signal-to-noise ratio.

Table 15.1: Source strength and location derived from the Bayesian inverse modelling (Luhar et al. 2013).

Source Parameter	CH$_4$		CO$_2$	
	Actual	Predicted	Actual	Predicted
q (g s^{-1})	8.4	7.8 ± 1.5	96.2	180.6 ± 57.2
x_i (m)	−49	−59.9 ± 7.3	−49	−49.3 ± 12.7
y_i (m)	−136	−172.2 ± 9.1	−136	−87.5 ± 37.0
z_i (m)	2.5	14.2 ± 1.7	2.5	10.2 ± 7.3

15.7 Conclusions

A suite of atmospheric monitoring techniques has been applied to the Otway Project over a period of six years, incorporating measurements of concentrations, tracers and fluxes, and interpreted with the help of transport modelling. The Otway Project provided for the first time a real world setting for the development and application of these techniques to geological storage of CO$_2$. The flat, homogeneous terrain and sparse development of the Otway region were advantageous for atmospheric monitoring. However, the ecosystem CO$_2$ flux, the main source of background "noise", was relatively high, requiring a number of measurement and interpretive methodologies for emissions to be detected and quantified.

It is likely that leakage of about 1000 t per year of the injected CO$_2$ would be detectable by the atmospheric strategy deployed at the Otway site, a relevant upper limit for large-scale commercial CCS projects to achieve effective climate abatement. This detection limit was estimated from model-simulated emission signals that could be detected in one or several of the atmospheric measurements and which was confirmed by the detection of actual emissions (though not leakage) of comparable magnitude.

Atmospheric monitoring was demonstrated at the Otway Project to be non-invasive, continuous and relevant to the needs of regulators, the public and project operators. It was well suited to the practical and economic constraints of long-term monitoring of other storage sites. Importantly, it tested the ability of geological storage to achieve its main purpose of reducing CO$_2$ emissions to the atmosphere.

Based on the Otway Project experience, it is possible to identify a suite of atmospheric methods that could be tailored for the characteristics of other storage sites. An essential requirement is for continuous, precise and well-calibrated measurements to ensure the accurate and reliable determination of differences across the storage site and over time.

A revised monitoring scheme should include at least two measurement locations, as was used for the Stage 2 controlled release experiment, which could more readily detect emissions across the site, and be used in inverse modelling to infer the source location and strength. The savings in initially having only the one-point monitoring were offset by the extra effort required to interpret the measurements and the lower certainty of the results. The eddy covariance flux station provided excellent information on wind fields, atmospheric stability and fluxes. However, for the purposes of selecting data, measurements of incoming shortwave radiation can be used as a good proxy for suitable CO$_2$ flux and stability conditions. Measurements of tracers, particularly if continuous and precise such as for CH$_4$, adds significant signal-noise over CO$_2$ measurements alone. However, there is a need for caution when including the monitoring of unknown systems such as the water well gases, if there is limited understanding of the system and no baseline record to judge potential variations in the observations.

The atmospheric monitoring at Otway produced a wealth of data on atmospheric composition, micrometeorology and carbon cycling that are available for research into monitoring geological CO$_2$ storage and for wider applications. For example, the local ecosystem carbon budget during the monitoring period was influenced by years of drought in south-east Australia followed by two consecutive La Niña high rainfall events. These variations were clearly observed in the measured CO$_2$ concentrations and fluxes (significantly more CO$_2$ uptake during the wet years of 2010 and 2011). Otway provided a field laboratory for the development of measurement technologies, including CRDS instruments for concentrations and isotopes, open and closed path CO$_2$ monitors for eddy covariance fluxes and for testing models such as the CABLE ecosystem carbon model. Model-data interpretation techniques were successfully developed to quantify the potential emissions from geological storage of CO$_2$ and could also be applied to the monitoring of other types of gas sources to the atmosphere.

15.8 References

Benson, S., Cook, P., Anderson, J., Bachu, S., Nimir, H., Basu, B., Bradshaw, J., Deguchi, G., Gale, J., von Goerne, G., Heidug, W., Holloway, S., Kamal, R., Keith,D., Lloyd, P., Rocha, P.,Senior, B.,Thompson, J., Torp, T., Wildenborg, T., Wilson, M.,Zarlenga, F., Zhou, D., Celia, M., Gunter, B., King, J.E., Lindegerg, E., Lombardi, S., Oldenburg, C., Pruess, K., Rigg, A., Stevens, S., Wilson,E. and Whittaker, S. 2005. Underground geologic storage, Chapter 5 in IPCC Special Report on Carbon Dioxide Capture and Storage. Cambridge University Press.

Enting, I.G., Etheridge, D.M. and Fielding, M.J. 2008. A perturbation analysis of the climate benefit from geosequestration of carbon dioxide. International Journal of Greenhouse Gas Control 2(3), 289–296.

Etheridge, D., Leuning, R., de Vries, D., Dodds, K., Trudinger, C., Allison, C. and Luhar, A. 2005. Atmospheric monitoring and verification technologies for CO_2 storage at geosequestration sites in Australia. CRC for Greenhouse Gas Technologies (CO2CRC), RPT05-0134 [O1].

Etheridge, D., Luhar, A., Loh, Z., Leuning, R., Spencer, D., Steele, P., Zegelin, S., Allison, C., Krummel, P., Leist, M. and van der Schoot, M. and 2011. Atmospheric monitoring of the CO2CRC Otway Project and lessons for large scale CO_2 storage projects: Energy Procedia 4, 3666–3675.

Foken, T., Aubinet, M. and Leuning, R. 2012. The Eddy covariance method. In: Aubinet, M., Vesala T. and Papale D. (Eds.), Eddy covariance: a practical guide to measurement and data analysis. Springer, ISBN 978 94 007 2350 4, 1–19.

Francey, R.J., Steele, L.P., Spencer, D.A., Langenfelds, R. L., Law, R.M., Krummel, P.B., Fraser, P.J., Etheridge, D.M., Derek, N., Coram, S.A., Cooper, L.N., Allison, C.E., Porter, L. and Baly, S. 2003. The CSIRO (Australia) measurement of greenhouse gases in the global atmosphere, Report of the 11th WMO/IAEA Meeting of experts on carbon dioxide concentration and related tracer measurement techniques, World Meteorological Organization.

Haugan, P.M. and Joos, F. 2004. Metrics to assess the mitigation of global warming by carbon capture and storage in the ocean and in geological reservoirs. Geophysical Research Letters 31(18).

Horst, T.W. and Weil, J.C. 1992. Footprint estimation for scalar flux measurements in the atmospheric surface layer. Boundary-Layer Meteorology 59, 279–296.

Humphries, R., Jenkins, C., Leuning, R., Zegelin, S., Griffith, D., Caldow, C., Berko, H. and Feitz, A. 2012. Atmospheric tomography: A bayesian inversion technique for determining the rate and location of fugitive emissions. Environmental Science and Technology 46(3), 1739–1746.

Hurley, P.J., Physick, W.L. and Luhar, A.K. 2005. TAPM: a practical approach to prognostic meteorological and air pollution modelling: Environmental Modelling and Software 20(6), 737–752.

Jenkins, C.R., Cook, P.J., Ennis-King, J., Underschultz, J., Boreham, C., Dance, T., de Caritat, P., Etheridge, D. M., Freifeld, B.M., Hortle, A., Kirste, D., Paterson, L., Pevzner, R., Schacht, U., Sharma, S., Stalker, L. and Urosevic, M. 2012. Safe storage and effective monitoring of CO_2 in depleted gas fields. Proceedings of the National Academy of Sciences of the United States of America 109(2), E35–E41.

Krevor, S., Perrin, J.C., Esposito, A., Rella, C. and Benson, S. 2010. Rapid detection and characterization of surface CO_2 leakage through the real-time measurement of delta(13) C signatures in CO_2 flux from the ground. International Journal of Greenhouse Gas Control 4(5), 811–815.

Leuning, R., Etheridge, D.M., Luhar, A.K. and Dunse, B.L. 2008. Atmospheric monitoring and verification technologies for CO_2 geosequestration. International Journal of Greenhouse Gas Control 2(3), 401–414.

Loh, Z., Leuning, R., Zegelin, S.J., Etheridge, D.M., Bai, M., Naylor, T. and Griffith, D. 2009. Testing Lagrangian atmospheric dispersion modelling to monitor CO_2 and CH_4 leakage from geosequestration. Atmospheric Environment 43(16), 2602–2611.

Loh, Z., Etheridge, D., Spencer, D., Gregory, R., Allison, C., Leist, M. and Krummel, P.B. 2011. Data framework report CO2CRC Otway Project atmospheric monitoring data (CSIRO). Canberra, Australia: Cooperative Research Centre for Greenhouse Gas Technologies, RPT11-2885 [O1].

Luhar, A.K., Etheridge, D.M., Leuning, R., Steele, L. P., Spencer, D.A., Hurley, P.J., Allison, C.E., Loh, Z., Zegelin, S.J. and Meyer, C.P. 2009. Modelling carbon dioxide fluxes and concentrations around the CO2CRC Otway geological storage site. CASANZ 2009 Conference : 19th International Clean Air and Environment conference, 6–9 September 2009, Perth Convention Exhibition Centre Perth, Western Australia, Clean Air Society of Australia and New Zealand.

Luhar, A., Etheridge, D., Leuning, R., Loh, Z., Jenkins, C., Spencer, D. and Zegelin, S. 2013. Bayesian modelling for source estimation, and application to a geological carbon storage site. 21st International Clean Air and Environment Conference, Sydney, September 2013.

Miller, B.R., Weiss, R.F., Salameh, P.K., Tanhua, T., Greally, B.R., Muhle, J. and Simmonds, P.G. 2008. Medusa: A sample preconcentration and GC/MS detector system for in situ measurements of atmospheric trace halocarbons, hydrocarbons, and sulfur compounds. Analytical Chemistry 80(5), 1536–1545.

Shaffer, G. 2010. Long-term effectiveness and consequences of carbon dioxide sequestration: Nature Geoscience 3(7), 464–467.

Jonathan Ennis-King, Lincoln Paterson

16. RESERVOIR ENGINEERING FOR STAGE 1

16.1 Introduction

The reservoir engineering effort for CO2CRC Otway Project Stage 1 extended back to 2004 and involved a long period of development of geological models and the corresponding dynamic models. The purpose of this chapter is to summarise the numerical modelling with particular application to the simulations done during 2008–12. Sections 16.2 to 16.14 describe the available field data and the sources from which it came. Section 16.15 shows how these data were analysed and incorporated into the numerical simulations, and the strategies employed in history matching. Section 16.16 outlines the history of the various generations of reservoir engineering models and the questions the models were designed to address. Sections 16.17 to 16.20 summarise the results of the simulation work.

16.2 Description of field data

16.2.1 Geological model

The geological model developed for the Waarre Sandstone is discussed in some detail in Chapter 5. In summary, a "static" geological model gives a gridded representation of the reservoir unit with rock properties (e.g. facies, porosity and permeability) for each grid block. This incorporates surfaces from seismic records, data from wire-line logs and core analysis, and builds in geological concepts of the depositional environment (Dance 2009; Dance et al. 2009; Chapter 5). This is distinct from a "dynamic model", which takes the exported geological model and creates a model ready for numerical simulation, adding fluid properties (initial fluid contacts and composition, pressure and temperature) and fluid-rock properties (relative permeability and capillary pressure), combined with operational data (e.g. injection or production rates) and field observations (e.g. downhole pressure). Section 16.16 summarises the history of such models over the life of the Project.

Table 16.1: Facies properties in the geological model.

Facies	Porosity Range (%)	Permeability Range (md)	k_v/k_h
Tidal fluvial (15%)	8–21 (17)	440–6000 (1553)	0.804
Wave reworked (4%)	9–14 (12)	1–281 (38)	0.305
Gravel-dominated channels (6%)	6–28 (14)	3–3751 (175)	0.4
Amalgamated channels (34%)	9–22 (17)	8–2428 (453)	0.388
Transgressive sand (16%)	10–19 (16)	62–2795 (554)	0.630
Abandoned channel fill (13%)	1–3 (2.5)	0.002–0.3 (0.01)	

After the ranges for porosity and permeability, the average value is in brackets. For each facies, the percentage of grid blocks in the reservoir model with that facies is given in brackets in the first column. The final column is the ratio of vertical to horizontal permeability at the core plug level.

Table 16.2: Relative permeability data from Perrin and Benson (2009), relative permeability to water k_{rw} and to gas k_{rg} are given as a function of water saturation S_l.

S_l	k_{rw}	k_{rg}
1.00	1.00	0.00
0.861	0.514	0.002
0.823	0.379	0.009
0.803	0.325	0.014
0.765	0.246	0.025
0.732	0.222	0.034
0.722	0.186	0.044
0.650	0.136	0.055
0.586	0.149	0.086
0.549	0.081	0.157
0.444	0.0	0.608

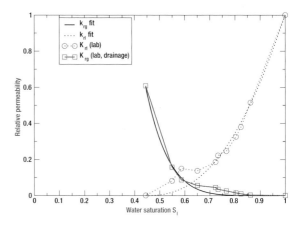

Figure 16.1: Relative permeability curves from tests on Waarre C cores after Perrin and Benson (2009).

The current version of the geological model incorporates surfaces from seismic and wire-line log data from Naylor-1 and CRC-1, and information from core analysis from CRC-1. It includes the whole area of the Naylor Field at the level of the Waarre C, and also includes some parts of the overlying Flaxman Formation. Two different cases were created: case 1 with a shorter range correlation (60–80 m) in the shale baffles and case 2 with long correlations (120–240 m). Five geostatistical realisations were created for each of these cases. Some details on the background to the geological model is given in other papers and reports (Cook and Sharma 2006; Dance 2009; Dance et al. 2009; Gibson-Poole 2005; Hortle et al. 2009; Spencer et al. 2006) and in Chapter 5.

The distribution of facies in the model is summarised in Table 16.1. The ratio k_v/k_h of vertical permeability k_v to horizontal permeability k_h is from core plugs, and doesn't take into account the lower ratios that usually result from upscaling.

16.2.2   Relative permeability data

Relative permeability data for the Waarre C Formation in the Naylor Field were limited, and came from three sources. The first was laboratory work on a core from the Waarre C in CRC-1 carried out at Stanford University,

Table 16.3: Rock properties and relative permeability endpoints for water-saturation after drainage and imbibition cycles with CO_2 and brine in CRC-1 cores analysed by Saeedi (2011).

Sample No.	ID	Porosity (%)	Permeability (md)	S_{lr} (drainage)	S_{lr} (imbibition)
1	CO2CRC-1-V	20.01	751.1	0.447	0.822
2	CO2CRC-2,3-H	17.31	788.3	0.501	0.804
3	CO2CRC-7-V	21.35	1899.6	0.469	0.906
4	CO2CRC-11,14-H	22.03	3409.8	0.524	0.910
5	CO2CRC-36-V	17.26	108.2	0.508	0.791
6	CO2CRC-39,42-H	15.49	252.6	0.546	0.802

California USA (Perrin and Benson 2009), giving a residual water saturation S_{lr} of 44.4% and a relative permeability to gas at this saturation (k_{rgmax}) of 0.608. The data are given in Table 16.2. Figure 16.1 shows the data and relative permeability curves fitted to it (the functional forms are discussed below). From the Stanford analysis, it is clear that a low permeability band across the core was largely responsible for poor sweep and larger residual water saturation. The Stanford sample was taken from the bottom of the Waarre C in CRC-1, in a tidally-influenced facies, with frequent small-scale laminations.

The second source of data was from special core analysis carried out for the Iona gas storage reservoir (also in the Waarre C formation) for TRUenergy, and history matching of reservoir models for that site.[1] For eight core plugs analysed from Iona-4, the average S_{lr} was 9.5%, the average residual gas saturation (S_{gr}) was 34.2%, the average relative permeability k_{rgmax} was 0.86, and the average relative permeability to water at residual gas saturation was 0.11. Matching of reservoir models suggests a likely S_{gr} value of 25–30%, although core analysis and logs gave values up to 40%. In the highest porosity sands, S_{lr} was as low as 5–10%, but reservoir models used an average value of around 24%. For the nearby North Paaratte field, average S_{lr} values were around 29% and average S_{gr} in the range 20–36%. For the Wallaby Creek field, S_{gr} values used ranged from 22–26.5%, and S_{lr} values average 26%.

The third source of data was from measurements undertaken at Curtin University as part of a PhD thesis (Saeedi 2011). The samples were taken from the top of the Waarre C in CRC-1, around 2059 mRT, from a fluvial dominated channel sands with fewer shale laminations. Table 16.3 shows the end-points found for different samples with drainage and imbibition measured under dynamic conditions. In the sample ID, the presence of two numbers (e.g. CO2CRC-2, 3-H) indicates that two short cores were put together to form a longer core for testing. The "H" indicates cores were drilled horizontally and "V" indicates cores drilled vertically from larger samples. The average water saturation on imbibition was 84%, implying a residual gas saturation of 16%. The average water saturation on drainage was 50.0%, comparable to the Stanford result, which was measured at static conditions. However, capillary end-effects in these data make it unlikely they are representative of reservoir flow conditions.

For the Naylor Field, the WCR for Naylor-1 gave an estimate of 11% for the residual water saturation. From post-production logging of Naylor-1 in May 2006, the residual gas saturation below the gas-water contact (GWC) was estimated at 20%.

Further work has also been done using pore-network reconstruction from X-ray CT scans of small pieces of core, which gave further guidance on the ranges of residual gas saturation expected in each facies (Knackstedt et al. 2010). The key parameter here is the average coordination

[1] Personal communication from Alex Goiye, Senior Reservoir Engineer, TRUenergy Gas Storage (now EnergyAustralia).

number of the pore network, Z_n. If $Z_n < 4$ then the trapped non-wetting phase saturation (in this case the residual gas saturation) is high, whereas $Z_n > 4$ is correlated with lower residual gas saturation. Table 16.4 summarises the analyses.

Ideally there would be multiple relative permeability measurements from each of the major facies, but the difficulty and expense of the process means this is rarely done even in the oil field domain. Here the data are restricted to the one Stanford sample, data from Curtin (which has limitations due to the procedures used) from a different facies, analogues from Iona, log estimates, and the analysis from digital core measurements (for four facies). In summary, there is a significant spread in end-point values: S_{lr} can be as low as 5–10% but as high as 45–50%, while S_{gr} can be in the range 20–40%.

Table 16.4: Facies properties from digital core measurements (Knackstedt et al. 2010); permeability is measured in the lateral directions as k_x and k_y, while k_z is in the vertical direction.

Facies	Permeability (md) (x, y, z)	k_v/k_h	S_{gr}
Tidal fluvial	400–500	1	Low
Wave reworked	130, 165, < 1	< 0.01	High
Gravel-dominated channels	2700, 2900, 2100	0.78	Low
Transgressive sand	70, 90, < 1	< 0.1	High

Table 16.5: Capillary pressure data from Waarre C cores in Boggy Creek-1 and Flaxman-1 (Daniel 2005).

Well	Depth (m TVDss)	Entry Pressure (kPa)
Boggy Creek-1	1673	1217.1
Boggy Creek-1	1674.1	4899.8
Boggy Creek-1	1676.4	2.482
Boggy Creek-1	1676.7	201.8
Flaxman-1	2194.9	59.7
Flaxman-1	2195.97	357.63

16.2.3   Capillary pressure data

The capillary pressure measurements made for the Otway Project focused on seal capacity of both top seals and intraformational seals (Daniel 2005; Daniel 2007; and Chapter 6), and there were no measurements made on reservoir samples from the Waarre C in the Naylor Field. The study of seal capacity at Boggy Creek (Daniel 2005) did include samples from the Waarre C, and the capillary entry pressures for these (corrected for brine-CO_2) are given in Table 16.5. Although the intention was to target intraformational seals, the sample from Boggy Creek-1 at 1676.4 m TVDss had an entry pressure of only 2.48 kPa, which is more representative of a reservoir facies than a seal.

Other samples from a different part of the Waarre C Formation were analysed by Damte (2002). In particular there were seven cores in the Waarre C from the Minerva-2a well in the Minerva field, Shipwreck Trough and offshore Otway Basin. Table 16.6 gives the sample details and the conversion to brine-CO_2. The conversion corresponds to the formula (Daniel 2007).

$$P_{b,c} = P_{a,m} \frac{\sigma_{b,c} \cos(\theta_{b,c})}{\sigma_{a,m} \cos(\theta_{a,m})}$$

where the subscript b,c indicates brine-CO_2, and subscript a,m indicates air-mercury (the fluids used in the measurement). Based on typical gradients, the pressure

Table 16.6: Capillary pressure data from Waarre C cores in Minerva-2a (Damte 2002).

Depth (m TVDss)	Air/Mercury Entry Pressure (psia)	Brine/CO$_2$ Entry Pressure (kPa)
1842.9	2	1.0
1845	2	1.0
1881.6	2	1.0
1882.5	2.5	1.25
1915.3	2.0	1.0
1931.2	24	12
1966.7	2.5	1.25

at the indicated depths is estimated at 18–19 MPa, and the temperature at 80–85°C. Representations for brine-CO_2 interfacial tension then give $\sigma_{b,c} \approx 27$ mN/m. The other parameters are taken as $\theta_{b,c} = 0°$, $\theta_{a,m} = 140°$, and $\sigma_{a,m} \approx 485$ mN/m. Substituting gives $P_{b,c} = -0.07267\ P_{a,m}$. With the conversion from psia for $P_{a,m}$ to kPa for $P_{b,c}$ the factor on the right hand side becomes 0.50.

These data suggest that, for good quality reservoir rock, the Waarre C has capillary entry pressure around 1–2 kPa, while for intraformational seals (presumably with lower permeability) the capillary entry pressure can be 10^2–10^3 kPa.

16.3   Well history

The original production well, Naylor-1, located near the top of the structure, produced gas from the Waarre C between June 2002 and October 2003. Thereafter the well was recompleted in a lower formation, the Waarre A, thus sealing off the Waarre C. It was subsequently found that the casing patch had slipped down and was not entirely sealing off the Waarre C, and this was used as the location of the uppermost U-tube sampling point.

During the preparations for the CO2CRC Otway Project, the Naylor-1 well was logged in May 2006, and this gave information on the level of the post-production GWC. It also provided information on the reservoir pressure in Naylor-1. After pressure depletion during production, the reservoir pressure slowly increased again due to influx from aquifers that connect to this fault block.

A new well, CRC-1, was drilled in early 2007 (spud date 15/2/07, rig release 15/3/07) about 300 m downdip from Naylor-1, so that injected carbon dioxide would migrate up towards Naylor-1 due to buoyancy and be detected there. A more detailed history of well events is given by Underschultz et al. (2011).

16.4   Well locations

At the wellhead, the Naylor-1 well-completion report gives the wellhead location as 38° 31′ 47.26″ S and 142° 48′ 30.43″ E (GDA94), which in Eastings and Northings is 657634.24, 5733850.997 (Naylor-1 LAS file). At the

reservoir level, the well location is 657648.96, 5733824.9. CRC-1 at the wellhead is at 657899.10, 5733759.0, while at the reservoir level is at 657920.86, 5733729.0. Thus the distance in the model between the wells is 280.3 m at the wellhead and at reservoir level is 288.3 m (see Figure 1.3).

16.5   Well completions

In Naylor-1, the rotary table was at 51.09 m above sea level. The depths in mRT (depth below the rotary table) have been corrected here to true vertical depth below mean sea level (TVDss) by subtracting off the rotary table offset, and correcting for well deviation. The original perforations were from 1977.4 to 1981.4 m TVDss. The casing patch inserted in late 2003 (to shut off the Waarre C, and allow production from the deeper Waarre A) extended from 1977.8 to 1986 m TVDss. Thus the casing patch did not completely shut off the perforations, and this was important for the U1 sampling location.

The major shale break observed in the Naylor-1 log was at 1983 m TVDss. The post-production GWC was estimated at 1988.4 m TVDss in Naylor-1 in May 2006.

When the bottom hole assembly (BHA) was put into Naylor-1 for monitoring purposes, the depths of the U-tube sampling points were as follows. The intake for U1 was located at 1977.7–1978.3 m TVDss, the U2 intake was at 1990.7–1991.3 m TVDss, and the intake for U3 at 1995.2–1995.8 m TVDss.

In CRC-1, the rotary table was at 50.0 m above sea level. The well was perforated between 2002.5 and 2008.5 mSS, and between 2010.5 and 2013.5 mSS (above and below a low permeability layer). The top of the Waarre C was at 2002.02 m TVDss, while the top of the Waarre B was at 2032.52 m TVDss.

16.6   Initial pre-production conditions

The discovery pressure (before production) was 19.5928 MPa at 1993.34 m TVDss (close to the middle of the Waarre C), with a gradient of 1.65 kPa/m. The 2P estimate of gas-in-place was 147×10^6 m^3 (where the conditions for gas volumes are taken to be 15°C and 0.101325 MPa, although there is a slight variation in the "standard" conditions used by different operations and in different countries). The downhole temperature was estimated at 85°C. However during the MDT test in CRC-1, the temperature was measured at 82.02°C at 2016.3 m TVDss, in the middle of the Waarre C. This may be cooler than the reservoir due to fluid injection. When the first downhole gauges were placed in CRC-1, on 11 February 2008, at a depth of 2036 mSS (23 m below the bottom of the perforations), the stable temperature was 85.9°C before injection began. Using an average geothermal gradient, this would indicate a temperature at 2008 mSS (near the middle of the perforations) of 85.1°C.

The original pre-production GWC was estimated to be between 2015 mSS and 2020 mSS, although there was no well intersecting this before production began. There are several lines of evidence to consider. Analysis of the original seismic data indicated a GWC in the vicinity of 2020 mSS, although it was tenuous and not clearly demarcated. The location of the spill point for the structure is also important. The Naylor fault was modelled and the highest sand-on-sand contact across the fault was assessed as 2015 mSS, while the lowest was 2030 mSS.

The hydrodynamic analysis is complicated by the fact that there was no pressure measurement available in the water leg before production. Using the available pressure point in the Naylor gas cap and the gas pressure gradient, and extrapolating to 2015 mSS, gave a pressure of 19.629 MPa and an estimated fresh water hydraulic head of −10 m, which is well below the regional hydrostatic prediction of 25 m, suggest that there was regional drawdown as a consequence of basin-wide production (Hennig 2007a; Hennig 2007b; Hortle et al. 2009; Underschultz et al. 2011). Thus the hydrodynamics alone did not determine the GWC location, although a deeper location (e.g. 2020 mSS) would increase the implied hydraulic head to −17 m, which becomes increasingly hard to explain.

The Waarre C is thought to be regionally hydraulically connected with high transmissivity. The most likely source of the pressure drawdown at the time Naylor-1 was drilled was production from the nearby Wallaby Creek field between September 1996 and December 2001. There the pre-production hydraulic head was 48 m, while in November 2006, almost 5 years after production, it was at −98 m, indicating a slow pressure recovery (Hennig 2007a; Hennig 2007b). Since the Naylor Field pressure was already reduced at the time of discovery, this meant that the GWC would have moved downwards due to gas cap expansion. It also meant that the eventual pressure recovery of the Naylor Field over a period of years (beyond the CO_2 injection) will return it to a higher pressure than on discovery. However, on current evidence, that recovery is very slow. At the end of injection the reservoir pressure in CRC-1 peaked at 19.278 MPa on 28 August 2009. On 5 October 2010 it was 18.519 MPa, and on 8 July 2011 it was 18.535 MPa, indicating that aquifer repressurisation was still continuing.

Table 16.7: Composition of gas produced from Naylor-1 well. WCR denotes well completion report, and BRM denotes data from sampling before breakthrough in 2008 (Boreham et al. 2011).

Component	Mol. Weight	Mole % WCR	Mole % BRM	Mass Frac. WCR	Mass Frac. BRM
CO_2	44.009	1.02	1.32	0.0235	0.0294
CH_4	16.043	84.34	82.17	0.7094	0.6682
N_2	28.018	6.52	2.04	0.0958	0.0290
C_2H_6	30.07	4.62	7.15	0.0728	0.1091
C_3H_8	44.096	2.01	3.94	0.0465	0.0880
C_4H_{10}	58.12	0.92	1.60	0.0280	0.0472
C_5H_{12}	72.15	0.24	0.41	0.0091	0.0148
C_6^+	86.18 (C_6)	0.33	0.33	0.0149	0.0143

16.7 Initial fluid compositions

The original gas composition in the Naylor Field is given in Table 16.7. The average molecular weight is 19.1. Subsequent gas analysis was carried out in the course of the sampling programme from Naylor-1 during Stage 1, and the composition prior to the arrival of the injected gas can also be used to estimate the post-production gas composition. In Table 16.7, the average of the first five samples from U-tube 1 were used (up to and including 17 July 2008, 121 days after injection began, when there was breakthrough of dissolved CO_2 at U-tube 2), corrected to be air-free (these are denoted by BRM). Compared to the WCR composition, these samples had less nitrogen and slightly more of the heavier hydrocarbons beyond methane.

The water salinity was 0.52 molal (i.e. moles per kg of water), corresponding to a salt mass fraction of 0.0295.

16.8 Production data

The Naylor-1 well produced gas from the Waarre C between June 2002 and October 2003. Thereafter the well was recompleted in a lower formation, the Waarre A, thus sealing off the Waarre C. Information available from the production phase included well completion details, monthly production rates and monthly wellhead pressures.

Since only monthly figures were supplied, the nominal start date of production was taken to be 1 June 2002. This was used as the start date for the whole simulation process, through post-production to injection and beyond. Table 16.8 gives the monthly production and the flowing tubing head pressures (FTHP) from Naylor-1.

16.9 Post-production conditions

During the preparations for the Otway Project, the Naylor-1 well was logged in May 2006, and this gave information on the level of the post-production GWC (1988.4 m sub-sea (mSS)), about 11 m below the top of Waarre C Formation in Naylor-1 (1977 mSS). It also provided information on the reservoir pressure in Naylor-1. After pressure depletion during production, the reservoir pressure

Table 16.8: Gas production and flowing tubing head pressures (FTHP) from Naylor-1 well.

Month	Elapsed Days	Production (10^6 m³)	FTHP (MPa)
June 2002	30	4.809	15.541
July 2002	61	4.877	14.760
August 2002	92	5.218	14.534
September 2002	122	6.419	13.975
October 2002	153	3.544	13.608
November 2002	183	3.875	14.098
December 2002	214	4.812	13.625
January 2003	245	3.963	13.377
February 2003	273	5.467	13.064
March 2003	304	3.861	12.761
April 2003	334	7.630	11.485
May 2003	365	6.845	11.330
June 2003	395	9.839	9.710
July 2003	426	9.915	9.151
August 2003	457	7.057	9.190
September 2003	487	3.502	10.343
October 2003	514	3.561	9.912

slowly increased due to influx from aquifer units connected to this fault block. The PSP combined log from 10 May 2006 gave the pressure at 2028 mRT (1977 m TVDss, approximately the level of the top of the Waarre C), as 16.568 MPa. In some simulation modelling (Bouquet et al. 2009; Underschultz et al. 2011) this was given as 17.4 MPa, but the details of the calculation giving this offset were not documented. Thus this pressure point had a degree of uncertainty, and the CRC-1 downhole gauge data were taken as more reliable.

In CRC-1, the pre-injection pressure during the MDT test was 17.801 MPa at 2013.1 m TVDss, giving a gradient of 9.76 kPa/m. The temperature was 82.02°C at 2016.3 m TVDss, i.e. in the middle of the Waarre C

Table 16.9: Composition of injected gas sourced from Buttress-1 well (source: Chris Boreham, Geoscience Australia), standard deviations (std) of mole % over measurements are given.

Component	Mole %	Std	Molecular Weight	Mass Fraction
CO_2	76.72	1.14	44.009	0.88269
CH_4	20.02	0.52	16.043	0.08397
N_2	1.66	0.50	28.018	0.01216
C_2H_6	0.76	0.023	30.07	0.00597
C_3H_8	0.259	0.007	44.1	0.00299
C_4H_{10}	0.104	0.025	58.12	0.00158
C_5H_{12}	0.060	0.044	72.15	0.00113
C_6^+	0.422	0.413	86.18	0.00951
He	6.44×10^{-4}	1.28×10^{-4}	4.0026	6.7×10^{-7}
H_2	1.1×10^{-6}	7.66×10^{-6}	2.0	5.7×10^{-10}

Formation. Further information was also obtained from the downhole gauges (see Section 16.11). In the period between 12 February and 10 March 2008, the pressure increase due to aquifer influx was fitted as 0.299 kPa/day for the bottom gauge, and 0.383 kPa/day from the top gauge, indicating some drift between the gauges (Paterson et al. 2010).

The top of the Waarre C in CRC-1 (2002 m TDVss) was below the post-production GWC in Naylor-1, but CRC-1 did cross the presumed level of the original pre-production GWC in the Waarre C (2015 to 2020 mSS). However, an RST log that was run in CRC-1 did not show a clearly defined GWC at this level, with residual gas being observed to the bottom of the Waarre C in CRC-1. This was interpreted as a paleo-gas column, i.e. residual gas that existed below the pre-production GWC.

16.10   Composition of injected gas

The gas for injection was sourced from the nearby Buttress Field, and the composition is given in Table 16.9. Data were taken from Buttress-1 before the separator and at CRC-1, starting before injection, and then approximately every 10,000 t of injected gas. The high standard deviation for the heavier hydrocarbon fractions reflects the fact that the

separation of condensate or wax from the Buttress-1 gas was at a maximum from November 2008 to May 2009.

The original Santos data on Buttress-1 gas (from 21 January 2002) had higher CO_2 (84.47 mole %) and lower methane (13.04 mole %), but there were possibly some errors in the sampling procedure. The nearby Boggy Creek structure had higher CO_2 (86–91 mole %) and lower methane (6–11 mole %).

16.11   Downhole pressure and temperature during injection

Before injection began, two downhole pressure gauges were placed in CRC-1 on 11 February 2008. These were located at 2036 mSS, about 30 m below the perforations at 2003 to 2013 mSS. Thus the recorded pressures (taken every 12 seconds, but only available when the gauges were pulled up after 6 months), should be about 0.29 MPa greater than the pressure at the perforations, assuming a column of formation water in the bottom part of the well. Detailed analysis of the pressure record (Paterson et al. 2010) provided a wealth of information about the reservoir response during injection, but it also gave the pressure value and the rate of pressure increase due to aquifer recharge before injection. The temperature record from the gauge in this location was not useful, in that it did not reflect the temperature of the injected gas.

Injection first began on 18 March 2008, prior to the official opening. At CRC-1, wellhead pressures, injection rates and cumulative injected amounts were recorded daily at first, and later every 10 minutes for wellhead pressures. The relation of wellhead pressure to downhole pressure was difficult to analyse due to temperature variations at the surface, and for this reason it was simpler to use the downhole pressures when they became available (downhole gauges were retrieved in August 2008 and March 2009). After the first two 6-month periods, subsequent gauges were set higher, about 149 m above the perforations. The offset in pressure from here to the perforations depended on the density of the injected gas in that part of the well, which in turn depended on the assumed pressure and temperature profile between there and the perforations. The offset was estimated to be 0.596 MPa. Figure 16.2 shows the gauge data adjusted to the level of the perforations. The original

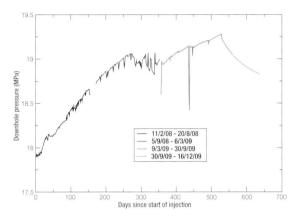

Figure 16.2: Daily average of the downhole pressure in CRC-1 (in MPa), adjusted to the level of the perforations, against days from the start of injection period.

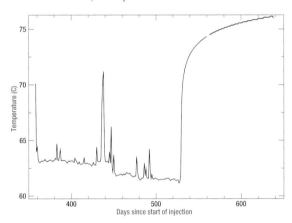

Figure 16.3: Daily average of the downhole temperature in CRC-1 (in °C), against days from the start of injection period.

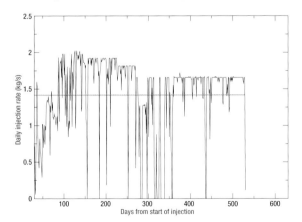

Figure 16.4: Daily average of the mass rate of injection at CRC-1 (in kg/s), against days from the start of injection (the red line is the average injection rate over the whole period).

data were filtered using a median filter with a width of 1 hour (which removed some of the extremes of the noise), then a running average with a width of 1 day, and then sampled at intervals of a day. The temperature data were only useful for the third and fourth sets of gauges, and are shown in Figure 16.3. These data were also filtered but not adjusted to the level of the perforations—an analysis of the temperature profile is given in Paterson et al. (2010). For the fourth set of gauges (30/9/09 to 16/12/09) the temperature was shifted up 0.55°C to match the trend from the third set of gauges, since the gauge location changed between the third and fourth sets of gauges. For non-isothermal simulations, the injected gas temperature at reservoir conditions could be set using Figure 16.3, which in the latter part of injection phase gave values of 61–63°C.

Figure 16.4 shows the daily injection rate in kg/s over the whole of the injection period. The red line is the average rate over the 528 days (1.419 kg/s). Brief periods of shut-in can be observed.

16.12   Tracer injection

The details of the tracer injection and monitoring programme are provided earlier in this volume (Chapter 12) and elsewhere (Stalker et al. 2009). In brief, for the first pulse, three different tracers were injected on 4 and 5 April 2008: 312 kg of sulphur hexafluoride (SF_6), 70 kg of krypton (Kr) and 1.6 kg of deuterated methane (CD_4). For the second pulse, two tracers were injected on 21 January 2009: 250 kg of SF_6 and 65 kg of 1,1,1,2-tetrafluoroethane (R-134a). Samples were taken at the Naylor-1 well at three depths within the reservoir, using the U-tube system (see Section 16.5 for details).

16.13   Gas and tracer sampling

Details of the procedures for sampling and analysis of tracers and the gas compositions are given in Chapter 12 and in other reports and papers (Stalker et al. 2009; Boreham et al. 2011). Samples were taken from the U-tubes in the Naylor-1 well on a weekly basis, starting in early 2008 before injection, and continuing on at a reduced frequency past the end of 2011. For comparison with simulation data, it is important to remember that U-tube 1, located at the top of the gas cap in Naylor-1, was always in a free gas phase; consequently, samples taken from here were always of a gas phase, and could be flowed up to surface directly through the tubing. By contrast, at the start of injection U-tubes 2 and 3 were below the post-production GWC at Naylor-1, and gas did not flow directly through the tubing, i.e. they were not 'self-lifting'. In order to get water samples under pressure from U-tubes 2 and 3, pressurised nitrogen was used in the U-tube system. After the water samples were collected they were then brought down to a lower pressure, so that dissolved gas would exsolve, and then be analysed.

Thus samples taken from U-tubes 2 and 3 were initially of exsolved gas. Since carbon dioxide is an order of magnitude more soluble than methane at downhole conditions, the concentration of carbon dioxide in the exsolved gas was proportionally much higher than in the residual but immobile gas phase. U-tube 2 transitioned fairly sharply to being self-lifting in the week leading up to 11 September 2008, i.e. that sample was self-lifting, but was not self-lifting the previous week. U-tube 3 had a more gradual transition to self-lift in the period leading up to 15 January 2009. Therefore, in summary, before these respective dates the gas compositions reported were of exsolved gas, whereas after these dates the compositions were of mobile free gas.

The other complication in the sampling was that it was not possible to separately pack off each sampling point within the Naylor-1 wellbore. Even though most of the space within the wellbore was occupied by the bottom hole assembly (BHA), there were sufficient gaps for fluid to migrate vertically. U-tube 1 was positioned near the leaking casing patch at the top of the reservoir unit, and was presumed to have communication with the reservoir through this route. The U-tube 2 sampling interval was 15 m deeper than the bottom of the U-tube 1 interval, and the U-tube 3 sampling interval was another 4 m deeper (see earlier).

On first glance this appears to complicate the analysis of the U-tube data significantly, in that if there was communication between the sample points then it would be hard to be certain of when the injected gas arrived at lower sample points. Modelling of the wellbore dynamics summarised by Underschultz et al. (2011) demonstrated that due to the low rate of extraction during U-tube sampling (around 20 kg/hour), the uppermost sample point, U-tube 1, was not affected by the composition lower down, although some of the denser fluid (containing injected gas) would be drawn up inside the wellbore. For the lower two U-tubes the situation was more complex. When U-tube 2 was self-lifting but U-tube 3 was not, the pressure drawdown during sampling was not enough to mix the samples from the two different depths, although U-tube 2 would have still sampled from a region below its screen location. When U-tube 3 also went to self-lift, then there was some mixing with the U-tube 2 samples. The amount of fluid withdrawn during a typical weekly sampling run was approximately equal to the estimated amount of free space in the wellbore around the BHA and below the packer. This also increased the likelihood of interference between U-tube 3 samples and U-tube 2 samples.

16.14   Post-injection conditions

After injection was concluded on 28 August 2009, downhole pressure and temperature gauge data were collected until 16 December 2009. Gas samples continued to be collected via the U-tube system at Naylor-1.

The other post-injection field observation was a cased hole resistivity log run in CRC-1 on 5 October 2010, which was compared to the open hole pre-injection logs run on 9 March 2007. Schlumberger's interpretation of the changes in resistivity is as follows. In the uppermost part of the perforations at 2004 mSS, the water saturation was around 80% (unchanged since the pre-injection log). In the middle of the perforations, around 2007 mSS there was an increase in gas saturation, with the water saturation going from a pre-injection 90% down to 70–80%. Towards the bottom of the perforations, around 2011 mSS, there had also been a substantial displacement of formation water, with water saturation changing from near 100% down to 70–80%. Finally, below the perforations in the range 2020 to 2028 mSS, there was a dramatic change in resistivity, with the water saturation going from 90–100% down to 60–70%. The instrument was not able to distinguish between residual methane-rich gas and CO_2-rich gas; consequently it was unclear if the increase in gas saturation was due to injected gas.

The eventual consensus was that it was difficult to compare different downhole instruments run in different conditions (open hole vs cased hole) and that a number of unexplained anomalies could result. The suggested interpretation was that the initial estimate of the pre-production GWC (2015 to 2020 mSS) was correct, and that the post-production gas saturation seen below this level was due to a paleo-gas column. This accords with the current interpretation of the pressure data, although it is hampered by not having a pressure measurement in the water leg in the Waarre C well. This paleo-gas column theory was supported by the fact that residual gas was encountered at Naylor South-1 and the nearby Croft Field which had a potential paleo-GWC at 2070 mSS, coincident with a seismic anomaly. It was concluded that this larger Naylor-Croft structure had once contained gas but activation of the Naylor-South fault after the late Cretaceous meant that methane gas leaked, leaving the smaller Naylor structure bound by the Naylor fault that terminates in the Belfast.

The cased hole resistivity run in CRC-1 (on 5 October 2010) and a pulsed neutron log run in CRC-1 on 8 July 2011 also gave pressure and temperature measurements in the Waarre Formation.

16.15   Simulation approach

16.15.1   Software

Two different simulation software packages were used in the reservoir simulation work. The first results were produced by the Earth Sciences Division of Lawrence Berkeley National Laboratory (LBNL) using TOUGH2. It used the EOS7c package developed by LBNL for carbon dioxide/methane/brine mixtures, and incorporates code for hysteretic relative permeabilities also developed by LBNL. TOUGH2 has been extensively used in modelling underground storage, and has had a significant role in two international code comparison projects (Pruess et al. 2002; Pruess et al. 2004; Class et al. 2009). It incorporates detailed representations of fluid properties and physical processes associated with carbon dioxide storage, and has capabilities similar to other simulation codes. The EOS7c package was extended for this project to include the effect

of brine salinity on carbon dioxide solubility. In general this is a salting-out effect, i.e. increasing salinity decreases carbon dioxide solubility. Importantly for the tracer work, the EOS7c package allows for one tracer component, with solubility set via a Henry's law relation, and diffusion of that tracer in both the gas and water phases.

As detailed below, from 2006 to 2009, Schlumberger's Eclipse 300 software was used extensively on the design of the field project, and on the initial history matching. The subsequent effort using TOUGH2, from 2009 onwards, build on the previous work in Eclipse, but focused specifically on tracer injection and monitoring. In this section the details are given specifically for representing the input data in TOUGH2, although they can be transferred without too much difficulty to other software.

16.15.2   Gridding

The geological model of the site is described in Chapter 5. It was created using Schlumberger's Petrel geological modelling software, and exported in a format suitable for Schlumberger's Eclipse simulator. Custom software was then used to convert the Eclipse format to one suitable for the TOUGH2 simulator. Among other changes, this involved changing the grid scheme from corner-point to PEBI (perpendicular bisector, where the line joining the centres of two connected blocks is perpendicular to the face). This conversion is not exact, but is accurate as long as the cornerpoint grid blocks are approximately rectangular in cross-section. The vertical grid spacing is about 1 m for the top part of the Waarre C. The lateral grid spacing is about 20 m in both directions.

16.15.3   Fluid properties

One of the limitations of the TOUGH2 EOS7c fluid module used in the Otway Project was that it does not allow for gas components in addition to carbon dioxide and methane, and thus cannot fully represent the composition of the original gas in the Naylor Field. The earlier work with Eclipse 300 allowed for a more complex gas composition.

There were two obvious choices for representing the original gas in the TOUGH2 simulations. The first was to match the density of the original gas at reservoir conditions but change its composition. The advantage of this approach was that the response of the gas cap to pressure changes would be accurately modelled. The matching could be achieved by using 10.3 mole % of carbon dioxide and 89.7 mole % of methane, using an implementation of the GERG2004 equation of state for natural gas mixtures. The obvious disadvantage of this approach was that the amount of carbon dioxide was actually much greater and this made it less useful for simulating changes in carbon dioxide concentrations during injection.

Table 16.10: Computed subsurface mass production rate and bottomhole pressure (BHP) for the Naylor-1 well.

Month	Elapsed Days	Production (kg/s)	BHP (MPa)
June 2002	30	1.2754	18.5754
July 2002	61	1.2485	17.6441
August 2002	92	1.3354	17.382
September 2002	122	1.6948	16.7184
October 2002	153	0.9046	16.2815
November 2002	183	1.0235	16.8651
December 2002	214	1.2283	16.3051
January 2003	245	1.0109	16.0061
February 2003	273	1.5426	15.6334
March 2003	304	0.9832	15.2719
April 2003	334	2.0009	13.7454
May 2003	365	1.7365	13.560
June 2003	395	2.5703	11.6204
July 2003	426	2.5043	9.9498
August 2003	457	1.7826	10.997
September 2003	487	0.9174	12.3964
1–11 October 2003	498	0.9130	
12–20 October 2003	507	0.0	
21–27 October 2003	514	1.1935	

The second approach was to retain the original carbon dioxide concentration (around 1.0%), or even set it to zero, for simplicity in simulating changes in concentrations at the monitoring well. The disadvantage was that the gas density was too low by about 16% at reservoir conditions, which affects the thickness of the gas cap. Also the amount of dissolved CO_2 was underestimated if there is no CO_2 in the gas phase. Since the monitoring results use the CO_2 concentration in the sampled gas, as well as tracer, the second approach was used in the simulations.

With the conversion of the production data (Table 16.8), there was a further choice—convert the produced amount to a mass of gas (using the full compositional data) and use this directly (with the altered composition in the reservoir). On the other hand it was important to take account of the volume removed at reservoir conditions, as this was more accurate for the movement of the GWC during and after production. The second option was the one used. Then the mass production in the simulation, M_{prod}^{sim}, is given by

$$M_{prod}^{sim} = V_{surf}^{field} \frac{\rho_{surf}^{field}}{\rho_{res}^{field}} \rho_{res}^{sim}$$

where V_{surf}^{field} is the reported volumetric field production data at surface conditions, ρ_{surf}^{field} is the density of the produced gas at surface conditions (using the full composition), ρ_{res}^{field} is the density of the produced gas at reservoir conditions (again using the full composition), and ρ_{res}^{sim} is the density of the gas in the simulation at reservoir conditions (using the simplified composition discussed above). It should be noted that although the subsurface density depends on subsurface pressure, this dependence approximately cancels out in the ratio of $\rho_{res}^{sim} / \rho_{res}^{field}$.

There is also a need to convert the reported flowing tubing head pressures during production to bottomhole pressures, for direct comparison with simulation results. This involved assuming a value for the surface temperature, and integrating the pressure change down the well due to the column of fluid, taking account of the geothermal gradient. Here, using the full composition of the produced gas is important for determining the pressure change. Frictional effects on the pressure change were investigated but found to be negligible at these flow rates. Information was not available on the gas temperature during production, so a surface temperature of 20°C was assumed (in fact the produced gas was most likely warmer than this). A more detailed calculation could be carried out of the likely wellbore temperature profile based on typical thermal properties of rock and well materials, but this was not done.

Table 16.10 gives computed subsurface mass production rate and bottomhole pressure (BHP) for the Naylor-1 well which in turn gives the converted mass flow rates and downhole pressures for the production phase. The GERG2004 equation of state has been used to compute gas density where the detailed composition of the gas is needed. Here the mass flow rates are the average for the month indicated. In the last month of production (October 2003) there was a brief shut-in period before the final stage of production, and data on this have also been incorporated.

16.15.4 Tracer properties and diffusion coefficients

The TOUGH2/EOS7c code characterises tracers by their molecular weight (as it affects diffusion), their diffusion coefficients in gas and water, and their Henry's law coefficient for dissolution into water. Table 16.11 gives the properties used in the simulations. The bulk diffusion coefficients (i.e. not including a tortuosity factor) in the gas phase D_{gas} are based on correlations for self-diffusion coefficients in carbon dioxide and methane (Liu and Macedo 1995) and relations for predicting binary gas diffusion coefficients (Fuller et al. 1996; Funazukuri et al. 1992; Shan and Pruess 2004).

The inverse Henry's law coefficients K_h are derived from Crovetto et al. (1982) and Shan and Pruess (2004) for

krypton and methane, and Mroczek (1997) for SF_6, for reservoir conditions of approximately 85°C. The definition is

$$K_h P_t = x_t$$

where P_t is the tracer partial pressure in the gas phase, and x_t is the tracer mass fraction in the aqueous phase.

16.15.5 Relative permeability

There is a range of plausible relative permeability parameters that might be used in simulations (Section 16.2). Five different relative permeability curves were used in simulations to explore the impact on history matching and breakthrough prediction. Table 16.12 gives the parameters, and the curves for drainage are shown in Figure 16.5. The production phase is dominated by the imbibition part of the gas relative permeability curve; in other words the value of the residual gas saturation is important there, whereas for injection it is the drainage part that is more important. Thus it is possible to use non-hysteretic relative permeability curves (standard in TOUGH2) with one S_{gr} value during production, and another value during injection. Hysteretic relative permeability curves allow more flexibility, and take care of both drainage and imbibition. Case 4 is based on the Iona data, whereas case 5 is based on the Stanford measurements.

For the non-hysteretic cases, Corey curves are used, which have the following forms for water relative permeability k_{rw}

Table 16.11: Properties of tracers used in simulations.

Tracer	Molecular weight	D_{gas} m² s⁻¹	D_{water} m² s⁻¹	K^h (Pa⁻¹)
SF_6	146.05	0.93×10^{-7}	1.0×10^{-9}	2.35×10^{-11}
Kr	84.0	1.33×10^{-7}	1.0×10^{-9}	2.2×10^{-10}
CD_4	20.0	1.57×10^{-7}	1.0×10^{-9}	9.1×10^{-11}
R-134a	102.03	0.9×10^{-7}	1.0×10^{-9}	2.1×10^{-9}

Table 16.12: Parameters for different relative permeability curves used in the simulations.

Case	Hysteretic	S_{gr}	S_{lr}	$k_{rg}(S_{lr})$	m	γ
1	No	0.325 (Im) 0.05 (drain)	0.11	1.0	1.99	10.3
2	No	0.325 (Im) 0.10 (drain)	0.11	1.0	1.99	10.3
3	No	0.325 (Im) 0.20 (drain)	0.11	1.0	1.99	10.3
4	Yes	0.300	0.11	0.86	0.737	0.5
5	Yes	0.250	0.444	0.605	1.03	3.99

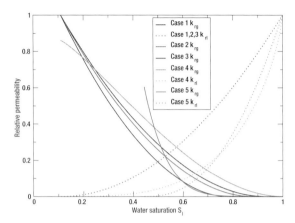

Figure 16.5: Relative permeability curves used. The gas relative permeability curves are the solid lines and the dotted lines are the water relative permeability curves. Note that the water relative permeability curves are the same for the first three cases.

$$k_{rw} = \overline{S}^m \qquad \text{where} \qquad \overline{S} = \frac{S_l - S_{lr}}{1 - S_{lr}}$$

Here S_l is the liquid saturation, and if $S_l < S_{lr}$ then $k_{rw} = 0$.

For gas relative permeability k_{rg} the form is

$$k_{rg} = (1 - \hat{S}^\gamma)(1 - \hat{S})^2 \qquad \text{where} \qquad \hat{S} = \frac{S_l - S_{lr}}{1 - S_{lr} - S_{gr}}$$

If $S_l > 1 - S_{gr}$ then $k_{rg} = 0$, while if $S_l < S_{lr}$ then $k_{rg} = 1$. The functional forms for the hysteretic relative permeability curves are described by Doughty (2007). The drainage curves are as follows:

$$k_{rw} = (S_l)^{1/2}(1 - (1 - S_l^{1/m})^m)^2$$

$$\text{and}$$

$$k_{rg} = k_{rgmax}(1 - S_l)^\gamma (1 - S_l^{1/m})^{2m}$$

When imbibition occurs, the trapped gas saturation depends on the saturation history in that location and the parameter S_{grmax} which sets the maximum trapped gas saturation. Other parameters also need to be supplied to describe the continuation of the relative permeability curves in the region $S_l < S_{lr}$.

16.15.6   Capillary pressure curves

Given the high permeability of most of the Waarre C Formation included in the geological model, capillary pressure is not as crucial a factor in the simulations, apart from the low permeability barriers that exist. As discussed in Section 16.2.3, no reservoir samples were tested for capillary pressure from Naylor-1 or CRC-1, but samples from other parts of the Waarre C suggest 1–2 kPa for capillary entry pressure for good quality reservoir, and 10^2–10^3 kPa for capillary entry pressures for intraformational seals.

The base case curve that was used was based on the Van Genuchten function (van Genuchten 1980; Pruess et al. 1999) with the form

$$P_{cap} = -P_0([S^*]^{-1/\lambda} - 1)^{1-\lambda}$$

with the restriction $-P_{max} \le P_{cap} \le 0$, and with

$$S^* = \frac{S_l - S_{lr}}{S_{ls} - S_{lr}}$$

where S_l is the liquid saturation, S_{lr} is the residual liquid saturation, P_0 is a capillary entry pressure, λ is a curvature parameter, and $S_{ls} \ge S_{lr}$. To avoid divergence in the capillary pressure, it is necessary that S_{lr} for the capillary pressure function is less than the corresponding S_{lr} value for the matching relative permeability function. Parameters for the capillary pressure curve give the values adopted in Table 16.13.

Also, the TOUGH2 option for block-by-block permeability modification allows for the scaling of capillary pressure with permeability changes (Pruess et al. 1999). The base permeability for a rock type is denoted by k_n and the capillary pressure curve for that rock type by $P_{cap,n}(S_l)$. When the permeability for a grid block of that rock type is changed by a factor ζ so the new permeability is $k_n\zeta$, then the capillary pressure is scaled as $P_{cap,n}/\zeta^{1/2}$. Thus, less permeable grid blocks have higher capillary entry pressures, as would be expected. For example, if the base curve corresponds to a 1 darcy permeability rock, then a 0.1 md rock has 100 times the capillary pressure threshold.

16.15.7 Vertical permeability

Although the geostatistical geological model gives block-by-block values for the porosity and absolute permeability, it does not readily give values for the vertical permeability k_v in relation to the horizontal permeability k_h. A common approach, adopted here, is to fix a ratio k_v / k_h. Table 16.1, discussed earlier, gives k_v / k_h ratios based on core plugs. However, even for relatively fine grids as used here (20 m laterally and 1 m vertically), the vertical permeability is still an upscaled quantity that is very sensitive to local heterogeneities. On a larger reservoir scale, it is possible to estimate vertical permeability based on shale distributions and depositional environment, which has consequences for the estimates of the typical lateral extent of shales (Begg and King 1985; Green and Ennis-King 2010). In most of these simulations, the ratio k_v / k_h has been treated as an additional parameter to adjust in the history matching. Early work on matching the production history of the Naylor Field by Dr Josh Xu led to a best fit value in the range k_v / k_h = 0.1–0.3 and this has been the base case adopted in subsequent work.

16.15.8 Comparison of gas composition and tracer with simulations

It is important to distinguish between the sample composition reported before self-lift in the U-tube, which represents dissolved gas, and the sample composition reported after self-lift, which represents free gas. Thus it is not sufficient to simply extract from the simulation the gas phase composition at the approximate depth of the U-tube sample points. One option is to treat the

sampling as a very slow production (at the rate of about 20 kg/hour for about half an hour). In TOUGH2, the default option is to weight the production according to the relative mobility of the phases. The composition of the produced fluid will then properly reflect the mobilities in the subsurface, and will transition from an aqueous to a gaseous phase when the injected free gas accumulates enough at the sample point. The inferred gas composition can then be computed from the mass fractions of gases in the produced sample. The other option is to output the compositions of the two phases in the U-tube grid block at that time, and externally compute a relative mobility for the phases, and derive an overall composition.

The slight drawdown during sampling means that U-tube samples cannot be considered to come exclusively from the grid-block closest in depth to the centre of the intake depth (which itself has a 0.6 m range). For U-tube 1 this means also monitoring an additional 1 m deeper. For U-tube 2, there was a similar effect once there was a transition to self-lift (i.e. monitor compositions 1 m deeper). For U-tube 3, once the transition to self-lift occurred, the effective sample range could also extend upwards by 1–2 m.

16.15.9 Aquifer response

As discussed earlier, the Naylor Field was most likely already pressure-depleted due to production from nearby fields at the time of discovery. During production, the field pressure fell from 19.5 MPa to almost 10 MPa by October 2003, and by the start of injection in March 2008 it had still only recovered to 17.8 MPa. This pressure change had a direct impact on the estimated size of the residual gas cap around Naylor-1, and consequently on the pre-injection GWC level. Since the downhole pressure in CRC-1 during injection was also an important set of observations, the aquifer properties used are correspondingly important. As is usual, there was little to guide the choice of parameters other than the demands of history matching. The aquifer response was clearly slow, and the permeability of the aquifer block controls this. The aquifer volume must also be large enough to repressurise the reservoir unit after production and injection. In one of the earlier phases of simulation, a two-aquifer model was proposed as one

Table 16.13: Parameters for capillary pressure curve.

Parameter	Value
λ	0.41
S_{lr}	0.0
$1/P_0$	5.0×10^{-4} Pa^{-1}
P_{max}	10^6 Pa
S_b	1.0

way to fit the observed pressure response during and after production (Hortle et al. 2009). As discussed previously the post-injection pressure declined from 19.278 MPa to 18.519 MPa, which was still below the discovery pressure.

16.16 Dynamic modelling process

16.16.1 Data to match

The observational data available for matching consisted of:

1. wellhead pressure at Naylor-1 during original production

2. downhole pressure data from logging in Naylor-1 in May 2006

3. saturation profile and location of GWC in Naylor-1 in May 2006

4. downhole pressure gauge data from CRC-1 from 11 February 2008 to 16 December 2009, plus additional points in October 2010 and July 2011

5. gas composition of U-tube samples from Naylor-1

6. tracer analysis of U-tube samples from Naylor-1.

The production rates from Naylor-1 and the injection rates at CRC-1 were known and used as inputs to the model (although it would be possible for example to take the bottom hole pressures as inputs, and try to match the injection rates). The use of different realisations of the geological model can be viewed either as a way of assessing the uncertainty of the model predictions, or as an additional way of fitting the observations.

16.16.2 History of Otway modelling

The numerical modelling of the Naylor Field for CO_2 storage, which commenced in early 2004, covered at least three different generations of static geological models, and at least four generations of dynamic models. For each version, it is important to note the data available and the observations that were being fitted by that model. Underschultz et al. 2011 gave a nomenclature for the various models described in that paper: DM1, DM2 and DM3. To that, two more cases have been added, namely DM0 denoting the earliest models and DM4 for the more recent work (Table 16.14). Section 16.16.1 gives a summary of the static and dynamic models, the data they were matching, and the questions that were being addressed.

The first simulation modelling effort was begun in 2004 by Dr Jonathan Ennis-King using TOUGH2 with a simple static models based on data supplied by Santos, the original operator of the Naylor Field. This static model, denoted as SM0, consisted of basic reservoir geometry (fault block extent and average slope), average rock properties and estimates for the GWCs. The dynamic model built from this, DM0, was designed to assist the decision of CO2CRC on whether to acquire the petroleum leases and whether the project concept was sound. DM0 simulations also fed into the first phase of site assessment, the first quantitative risk assessment of the Project, and the initial submissions to the Victorian Environment Protection Agency to obtain approval for the Project (Sharma and Robinson 2005; Sharma and Berly 2006).

Table 16.14: Dynamic models for the Otway Project; the geological models are described in the text, and the observations matched and predicted refer to the list in Section 16.16.1.

Name	Software	Date	Static Model	Match	Predict	Ref
DM0	TOUGH2	2004–5	SM0	None	Feasibility	Sharma et al. 2006
DM1	Eclipse	2005–6	SM1	1	Well location	Spencer et al. 2006; Xu 2006; Xu and Weir 2006; Xu et al. 2006a
DM2	Eclipse	2006–7	SM2	1, 2, 3	5, operations	Dance et al. 2009
DM3	Eclipse/ TOUGH2	2007–9	SM2 + Rates	1, 2, 3, 4	5	Underschultz et al. 2011
DM4	TOUGH2	2009–10	SM2 + Rates	1, 2, 3, 4	5, 6	Ennis-King et al. 2011

The second phase of simulation began in 2005 when Dr Josh Xu took over the majority of the modelling effort, using Schlumberger's Eclipse simulation software, with contributions from Dr Yildiray Cinar. This was initially based on the static model SM1—the first "proper" model—which was created in Schlumberger's Petrel software using detailed surfaces derived from seismic and populated with rock properties based on the existing wells (mainly Naylor-1). This effort explored many geological scenarios, including hypotheses about the depositional environment. The dynamic model DM1 was matched to production data from Naylor-1. The focus was on a series of important operational questions, such as where to locate the yet-to-be-drilled injection well. The aim was to ensure that the injected gas would arrive at the monitoring well within 6–12 months, in order to meet external constraints on project operation.

The static model was then updated to incorporate new data from the relogging of Naylor-1 in May 2006, and log data from the new CRC-1 well drilled in March 2007 (this is designated as SM2). The next dynamic model, DM2, used the SM2 model and the pressure measurements from Naylor-1 in May 2006 to improve the history match of production (by adjusting aquifer properties), and revise the predictions for the performance of the injection, in particular the breakthrough time at Naylor-1. There were several rounds of peer review of the models (static and dynamic) by representatives from some of the CO2CRC industry participants. An overview of the Eclipse modelling work is given by Underschultz et al. (2009; 2011). Earlier reports also gave more detail (Bouquet et al. 2009; Xu 2006b, 2007; Xu and Weir 2006; Xu et al. 2006a). In this period there was also work using TOUGHREACT to assess the impact of mineral reactions on long-term CO_2 storage in the Naylor Field (Ennis-King et al. 2006).

Once the Stage 1 injection commenced on 18 March 2008, field data such as U-tube sampling and injection rates began to be collected. Since memory gauges were used in CRC-1, the first downhole pressure data was only available after 6 months when the gauges were replaced. The dynamic model DM3 (still using the static model SM2) began the history match of injection data and observations, initially with Eclipse. In early 2008 DM3

was converted for use in TOUGH2. The initial aim was to complement the Eclipse work by allowing a more detailed prediction of tracer migration in the Naylor Field, and thus to decide on the timing of the tracer injections.

Injection ceased at the end of August 2009, with memory gauges being replaced on 30 September 2009, and then finally retrieved on 16 December 2009. U-tube sampling continued at a weekly rate. The next dynamic model, DM4, was aimed at a close history match of the downhole pressure record (as shown in Figure 16.2), and then comparing predictions for the U-tube sampling results with field observations (Ennis-King et al. 2011). This continued up until the end of 2010, with the detection of the second pulse of tracer at the monitoring well, and the repeat of the cased hole resistivity measurements at CRC-1. It was also the basis for seismic forward modelling of the expected response on time-lapse seismic, to compare to the field data. DM4 was used to generate the results referred to by Jenkins et al. (2012).

16.16.3 Approaches to history matching

As progressively more field data became available, the dynamic models were fitted to this data, a process that the petroleum industry refers to as "history matching". In that industry context, history matching is generally undertaken to address specific operational questions, e.g. how can production be maximised? Where should additional wells be drilled? What is the production revenue going to be?

In a CO_2 storage context there are operational questions to address such as: Where will the plume be located over time? How should a monitoring programme be designed to track the plume? Should water production wells be planned in case pressure relief is needed during injection? It is also useful to match only part of the field data, and then predict other observations, to test how well the model represents the storage site. This was done in DM3 (Underschultz et al. 2011) and DM4 (Ennis-King et al. 2011; Jenkins et al. 2012) to predict the fluid compositions at Naylor-1 and compare them to the U-tube sampling results.

For some CCS storage sites there will be an important issue around the potential transfer of long-term liability for the site, and when and under what conditions that liability might be transferred. For such a situation, a fully matched model of the storage operation would be important for demonstrating that the observed behaviour is understood, and the long-term behaviour can be predicted with confidence.

To date most of the history matching for the Otway models has been "manual", i.e. doing a set of runs, varying one parameter, and choosing a value that gives the best fit. Some of the Eclipse work involved automated matching, and there were some experiments with iTOUGH, the inverse modelling software package built on TOUGH2.

The key issue in history matching is to decide which model parameters can be varied to improve the fit with observations, without significantly distorting the geological model. As discussed above, the relative permeability at the Otway site is not very well constrained, and can be adjusted within some ranges, hence the need for the multiple models discussed in the previous section. The absolute permeability throughout the model is a geostatistical realisation conditioned to data at the wells. Even then, the well test at CRC-1 initially failed due to the perforations being blocked, so that there was a lack of a reservoir-scale average permeability around CRC-1, although there were measurements on cores. Thus the actual permeability throughout the reservoir could differ considerably from the model, as could the distribution of that permeability away from the wells (particularly the location and connectivity of high permeability streaks), and the ratio of vertical to horizontal permeability. The third major sensitivity was likely to be to aquifer characterisation, which is also typically poorly constrained in many oilfield models.

As one would expect, the observations are not uniformly sensitive to the same parameters. For example, the behaviour of the gas cap during production depends on the imbibition branch of the relative permeability curve. Thus the level of the GWC in Naylor-1 in May 2006 depended largely on the initial amount of gas and the residual gas saturation on imbibition. The original GWC is estimated to be between 2015 mSS and 2020 mSS. Based on earlier history matching with Eclipse, 2015 mSS was selected.

The value of S_{gr} = 0.325 matches the GWC in May 2006 (within the resolution of the grid block vertical thickness, which is around 1 m). The pressure response after production finished, referred to here as the "recovery" period, is sensitive to changes in aquifer parameters. Here the observations to match are the pressure in Naylor-1 in May 2006, and the pressure in CRC-1 in February 2008 i.e. just before the start of injection.

The downhole pressure history during injection was strongly affected by two factors. One was the probable existence of a paleo-gas column below the pre-production GWC. This increased the compressibility of the system and limited the pressure response due to injection. The other factor was the average permeability across the reservoir, which could likewise limit the pressure buildup at the well.

The numerical modelling of Stage 1 of the CO2CRC Otway Project has now reached a mature stage, in which the static geological models have been refined to take account of all the available geological information, and the dynamic models have incorporated all the field data up to late 2010, a year after the end of injection. Each piece of field data requires a number of interpretive steps before it can be properly used in a numerical simulation, whether as an input or an output.

So far, this chapter has described the field data that are relevant to numerical simulation of Stage 1 and the interpretative decisions that have been made in each case. It is important to now consider the actual modelling results.

16.17   Pre-injection modelling results

16.17.1   Earliest model DM0: testing the concept

As indicated above in Section 16.16.2, the earliest models were designed to explore whether the project concept was sound. The static model SM0 of the Waarre C reservoir in the Naylor Field was a sloping block, with the angle inferred from depth maps and the lateral dimensions chosen to match the estimated field area. The reservoir was assumed to be homogeneous, with a porosity of 17%

(from well logs) and a permeability of 4.5×10^{-12} m^2 (4.5 darcies), based on regional estimates. The amount of gas originally in place (OGIP) before production was taken as the 2P estimate of 5.6 billion standard cubic feet (Bscf), with a pre-production GWC of 2018 mSS. The total gas production of 3.4 Bscf was then simulated before the injection of CO_2. At this stage of planning it was assumed that pure CO_2 would be used, with the separation of the methane component from the Buttress gas. However this was predicated on information initially supplied that the CO_2 content of Buttress gas was around 90 mole %. This was inconsistent with the pressure gradient in the reservoir. When the composition of the Buttress gas was measured again at the time of a well test (see Xu et al. 2007), the CO_2 content was found to be about 80 mole %. This led to an operational decision to inject the full Buttress gas mixture rather than separate out the methane.

Figure 16.6 shows the distribution of the injected CO_2 after 87.6 days of injection, for a hypothetical well location of 680 m downdip, at a depth of 2060 mSS. The simulations indicated that migration up the edge of the gas cap would occur within 3 months, and was not very sensitive to injection location. Arrival at the wellbore would occur within 6 months, but not in the interval of the previous completion (top 4 m). A lower GWC (e.g. due to the greater OGIP) or the presence of faults or barriers would impede the arrival of the CO_2 at the monitoring well.

16.17.2 First detailed model DM1: refining the plan

As summarised in Section 16.6.2, the static model SM1 used surfaces from seismic interpretations, logs from Naylor-1 and Naylor South, and geological analogue information. Two geologically plausible models for the depositional environment were explored: a transgressive shoreline model, which was judged to be possible, and a regressive braided fluvial model which was judged to be most likely. The production history at Naylor-1 was matched, and this assisted in distinguishing which depositional models were most likely (Spencer et al. 2006). The median case for the average permeability was 7.5×10^{-13} m^2 (750 md), based on the production history match.

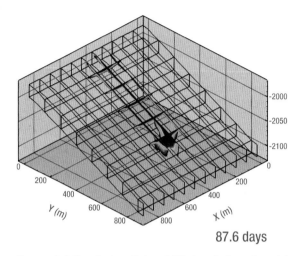

87.6 days

Figure 16.6: Distribution of injected CO_2 in early dynamic model (DM0), for injection location 620 m downdip at a depth of 2060 mSS, after 87.6 days of injection.

The aim of the DM1 model was to allow for better quantification of the test design, particularly the choice of well location and the design of the monitoring programme. It also investigated the possibility of an enhanced gas recovery option, where there could be renewed production at the observation well in order to speed up breakthrough at Naylor-1. The effect of spatial discretisation was checked, indicating that a grid block size of 20 m × 20 m × 1 m would capture the salient details. Figure 16.7 shows the shape of the CO_2 plume in the most geological likely case (without further production).

The base case simulations, with the injection well 300 m downdip, predicted breakthrough in 9 months. The key sensitivities were to permeability and well location. For an average permeability of 1125 md, the breakthrough time dropped to 6 months, while for a permeability of 375 md it was 14 months. For a well location 200 m downdip the time was 4 months, while for 400 m downdip it was 11 months, and for 500 m downdip 14 months. This analysis contributed to the decision to locate the new CRC-1 well 280 m downdip from Naylor-1, with a predicted breakthrough time in the range 6–14 months (Xu and Weir 2006). It was also anticipated that the shape and distribution of shales would have a significant impact on CO_2 plume shape and migration path.

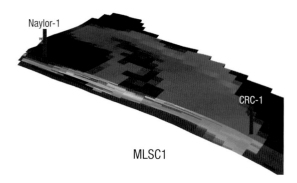

Figure 16.7: Predicted distribution of CO_2 plume for the most likely case of the DM1 model. The chosen well locations are given, with CRC-1 being the injector, and the model has been cut on a vertical section along the line of the wells.

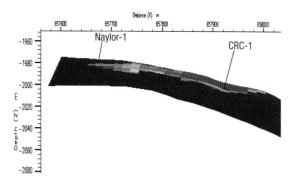

Figure 16.8: Predictions of DM2 pre-injection model for CO_2 arrival at the monitoring well after 7 months (base case): vertical cross-section along the line of the wells.

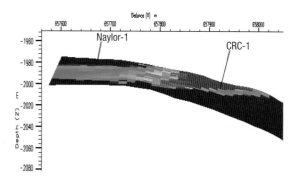

Figure 16.9: Predictions of DM2 pre-injection model for CO_2 distribution 18 months after injection ceased (Dec 2010): vertical cross-section along the line of the wells.

16.17.3 Detailed model with well data DM2

The drilling of the CRC-1 well drilled in March 2007 provided a wealth of new geological data, including well logs and cores (there being no core available from the Naylor-1 well). The geological model using this data, denoted here as SM2, was a significant improvement on SM1. The dynamic model DM2 therefore further revised the predictions of the breakthrough time of the injected gas. The choice had now been made to use the gas from the Buttress well directly, rather than separating out the CO_2, so the injection composition was no longer pure CO_2 as in previous dynamic models. The predicted breakthrough time was now estimated at 4–8 months, and the maximum bottom hole pressure as 19.2 MPa at the end of injection (assuming 100,000 t injected).

Figure 16.8 shows the predictions of the vertical distribution of CO_2 in the revised model for an arrival time of 7 months, and Figure 16.9 the distribution projected for December 2010, 18 months after the expected end of injection (assuming 100,000 t). Figure 16.10 shows the lateral distribution of the plume at the end of injection (projected for June 2009), and Figure 16.11 for December 2010. It is noticeable that across the first 4 years of modelling (2004–07), the overall features of the simulation results are the same: the injected gas migrates updip and accumulates beneath the residual gas cap, and is detected at the monitoring well. The predictions for the arrival time and the pressure increase improved as more geological data became available, but there was always an uncertainty in the timing due to uncertainties about the reservoir heterogeneity. The injected gas was always above the anticipated level of the spill-point (2015 mSS).

16.18 Injection and post-injection modelling results

16.18.1 DM3: use injection data and compare to gas composition

In February 2008, not long before the start of injection, the CRC-1 well was perforated over the interval 2010.5–

2013.5 mSS, in addition to the earlier perforation over 2002.5–2008.5 mSS. This was due to a concern over the possible effect of a shale break (in the 2008.5 to 2010.5 interval) which if extensive enough might prevent the CO_2 mixture from reaching the lower sampling intervals at the Naylor-1 monitoring well. Once injection began on 18 March 2008, field data began to accumulate—the first batch of pressure data would not be available until the first change of the memory gauges on 20 August 2008, and then the second batch on 6 March 2009. The U-tube sampling continued on a weekly basis, but although gas composition was available, the tracer concentration data was not available until laboratory analyses could be done. Meanwhile the dynamic models continued to be used for operational decisions. As noted above, the DM2 model was converted to a TOUGH2 format, in order to simulate tracer injection. The operational issue was timing: should the initial tracer pulse be added as soon as injection began, or at some later point?

While injection was still proceeding, models were improved to reflect the actual field conditions. The DM2 model was updated with the final perforation interval in the CRC-1 injection well (above and below the shale break), the actual injection gas composition (as measured from the Buttress-1 supply well), relative permeability data measured on CRC-1 core, and injection rate and/or downhole pressure data. This model was designated DM3. The first reported results from a version of DM3 were presented in September 2009 (Bouquet et al. 2009), as a collaboration with researchers from Schlumberger Carbon Services. This paper used the bottom-hole pressure (BHP) data from March to August 2008 as an input, then the actual rate data from August 2008 to November 2008, and compared the cumulative injection predicted from the simulation with the field data. Unfortunately, the input gas composition was taken as 80% CO_2 and 20% methane by mass, whereas these figures should be the proportions by mole, meaning that the injected gas was not dense enough, and buoyancy effects would have been overestimated. For conditions of 85°C and 18.0 MPa (appropriate to the start of injection), the density of the actual injected gas (approximated in composition as 80 mole % CO_2 and 20 mole % CH_4) is about 352 kg/m³, whereas for 80 mass % CO_2, equivalent to 59.3 mole %, the density is approximately 259 kg/m³. This compares with the gas-in-place, which has a density

of 120 kg/m³ at the same conditions. Thus the density difference between the injected gas and the native gas was underestimated by 40%.

The predictions of the DM3 model by Bouquet et al. (2009) for the composition at the U2 and U3 sampling points were compared to the early field data, using tracers in the numerical simulation to label the injected gas. Simulated tracer concentration increased from July to

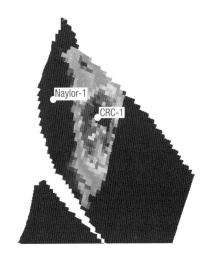

Figure 16.10: Predictions of DM2 pre-injection model for lateral CO_2 distribution at the end of injection.

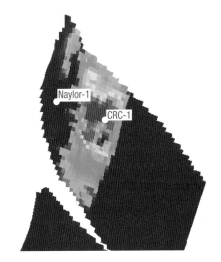

Figure 16.11: Predictions of DM2 pre-injection model for lateral CO_2 distribution 18 months after the end of injection (projected for December 2010).

September 2008, which was consistent with the field data for the first arrival of injected CO_2 at the monitoring well in July 2008. However, at this time the U-tube data had not been fully analysed. It should be noted that these simulations were only for a single realisation of the permeability distribution in the static model. An exploration of model sensitivity showed that the choice of relative permeability curves was important. If the injection rate was also kept fixed across the range of relative permeability curves (for constant BHP the relative permeability would affect injectivity), then the spread of breakthrough times was 3 months. Other parameters that were examined, such as compressibility and perforation depth, had a lesser impact on breakthrough time.

Further work using the DM3 model was carried out in 2009 (Underschultz et al. 2009; 2011), but with a different approach to history matching, and using the correct composition for the injected Buttress gas. The monthly average of the injection rate at CRC-1 was used as an input, and the downhole pressure was matched by varying aquifer parameters, using an automated process. In this model a paleo-gas column was assumed (i.e. residual gas below the pre-production GWC), and this had an impact on the rate of increase of pressure during injection. With the compositional formulation of the Eclipse 300 software, it was possible to model the gas-in-place in detail as having components of CO_2, N_2, C1, C2 and C3 (see Section 16.7 for proportions). Tracer injection was not modelled, so the main focus was on the gas

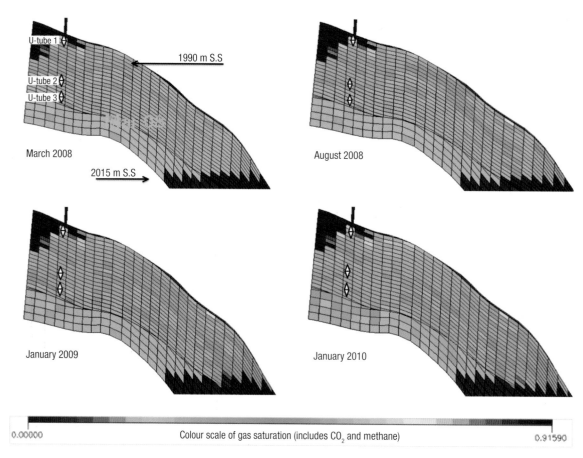

Figure 16.12: Vertical cross-sections along the line of the wells for the gas saturation in the DM3 model. Top left: March 2008, before the start of injection. Top right: August 2008, near the time of breakthrough at U2. Bottom left: January 2009, near the time of U3 breakthrough. Bottom right: January 2010. The pairs of white triangles represent the U-tube sampling points.

composition at the U1, U2 and U3 sampling points at the Naylor-1 monitoring well.

The evolution of gas saturation in the DM3 model is shown in Figure 16.12. The top left shows the situation just before the start of CO_2 injection. The red-green area is the methane-dominated gas cap, with the post-production GWC at 1990 mSS indicated. Later analysis of the depth offsets adjusted the measured GWC to 1988.4 mSS, as discussed in Section 16.9. Underlying this is a region (coloured light blue) that contains residual gas from the production phase, down to the original pre-production GWC at 2015 mSS. Below this is the dark blue of the water leg. In this version of the model, the paleo-gas column below 2015 mSS is not included. After injection began, the GWC moved downwards (while not being horizontal, due to dynamic fluid movement), until in August 2008 it was at the level of U2, near to the observed breakthrough time (top right in Figure 16.12). With further

injection, in January 2009 (bottom left in Figure 16.12) the GWC was at the level of U3, around the time when breakthrough was observed. Finally the post-injection gas saturation is shown in the bottom right of Figure 16.12.

The gas saturation includes both residual methane and the injected CO_2. Figure 16.13 gives a closer view of the CO_2 saturation alone. An interesting feature observed in the August 2008 picture (top left), is a finger of CO_2-rich gas near to U1, as well as lower down near U2. By December 2008 (top right), the upper finger was less prominent. Since the injected gas was much denser than the native gas (352 kg/m^3 vs 120 kg/m^3 at the start of injection), gravity segregation pushed the injected gas down below the free gas in the Naylor gas-cap. This finger of CO_2 at the U1 level turns out to be significant for later models for explaining the field observations at U1.

The predictions of the DM3 model for the gas composition

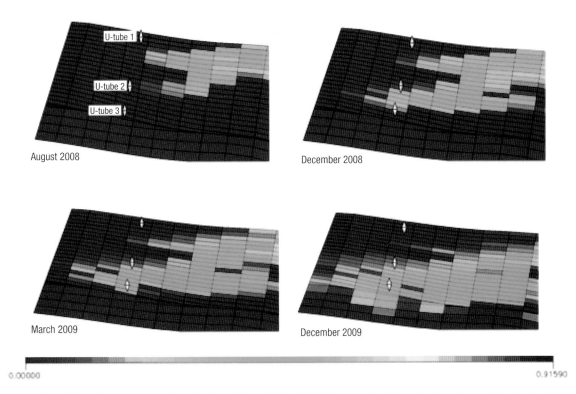

Figure 16.13: Vertical cross-sections along the line of the wells for the CO_2 saturation (shown by the colour scale) alone in DM3 model. Top left: August 2008, near the time of breakthrough at U2. Top right: December 2008. Bottom left: March 2009. Bottom right: December 2009.

at the U1, U2 and U3 monitoring levels are shown in Figure 16.14, Figure 16.15 and Figure 16.16. The vertical grid spacing (approximately 1 m) restricts the spatial resolution, so that the simulation results are from the nearest grid cell in depth to the U-tube intake. In turn, that intake samples fluids over a vertical distance of at least 0.6 m, without taking into account mixing within the wellbore. The combination of these factors limits the possible accuracy of simulations for predictions even if every other aspect of the model was correct. The DM3 results for U1, given in Figure 16.14, show a slight rise to 4 mole % around 200 days, and then a decline. This appears to correspond to the presence of a finger of CO_2-enriched gas shown in Figure 16.13, and probably includes diffusion. However the model results do not clearly distinguish between gas-phase CO_2 and dissolved CO_2, which is important for the interpretation. By contrast the field observations show a sharp increase in CO_2 concentration in the gas phase to 20 mole % after 255 days of injection.

The predictions of DM3 for the deeper sampling levels at U2 and U3 were more in line with observations. Figure 16.15 shows the U2 field results, with a transition to self-lift after 121 days of injection and a plateau at 177 days. The simulations predict that the gas composition increases to 40 mole % CO_2 (slightly below the observed level), and the predicted rise begins approximately 50 days after the field observations. Figure 16.16 shows that the DM3 predictions are similar to the U3 field data, but with the breakthrough being delayed by about 50 days.

The strategy in the DM3 modelling was to use the production and injection rates as inputs, fit the pressure history throughout, and then predict the observations at the monitoring well. The results shown here are for a single "base case" realisation of the geological heterogeneity in the static model, and for a single choice of relative permeability curves (although the sensitivity to relative permeability curves was explored in Bouquet et al. 2009). Thus it is not clear how sensitive the predictions might be to the remaining geological uncertainties. Also, the tracer pulses have not been modelled, and the additional insight from the tracer concentrations at the U-tube has not been included. Nevertheless, the original concept for Stage 1 was again confirmed by the later and more

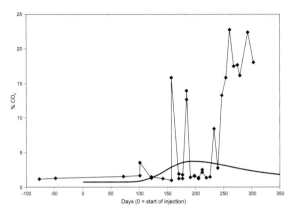

Figure 16.14: Predictions of the DM3 model for the gas composition at U1 sampling point in the upper part of the gas cap (solid line), compared to field data available at that time (diamonds).

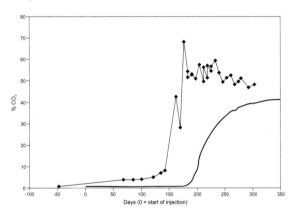

Figure 16.15: Predictions of the DM3 model for the gas composition at U2, initially below the GWC (solid line), compared to field data available at that time (diamonds).

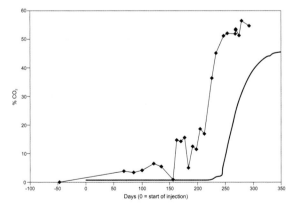

Figure 16.16: Predictions of the DM3 model for the gas composition at U3 (solid line), compared to field data available at that time (diamonds).

detailed DM3 simulations, which already predicted many aspects of the field observations during injection. The remaining puzzles at this stage of modelling were the CO_2 concentrations observed at U1, and the timing of the breakthroughs at U2 and U3.

16.18.2 DM4: multiple realisations, relative permeability and tracers

The conversion of DM3 to TOUGH2 format opened up the possibility of using the tracer options present in some TOUGH2 equation-of-state modules. The EOS7c module used here allows for a single tracer, with a Henry's law coefficient describing the partitioning between the gas phase and the aqueous phase. It is assumed that the tracer concentrations remain small, and their effect on the bulk phase properties is not considered. For tracers that are less soluble than CO_2 (all of those considered), there will be a slight enrichment of tracer concentration near the front of the first pulse of tracer, as the more soluble CO_2 is dissolved faster than the tracer. This effect is limited because the original gas in the Naylor Field contained CO_2 (at about 1 mole %), so the formation water would be relatively well-saturated with dissolved CO_2 at the beginning of injection, depending somewhat on the incoming water from the boundary aquifers. If the hypothesis about the paleo-gas column is correct, then the aquifer water would already have dissolved CO_2, whereas if there were no residual CO_2 below 2015 mSS, then the inflowing aquifer water could have more capacity to dissolve injected CO_2. Note that the enrichment effect at the front due to differing solubility also applies to the methane in the injected gas (and other components). Here again this effect is limited by dissolved components (e.g. methane) in the formation water, but also because the injected gas at the front is mixing with the residual original gas and mobilising it, so there is some dilution of the higher CO_2 concentrations that are injected, and some dispersion at the front.

Another dispersion effect also applies to the tracers, which are injected in brief pulses. The TOUGH2 version used, does not have explicit hydrodynamic dispersion, which is known to be important for tracer modelling. The lateral spacing of the grid, around 20 m, will on its own cause a spreading of the tracer pulse. In simpler models it is possible to calibrate the lateral grid spacing to match the expected physical dispersion, but the complex nature of the static model makes it difficult to regrid the full field model solely for that purpose (especially as a finer lateral grid would change the expression of the heterogeneity).

To consider the effect of geological heterogeneity, multiple realisations were generated for the porosity and permeability distribution by geostatistical techniques, while honouring the known log and core data in the wells. As explained previously, two different cases were created: case 1 with a shorter range correlation (60–80 m) of the variograms for the geostatistical model, and case 2 with a longer range (120–240 m). Five geostatistical realisations were created for each of these cases, but only two instances of each were used in the history matching process.

The DM4 model in TOUGH2 used the production and injection rates as input, and adjusted the model parameters to match the bottom-hole pressures (as was done for DM3 in Eclipse). Figure 16.17 shows the comparison for case 1 (shorter-range heterogeneity) and realisation 1. As discussed in Section 16.9, there is an issue with the interpretation of the data point for May 2006 in Naylor-1. On the other hand the data from the downhole gauges at CRC-1 are of high quality, and should be the primary constraint on the post-production pressure recovery. The greater sensitivity was to the permeability of the bounding aquifer blocks, and this value was adjusted to match the pressure recovery. It is noticeable that the largest part of the pressure recovery had already occurred by the time injection began, which corresponds to a significant influx of formation water from the boundary aquifer. Another concept that was explored earlier (with the DM2 model) was whether the leaking casing patch in Naylor-1 at the top of the Waarre C interval could have allowed either gas or water from the deeper Waarre A Formation to enter, and thus contributed to the repressurisation. The earlier work found this was not sufficient to explain the field observations on the movement of the GWC and the pressure response during and after injection, so it was not investigated again.

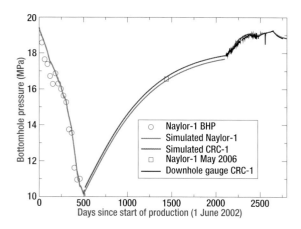

Figure 16.17: Matching of bottomhole pressure at Naylor-1 and CRC-1 over both production and injection phases, for case 1 (shorter-range heterogeneity) and realisation 1 of geological model DM4. Green curve: the simulated reservoir pressure at Naylor-1 (the production and observation well). Red curve: the simulated reservoir pressure at CRC-1 (the injection well). Black curve: actual reservoir pressure at CRC-1 during injection.

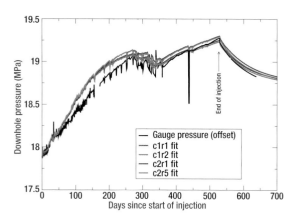

Figure 16.18: Matching of downhole pressure during injection at CRC-1 to four realisations of the geological model, in DM4 simulations.

A detailed view of the match to the downhole pressure during injection is shown in Figure 16.18. Although many metres of core were obtained during drilling of CRC-1 (so that permeability measurements could be tied to well logs), the well-test that was performed failed due to blocked perforations. Thus the pressure changes during injection provide vital information about the average permeability on a reservoir scale. Among the many parameter variations explored, adjustments to the

overall reservoir permeability were the most effective at improving the history match with the pressure data. Non-isothermal effects (due to the injected gas being cooler than the reservoir), near-well permeability enhancements and relative permeability variants were explored, but all had only a weak effect on predicted downhole pressure. In this matching process, each realisation of the geological model had to be fitted separately. It should be noted that the fitting process was non-unique, and there was a parameter space of variations which would fit equally well. The quality of the match achieved in Figure 16.18 is worst in the stage between 100 and 250 days after the start of injection, which is the time period of the breakthrough at the monitoring well. Nevertheless the key features are captured, including the initial pressure fall-off when injection ceased. Later pressure data points (from 5 October 2010 and 8 July 2011) were slightly below the simulation predictions, indicating that the aquifer pressure recovery was weaker than so far fitted.

Having matched each separate realisation of the DM4 to the pressure data, the predictions of the models for the observations at the monitoring well can now be examined. Figure 16.19 shows the predictions of these fitted models for the gas composition at U2, the sampling point just below the post-production GWC, compared to the field data. Four realisations of the model are shown, with two different relative permeability curves for each, one non-hysteretic (NH) and one hysteretic (HY). The timing of self-lift is indicated at 177 days. It is important to note that so far the field data from U2 had not been used in the fitting process. For the simulations, the two phases were sampled from a grid block in proportion to their relative mobility, and for the liquid phase the composition of exsolved gas was deduced from the mass fractions of dissolved gas, assuming complete equilibrium. The simulation results gave an indication of the inherent uncertainty coming from the model uncertainty about the heterogeneity of the permeability distribution, and on top of this, the differences due to changes in assumptions about the relative permeability curves. Without the fitting to the CRC-1 pressure data, predictions for the arrival time at U2 were in the range 140 to 290 days, with an average of 220 days. After adjustments to the bulk permeability in each realisation to fit the pressure data (as shown in Figure 16.18), predictions for the arrival time at

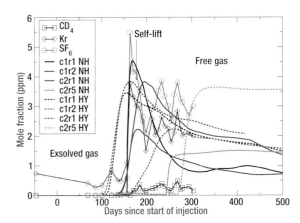

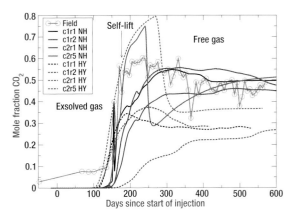

Figure 16.19: Comparison of DM4 simulation (lines) and field data (open symbols) for SF6 tracer at U2. The field data for Kr and CD4 tracers are also shown for comparison. Curves as in Figure 16.19.

Figure 16.20: Comparison of DM4 simulation (lines) and field data (open circles) for gas composition at U2. There are two cases (case 1 is short-range heterogeneity, and case 2 is longer-range heterogeneity), and two realisations of each case. "NH" indicates non-hysteretic relative permeability curves and "HY" indicates hysteretic ones.

U2 (here taken as self-lift) have a range of 140–340 days and an average of 180 days. This confirms the usefulness of the matching procedure in constraining the model.

Tracers were important for confirming that the increase in CO_2 in the U-tube samples was indeed due to the arrival of the injected gas. Figure 16.19 shows the predictions of the fitted simulations (i.e. fitted to downhole pressure) against the field data for SF_6 at U2. The arrival time of the tracers was within the range of uncertainty of the forward predictions, with different assumptions about the relative permeability curves (hysteretic vs non-hysteretic) contributing as much variation as the heterogeneity of the geology.

All the previous simulation models indicated that detection of the injected gas at U3 was expected to be later than at U2, as the denser injected gas filled the reservoir downwards beneath the residual methane-rich gas cap. Figures 16.20 and 16.21 shows the comparison of the DM4 simulation predictions with field data. After U2 was self-lifting, and before U3 went to self-lift at 303 days after injection, there was a period from 233–303 days where the exsolved gas from the U3 sample gradually became richer in CO_2. This was also found in the simulation curves, although the simulations typically predict a maximum in CO_2 content in the dissolved gas just before self-lift occurs. Even with a small pressure drawdown during U-tube sampling, the composition of U3 samples will be influenced by the

gas properties slightly above the U-tube location. It is difficult to reproduce this fully in the simulation, and this may account for the broader transition in the field data between the composition of the exsolved gas and the free gas, and the slightly earlier arrival of the injected gas at U3 in the field results. As was the case for U2, the geological heterogeneity gave a spread of forward predictions for the arrival time of the plume at U3. Figure 16.22 shows the corresponding predictions for the arrival of SF_6 tracer at U3, compared to the field data, with Kr and CD4 data given, and the timing of self-lift. It is noticeable that in the field data the peak in tracer arrival was before self-lift. The DM4 simulations predict a much later arrival of tracer, which matches to the predictions for gas composition in Figure 16.20. One of the factors involved here is the potential communication between the sampling points within the wellbore, so that the U3 samples may be partly influenced by the gas composition at U2. Although it is possible to devise a model for the near-wellbore fluid movement, it is difficult to incorporate this into the field scale simulations. The key result is that the field data confirm the essential simulation picture of the depleted gas field filling downwards below the residual gas cap as the injected gas migrates there.

Figure 16.23 shows vertical cross-sections (along the line of the wells) for one realisation of SM2 (case 1—shorter-range heterogeneity—and realisation 1) at the end of injection. Note that the vertical exaggeration of these plots is 5:1, and the lateral distance between the wells is 300 m. The plot of gas saturation (Figure 16.23(a)) shows the location of the residual methane-rich gas cap, the region of high saturation (red). The plot of CO_2 mass fraction (Figure 16.23(b)) indicates the location of the injected CO_2-rich gas, which due to greater density stays beneath the original methane-rich gas. The dispersion of the first tracer pulse can be seen in the plot of tracer mass fraction for SF_6 (Figure 16.23(c)). There are thermal effects since the injected gas had a temperature of around

63°C at the perforations, which was 20°C colder than the reservoir. The plot of temperature (Figure 16.23(d)) shows that the cooling effect is limited to a small region around the wellbore.

Figure 16.23(e) and Figure 16.23(f) compare the effect of different realisations and assumptions about the heterogeneity on the distribution of injected CO_2 (see also Figure 16.23(b)). Within the same case 1 for shorter-range heterogeneity (Figure 16.23(f) and Figure 16.23(b)), changing the realisation does not have a striking impact on the injected CO_2, although the arrival time at U3 is obviously affected (see Figure 16.20). On the other hand, changing the assumptions about the correlation length of the heterogeneity (Figure 16.23(e) vs Figure 16.23(f)) has an obvious effect, with higher concentrations of CO_2 towards the outer edge of the residual gas cap in case 2 than in case 1.

While the comparison with field data for U2 and U3 confirms the general picture from simulations, there is an issue with the data from U1—the sampling location at the top of the residual gas cap. The mole fraction of CO_2 from U1 (see the open squares in Figure 16.20) also rises from 200 days, and plateaus at about 25 mole %. Similarly tracers are detected there, signalling the arrival of the injected gas. The DM4 simulations do not predict this observation—it is clear from Figure 16.23 that the injected CO_2 stays in the lower part of the residual gas cap.

Further field observations on the sampling procedure, and numerical modeling of the near wellbore environment have shown that U1 is independent of U2 and U3 during sampling. This implies that the gas sampled at U1 does not migrate inside the wellbore, although there remains a possibility of a higher permeability pathway in the near wellbore environment (e.g. between casing and wellbore). Of the multiple scenarios investigated, the most plausible again relates to heterogeneity: there is a major shale break in the Waarre-C at Naylor-1 about 8 m from the top of the formation, below the perforations used for production. It is possible that production preferentially depleted the gas above the shale break, until water breakthrough. In this case some of the initial injection gas would fill this depleted region above the shale interval, with breakthrough at U1, as well as filling downwards below the GWC underneath the shale.

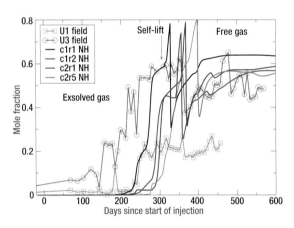

Figure 16.21: Comparison of DM4 simulation (lines) and field data (open circles) for gas composition at U3. The gas composition at U1 is also shown (open squares).

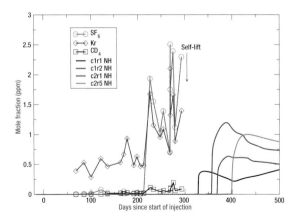

Figure 16.22: Comparison of DM4 simulation (lines) and field data (open symbols) for SF6 tracer at U3. The field data for Kr and CD4 tracers are also shown for comparison. Curves as in Figure 16.19.

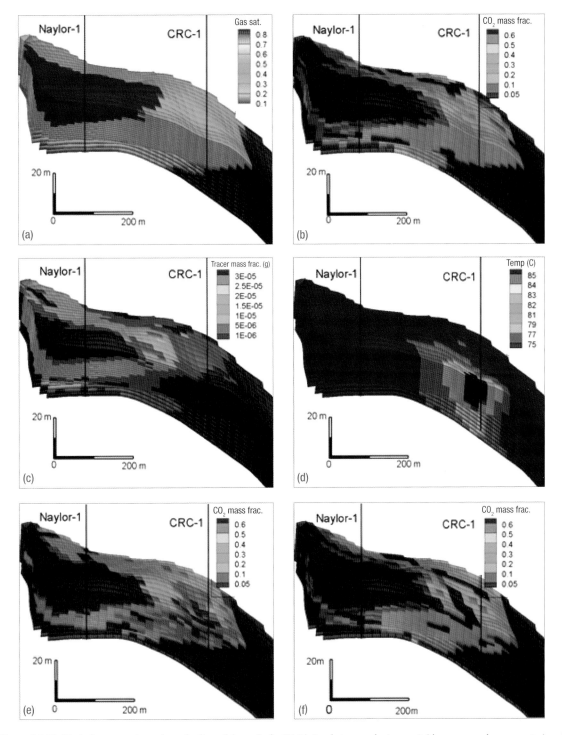

Figure 16.23: Vertical cross-sections along the line of the wells for DM4 simulation results in case 1 (shorter-range heterogeneity) and realisation 1, at the end of injection. The vertical exaggeration is 5:1. (a) Gas saturation.(b) Total CO_2 mass fraction. (c) Tracer mass fraction in gas. (d) Temperature (°C). (e) Case 2, realisation 1, total CO_2 mass fraction. (f) Case 1, realisation 2, total CO_2 mass fraction.

16.19 Dynamic storage capacity of a depleted gas field

Depleted gas fields represent potential geological storage capacity for carbon dioxide (although there may be competing options for use, such as natural gas storage). There is uncertainty regarding the extent to which the pore space previously occupied by gas (primarily CH_4, accumulated over geological time) can be reoccupied by CO_2 injected in only a few years. Global estimates of capacity depend on applying average assumptions to a wide range of different circumstances and need to be checked by experience. The results from Stage 1 of the Otway Project present an opportunity for estimating the efficiency of storage capacity in a depleted gas field.

Storage capacity itself has a range of possible definitions. "Theoretical" storage capacity for a depleted gas field could be calculated by using the known produced volume of hydrocarbons, or the pore volume from a static geological model, and converting that value into an equivalent volume of injected CO_2. However, there are technical constraints on the efficiency of filling the pore space, which give an "effective" capacity. This still does not consider economic constraints or source-sink matching.

The reduction in storage capacity from the theoretical capacity to the effective capacity depends among other things on the strength of the aquifer drive—the rate at which formation water is able to re-enter the reservoir. For a field with a weak aquifer drive, the reservoir behaves more or less like a "tank". In depleted fields with weak or non-existent aquifer support, the same pore space in the reservoir is still available for gas, and consequently in returning to the original reservoir pressure, it is possible to store the same subsurface volume as originally produced. This would be the case for depletion drive reservoirs, where the gas cap provides the needed pressure for production.

In depleted gas fields with strong aquifer support, two related phenomena limit the storage capacity. The first is that, as production proceeds, formation water occupies some of the pore space previously occupied by gas. If the injection pressure exceeds the aquifer pressure, then it should be possible to displace some of this formation water and recover some of the pore space. Also, during gas injection, the residual water saturation after drainage will generally be larger than it was pre-production, and this is observed as hysteresis in the relative permeability curves. Laboratory studies show that with repeated cycles of imbibition (water inflow) and drainage (gas displacement of water), more formation water is trapped in the pore space. Thus not all the pore space previously occupied by gas can be reoccupied.

For a depleted gas field with strong aquifer support, the amount of influx of formation water will also be influenced by the timing of injection. For example, production from the Waarre-C in the Naylor Field ceased at the end of October 2003, when the formation pressure was around 10 MPa. For the CO2CRC Otway Project, injection began in March 2008, by which time the reservoir pressure had recovered to around 17.8 MPa. This indicates a substantial influx of formation water from the aquifer system. Therefore effective storage capacity of a depleted gas field with a strong aquifer drive will depend upon the timing of injection relative to production, and the limitations on the maximum injection pressure.

The rate of filling of the Waarre-C reservoir can be assessed from the timing of self-lift at the two lower sampling points in the Naylor-1 observation well. U-tube-2 transitioned to self lift on 11 September 2008 after 21,100 t injection. U-tube-3 transitioned to self lift on 15 January 2009 after 38,100 t injection, although there was a period of some weeks beforehand in which sampling produced a mixture of gas and water. However, 21,100 t is not equal to the effective storage capacity of the reservoir between the GWC and U-tube-2 elevation because there is a portion of that volume which is "in-transit" between the injection and observation wells. Since the daily injection rate was reasonably constant, the volume of gas injected between the times that U-tube 2 and 3 transitioned to self lift (17,000 t) should represent a reasonable estimate of the effective storage capacity of the pore space between the elevations of U-tube 2 and U-tube-3. The elapsed time interval was 126 days, and the sampling interval was 7 days, so the uncertainty in the timing is at least 10–15%, and perhaps more due to the transition period at U-tube-3. Thus the volume of gas has a proportional uncertainty of 17,000 ± 2000 t.

Although the depths of the U-tubes was accurately known, each samples from an interval of finite width, of around 0.6 m. U-tube 2 and U-tube 3 were not packed off from each other, and were adjacent to the same interval of perforations. Sampling from a U-tube induces a flow into the U-tube at the reservoir level, and the origin of this fluid will depend on the productivity (i.e. the product of permeability and thickness) of the reservoir section near the perforations, as well as possible flow within the wellbore. Since each sampling operation removes a fluid volume comparable to the free volume within the wellbore, there is also some dependence on the history of fluid withdrawal due to sampling. The difference in depth between the two U-tubes is 4.5 m. Detailed modelling of the wellbore flow suggests that the movement of the GWC between the two transitions to self-lift was in the range 3.5 ± 0.5 m. Here the GWC is used as shorthand for the complex distribution of saturation as the reservoir fills, which depends on local porosity, permeability and fluid-rock properties such as capillary pressure and relative permeability.

The effective storage capacity in a region of the reservoir depends on the residual gas saturation S_{gr} upon imbibition (when the field is depleted and is being repressurised by water influx), and the maximum gas saturation S_{gmax} upon drainage (when the injected gas displaces some of the water that has occupied the pore space). S_{gmax} should be less than the $1 - S_{lr}$, where S_{lr} is the original residual water saturation. The efficiency of filling due to hysteresis, which will be denoted by C_{hyst}, is then given by

$$C_{hyst} = \frac{S_{g\,max} - S_{gr}}{1 - S_{lr} - S_{gr}}$$

If there is no hysteresis, then $S_{gmax} = 1 - S_{lr}$ and $C_{hyst} = 1$. Note that this is only the local hysteresis, and does not take account of reservoir pore volume that is invaded by formation water that is not ultimately displaced. This calculation also implicitly assumes that the end-points of the relative permeability curves are independent of pressure, and that the reservoir is nearly back to its discovery pressure. In fact there are more complex considerations, where gas that is trapped at residual saturation will increase in density as the reservoir pressure recovers, reducing the saturation further.

From the static geological model, the pore volume that is filled between the two transitions to self-lift can be estimated at 0.7×10^5 m³ – 1.3×10^5 m³ using the estimated range of movement of the GWC between the transitions. It has been estimated from logs that $S_{lr} = 0.11$ and $S_{gr} = 0.20$. Using the lower value for the vertical movement of the GWC (3.0 m) gives $C_{hyst} = 0.75$ to 0.84. These in turn imply $S_{gmax} = 0.72 - 0.78$. For the upper value of vertical movement (4.0 m) the range is $C_{hyst} = 0.56$ to 0.62, and so $S_{gmax} = 0.58 - 0.63$. Although there is still a considerable range of values, these measurements provide evidence that in this case $C_{hyst} < 1$. The values of the relative permeability endpoints (S_{lr}, S_{gr} and S_{gmax}) are a fluid-rock property, and consequently specific to particular sites. However it is likely that hysteretic effects, on top of pore volume loss due to aquifer influx, will further reduce the effective storage capacity of depleted gas fields with moderate aquifer drive.

For the Waarre-C in the Naylor Field, the observations have so far only constrained the efficiency of filling between U-tube 2 and U-tube 3. An estimate of the effective storage capacity of the whole reservoir unit depends on taking some of the ensemble of numerical models matched to the observations, and simulating the effect of continued injection until a suitable stopping criterion is reached. This criterion might be economic or technical in nature.

The calculations described above predict that 56–84% of the space originally occupied by recoverable CH_4 is reoccupied by CO_2, taking account of uncertainties in the pore volume, post-production residual CH_4 saturation and the possible range of movement of the GWC. Generic estimates used to calculate global capacity in depleted gas fields have assumed 75%. While dependent on detailed circumstances, e.g. rock type or timing of storage after production, the Stage 1 result is consistent with the generic estimate, and indicates that depleted gas fields with strong aquifer support have the potential to store injected carbon dioxide.

16.20   Conclusions

Dynamic modelling of Otway Stage 1 began in 2004, and extended through three major generations of static models, five major generations of dynamic models, and two entirely different software packages. Each generation of model was developed to answer fairly different questions, beginning with the project concept (before acquisition of the petroleum licences), then the design, the operations and the monitoring. Thus the dynamic models were the place where much of the project data—geological models, operational history, downhole sampling and monitoring results—were integrated into a unified representation of the behaviour of the injected gas in the subsurface.

Across the generations of models, a few features were consistently observed. The injected gas was expected to move updip from the injection well and fill downwards beneath the post-production gas cap, but remain above the pre-production GWC, which was presumed to be the spill point for the reservoir. This general picture was confirmed by the field data, particularly the composition of gas samples from the three U-tubes in the monitoring well.

Some predictions changed as the models were improved over time, most notably the arrival times at the U-tube sampling points, which were affected by the estimates of average reservoir permeability, choice of relative permeability curves and injection details. The changes in gas composition at the two lower sampling points agreed with the ensemble of predictions from the final generation of dynamic models (using the actual injection rates and downhole pressure), although the actual arrival times were towards the early end of the predicted range, which may be the result of areas of higher permeability than anticipated in the geological model. The timing of the arrival at the highest U-tube was explained by heterogeneity in the permeability, in particular the effect of a shale break in the vicinity of the monitoring well.

Even when all the field data were used, a significant uncertainty still remained in the modelling due to heterogeneity of rocks in the subsurface, and insufficient relative permeability data. Thus long-term predictions of how the trapped CO_2 will behave are still only probabilistic. However, there is strong evidence that the CO_2 in Otway Stage 1 is "behaving as predicted", and will remain trapped for a long period of time.

Finally, it is apparent from the Otway results that in the future depleted gas fields with strong aquifer support could provide valuable opportunities for CO_2 storage.

16.21   References

Begg, S.H. and King, P.R. 1985. Modelling the effects of shales on reservoir performance: Calculation of effective vertical permeabilities. In: SPE Reservoir Simulation Symposium, Dallas, Texas, U.S.A., February 10–13. SPE 13529.

Boreham, C., Underschultz, J., Stalker, L., Kirste, D., Freifeld, B., Jenkins, C. and Ennis-King, J. 2011. Monitoring of CO_2 storage in a depleted gas reservoir: gas geochemistry from the CO2CRC Otway Project, Australia. Int J Greenhouse Gas Control 5(4), 1039–1054.

Bouquet, S., Gendrin, A., Labregere, D., Nir, I.L., Dance, T., Xu, J. and Cinar, Y. 2009. CO2CRC Otway Project, Australia: Parameters influencing dynamic modeling of CO_2 injection into a depleted gas reservoir. SPE Offshore Europe Conference, held in Aberdeen, UK, 8 October – 11 September. SPE 124298.

Class, H., Ebigbo, A., Helmig, R., Dahle, H.K., Nordbotten, J.M., Celia, M.A., Audigane, P., Darcis, M., Ennis-King, J., Fan, Y., Flemisch, B., Gasda, S.E., Jin, M., Krug, S., Labregere, D., Naderi Beni, A., Pawar, R.J., Sbai, A., Thomas, S.G., Trenty, L. and Wei, L. 2009. A benchmark study on problems related to CO_2 storage in geologic formations. Computational Geosci 13(4), 409–434. URL http://www.springerlink.com/index/10.1007/s10596-009-9146-x.

Cook, P.J. and Sharma, S. 2006. Detailed geological site characterization of a CO_2 storage site: Otway Basin Pilot Project, Australia. Proceedings of the CO2SC Symposium 2006, Lawrence Berkeley National Laboratory, Berkeley, California, 20–22 March 2006. CO2CRC Report No. CNF06-0050.

Crovetto, R., Fernández-Prini, R. and Japas, M.L. 1982. Solubilities of inert gases and methane in H_2O and in D_2O in the temperature range of 300 to 600 k. J Chem Phys 76(2), 077–1086.

Damte, S.W. 2002. Analysis of the sealing capacity of the Flaxman Formation and the Belfast Mudstone in the vicinity of the Shipwreck Trough, Otway Basin, Victoria. Honours Thesis. National Centre for Petroleum Geology and Geophysics, University of Adelaide.

Dance, T. 2009. A workflow for storage site characterisation: A case study from the CO2CRC Otway Project. AAPG/ SEG/SPE Hedberg Conference on Geological Carbon Sequestration: prediction and verification, 16–19 August 2009, Vancouver, BC, Canada.

Dance, T., Spencer, L. and Xu, J.Q. 2009. Geological characterisation of the Otway Project pilot site: What a difference a well makes. Energy Procedia 1(1), 2871–2878. http://linkinghub.elsevier.com/retrieve/pii/S187661020 9007048.

Daniel, R.F. 2005. Carbon dioxide seal capacity study, Boggy Creek, Otway Basin, Victoria. Tech. rep., CO2CRC. CO2CRC Report No. RPT05-0045.

Daniel, R.F. 2007. Carbon dioxide seal capacity study, CRC-1, CO2CRC Otway Project, Otway Basin, Victoria. Tech. rep., CRC for Greenhouse Gas Technologies (CO2CRC). CO2CRC Report No. RPT07-0629.

Doughty, C. 2007. Modeling geologic storage of carbon dioxide: Comparison of non-hysteretic and hysteretic characteristic curves. Energy Convers Mgmt 48, 1768–1781.

Ennis-King, J., Dance, T., Xu, J., Boreham, C., Freifeld, B., Jenkins, C., Paterson, L., Sharma, S., Stalker, L. and Underschultz, J. 2011. The role of heterogeneity in CO_2 storage in a depleted gas field: history matching of simulation models to field data for the CO2CRC Otway Project, Australia. Energy Procedia 4, 3494–3501. Proceedings of the Tenth International Conference on Greenhouse Gas Control Technologies, 19–23 September 2010, Amsterdam, The Netherlands.

Ennis-King, J., Perkins, E. and Kirste, D. 2006. Reactive transport modelling of the Otway Basin CO_2 storage pilot. Eighth International Conference on Greenhouse Gas Control Technologies, 19–22 June, Trondheim, Norway.

Fuller, E.N., Schettler, P.D. and Giddings, J.C. 1966. A new method for prediction of binary gas-phase diffusion coefficients. Ind Eng Chem 58(5), 19–27.

Funazukuri, T., Ishiwata, Y. and Wakao, N. 1992. Predictive correlation for binary diffusion coefficients in dense carbon dioxide. AIChE J 38(11), 1761–1768.

Gibson-Poole, C. 2005. Sedimentology of Waarre Formation from the Boggy Creek-1 and Flaxman-1 Wells, Otway Basin, SE Australia. Tech. rep., CRC for Greenhouse Gas Technologies (CO2CRC). CO2CRC Report No. RPT05-0029.

Green, C.P. and Ennis-King, J. 2010. Vertical permeability distribution of reservoirs with impermeable barriers. Transport Porous Med 83(3), 525–539.

Hennig, A. 2007a. Hydrodynamic interpretation of the formation pressures in CRC-1: Vertical and horizontal hydraulic communication. Tech. rep., CRC for Greenhouse Gas Technologies (CO2CRC). CO2CRC Report No. RPT07-0626.

Hennig, A. 2007b. Hydrodynamic interpretation of the Waarre Formation Aquifer in the onshore Otway Basin. Implications for the CO2CRC Otway Project. Tech. rep., CRC for Greenhouse Gas Technologies (CO2CRC). CO2CRC Report No. RPT07-0634.

Hortle, A., Xu, J. and Dance, T. 2009. Hydrodynamic interpretation of the Waarre Fm Aquifer in the onshore Otway Basin: Implications for the CO2CRC Otway Project. Energ Procedia 1(1), 2895–2902. http://linkinghub.elsevier.com/retrieve/pii/S1876610209007073.

Jenkins, C.R., Cook, P.J., Ennis-King, J., Underschultz, J., Boreham, C., Dance, T., de Caritat, P., Etheridge, D.M., Freifeld, B.M., Hortle, A., Kirste, D., Paterson, L., Pevzner, R., Schacht, U., Sharma, S., Stalker, L. and Urosevic, M. 2012. Safe storage of CO_2 in a depleted gas field—the CO2CRC Otway Project. Proc Nat Acad Sci USA 109(2), E35–E41.

Knackstedt, M., Dance, T., Kumar, M., Averdunk, H. and Paterson, L. 2010. Enumerating permeability, surface areas, and residual capillary trapping of CO_2 in 3D: Digital analysis of CO2CRC Otway Project core. SPE Annual Technical Conference and Exhibition, held in Florence, Italy, 19–22 September. SPE 134625.

Liu, H. and Macedo, E.A. 1995. Accurate correlations for the self-diffusion coefficients of CO_2, CH_4, C_2H_4, H_2O and D_2O over wide ranges of temperature and pressure. J Supercrit Fluid 8, 310–317.

Mroczek, E.K. 1997. Henry's law constants and distribution coefficients of sulfur hexafluoride in water from 25°C to 300°C. J Chem Eng Data 42(1), 116–119.

Paterson, L., Ennis-King, J. and Sharma, S. 2010. Observations of thermal and pressure transients in carbon dioxide wells. SPE Annual Technical Conference and Exhibition, held in Florence, Italy, 19–22 September. SPE 134881.

Perrin, J.C. and Benson, S. 2009. An experimental study on the influence of sub-core scale heterogeneities on CO_2 distribution in reservoir rocks. Proceedings of the Ninth International Conference on Greenhouse Gas Control Technologies, 16–20 November, Washington D.C., USA.

Pruess, K., Garcia, J., Kovscek, T., Oldenburg, C., Rutqvist, J., Steefel, C. and Xu ,T. 2002. Intercomparison of numerical simulation codes for geologic disposal of CO_2. Tech. rep., Lawrence Berkeley National Laboratory, Earth Sciences Division. LBNL-51813.

Pruess, K., García, J., Kovscek, T., Oldenburg, C., Rutqvist, J., Steefel, C. and Xu, T. 2004. Code intercomparison builds confidence in numerical simulation models for geologic disposal of CO_2. Energy 29(9–10), 1431–1444.

Pruess, K., Oldenburg, C. and Moridis, G. 1999. TOUGH2 User's Guide, Version 2.0. Tech. rep., Lawrence Berkeley National Laboratory, Earth Sciences Division. LBNL-43134.

Saeedi, A. 2011. Experimental study of multi-phase flow in porous media during CO_2 geosequestration processes. PhD thesis, Curtin University Department of Petroleum Engineering.

Shan, C. and Pruess, K. 2004. EOSN—a new TOUGH2 module for simulating transport of noble gases in the subsurface. Geothermics 33, 521–529.

Sharma, S. and Berly, T. 2006. Otway Basin Pilot Project (OBPP). Submission to the Victorian Environment Protection Agency (EPA). Seeking approval to demonstrate the viability of carbon capture and storage through a demonstration project in the Victorian Otway Basin. Tech. rep., CRC for Greenhouse Gas Technologies (CO2CRC). CO2CRC Report No. RPT05-0043.

Sharma, S., Cook, P., Robinson, S. and Anderson, C. 2006. Regulatory challenges and managing public perception in planning a geological storage pilot project in Australia. Eighth International Conference on Greenhouse Gas Control Technologies, 19–22 June 2006, Trondheim, Norway. CO2CRC Report No. ABS06-0089.

Sharma, S. and Robinson, S. 2005. Otway Basin Pilot Project (OBPP) discussion paper of risks associated with the proposed Otway Basin Pilot Project to demonstrate CO_2 capture and storage. Tech. rep., CRC for Greenhouse Gas Technologies (CO2CRC). CO2CRC Report No. RPT05-0084.

Spencer, L., Xu, J.Q., LaPedaline, F. and Weir, G. 2006. Site characterisation of the Otway Basin carbon dioxide geo-sequestration pilot project in Victoria, Australia. Eighth International Conference on Greenhouse Gas Control Technologies, 19–22 June 2006, Trondheim, Norway.

Stalker, L., Boreham, C., Underschultz, J., Freifeld, B., Perkins, E., Schacht, U. and Sharma, S. 2009. Geochemical monitoring at the CO2CRC Otway Project: Tracer injection and reservoir fluid acquisition. Energ Procedia 1(1), 2119–2125. http://linkinghub.elsevier.com/retrieve/pii/S187661020900277X.

Underschultz, J., Boreham, C., Dance, T., Stalker, L., Freifeld, B., Kirste, D. and Ennis-King, J. 2011. CO_2 storage in a depleted gas field: An overview of the CO2CRC Otway Project and initial results. Int. J. Greenhouse Gas Control 5(4), 922–932.

Underschultz, J.R., Freifeld, B., Boreham, C., Stalker, L. and Kirste, D. 2009. Geochemical and hydrogeological monitoring and verification of carbon storage in a depleted gas reservoir: examples from the Otway Project, Australia. AAPG/SEG/SPE Hedberg Conference "Geological carbon sequestration: prediction and verification", 16–19 August 2009, Vancouver, BC, Canada. CO2CRC Report No. ABS09-1508.

van Genuchten, M.T. 1980. A closed-form equation for predicting the hydraulic conductivity of unsaturated soils. Soil Sci Soc 44, 892–898.

Xu, J. 2006. CO2CRC Otway Basin Pilot Project: Naylor-1 history matching. Tech. rep., CRC for Greenhouse Gas Technologies (CO2CRC). CO2CRC Report No. RPT06-0035.

Xu, J. 2007. Displacement mechanisms and evaluation of enhanced gas recovery (EGR) potential in the Otway Basin Pilot Project. Tech. rep., CRC for Greenhouse Gas Technologies. CO2CRC Report No. RPT07-0559.

Xu, J. and Weir, G. 2006. Otway Basin Pilot Project CO_2 injection modelling. Tech. rep., CRC for Greenhouse Gas Technologies (CO2CRC). CO2CRC Report No. RPT06-0068.

Xu, J.Q., Weir, G., Ennis-King, J. and Paterson, L. 2006a. CO_2 migration modelling in a depleted gas field for the Otway Basin CO_2 storage pilot in Australia. Eighth International Conference on Greenhouse Gas Control Technologies, 19–22 June 2006, Trondheim, Norway. CO2CRC Report No. ABS05-0096.

Xu, J.Q., Weir, G., Paterson, L. and Sharma, S. 2006b. Dynamic modelling in carbon dioxide geosequestration. Proceedings of the AAPG 2006 International Conference and Exhibition, Perth, Australia, 5–8 November 2006. CO2CRC Report No. CNF06-0140.

Xu, Q., Weir, G., Paterson, L., Black, I. and Sharma, S. 2007. A CO_2-rich gas well test and analysis. SPE Asia Pacific Oil and Gas Conference and Exhibition, Jakarta, Indonesia, 30 October – 1 November, SPE 109294.

Lincoln Paterson, Chris Boreham, Mark Bunch, Tess Dance,
Jonathan Ennis-King, Barry Freifeld, Ralf Haese, Charles Jenkins,
Matthias Raab, Rajindar Singh, Linda Stalker

17. CO2CRC OTWAY STAGE 2B RESIDUAL SATURATION AND DISSOLUTION TEST

17.1 Introduction

Key risks for commercial-scale CCS projects include uncertainties regarding storage capacity and CO_2 containment. The CO2CRC Otway Project Stage 2B sought to address some key factors related to both these risks and also provide the basis for a cost-effective aquifer appraisal process.

Residual capillary trapping and dissolution trapping are important mechanisms in geological storage (Figure 17.1) because they provide secure containment without necessarily the need for an overlying seal rock according to Benson, Cook et al. (IPCC 2005). Residual carbon dioxide saturation (the fraction of the pore space in which carbon dioxide is trapped) is the end-point of a CO_2-water relative permeability curve. The relative permeability of the reservoir plays a dominant role in determining the migration distances and migration rates of injected carbon dioxide plumes. Dissolution trapping refers to the portion of CO_2 that dissolves into formation water.

The relative permeability of CO_2 and water, and the degree of residual and dissolution trapping, are major uncertainties in the development of carbon dioxide storage sites within aquifers. There has been particular emphasis in recent years on the nature of relative permeability hysteresis and the role it plays in storing CO_2 by residual saturation trapping. This phenomenon can be exploited to increase trapping by co-injecting brine with carbon dioxide, or by injecting chase brine after the carbon dioxide. This affects both the amount of residual trapping and the rate of CO_2 dissolution into the formation brine. The possibilities raised by these injection strategies make it imperative to obtain a reliable characterisation of relative permeability behaviour at sites being considered for underground storage within aquifers.

It is difficult to obtain core samples suitable for the laboratory determination of relative permeability. Core samples are usually oriented in the vertical direction whereas relative permeability measurements are also required in the horizontal direction. Even then, laboratory results from cores only represent a small part of the reservoir and may not be representative of larger scale heterogeneities and facies transitions.

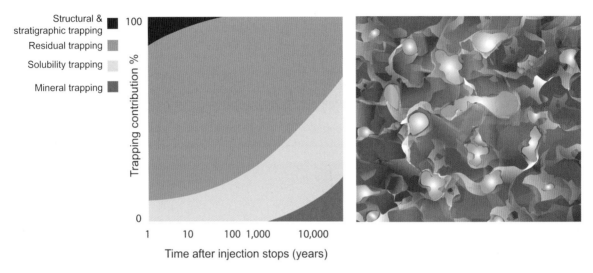

Figure 17.1: Temporal concept of trapping mechanisms of CO_2 (left), concept of residual trapping of CO_2 (right).

There is great value in conducting a short-term test in the vicinity of an injection well prior to large-scale injection, to obtain reliable estimates of reservoir-scale residual saturation that can be used to estimate reservoir-scale CO_2-water relative permeability. This reduces risk and uncertainty in the modelled performance of a potential carbon dioxide storage site, enhancing the "bankability" of a commercial project at an early stage. The Otway site provided an outstanding location to put this concept to the test. Following on the success of the Otway Project Stage 1, it was decided to embark on Stage 2 (see Chapter 1).

17.2   Test concept

The Otway Stage 2B test was conceived as a single-well test that could be potentially applied at commercial sites in order to reduce uncertainty in estimating field-scale residual trapping. Several methods already exist for single-well measurements of residual oil and non-aqueous phase liquids (NAPL—essentially petroleum contamination) but these have not yet been applied in the context of carbon dioxide. Carbon dioxide presents some particular challenges, mainly due to the high solubility of carbon dioxide compared to petroleum. Furthermore, the CO_2 residual saturation has to be established prior to measurement. For residual oil, the original oil in place

commences with an accumulation that has stabilised under gravity but this is not the case for most situations with carbon dioxide. For these reasons there was seen to be value in comparing multiple approaches, especially when the incremental cost of each additional testing method is relatively minor. The CO2CRC Otway site presented an ideal research opportunity to test a variety of methods.

The methods for measuring residual saturation were as follows:

1. Pressure test (or hydraulic test). Essentially this is a multi-phase well test where numerical simulation is used to inverse model the reservoir response to derive the residual saturation. This method is described in detail by Zhang et al. (2011). Further development of analytical methods may allow a formula to be derived and used similar to conventional well testing.

2. Measuring the hydrogen index using pulsed neutron logging. Commercial pulsed neutron logging is an established practice although experience with CO_2 is limited. The major drawback is that the pulsed neutron capture tool only penetrates to a depth of about 20 cm, so this method samples the smallest rock volume of all the methods.

Figure 17.2: CRC-2 site facility. The CRC-2 wellhead is left of centre, the CO_2 storage tank is at right and water pipes are in the foreground. The water storage tanks are out of the frame to the right. A man to the left of the tank provides a scale.

3. Thermal test. In the thermal test the borehole is heated and temperature is recorded, using a fibre-optic distributed temperature sensor (DTS). The depth of penetration into the formation is 1–2 m—significantly greater than pulsed neutron logging (Zhang et al. 2011).

4. Organic tracer test. Organic reactive tracers have been used to measure residual oil (Tomich et al. 1973); however, ethyl acetate that is used for measuring residual oil is not optimal for carbon dioxide. The application of this test requires identification of new tracers that perform the same role. Three tracers were identified and used in Otway Stage 2B. The lack of prior experience with these tracers meant that there was a strong research component to this test.

5. Dissolution test. This test is based on a method for residual oil described by Bragg et al. (1976). It is a relatively simple test involving the injection of water, dissolving residual fluid, and back-producing the water to measure the dissolved fluid. It was easy to implement at Otway because of the stored water at the surface and the installed downhole fluid sampling. The downside of the test was that, unlike the other tests, it reduces the residual fluid, removing the possibility of further testing.

A brief overview of the Otway Stage 2B test methods has been reported by Paterson et al. (2013). A photograph of the CRC-2 site facility is shown in Figure 17.2 and the facilities are illustrated schematically in Figure 1.11.

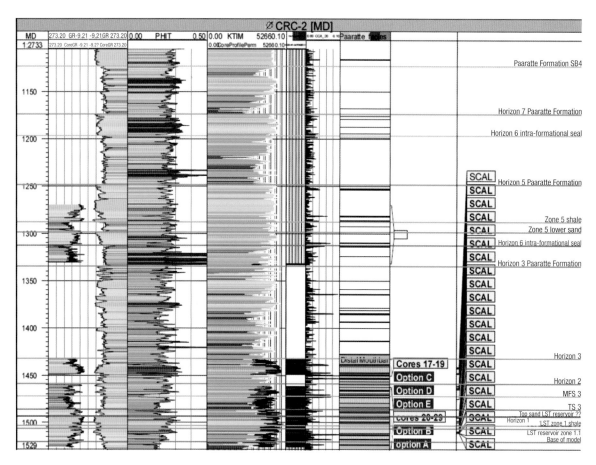

Figure 17.3: Paaratte Formation well composite with log porosity, permeability, core porosity and lithology characterisation indicated.

17.3   Injection target

The top of the Paaratte Formation succession occurs at 1123 m MD[1] (1119 m below ground level) and bottom at 1520 m MD (1516 m below ground level). The Paaratte Formation well composite (Figures 17.3 and 17.4), shows well-log porosity, permeability, core porosity and lithology characterisation in terms of depositional-diagenetic facies associations.

Following core and well log interpretation, five potential CO_2 storage systems were identified within the lower part

of the Paaratte Formation (1440–1520 m MD) formerly referred to as Zone 1. These five storage systems, labelled as options A to E, were assessed for homogeneity in the reservoir (injection) interval and sealing capability of any non-reservoir layers immediately above them (Figures 17.3 and 17.4).

The deepest option (option A) had no sealing layer, which would have made the test more difficult to interpret owing to likely vertical pressure communication and migration of CO_2. The aim was to reduce unknown effects on fluid movement in vertical and lateral directions due to the ratio of vertical and horizontal permeability, (k_v/k_h). A good sealing cap rock was preferred to isolate the CO_2 plume migration so that it remained within the reservoir interval only.

[1] Note on depths: Unless otherwise specified, depths quoted in this chapter are measured depths in the CRC-2 well (MD), so that the perforations are at 1440 m – 1447 m MD, 1392.1 m – 1399.1 m TVD SS (sub mean sea level), or 1436.1 m – 1443.1 m below the surface (ground level) to the nearest 0.1 m. The datum for CRC-2 is 47.9 m above MSL.

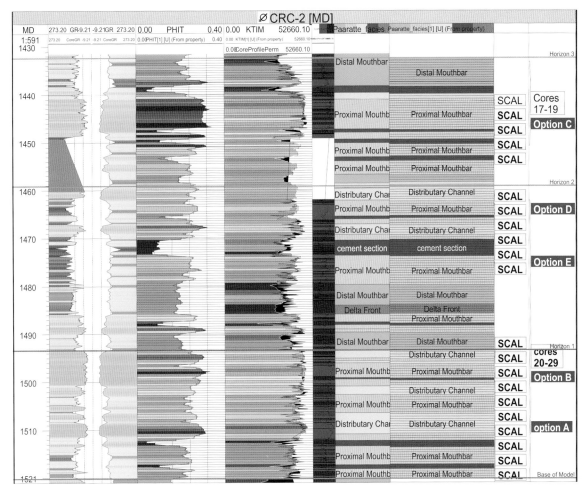

Figure 17.4: Log compilation of the lower part of the Paaratte Formation showing five sections (labelled as options A to E) that were regarded as potentially suitable for the residual saturation and dissolution test.

Option B (above option A in Figure 17.4) contained cemented layers and variable porosity and permeability, which made the test interval more heterogeneous. This would have introduced uncertainty on the pressure response, that would have been difficult to differentiate from possibly vertical migration of the CO_2. However, MCIP test results showed a high potential to retain CO_2 (Figure 17.5).

Option E had no confirmed capping interval. A relatively thick cemented layer centred on 1472 m MD demonstrated poor sealing capacity. The risk of vertical migration would have introduced uncertainty into interpretation of the pressure response.

Like option B, option D contained a thin cemented interval and a mixture of reservoir lithofacies. A thin sealing interval centred on 1457.5 m MD was unproven and was interpreted from well logs alone, as core was not cut between 1449–1460 m MD.

The reservoir interval of the shallowest option, option C (near the top in Figure 17.3) was the most homogenous when compared with the other options. Porosity (average approximately 28%) and permeability (average approximately 2196 md) values from logs and cores were high and relatively consistent. Option C had a cemented interval and a thick non-reservoir lithofacies interval above with a high sealing capacity for withholding CO_2 (Figure 17.5).

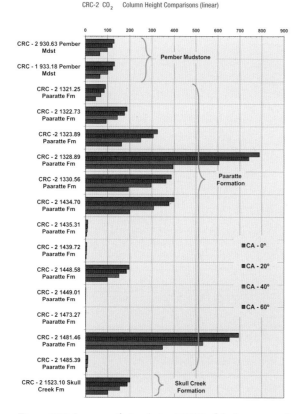

Figure 17.5: Log compilation (using MICP) of the lower part of the Paaratte Formation. Division in five sections (options A to E).

For these reasons, option C was chosen as the injection interval for the new residual gas saturation and dissolution test, and was perforated for the test within the depth range of 1440–1447 m MD (1436–1443 m below ground level).

17.4   Test sequence

The planned test sequence is shown in Figure 17.6; the actual test sequence is shown in Figure 17.7. The main difference was due to operational delays, but otherwise all the activities in the original sequence were implemented. The process flow diagram is shown in Figure 17.9. Following an initial water production phase, the residual gas saturation test sequence commenced on the 27 June 2011 and lasted for 76 days (see Figure 17.7 and Table 17.1). The initial water production phase comprised 10 days of water production, where 505 t of water were produced. Three surface tanks were installed to handle the water storage. Since the gas

lift method was employed with Buttress gas (CO_2/CH_4 mixture) via the bottom gas mandrel mounted on the tubing at 996.6 m, provisions were made to safely vent the CO_2 that was both in free phase and exsolved from the water and the free-phase CH_4.

The water was treated with an ultraviolet system to prevent microbial activity that could have potentially entered the system during reinjection into the formation, thereby blocking the entry pore space at the Paaratte injection zone. Water production provided a measure of the bulk permeability of water in the Paaratte Formation, derived by the pressure drawdown that was recorded by downhole pressure gauge measurements.

A maximum of 590 t (590,000 L) of formation water was produced from the reservoir. At various stages, portions of this water volume were reinjected and re-produced as shown in Figure 17.7. In total there were 1003 t of pumping, both in and out. After the initial surface stock of 505 t was established in Phase 1, 104 t were injected in Phase 2 followed by 190 t of production. This took the stored surface volume up to the maximum inventory of 590 t.

In Phase 3, the largest water injection period (454 t) was used to drive the CO_2 to residual saturation where the pressure would have reached steady state. The remaining phases had four periods of production (104 t, 6 t, 75 t and 251 t final disposal) and three periods of production (182 t, 24 t and 131 t).

Phase 2 consisted of the baseline response tests when the system was in initial condition and water saturated (Figure 17.7). It began with the operation of the distributed thermal perturbation sensor (DTPS) to produce baseline thermal conductivity estimates. The formation was heated above 2°C of formation temperature within 2 days. The temperature fall off was recorded over a period of 3 days. During that period, the borehole remained unperturbed and no tools entered the well. The water-saturated reference test was also conducted with tracers (krypton and xenon). This involved co-injecting 104 t of water with 1200 L of noble gases. They were injected at 5.0 L/min (totalling 600 L of Kr and 600 L of Xe), maintaining a Kr:Xe ratio of 1. The downhole pressure increase was recorded for a pressure-buildup well test interpretation.

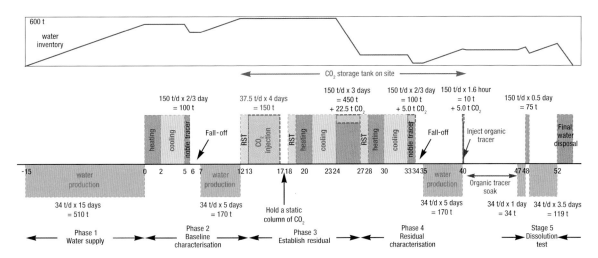

Figure 17.6: Otway Project Stage 2B planned residual gas saturation test sequence, 18 April 2011, prior to the field test.

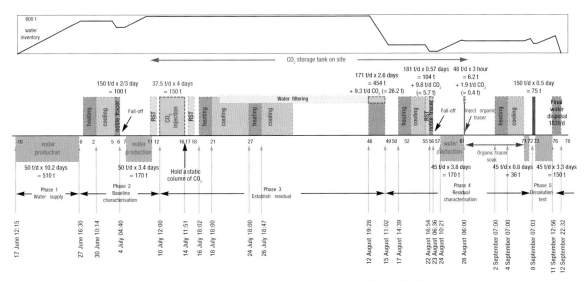

Figure 17.7: Otway Project Stage 2B actual residual gas saturation test sequence, 15 September 2011.

Surface installations were provided that facilitated co-injection of gaseous tracer material with the water. The tracers were introduced in a discrete pulse at the start of injection to focus and concentrate the front so that the breakthrough curve was at its most concentrated on its return. Subsequently, 190 t of the injected water and tracers were back-produced over 3.4 days and samples were analysed for tracer concentration. At the end of Phase 2 the well was logged with pulsed neutron measurements (using Schlumberger's RST tool) to establish the baseline at fully water-saturated conditions.

Phase 3 commenced with the injection of 37.5 t CO_2/day over a period of 4 days, for a total of 150 t injected. Carbon dioxide was brought into the stream by direct injection into the 2⅜″ tubing. The expected extent of the plume and gas saturation is illustrated in Figure 17.10.

The formation was left to equilibrate for one day. Another pulsed neutron log run in the well showed CO_2 saturation close to the borehole, averaged 30% in the lower part of the perforated interval and 45% in the upper part. The

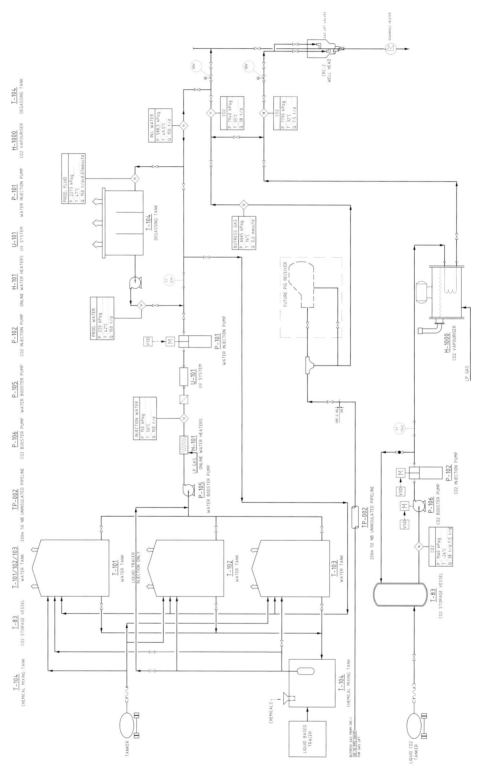

Figure 17.8: CRC-2 process flow diagram for water and CO$_2$ injection.

Table 17.1: Injection and production quantities used during the five phases of the residual saturation test.

Description	Production	Injection
Initial water supply and pressure drawdown test	505 t water.	
Noble tracer injection and pressure build-up test		104 t water plus noble tracer.
Noble tracer analysis and additional water supply	190 t water plus noble tracer.	
CO_2 injection		150 t CO_2.
Water to push to residual saturation plus pressure response to analyse		454 t water plus 26.2 t CO_2 to prevent dissolution.
Noble tracer injection		104 t water plus 5.4 t CO_2 plus noble tracer.
Noble tracer analysis	182 t water plus noble tracer.	
Organic tracer injection		12 kg of triacetin, 10 kg of propylene glycol diacetate and 8 kg of tripropionin by 8 t water plus 0.3 t CO_2.
Organic tracer analysis	23.7 t water plus ester tracers and hydrolysis products.	
Dissolution test water		75 t water without CO_2. Methanol was added in the water at a concentration of 1091 ppm.
Dissolution test analysis	131 t water plus dissolved CO_2.	
Final water disposal		251 t water.

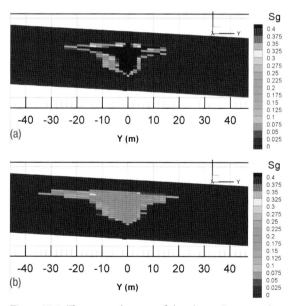

Figure 17.9: The expected extent of the plume. Reservoir slice in dipping direction (2° dip) of contours of gas saturation (a) after gas injection and (b) after residual saturation (after drive to residual to S_{grmax} = 20%).

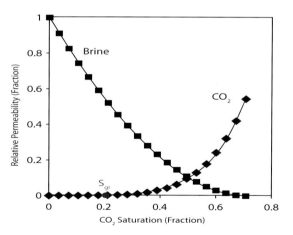

Figure 17.10: Relative permeability curve from Basal Cambrian sandstone drainage from tests by Bennion and Bachu (2005). S_{gr} = 20%.

DTPS measurement was then repeated, using changes of observed thermal conductivity as a proxy measurement for CO_2 saturation. Heating for 2 days followed by a quiescent cooling period of 3 days was used to measure thermal conductivity. After a delay, this was followed by an injection of 171 t/day of water for 2.6 days at reservoir CO_2 saturation, to drive the reservoir to residual gas saturation conditions. The pressure response and flow rate were used to reconstruct the CO_2-water relative permeability curves based on the concepts of multiphase well testing. The relative permeability curve provided the residual gas saturation at zero gas mobility (Figure 17.8), although this curve is for drainage (CO_2 displacing water) rather than imbibition (water displacing CO_2) and doesn't consider hysteresis.

The injected water that drives the CO_2 to residual gas saturation was saturated with CO_2 to avoid dissolving the residual trapped CO_2. The water was mixed with pure CO_2 downhole, by pumping the CO_2 that had been vaporised and compressed through the upper gas injection mandrel at 846 m depth: 26.2 t of CO_2 were used to saturate the 454 t of water. Pressure was monitored throughout the CO_2 and CO_2-saturated water injection. The pressure response during the water injection period provided information about residual gas saturation for history matching with a flow simulator. Figure 17.11 shows the expected sensitivity of pressure to residual gas saturation, demonstrated for 10% and 20% S_{gr}. It should be noted that the timeline in Figure 17.11 only represents the CO_2 injection and subsequent creation of the residual gas saturation field.

Depending on the heterogeneity of the formation, the response will be more or less pronounced. This was tested and studied in several simulation runs (Figure 17.12).

After the creation of the residual gas field through water injection, the formation was left to equilibrate for one more day before thermal logging and the third and last pulsed neutron log was run to establish an estimate of the residual gas saturation. The pulsed neutron test was used to estimate the residual gas saturation close to the borehole, whereas the history matching interpretation of the residual gas saturation indicated the extent of the CO_2 plume (approximately 15 m radius). The combination of both history matching and pulsed neutron logging provided complementary information on the estimate of residual gas saturation.

A third DTPS dataset was collected at the beginning of Phase 4 to further reduce the uncertainty of the residual gas saturation estimation. Six days after Phase 4 was commenced, the second tracer injection took place in an identical manner to Phase 2, where water was spiked with noble gas tracers, and saturated with CO_2 (the noble gas tracers partitions according to the gas phase present, and the tracer recovered is inversely proportional to the gas saturation (residual gas saturation).

The reactive ester tracer partitioning test was based on the single-well tracer method to measure residual oil saturation described by Tomich et al. (1973), except that different tracers were required for measuring residual carbon dioxide. Triacetin (12 kg) with formula $C_9H_{14}O_6$, tripropionin (8 kg) with formula $C_{12}H_{20}O_6$ and propylene glycol diacetate (10 kg) with formula $C_7H_{12}O_4$ are benign liquids (food grade products) and were added undiluted at the start of this section of testing followed by 6.2 t of water saturated with 380 kg of CO_2 to prevent dissolution of the residual CO_2. The organic tracers were allowed to hydrolyse/soak for 10 days in the reservoir. Then 23.7 t of water was produced and analysed for tracer composition and concentrations.

The last step was the dissolution test based on the Bragg-Shallenberger-Deans (1976) method. This involved injecting 75 t of water, this time without any CO_2, and deliberately dissolving the residual CO_2. Approximately 78 L of methanol was uniformly added to the injected water as a tracer (giving a concentration of 1091 ppm). Then 131 t of this water was back-produced from the formation and the CO_2 content was measured to calculate the residual saturation. Unlike the other methods this method dramatically alters the residual saturation hence was conducted last. CO_2 was brought back to surface in this final step. Provisions were made to safely vent the CO_2 that is both in free phase and exsolved from the water.

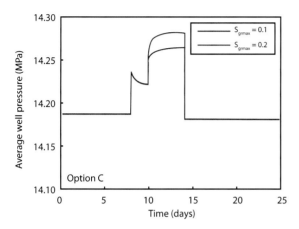

Figure 17.11: Pressure response and sensitivity to residual gas saturation (option C).

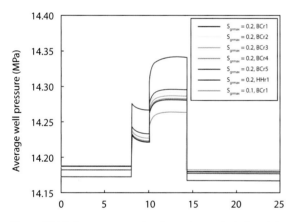

Figure 17.12: Pressure versus time for various geological models, BCr1 BCr2, BCr3, BCr3, BCr4, BCr5 , HHr1. Note 1: BC is base case heterogeneity HH is high heterogeneity (with shorter correlation length scale). Note 2: r1, r2 etc. is realisation 1, 2, etc. (which are all equally probably geological models).

17.5 Downhole completion

The CRC-2 well was cased and 0.14 m (5½″) production casing run from surface to the total depth of the well of 1565 m.

In February 2011, the downhole completion was installed in the well after perforating the interval from 1440 to 1447 m (7 m). The completion consisted of a 2⅜″ Fox 13 Cr80 tubing, an inflatable packer configuration, set to straddle the test zone and minimise the sump area, and several steel lines for gas lift, pressure gauges, distributed

Table 17.2: Control lines in the CRC-2 completion.

Control Line Description	Diameter	Termination Depth MD (m)
Packer inflate system	⅜″	1435 to below perforated interval
U-tube #1 and #2	⅜″	1440
Gas injection mandrel	⅜″	846 with a ¼″ orifice valve
Gas lift mandrel	⅜″	996 with a ¼″ orifice valve
DTS/heater	⅜″	1460
Panex TEC #1	¼″	1439.8
Panex TEC #2	¼″	1448.3

temperature perturbation sensor (DTPS), heating wire, and U-tube sampling (Figures 17.13 and 17.14). Table 17.2 describes the various control lines at the wellhead. All lines were bare 316 L stainless steel welded tubes. The seven steel lines housing the in-line P/T gauge coax cables, DTS, heater lines, and U-tube samplers lines and packer inflate lines penetrated the top packer, which had been set at 1435.48 m. There were four real-time high-accuracy surface read-out P/T gauges on two TEC cables, U-tube, and DTPS measure fluid properties beneath the packer. The memory P/T gauges were located in X-nipples above the packer and below the gas lift mandrels in the tubing string, when required. The perforated interval was isolated from the bottom of the well by a lower packer set at 1449.55 m (see Figure 17.13).

After the completion was installed, the well was flowed to retrieve a pristine formation sample for the baseline water geochemistry (salinity, type of dissolved solids, etc.) and microbial activity (original microbe content in formation water) of the Paaratte Formation by swabbing. Preliminary analysis of the samples showed that the swabbing operation was not long enough and consequently the completion fluid KCl brine and fluorescein were not fully retrieved. At the start of the water production phase the well was flowed until pristine formation water conditions were obtained, and this water was disposed of separately. Formation fluids from MDT sampling in CRC-2 also showed mud contamination (fluorescein). The perforated

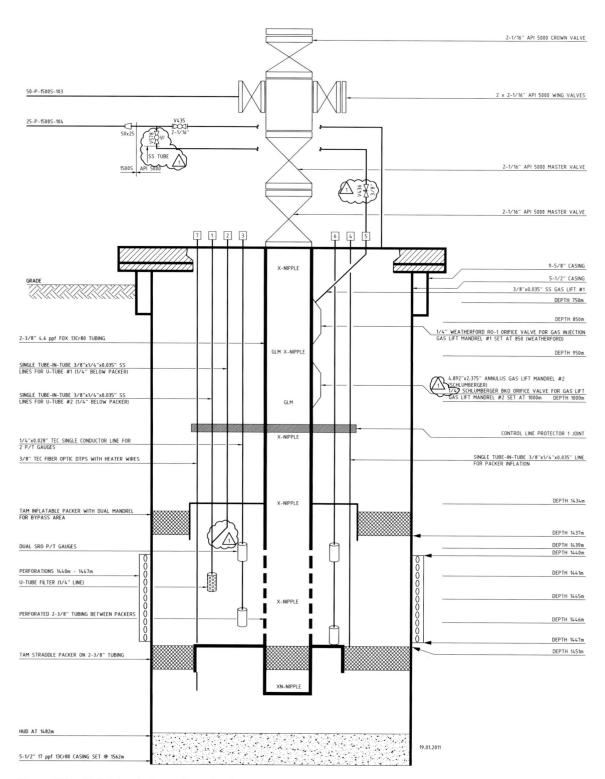

Figure 17.13: CRC-2 downhole completion details.

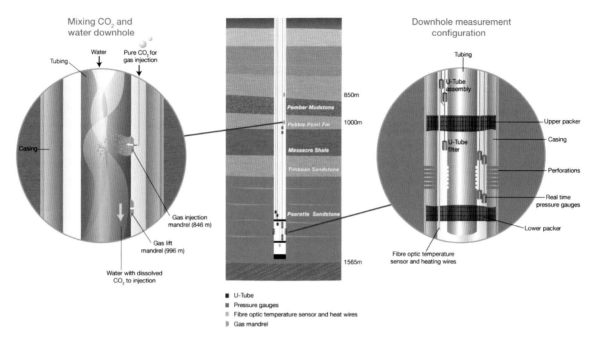

Figure 17.14: Downhole completion of CRC-2 for the residual gas saturation and dissolution test. The diagram on the left shows the gas mandrel above the packer for mixing CO_2 and water downhole. The diagram on the right shows instruments below and above the packer.

formation was at hydrostatic pressure and contained only formation water, consequently the well would not flow by itself. To produce water, a gas lift method using Buttress gas was employed to lift the water column in CRC-2 via the casing annulus and up the 2⅜″ tubing. The Buttress gas entered the 2⅜″ tubing via an orifice valve located in the gas lift mandrel at 996.6 m. The gas lightened the water column in the well thus, allowing flow to reach the surface. The Buttress gas and water mixture from the gas lift was first flowed into a degassing tank where any excess gas that came out of solution was vented.

The well was initially flowed for 10 days, and subsequently for 3.4 days during the water reference test and 3.8 days at residual gas saturation to produce the injected tracers back to surface (see Table 17.1 and Figure 17.7).

Facilities were provided to hold the produced water at the surface. A sampling protocol was established and the formation fluid composition analysed in real-time via the U-tube while flowing the well. Some pristine formation samples were taken using the U-tube system at the perforation interval, while the fluid column was lifted to surface with the CO_2/CH_4 gas mixture from Buttress.

A second gas injection mandrel complete with a ¼″ orifice valve located at 846 m MD provided the inlet for pure CO_2 to be injected into the tubing for a later phase in the residual gas saturation test. The CO_2 was mixed with water for the "water injection with dissolved CO_2," activity after the pure CO_2 injection (day 46 to 49 in Phase 3 of 7). The solubility of CO_2 in water at 846 m was calculated for three different temperatures,[2] assuming a hydrostatic pressure gradient that included a pressure increase during injection, of about 0.1 MPa for 1 darcy rock, and about 1 MPa for 100 md rock. The distance from 846 m to 1435 m was deemed adequate to dissolve the CO_2 before it reached the perforated formation.

Four real-time P/T gauges and two retrievable memory gauges were located beneath the lower gas mandrel to confirm optimum mixing of CO_2 with water to prevent CO_2 under- or over-saturation (Figure 17.14). The

---

[2] The first is in equilibrium with the geothermal gradient (an initial state, or at very low rates). The second assumes a steady state water injection over 5 days with the overall wellbore heat transfer coefficient set to 50 J/(s m² K) (a moderate to high value), giving a thermal equilibrium length (1/LR) of about 1000 m. The third case assumes a lower value of the overall wellbore heat transfer coefficient of 20 J/(s m² K).

Table 17.3: Measurement types for formation fluid sampling and residual gas saturation test and timeline.

Surface	Measurement	Comments	Time
	Flow rate in	Rosemount 1700 series Coriolis flow meter for CO_2 injection. Rosemount 8732 magnetic flowmeter for water.	From day 15 onwards, with start of Phase 2
	Flow rate out	Rosemount 8732 magnetic flowmeter for water.	From day 1 of Phase 1
	Water cut	No separator was required.	From day 1 of Phase 1
	Wellhead pressure and temp at CRC-2	This was connected to the DCS providing a continuous record.	From day 1 of Phase 1
Downhole	**Measurement**	**Comments**	**Time**
	Pressure and temperature (P/T)	Four real-time surface read out Panex gauges across perforated interval.	Baseline recording for 2 days before day 1 of Phase 1, recording continuously until end of test.
			Real-time gauges were critical during the entire test.
	P/T in tubing	Two Metrolog memory gauges below top gas mandrel and above top packer.	Memory gauges were only deployed when injecting water with saturated CO_2.
Downhole	**Measurement**	**Comments**	**Time**
	Formation gas and water and tracer detection from test via U-tubes	At surface, sampling lab was installed at CRC-2 wellsite with LBNL mass spectrometer. GA mobile field lab was equipped with 2 GC-MS wet chemistry equipment and DIC tester.	Baseline sampling for 2 days before day 1 of Phase 1.
			Sampling during back production in Phase 2, day 6 to 13, for tracers. Sampling during CO_2 injection in Phase 3, day 14 to 18.
			Sampling during water with diss. CO_2 injection in Phase 3, day 24 to 27, to confirm mixing.
			Sampling during back production in Phase 4, day 37 to 44, for tracers.
	DTPS and heating wires.	Standalone measurement	Heating formation and DTPS baseline recording for 2 days before day 1 of Phase 1.
			Continuous DTPS recording until the end of test. Heating formation from day 18 to 20 in Phase 3. Heating formation from day 28 to 30 in Phase 4.
	Pulsed neutron log	Logging tool, time lapse log	Evaluation of 100% water saturated conditions, day 12. Record near-borehole 100% CO_2 saturation, day 23
			S_{gr} measured day 27

memory gauges were installed at 948 m and 1095 m MD and retrieved using Slickline. A set of mixing tables for water versus CO_2 rates were made available to the pump operator in order to ensure that the correct rate ratio was transmitted during injection.

17.6   Measurements

A range of instruments were installed downhole and at the surface. Measurement types for formation fluid sampling and residual gas saturation test and timeline are listed in Table 17.3.

17.7   Surface data

The flow rates of fluids in and out of the well CRC-2 are the basic elements of the test and of subsequent simulations. The rates recorded were for CO_2 injection and water injection, and the production of water, at 1 minute intervals. At some stages there was much short term variation in the rates and therefore for modelling purposes, the rates have been approximated by a piecewise constant injection rate.

Figure 17.15 shows the water rates for injection and production throughout the test. Water injection could be carried out at 150–180 t/day, whereas average water

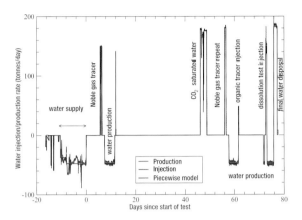

Figure 17.15: Water production rates (negative, blue lines) and water injection rates (positive, red lines) over the whole test. In black in shown the piecewise constant model for the injection rates. Time origin: 27 June 16:30, production rates use 4 hour average.

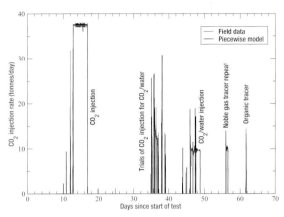

Figure 17.17: CO_2 injection rates over the whole test, compared to a piecewise constant model. Time origin of test is 27 June 16:40.

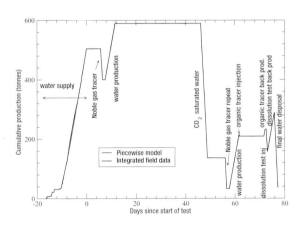

Figure 17.16: Cumulative water injection and production in tonnes, from the beginning of the gauge data. The time origin of the test is chosen as 27 June 16:30.

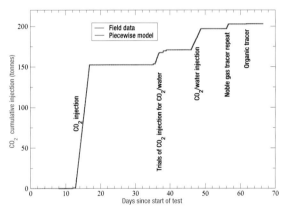

Figure 17.18: Cumulative CO_2 injection during the whole of the test. The piecewise constant simulation model is shown for comparison. The time origin of the test is chosen as 27 June 16:30.

production rates, limited by gas lift, were around 50 t/day. Therefore by design the production periods were longer than the corresponding injection periods. Figure 17.16 shows the cumulative sum of water production (here taken as positive) and water injection (here taken as negative) over the test cycle. At the end of the test there was some water remaining in the tanks.

The CO_2 injection rates for the whole test are shown in Figure 17.17, along with a piecewise constant model for simulation purposes. Figure 17.18 shows the cumulative amount of CO_2 injection over the test. The main CO_2

injection was relatively straightforward. Subsequent CO_2 injections had to be balanced with simultaneous injections of water to produce CO_2-saturated brine at bottom-hole conditions. The design rate for these sections was 7.5 t/day of CO_2, but this proved difficult to achieve without the CO_2 pump gassing up. After much experimentation with surface equipment, a way was found to get a relatively stable rate of around 10 t/day, balanced with a suitable water rate.

The overall mass ratio for this part of the test was 0.0578, whereas the target rate was 0.0548. Design calculations were made based on CO_2 solubility in brine at downhole conditions (the mixing of the CO_2 and brine occurred at the check valve depth for the instrument tubing at about 840 m depth). The aim was for the brine to be very slightly undersaturated at bottomhole conditions (to avoid introducing any further gas phase CO_2), and very slightly oversaturated far from the well. As discussed later, injection of brine lowered the downhole temperature to about 49°C, in other words as brine flowed from the well in the target formation, the temperature increased back to the formation temperature of 59°C, while the pressure decreased. Given the high reservoir permeability, the pressure change was small; in other words, the temperature effect on solubility was much more important.

After the injection of the CO_2-saturated brine, which drove the near-well environment to residual gas saturation, tests were then conducted to characterise this residual saturation. The average mass ratio of CO_2/brine in this

part of the test was 0.0548, which agreed with the target ratio of 0.0548.

The organic tracer test similarly involved the injection of CO_2-saturated brine. As this was the last test using CO_2, the amount remaining was not sufficient for the planned test, resulting in a number of operational issues. Therefore the CO_2/brine mass ratio was not calculated for this part of the test.

The last residual characterisation test was the dissolution test, using unsaturated brine, where the explicit aim was to dissolve the residual gas, and monitor the dissolved CO_2 in the back-produced water.

The overall amounts water injected and produced in each part of the test are summarised in Table 17.4, along with the average flow rates. In cases where there were

Table 17.4: Summary of water injection and production.

Time (Days Since Start of Injection)	Purpose	Amount of Water (t) (+ for Production, – for Injection)	Average Rate (t/day) (+ for Injection, – for Production)
–16.358–0.000	Water supply	+504.4	–30.8
5.744–6.507	Initial noble gas tracer injection	–103.9	+136.2
7.538–11.623	Water back production for noble gas tracer	+190.2	–46.6
46.123–48.773	CO_2 saturated brine injection	–453.7	+170.6
56.016–56.590	Repeat noble gas tracer injection	–104.1	+181.3
Time (Days Since Start of Injection)	**Purpose**	**Amount of Water (t) (+ for Production, – for Injection)**	**Average rate (t/day) (+ for Injection, – for Production)**
57.740–61.641	Back production for repeat noble gas tracer	+182.4	–46.8
61.670–61.797	Organic tracer injection	–6.18	+48.4
71.674–72.158	Back production for organic tracer test	+23.7	–48.9
72.605–73.106	Dissolution test: injection	–75.0	+149.7
73.139–75.852	Back production for dissolution test	+130.86	–48.2
75.881–77.252	Final water disposal	–250.6	+182.8

Table 17.5: DTPS operational tests conducted.

Date (Heating)	Comments	Acceptable for Thermal Analysis
27–29 June 2011	Baseline dataset: fully water saturated formation	Yes
16–18 July 2011	Post-CO_2 injection: large thermal background trends	No
24–26 July 2011	Post-CO_2 injection: moderate background thermal trend	Yes
17–19 August 2011	Post-H_2O sweep: large thermal background trends	No
2–4 September 2011	Post-H_2O sweep: moderate background thermal trend	Yes

many interruptions to injection, the latter average was not representative of the typical rates when flow was occurring. In the tables, the time origin of the test was taken as 16:30 on 27 June 2011.

17.8   Thermal logging

A DTPS instrument can provide estimates of formation thermal parameters (Freifeld et al. 2008) and can be used to infer hydrological properties and fluid phase saturation using changes in thermal properties as proxy measurements. The DTPS consists of a borehole length heater and a DTS. These DTSs are commercially-available temperature measurement devices, capable of high spatial resolution (approximately 1 m) with measurement precision as good as 0.01°C. Typical operational accuracy is closer to ± 0.1°C. DTS systems operate using standard telecommunications grade multimode fibre-optic sensors as the sensing medium.

For the Otway Stage 2B residual saturation test, the DTPS was deployed to monitor changes in CO_2 saturation. Table 17.5 shows when the heating phase of DTPS testing was conducted. To observe changes in CO_2 saturation, the heater created a thermal pulse using fixed power input (30 kW), and then, after a fixed duration of heating, the thermal decay was monitored. A snapshot of the temperature profile along the borehole was recorded every 15 minutes. For CRC-2 testing, after 2 days of heating, cooling trends were monitored for an additional 3 days. Since supercritical CO_2 has a thermal conductivity roughly 20 times less than that of formation fluid, the temperature was hotter in CO_2-rich regions of the reservoir and the thermal decay was slower, as the CO_2 acted to reduce the formation thermal diffusivity.

17.9   Noble gas tracer tests

The noble gases krypton (Kr) and xenon (Xe) were included in the design of a single-well injection-withdrawal test in the Paaratte Formation sandstone reservoir. Their ability to partition between the mobile water phase and immobile gas phase was used as a potential measure of residual CO_2 trapping (Zhang et al. 2011). To determine the residual gas saturation (S_{gr}) the noble gases were co-

injected with water before the injection of CO_2 when the pore space was fully water-saturated (reference test) and once again when the residual CO_2 field had been created (characterisation test). The push-pull partitioning of tracers in combination with the hydraulic and thermal tests reduces the estimation uncertainty in the S_{gr} field (Zhang et al. 2011). Furthermore, performing a reference test provided information on formation dispersivity, resulting in reduced parametric and geological uncertainty and helped constrain two-phase parameters used in the synthetic models (Zhang et al. 2011). Subsequent to the proposed simulated test sequence by Zhang et al. (2011) an organic tracer test and a dissolution test were added, with the latter test also involving the analysis of recovered Kr and Xe.

To collect suitable fluid samples for analysis, the U-tube assembly (Figure 17.19) enabled a pressurised water sample to be taken directly from the reservoir level and transferred to the surface into a 4.7 L high pressure stainless steel cylinder under reservoir conditions of around 14 MPa. Subsequently, a 150 mL pressurised water sample was taken for wet chemical analysis (pH, alkalinity, electrical conductivity and salinity) in the purpose-built field laboratory. The controlled pressure release of the storage cylinder resulted in the exsolution of dissolved gas, which flowed to vent. An instantaneous gas sample was collected in both an isotube and gas-bag once the initial pressure had dropped to around 4.8 kPa. The gases were analysed in the field laboratory for Kr, Xe, CO_2, O_2, N_2 and CH_4 by gas chromatography and mass spectrometry.

17.10   Testing phases

17.10.1   Phase 1 to initial Phase 2: background test

Periodic collection of U-tube-derived gases showed baseline concentration of gases for Kr and Xe were below GC detection limits (< 1 ppm) while CO_2 and CH_4 concentrations averaged 202 and 56 ppm, respectively. The balance of the gas was N_2 resulting from the initial 4.7 L N_2 in the SS storage cylinder at atmospheric pressure before filling with Paaratte water and its associated dissolved gases.

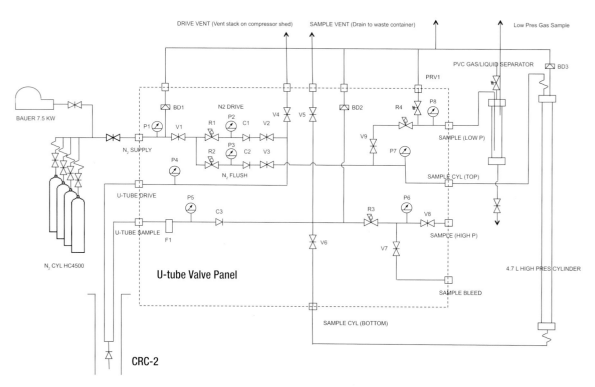

Figure 17.19: Schematic of U-tube surface assembly.

17.10.2   Phase 2: reference (baseline) test

At the start of the water injection phase, 2.46 kg of Kr (658 L @ STP) and 3.62 kg of Xe (618 L @ STP) were added at a constant flow rate over 120 minutes to the Paaratte Formation water, which was being reinjected from the surface storage tanks at an average rate of 136 t/day. Water injection continued for another 14 hours (push) with the injection of 104 t of water (Figure 17.8). After a further 24 hours, water production continued (pull) uninterrupted for another 3.4 days and 190 t of water. Round-the-clock U-tube-sampling produced splits of derived water and gas at 90 minute intervals for near real-time aqueous and gas analysis.

17.10.3   Phase 4: characterisation test

At the start of the water injection phase, 2.96 kg of Kr (791 L @ STP) and 4.30 kg of Xe (734 L @ STP) were co-injected at a constant flow rate over 100 minutes into the Paaratte Formation, with water being reinjected from the surface storage tanks at a flow rate of ~181 t/day. Carbon dioxide was injected through a ⅜″ capillary line and mixed downhole with the water at a position high above the Paaratte Formation to maintain near CO_2-saturated water (~1 mol CO_2/kg water) under reservoir conditions. Water and CO_2 injection continued for another 12 hours (push) with the injection of 104 t of water. After a further 24 hours, continuous water production (pull) continued for close to 4 days and 182 t of water. During this time, U-tube-derived water and gas samples were continuously sampled with a U-tube sampling period of ~90 minutes. Figure 17.20 shows the concentration of Kr and Xe plotted against the cumulative water production for the characterisation test. The balance of the gas was CO_2 and N_2 with an average CO_2/N_2 ratio of 5.65 ± 1.17 σ.

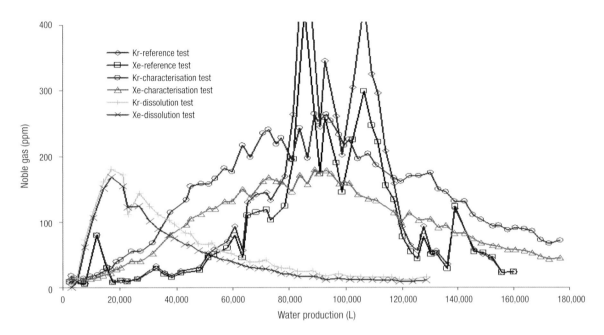

Figure 17.20: Breakthrough curve (BTC) with the concentration of Kr and Xe plotted against the cumulative water production for the reference test. The balance of the gas was N_2.

17.10.4 Phase 5: dissolution test

In this test, CO_2-free formation water was injected into the Paaratte Formation, resulting in dissolution of the residual CO_2 phase and remobilisation of the noble gas tracers back into the water phase. Subsequent water production and gas sampling produced the BTCs for Kr and Xe in the dissolution test in the following manner: (1) the field laboratory provided real-time analytical results on gas composition and enabled sound operational decisions to be made (e.g. when to stop water production and sampling); (2) noble gas tracers were successfully injected, sampled and analysed; (3) noble gas tracer BTCs were observed in both the reference and characterisation tests; (4) the noble gas tracer BTCs were quite different to the characterisation test, showing a much broader elution profile, in accordance with tracer partitioning between the water and residual CO_2 phases; and (5) in the dissolution test, noble gas tracers that had partitioned into the residual CO_2 phase during the characterisation test were remobilised as a result of dissolution of the residual CO_2 phase with the injection of CO_2-free formation water.

17.11 Downhole data (memory gauges)

For part of the test, pressure and temperature (P/T) memory gauges were run on battery and retrieved by slick line. These gauges were placed at 948 m and 1095 m. The pressures are shown in Figure 17.21 and the temperatures in Figure 17.22. This covered the period after the trial CO_2 injection (when there were numerous attempts to stabilise the CO_2 rate at a suitable level) and the whole of the injection of CO_2-saturated brine, and the subsequent heating/cooling test.

17.12 Downhole data (permanent gauges)

Four real-time pressure and temperature (P/T) gauges were placed around the perforated interval—two above and two below. The entire sequence of measurements from these gauges is shown in Figures 17.23 and 17.24. While the pressure measurements were primarily used as input to numerical history matching using reservoir simulation, there were parts to the test sequence that were

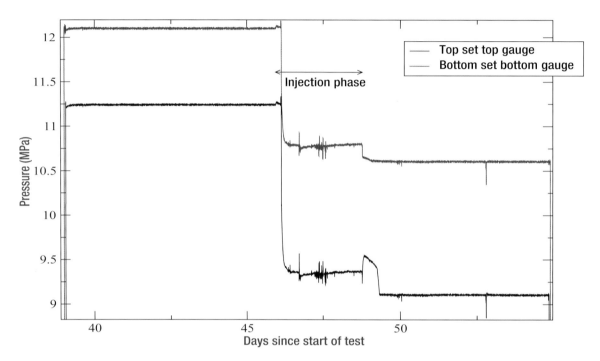

Figure 17.21: Memory gauge pressure during CO_2-saturated brine. Time origin of the test is 27 June 16:30. Gauge depths at 948 m and 1095 m.

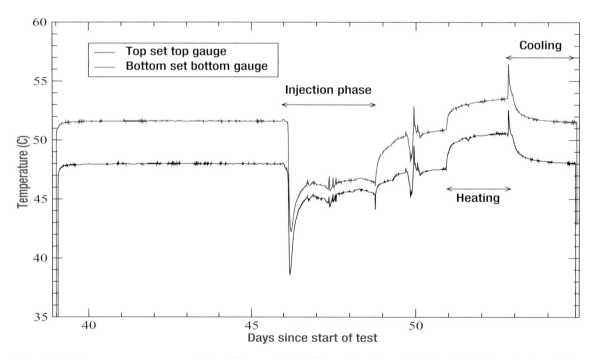

Figure 17.22: Memory gauge temperature during CO_2 saturated brine. Time origin of the test is 27 June 16:30. Gauge depths at 948 m and 1095 m.

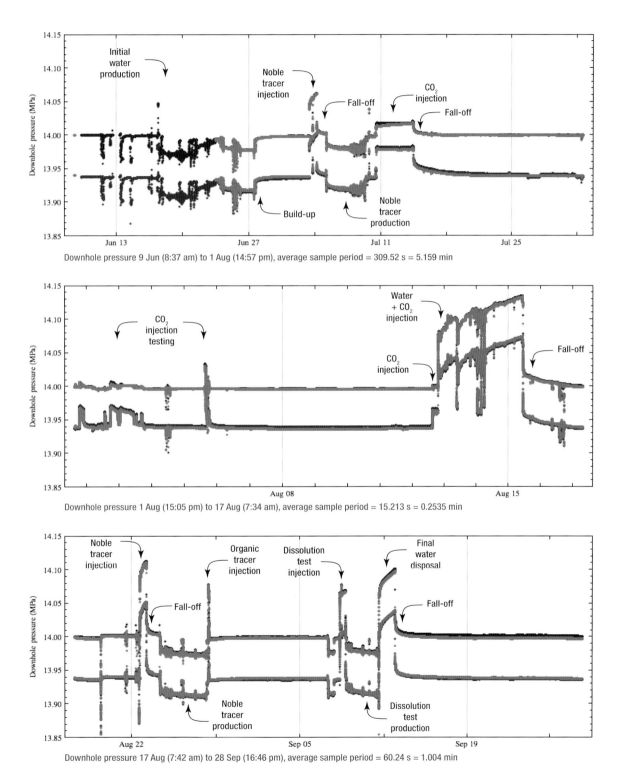

Figure 17.23: Downhole temperature measured by the four permanent gauges.

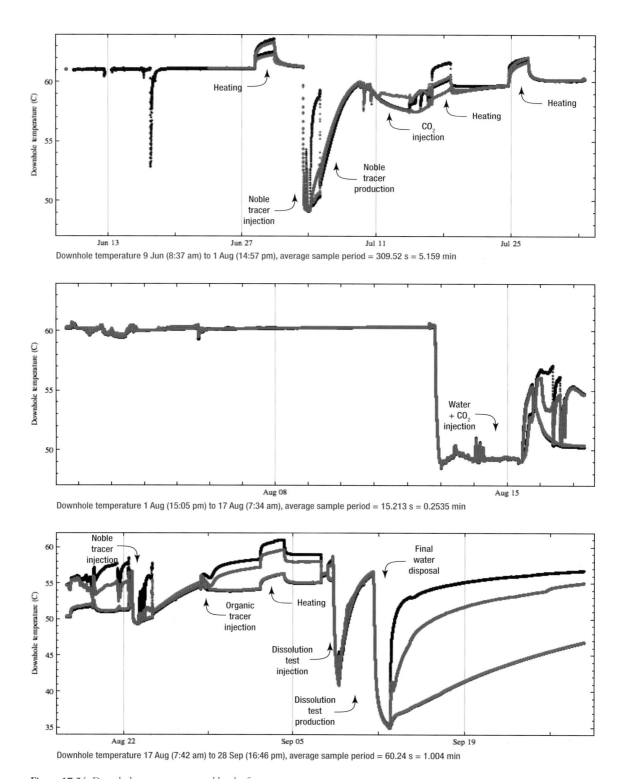

Downhole temperature 9 Jun (8:37 am) to 1 Aug (14:57 pm), average sample period = 309.52 s = 5.159 min

Downhole temperature 1 Aug (15:05 pm) to 17 Aug (7:34 am), average sample period = 15.213 s = 0.2535 min

Downhole temperature 17 Aug (7:42 am) to 28 Sep (16:46 pm), average sample period = 60.24 s = 1.004 min

Figure 17.24: Downhole pressure measured by the four permanent gauges.

amenable to standard well test analytical analysis. These pressure test data could be analysed using the methods described in Earlougher (1977) and Horne (1995). The first step in a simple semilog analysis was to produce a log-log plot of Δp vs t, shown in Figure 17.25. The next step was to determine the time at which the unit slope ended. To perform initial analysis, the period with water injection containing noble tracers was chosen, along with the subsequent pressure decline. With measurements at 5-minute intervals it was difficult to find the unit slope, but from the swabbing analysis, the orange line with unit slope in Figure 17.26 was consistent with 7 minutes. The semilog straight line could be expected to start 1.5 log cycles ahead of that point at around 3.7 hours. A Horner plot is shown in Figure 17.25 with the decline commencing Monday 4 July 2011 at 04:33:25. A 10.0 kPa/cycle slope for the solid line fit corresponds to:

$$k = 0.183234 \frac{q B_w \mu_w}{mh}$$

$$= 0.183234 \frac{(0.00174)(1)(0.48 \times 10^{-3})}{10.0 \times 10^3 \times 8} = 1.9 \text{ darcy}$$

This indicates the bulk permeability of the injection interval was about 2 darcies, consistent with the earlier estimate of 2.2 darcies when option C was chosen as the injection interval.

17.13   Pulsed neutron logging

Pulsed neutron logs were run in CRC-2 on 7 July, 16 July and 22 August in the interval 1200 m to 1455 m MD. The tool was run from a Schlumberger logging truck (Figure 17.27) through the tubing (tool diameter: 43 mm) below the packer to measure the saturation in the 7 m perforated interval. Three measurement points were established, one at 100% water saturation, the second at 100% CO_2 saturation, and last at residual gas saturation. Prior to completion, a baseline log was recorded in the cased borehole at 100% water saturation. This helped to analyse any variations caused by the completion. The tool's depth of investigation was 0.25 m and gave a detailed fluid saturation profile along the borehole, with 0.38 m vertical resolution. Output from the three logs is summarised in Figure 17.28. In this figure, the blue shaded region is CO_2 saturation after CO_2 injection; the green shaded region is after water injection.

Three pulsed neutron logging runs were run in CRC-2 during this test. The first pulsed neutron logging was run on 7 July 2011 to characterise the state of the water-filled formation. As there was no pressure at the wellhead, this job was conducted using a standard pack-off at surface. After the pure CO_2 was injected, a second pulsed neutron logging run was conducted to log the CO_2 saturated formation. A third pulsed neutron logging run was completed on 22 August 2011 to record the formation at residual state.

To measure formation water saturation the tool was run in sigma mode, thus outputting formation neutron capture cross section, sigma (SIGM) and thermal decay porosity (TPHI), a measure obtained from the ratio of the "near to far" detector capture count rates. The interpretation of the data was complicated by the fact that the TPHI computation from pulsed neutron logging acquisition is well defined for conditions when fluid in the wellbore has a hydrogen index (HI) of approximately 1 but is not well defined for conditions when the fluid does not have a HI of 1. Efforts were made to ensure the presence of formation water around the tool between the dowhole packers, but a small "dead zone" in the annulus just below the top packer created difficulties and the post-CO_2 injection log was completed with CO_2 in the wellbore, which has a HI $\approx$ 0. Consequently a shift was made to the TPHI log when CO_2 was in the borehole to account for the changed pulsed neutron logging response in the near-tool region.

Due to the low salinity of the formation water (below 800 ppm), the TPHI output was used in preference to the SIGM output (sigma and TPHI can both be used to invert for CO_2 saturation). Therefore TPHI was used for CO_2 saturation computation; both outputs are discussed below.

The results using the TPHI output are shown in Figure 17.28. The TPHI output responded to the hydrogen index in the formation. The log shows a small change in S_{gr} within the perforation zone. It is noted that the formation was not 100% saturated after injection of pure CO_2 (blue curve) when driven to full saturation. It was observed that formation water re-entered the well soon after the injection and this could be the reason why the saturation values were lower for the CO_2-saturated case. Residual state S_{gr} had a value of approximately 18% in the bottom half of the perforated interval and 23% (average) in the top half.

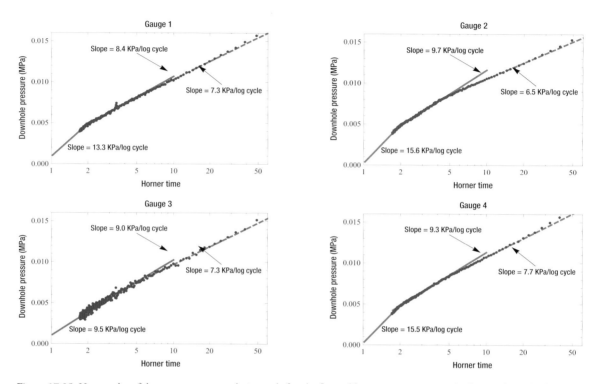

Figure 17.25: Horner plot of the pressure response during and after the first noble tracer injection period. Plots are for each of the four permanent gauges. The slope can be used to calculate permeability.

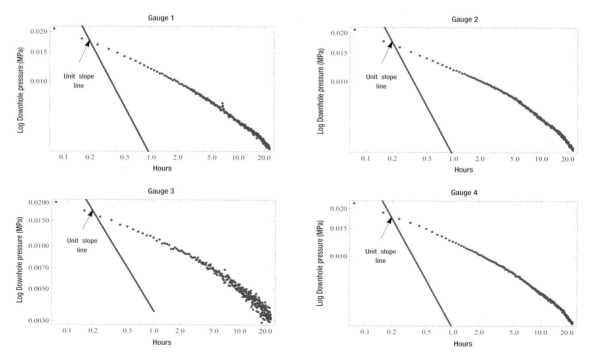

Figure 17.26: Log-log plot of Δp vs t after the first noble tracer injection for each of the four permanent gauges (see Horne 1995, p. 25), confirming that the analysis is beyond wellbore storage effects.

The results using SIGM output are shown in Figure 17.29. Residual S_{gr} is 20% on average in the perforated interval. As indicated earlier, due to the low salinity of the formation water, TPHI was preferred as a saturation measurement.

Figure 17.27: (a) Schlumberger logging truck, and (b) the reservoir saturation tool (RST).

17.14   The organic tracer test

Three reactive ester tracers (triacetin, tripropionin and propylene glycol diacetate) were identified and assessed in preliminary laboratory tests as potentially suitable for measuring residual saturation. This approach was pursued as there was concern that limited differences in the breakthrough profiles of the inert tracers would be observed (based on a general understanding of the chromatographic behaviour of krypton and xenon). This was further exacerbated by the difficulties in obtaining highly accurate and reproducible results from field experiments. The modelling of inert tracer behaviour is particularly sensitive to drift, which may add further uncertainties to interpretation. The organic reactive tracers were designed to complement the other tests and the theory behind the reactive ester tracers is based on earlier work by Deans et al. (1971) who designed a series of experiments using ethyl acetate to determine, from a single well test, the residual oil saturation of an oil field. There are obvious similarities between their approach and the approach used for the single well test at CRC-2.

All three ester tracers (30 kg total) were food grade liquids and were injected undiluted. The ester tracers were allowed to partially hydrolyse in the subsurface for 10 days. During water production 621 water samples (approx. 50–80 mL each) were collected, immediately frozen on site using dry ice and then shipped to the National Measurements Institute in Perth for analysis using standard chromatographic techniques. The initial results of the concentration breakthrough profiles for the parent tracers are given in Figure 17.30 and hydrolysis products of the parent tracers are given in Figure 17.31.

The amounts of tracers in the samples were sufficiently high to be analysed and in some cases required dilution. The results also confirmed the earlier calculations of kinetic behaviour where a 10-day soak period was preferred for an ideal mix of parent and daughter compounds as proposed by Myers et al. (2012). Further samples were chosen for analysis in order to clarify the early parts of the curves and better understand the arrival of the tracers at the start of water production. The results were compared with theoretical models developed by Myers et al. (2012). Using the data, the best suite of compounds were used to calculating the residual CO_2 saturation for that part of the overall Stage 2B test.

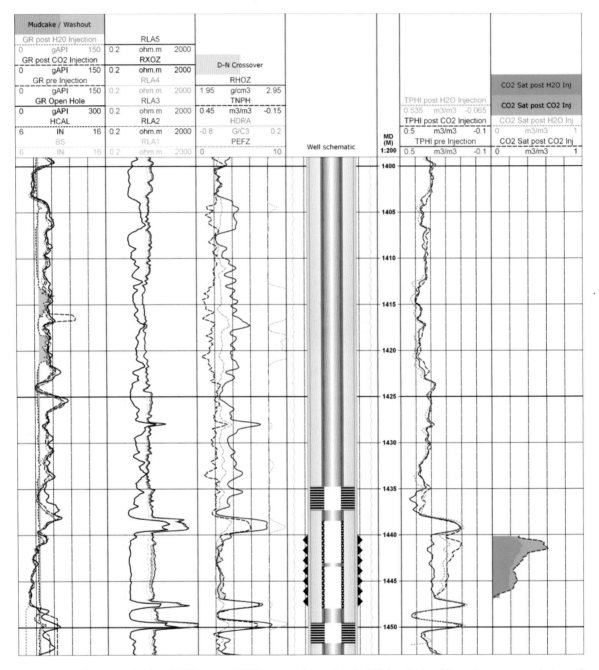

Figure 17.28: Comparing the three RST logs using TPHI output indicates that the CO_2 has displaced formation water over the interval from 1440 m to 1447 m. The blue shaded region is CO_2 saturation after CO_2 injection; the green shaded region is after water injection.

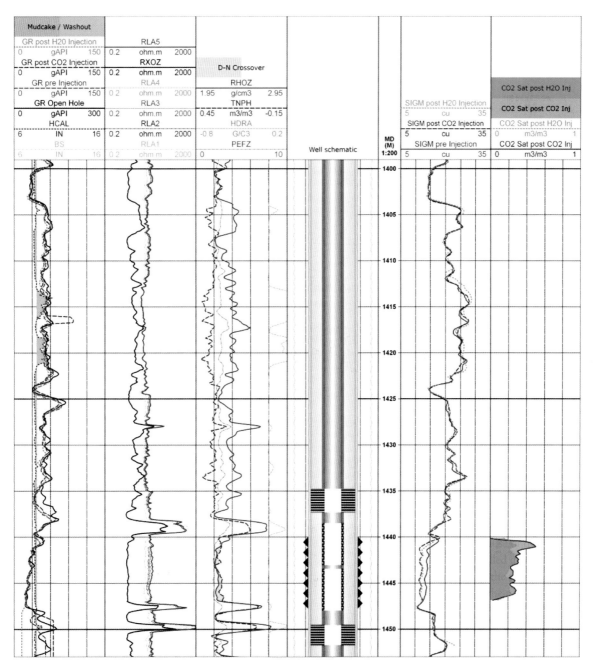

Figure 17.29: Comparing the three RST logs using SIGM output also indicates that the CO_2 has displaced formation water over the interval from 1440 m to 1447 m. The blue shaded region is CO_2 saturation after CO_2 injection; the green shaded region is after water injection.

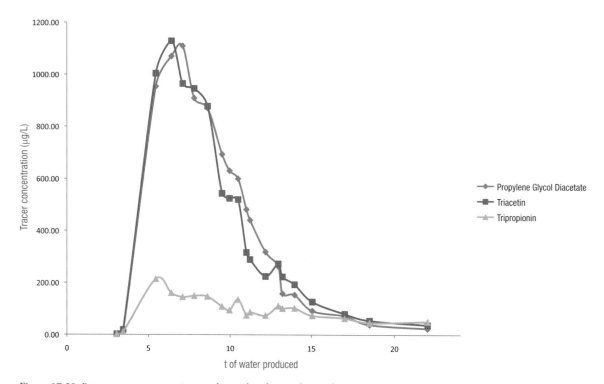

Figure 17.30: Parent tracer concentrations in the produced water during the organic tracer test.

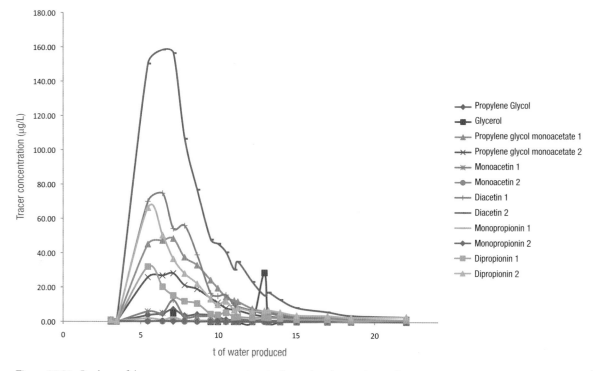

Figure 17.31: Products of the parent tracer concentrations in the produced water during the organic tracer test.

17.15   The dissolution test

The dissolution test has been described by Haese et al. (2013). Following a period of injecting water without CO_2, the dissolution test measured the dissolved CO_2 during production. Using the method described in the original Bragg et al. (1976) paper, it was established that the measurement metric (XCO_2aq) had sensitivity primarily to one parameter: residual gas saturation, S_{gr}. Other parameters have only minor effect on the measured metric.

For the test, 75 t of water without CO_2 was injected over 12 hours. Methanol was added to the injection water at a concentration of 0.1 vol % as a marker to estimate the aquifer drift during subsequent water production.

Formation water and its dissolved gas were sampled using the U-tube system, with (1) fluid samples taken under reservoir pressure (~14.0 MPa) and (2) gas samples derived from the exsolved gas at reduced pressure (6.2 MPa). U-tube samples were taken continuously, approximately every 1.5 hours during the 62.5 hours of the water production period leading to the collection of 40 samples. Water production commenced immediately after the injection of 75 t of water and production was continuously at a rate of 49 t per day.

Fluid samples were collected in two 150 mL stainless steel cylinders at reservoir pressure. One sample cylinder was used for the analysis of the dissolved inorganic carbon (DIC, DIC = ΣCO_2, HCO_3^-, CO_3^{2-}) concentration and its $\delta^{13}C$ and $\delta^{18}O$ isotopic composition. The other sample cylinder was used for general water properties (temperature, pH, salinity, alkalinity) analysed on site and for subsamples taken for the analysis of cations, anions and methanol at the Geoscience Australia laboratories.

The cylinder filled with formation water for DIC and isotopic analysis was connected to another cylinder filled with 100 mL of 0.5 M NaOH and valves were opened to allow the mixing of the two fluids. The aim was to make the sample fluid alkaline, so that all dissolved CO_2 would be converted into CO_3^{2-}, which would not escape from the fluid (as CO_2 gas) when pressure was reduced. The fluid from inside the two cylinders was collected over 15 to 20 minutes. DIC was analysed on site using

the AS-C3 model by Apollo SciTech, which included an infrared-based CO_2 detector (LiCor 7000). Carbonate was precipitated from solution by adding excess of SrCl leading to the formation of $SrCO_3$, which was filtered and dried in preparation for the carbon and oxygen isotopic analysis by isotope ratio mass spectrometry.

The second sample cylinder was opened and immediately depressurised. Temperature, pH and total dissolved solid concentration were measured by standard electrodes. Total alkalinity was determined by titration using 1 mL of sample, 0.02 M HCl as titration solution and a colour indicator. From every alternate cylinder, 4, 30 and 30 mL of fluid was subsampled for methanol, anion and cation analysis, respectively. The cation sample was acidified using 1 or 2 drops of concentrated HNO_3. Methanol was analysed by gas chromatography, anions by ion chromatography and cations were analysed by inductively coupled plasma optical emission spectrometry. Gas samples were collected in Isotubes® and gas bags and analysed on site for CO_2, N_2, Kr and Xe on the gas chromatograph and quadrapole mass spectrometer (see Section 17.9).

The DIC concentrations show a clear pattern, with a steep increase in concentration from the start of the experiment to 26.9 t of water production followed by relatively high concentrations to the end of the experiment (Figure 17.32). Concentrations ranged from 35 to 456 mmol/L DIC with significant variation from one sample to another suggesting variable loss of CO_2 during sampling despite the attempt to convert all CO_2 into CO_3^{2-}. The loss of CO_2 was also suggested by the difference between observed and predicted maximum CO_2 concentration. The CO_2 saturation concentration under reservoir conditions was close to 1000 mmol/L and this concentration was expected once the plateau of high concentrations was reached. However, maximum measured concentrations of only ~450 mmol/L were achieved. It appears the trend in DIC concentration follows the predicted pattern, but measured DIC concentrations underestimated the true value, particularly in the high concentration range.

Total alkalinity is a measure for the buffering of excess positive charge resulting from the cation–anion balance in solution. Concentrations of HCO_3^- (bicarbonate) and CO_3^{2-} (carbonate) usually determine total alkalinity and

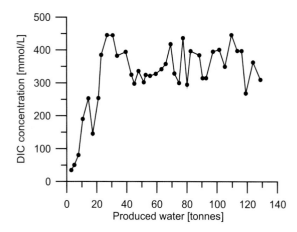

Figure 17.32: Measured dissolved inorganic carbon concentrations.

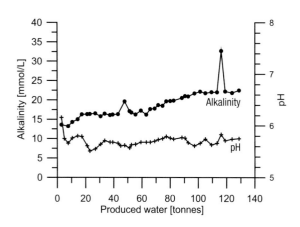

Figure 17.33: Measured pH (crosses) and total alkalinity concentrations (dots).

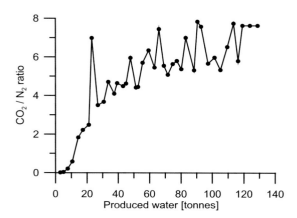

Figure 17.34: Measured CO_2/N_2 ratios after CO_2 exsolved from the fluid sample at 62 bars.

contribute to the DIC concentration. As the pH of the formation water of the Paaratte Formation was consistently low with a value of approximately 5.6 (Figure 17.33), CO_3^{2-} would not stabilise. Consequently, total alkalinity represented the HCO_3^- concentration. Total alkalinity concentration showed an increase from 13 to 22 mmol/L during the course of water production (Figure 17.33). Such low concentrations demonstrated that HCO_3^- only contributed significantly to the DIC concentration at the beginning of the experiment.

Exsolved CO_2 was collected at a pressure of 6.2 MPa in a 4.7 L stainless steel cylinder, which was initially filled with pure N_2. The gas mixture was sampled in an Isotube and the CO_2/N_2 ratio determined by gas chromatography. Similar to the DIC concentration curve, a steep increase in the CO_2/N_2 ratio was evident from the beginning of the experiment to about 26.9 t of water production followed by variable, but overall high values at the end of the experiment (Figure 17.34).

17.16   Conclusions

The Otway Stage 2B residual saturation and dissolution test is believed to be the first field test of this nature to be undertaken. Residual trapping can immobilise carbon dioxide within a reservoir, preventing buoyant migration where structural traps are absent or if seal integrity is lost. Even for storage sites that are based on secure structural trapping, potentially a large component of the carbon dioxide can be retained by residual trapping. Hence estimation of the amount of residual trapping is a very important consideration in storage site evaluation. With this in mind, the goal of CO2CRC Otway Stage 2B was to measure large-scale residual trapping of CO_2 in an actual field project using five different methods, compare the methods and make recommendations. It was realised during design that substantial information would be collected on dissolution trapping also allowing for analysis of this mechanism.

Measurement of residual trapping, S_{gr}, was undertaken by several independent techniques in the test (order indicates priority):

1. history matching injection and production pressures and flow rates by developing a two-phase relative permeability test in a well, based on Zhang et al. (2011)

2. measure hydrogen index using a pulsed neutron log

3. measure the thermal conductivity of fluids in the formation by thermal logging

4. reactive ester tracer partitioning according to residual saturation, based on the Tomich et al. (1973) method for residual oil

5. a dissolution test, based on the Bragg et al. (1976) method for residual gas (methane).

Following a lengthy design period, the field test programme for Otway Stage 2B commenced on 17 June 2011 and finished on 12 September 2011. All the planned components of the field test were completed including the five methods for measuring residual trapping. Extensive high quality data were obtained throughout the programme that allows detailed analysis within each method. The responses at each stage showed that the injected CO_2 was driven to residual saturation and was detected by each of the five measurement methods.

The conclusions can be summarised as follows:

1. Excellent quality downhole pressure data were acquired throughout the field programme from the permanent gauges. These gauges are more accurate than the memory gauges used in the Stage 1 injection in the Waarre Formation. The high permeability of the injection interval did not cause problems with getting a sufficient pressure response to give a good signal-to-noise ratio.

2. Excellent downhole temperature data were also acquired. The instantaneous readout of downhole temperatures proved extremely useful in diagnosing operational issues at several stages. Elements of the temperature records will take some effort to interpret due to persistent heat responses.

3. All three pulsed neutron logs were run as intended. Two of the logs were run in a standard water environment, the other log was run when CO_2 was in the tubing and hence was run in a CO_2 environment. The latter log will require some additional considerations for a complete interpretation. Current interpretation has residual CO_2 saturation around 0.18 in the lower half of the perforated interval and around 0.23 (average) in the upper half.

4. Fluid sampling was completed consistent with the test plan. The noble and organic tracers were added to the injection stream as planned.

5. The reactive ester tracer test has not previously been conducted in the field. The concentrations of added ester tracers were more than sufficient to be detected by standard chromatographic methods. The results to date show the partial breakdown of the three parent compounds (triacetin, tripropionin and propylene glycol diacetate). The 10-day soak period was a good fit with the theoretical breakdown kinetics. Further work will be done to fill in more data points and begin the interpretive stage of the work.

6. Sampling for the dissolution test including the added methanol, was performed as intended.

7. Despite initial testing, it proved difficult to control the low CO_2 flow rates that were required when injecting water with dissolved CO_2. Modifications led to delays and, although the problem was largely controlled, difficulties did persist during the shorter injection intervals.

8. Experience with the Stage 2B testing shows that it would be helpful in future work to have a more detailed well model that takes into account two-phase flow movement within the casing. "Off-the-shelf" software may have problems correctly implementing CO_2 properties in the vicinity of the critical point, and a code comparison study of wellbore flow codes would be useful.

Other results from the Otway Stage 2B test sequence will be progressively documented in separate publications.

17.17   References

Bennion, B. and Bachu, S. 2005. Relative permeability characteristics for supercritical CO_2 displacing water in a variety of potential sequestration zones in the Western Canada Sedimentary Basin. SPE 95547. SPE Annual Technical Conference and Exhibition held in Dallas, 9–12 October.

Bragg, J.R., Shallenberger, L.K. and Deans, H.A. 1976. In-situ determination of residual gas saturation by injection and production of brine. SPE 6047. 51st Annual Fall Technical Conference of the Society of Petroleum Engineers, New Orleans, 3–6 October.

Deans, H.A. 1971. Method of determining fluid saturations in reservoirs, U.S. Patent #3,623,842.

Earlougher, R.J. 1977. Advances in well test analysis. SPE, Richardson, Texas.

Freifeld, B.M., Finsterle, S., Onstott, T.C., Toole, P. and Pratt, L.M. 2008. Ground surface temperature reconstructions: Using in situ estimates for thermal conductivity acquired with a fiber-optic distributed thermal perturbation sensor. Geophysical Research Letters 35(14) article number L14309.

Haese, R.R., LaForce, T., Boreham, C., Ennis-King, J., Freifeld, B.M., Paterson, L. and Schacht, U. 2013. Determining residual CO_2 saturation through a dissolution test—results from the CO2CRC Otway Project. Energy Procedia 37, 5379–5386.

Horne, R.N. 1995. Modern well test analysis. Petroway, Palo Alto, California, 2nd edition.

IPCC 2005. IPCC Carbon dioxide capture and storage: Special report of the Intergovernmental Panel on Climate Change. Cambridge University Press, London.

Myers, M., Stalker, S., Ross, A., Dyt, C. and Ho, K-B. 2012. Method for the determination of residual carbon dioxide saturation using reactive ester tracers. Applied Geochemistry 27, 2148–2156.

Paterson, L., Boreham, C., Bunch, M., Dance, T., Ennis-King, J., Freifeld, B., Haese, R., Jenkins, C., LaForce, T., Raab, M., Singh, R. and Stalker, L. 2013. Overview of the CO2CRC Otway residual saturation and dissolution test. Energy Procedia 37, 6140–6148.

Tomich, J.F., Dalton Jr., R.L., Deans, H.A. and Shallenberger, L.K. 1973. Single-well tracer method to measure residual oil saturation. Journal of Petroleum Technology 25(2), 211–218.

Zhang, Y., Freifeld, B., Finsterle, S., Leahy, M., Ennis-King, J., Paterson, L. and Dance, T. 2011. Single-well experimental design for studying residual trapping of supercritical carbon dioxide. International Journal of Greenhouse Gas Control 5, 88–98.

18

Peter Cook

18. WHAT WAS LEARNED FROM THE OTWAY PROJECT?

18.1 Introduction

Michael et al. (2010) reviewed some of the lessons learned from saline aquifer storage projects; Stalker et al. (2012) have summarised monitoring methodologies for a range of projects; Cook et al. (2013) have summarised the current state of small-scale CO_2 injection projects of less than 100,000 t around the world and some of the lessons learned from them. Cook et al. (op cit) found that the Otway Project was one of the largest of the projects falling into the class of "less than 100,000 t" (Figure 18.1). Otway Stage 1 is in the top three in terms of amount of CO_2 injected, not that this necessarily is an indicator of greater or lesser importance, although it could reasonably be argued that the larger the injection the more commercially relevant it is likely to be. It is also relevant to note that, in terms of total cost, it is also in the top five. In other words by comparison with other pilot or demonstration projects, Otway 1 was expensive but it was large—a not unexpected correlation. More importantly perhaps, it was comprehensive by comparison with many other projects. Did it deliver value for money? The reader can make a judgement perhaps based in part on this book, but the Board, the Project team and the key stakeholders are firmly of a view that it did.

Does this then translate into "best practice"? A number of reports have been published which provide an indication of "best practice" and these have been reviewed by Soroka (2011) and Cook et al. (2013). The best practice reports range from very topic-specific manuals, to those covering the entire CCS chain including transport. They vary in the level of detail provided, with some offering overviews of concepts, others offering highly detailed discussions and some providing the technical operations, calculations and geological parameters that went into projects. Valuable examples of comprehensive publications on important small-scale storage projects which contain elements of "best practice" include those compiled for projects undertaken through the US Regional Partnerships Program (Lytynski et al. 2013), a number of Canadian projects (Hitchon 2012, 2013), the Nagaoka Project (Mito et al. 2013), the Ketzin Project (Wickstrom et al. 2009; Würdeman et al. 2010), the Altmark Project (Kuhn and Munch 2013) and for a range of projects (Gluyas and Mathias 2013).

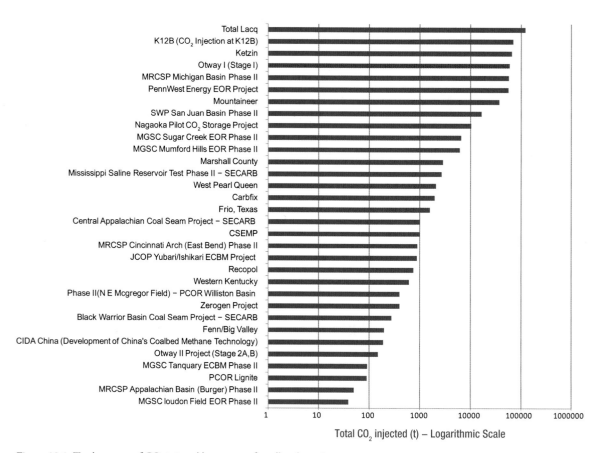

Figure 18.1: Total amount of CO_2 injected by a range of small-scale projects.

However, Cook et al. 2013 make the point that "There is no such thing as a 'perfect project' that can be used as a template for establishing the technical or governance parameters for a future project and there is no recipe that ensures a successful project".

This book on the CO2CRC Otway Project does not claim to be a best practice manual that is applicable to every CCS project, large or small. Nor was it always possible to adopt "best practice" for the Otway Project. In some areas, particularly those relating to health, safety and the environment (HSE), the Project always endeavoured to attain best practice and its outstanding record in HSE is well documented in Chapter 1. Similarly, to the extent that there is "best practice" in communications, the Project attained this, if the measure is that the Project was able to go ahead with no major hold-ups arising from community concerns, as documented in Chapter 2 and as discussed by Ashworth et al. (2010). The procedures

relating to regulation, developed through close cooperation between the Project and the State Government and its regulators, were exemplary in that they worked well, they enabled the Project to proceed in a very satisfactory manner and they were subsequently used to guide more formal regulation for large-scale CCS projects (Chapter 3). Similarly, project management "best practice" was followed, to the extent that project managers were able to identify "best practice".

In the case of science and technology, it was often less clear what might constitute "best practice". In some instances, e.g. in the preferred choice of technology for transport, it was decided not to compromise, and to use stainless steel piping, because the water content of the CO_2 could have constituted a potential hazard to the long-term integrity of a pipeline. However, in other cases, while the underlying principle was *always* that HSE would not be compromised, it was not uncommon to compromise on non-HSE matters, most commonly for financial reasons. A good example of this is documented in Chapter 4 where the initial idea was to separate methane from the Buttress gas. Obviously it would have been preferable to inject pure CO_2 in that this would have removed any atypical behaviour due to the presence of the methane impurity. However, the high cost made this impossible and therefore it was decided to inject a CO_2-methane mixture. Was this best practice? Perhaps not, to the extent that most CCS projects will not store comparable gas mixtures. Did this matter, in terms of the methodologies adopted or the scientific results obtained? No, and it saved the Project several million dollars! A second example of compromise is offered by the experience with the Naylor monitoring well. There were two compromises involved here. First, there was the compromise of only having one monitoring well, when clearly two or more would have been far preferable in terms of being able to better document the sub surface behaviour of the injected CO_2. After all, the likely trajectory of the CO_2 was known with a degree of confidence, but not with certainty. Given the heterogeneous nature of the reservoir it would not have been impossible for the CO_2 to never reach the single monitoring well. Fortunately the CO_2 did arrive at the monitoring well within the anticipated time and therefore this compromise worked and saved the cost (several million dollars) of an extra well. A less successful compromise perhaps, in terms of the technology, was to use a small diameter monobore well (again, Naylor-1) as the monitoring well, in that it proved very difficult to place all of the downhole instrumentation into such a narrow well. In the event, not all of the instrumentation worked and perhaps it would have been better to have gone to the expense of drilling a new larger diameter well. Bearing all this in mind, the claim is not made that this book is a "best practice manual". Nonetheless we hope it provides a comprehensive guide to what worked and what did not work. With this in mind, it is useful to briefly review some of those lessons.

18.2 Organising a project

The organisational and governance structures of the entities relating to the Otway Project were outlined in Chapter 1. Throughout the period 2003–09, there were three entities:

> an unincorporated CO2CRC Joint Venture, which provided the "umbrella" for initiating the Project, obtaining the funding and developing the science

> CO2CRC Management Pty Ltd (CMPL), a company limited by shares, which acted as the Centre Agent under a Management Agreement

> CO2CRC Technologies Pty Ltd (CO2TECH), a company limited by shares, responsible for protection, management and commercialisation of Centre IP.

In 2005 a fourth entity was formed:

> CO2CRC Pilot Project Ltd, a special purpose company Limited by Guarentee, responsible for undertaking the operational aspects of the Otway Project.

In 2010, the structure was revised again, with the termination of the CO2CRC JV and CMPL. IP and commercial issues continued to be handled by CO2TECH, the company limited by shares. The primary organisation for all aspects (operational, financial and scientific) now became the company limited by guarantee (CLG), CO2CRC Ltd (now renamed from CPPL) with a revised mandate and Board structure.

Why was this course of action followed and could there have a more satisfactory structure established right from the start? The original decision to have an unincorporated joint venture (UJV) in 2003 was a consequence of the transition from the Petroleum CRC (APCRC), which was also an UJV, to the new CO2CRC. The key stakeholders were comfortable at that time with the UJV. Would it have been better to have CO2CRC Ltd established right from the start? Yes it would, but it was necessary for this to become evident to the stakeholders. Therefore the initial structure was a practical compromise. The intermediate step of establishing the "Coalition of the Willing" and then CPPL enabled the Members to become comfortable with a CLG and with their responsibility for operational aspects of the Otway Project. CPPL played a crucial role in enabling the Project proposal to move ahead. Were there alternatives? The option of a company limited by shares was considered but was a less suitable option on a number of grounds.

In the absence of a commercial company willing to undertake the Otway Project on behalf of the CO2CRC JV, the CPPL arrangement, and the subsequent CO2CRC Ltd arrangement was the only realistic option. More importantly, it worked very well and is probably an arrangement that could be more generally applied to similar projects. The advantages of the CPPL/CO2CRC arrangement compared to a commercial operator taking over everything, was that science was the primary driver, with no commercial considerations (other than the obvious one of seeking to do things in a cost effective way) and where for example showing visitors around was not seen as interrupting a commercial operation, but as one of the important activities of the Project. Similarly the science was not seen as interrupting a commercial driver such as getting more oil out of the ground; science was the whole raison d'etre of the Project. This had major advantages in terms of interacting with the community and with the regulators.

At the same time, because the Board of CPPL was composed of industry Directors with a high level of consciousness regarding HSE issues and management of risk, industry standards were adopted throughout the operations of the Otway Project. Similarly, industry processes were followed in terms of project management and approvals. Therefore in summary, the board arrangements and the overview of the Project by CPPL/CO2CRC Ltd worked very well and is a model which other projects of similar scope may wish to consider.

Was it necessary to continue to have a commercial arm (CO2TECH) alongside CO2CRC Ltd or could this commercial entity have been subsumed into CO2CRC Ltd? Overall there were seen to be advantages in retaining the two companies, in that it provided a clear separation of research activities from commercial activities, but at the same time provided a mechanism for capturing monetary value arising from the Otway Project, should there be a wish to do so. However such an arrangement did result in additional management costs and was not an absolute necessity, whereas CO2CRC Ltd, as a CLG, was absolutely essential to the safe and effective conduct of the Project.

18.3　Managing a project

As pointed out in Chapter 1, the template commonly used by the oil industry for the project development, was very directly applicable to the management of the Otway Project. However the template needed to be underpinned by a number of other measures, including:

> Ensuring that the objectives of the project were agreed and clearly defined in a manner that addressed the expectations of all key stakeholders, while recognising that as a research project, not all objectives would necessarily be achieved. It was also important to prioritise objectives as well as retain the necessary degree of flexibility, so that when there were unexpected outcomes, objectives could be modified or re prioritised in a sensible manner.

> As the Project was undertaken by a consortium (and most projects are), it was necessary to ensure that there was full alignment between all participants on issues such as budgets, funding, responsibilities, liabilities, governance, confidentiality, communications, operations and management and board responsibilities.

> An agreed and well defined work flow with clear decision points was put in place to enable the Project to be logically taken from identification of the opportunity through to the operate stage and in the future, to the abandonment stage.

> To the extent possible, the project team ensured that there was alignment of funding, budgets, and the expenditure profile over the life of the Project, though given the research environment, this was not always possible.

> An agreement was in place to ensure a clear understanding of the criteria to be applied for approving the commencement of CO_2 injection, the suspension of injection and the closure of the Project, including ensuring that there was adequate funding available to meet abandonment (including remediation) requirements.

> Key performance indicators were agreed upon with the regulators, so that the objectives of monitoring were clear and the Project could confidently move forward in the knowledge that the "ground rules" would not change in the middle of the Project.

> HSE and other protocols were in place and steps taken to ensure that they were enforced with all staff and all contractors and subcontractors. Industry participants from the oil and gas sector involved in CO2CRC were well experienced in this area and therefore were able to provide valuable advice.

> Risk assessment and risk management were embedded within all aspects of the Project including research, monitoring and operations.

Along with all of these was the need to have systems in place for effective resource management, whether for staffing, contracting, finances, equipment, travel, auditing etcetera. In the case of the Otway Project, it was necessary to ensure that these systems were compatible with the requirements of CO2CRC and its funders.

18.4   Funding a project

Every project will encounter its own unique set of funding issues and the Otway Project was no exception to this. Chapter 1 outlines some of the funding challenges, particularly the uncertainties of funding major operational activities such as drilling a well. This undoubtedly resulted in some activities being more costly than they otherwise would have been. However, the one very important thing which CO2CRC had and which made the Project possible in the first instance was the certainty of underpinning funding (though not major capital funding), to cover the entire period 2003–10. The global review of small-scale injection projects by Cook et al. (2013), clearly showed an average time span of 3 or 4 years from development of the first idea of a project to the start of CO_2 injection. This suggests an absolute minimum of 5 or 6 years of funding certainty is necessary to successfully complete a storage project.

The Otway Project showed a tendency on the part of the Project proponents within the Centre to be overly optimistic about costs. At the same time, had they known what the final costs would be, it is not certain they would have started down the project path in the first place and even more doubtful that the necessary funds would have been forthcoming! Therefore an incremental approach to funding was perhaps inevitable, but the clear message from the Otway Project experience is to recognise, right from the start, the need for an adequate contingency of perhaps as much as 30–40% on initial cost estimates.

The Otway experience showed that there was usually only limited scope for compromising on engineering and related works—and of course zero scope for compromise where HSE issues were involved. Therefore the financial compromises had to originate, for the most part, within the science programme. At the same time it was evident that, even if no science whatsoever was undertaken at the Otway site, the costs for maintenance, security, fees etc. were of the order of one million dollars per annum. Depending on the circumstances and the regulatory regime, there was also a need to make adequate provision for abandonment and remediation of the site. In the case of Otway it is estimated that this will cost approximately $3.5 million. Again this is a cost that has to be borne and does not necessarily contribute directly to the science, but is nonetheless an integral part of the programme.

The total cost of small-scale storage projects is extremely variable, as shown by Cook et al. (2013), with the variability depending not only on the particular features of the site and the scope of the project, but also on the manner in which researchers are charged out to the project and particularly the cost of undertaking operations such as drilling a well. The total cost of undertaking Otway 1 was $41 million. The cost of undertaking a comparable project in an active oil exploration area in Texas would probably be of the order of $20–30 million or even less. Otway Projects 2A and 2B cost a total of approximately $30 million, but again the cost of undertaking comparable projects in Texas would probably be less than $20 million. Costs also show great variability depending on whether a component of the cost relates to the purchase of CO_2. In the case of Otway it cost several million dollars to complete the Buttress well, lay a pipeline and put in the gas processing facility. However, this also secured abundant CO_2 for Otway 1 and beyond, and compared to buying commercial CO_2 saved several million dollars.

18.5 Project communications and collaboration

There were a number of aspects of the communications programme that contributed to the success of the Otway Project, and from which lessons can be learned. The decision to inform the local community of the Project at a very early stage (before there was any certainty that a Project would go ahead), and to involve senior management in that first step, was very important. It helped to set the scene for the open non-hierarchical way in which things would be done throughout the Project. At the same time it was important to have a clear structure for the communication activities, as outlined in Chapter 2.

Throughout the period of the Project, as many opportunities as possible were taken to enable the local community to visit the site and for scientists to explain exactly what was going on. Along with this openness, a wide range of stakeholders from throughout Australia and internationally were encouraged to visit the site and, in the time that Otway Stage 1 was underway, many hundreds of people visited. There was a cost involved in this but it was money well spent.

One of the decisions that was critical to the success of the Project was to employ a local person to act as the liaison officer: the first local point of contact for landholders if they had any issues with any aspect of the Project. The Project was especially fortunate to be able to employ a high calibre person who knew all the local people, understood the science and had an open approach to problem solving. Coupled with this was a commendable willingness to tell scientists or technical staff if they neglected to follow the code of practice when undertaking field activities. Every project would benefit from such a person.

Collaboration was an essential feature of the Otway Project at all levels and across all topics—scientific, technical, operational, etc. Collaboration is a vital component of a Cooperative Research Centre, and not only served to enrich the science that was undertaken as part of the

Otway Project, but also helped to ensure that the best researchers could be brought in as needed, from more than a dozen organisations throughout Australia and around the world. At the same time, it was critical to have project leadership and coordination undertaken by people who were focused solely on the Otway Project. Without this, many excellent researchers with numerous calls on their time and their skills would have encountered difficulties with ensuring that they were available to the Project when required. Nonetheless it was at times challenging to coordinate the input of a large number of people who were not necessarily directly employed by CO2CRC.

It is essential to have effective time management systems in place for any project as complicated as the Otway Project. It has to be said that such a system was not in place from the start of the Project; this might have resulted in some increased costs and some missed deadlines on occasions, particularly in the writing up of results. In addition it has to be recognised that CO2CRC did not adequately budget for the final writing-up stage.

18.6 Regulating a project

As outlined in Chapter 3, there were no specific CCS regulations in place, when Otway Project planning got underway. It was therefore critically important that CO2CR worked closely with government officials to establish an appropriate regime. The lack of specific CCS regulations did not turn out to be a major obstacle, in large measure because of the willingness of pragmatic officials to find ways to progress the Project. The existence of regulations permitting R&D was particularly important to the Project. As mentioned in Chapters 1 and 3, the development of specific CCS regulations later in the life of the Project did not help to facilitate research, in that it potentially applied much more onerous conditions to an injection and storage licence. Fortunately the Act was amended to allow for small-scale projects to be more appropriately regulated. The lessons arising from that experience were twofold: first, regulatory authorities should ensure regulations for commercial large-scale storage projects are not seen as the basis for regulating research or small-scale injection projects (the European limit of up to 100,000 t of CO_2 is useful in this regard), and, second, small-scale projects should be seen as providing a pathway to sensible regulation of CCS and therefore an important part of any roadmap to accelerate the application of CCS. Further, the Otway Project proved to be a valuable tool in providing training for CCS regulators. It is likely that other small-scale projects could fulfil a similar role.

18.7 Identifying a suitable project site

Earlier in this volume, the process that led to the identification of the Otway site as being suitable was outlined. It is clear that a while a seriatim of decision points was followed in the process, there were also single decision points at which a practical consideration became the deciding factor, overwhelming in an otherwise logical progression of decision-making. For example, at the start it was clear that whatever the geological merits of a potential site, it had to be logistically feasible. It would not have been sensible to propose a small-scale offshore project (whatever the merits of the geology), because it would have been logistically very difficult and very expensive. It was necessary to identify a site that would be geologically viable, not a site that was geologically "ideal". In the same way, it was necessary to place a higher priority on a site that had a ready supply of CO_2 nearby, rather than one which had "better" geology. Indeed, as pointed out earlier, the starting point for the search for the site became dominated by CO_2 supply rather than geology. It was for this reason that a site was favoured in the Otway Basin. Subsequent events and the experience of other projects, including projects in Australia, have shown the wisdom of this decision.

The process of geological site characterisation is discussed in considerable detail in Chapter 5 and provides a fine example of what might reasonably be described as geological best practice. Cook (2006) defines site characterisation as "The collection, analysis and interpretation of subsurface, surface and atmospheric data (geoscientific, spatial, engineering, social, economic, environmental) and the application of that knowledge to judge, with a degree of confidence, if an identified site will geologically store a specific quantity of CO_2 for a defined period of time and meet all required health, safety, environmental and regulatory standards".

This then serves to summarise all the criteria that were drawn together and which enabled CO2CRC to conclude that the Otway site would be suitable for the proposed project. This definition could perhaps also serve as a useful guide to the choice of other sites, not just for small-scale projects but perhaps also for large commercial projects.

The Otway site could be referred to as a "brown field site" in that there had already been hydrocarbon exploration and production at the site. This had major benefits for the Project in that there was a great deal of pre-existing information as well as some useful facilities and considerable cost savings, and it served to advance the project timetable by 1 or 2 years. It also raised the level of confidence that the site was geologically suitable and decreased the level of risk (Chapter 8). For these same reasons there have been few onshore "green field" storage sites anywhere and this is likely to continue to be the case for most future small-scale projects. Finally the point needs to made that, based on the Otway experience, site characterisation is not a one-off exercise; the geological model developed as part of the characterisation process needs to be constantly updated as new information comes to hand.

18.8 Deciding on project science

The broad objective of the Otway 1 science was defined at the start of planning, but only in the most general terms, namely to demonstrate safe geological storage of CO_2. The science needed to demonstrate that this was so was influenced in no small measure by the regulatory requirements (albeit requirements arrived at in discussions with the scientists regarding what was possible) and the key performance indicators that underpinned those requirements. This essentially iterative process revolved around the integrity monitoring and assurance monitoring undertaken as part of Otway Stage 1. This is discussed in Chapter 9 and in somewhat greater detail in Chapters 10–15 and also in Section 18.9 of this chapter. However, there was also a need for outstanding science in order to characterise the site and this is demonstrated in Chapters 5–8. High quality science was also essential in determining the likely behaviour of CO_2 within the reservoir (Chapter 16) and in demonstrating that the capacity of the Naylor structure was sufficient to store 100,000 t of CO_2.

In all, these added up to the drivers for the science needed to demonstrate capacity, containment and injectivity. So how were research priorities established between or within these three? If there was a regulatory requirement for a measurement to be made, then this obviously had the highest priority within Otway 1. Beyond that, the issue of cost obviously had to be borne in mind. But overall the incremental cost of the science was relatively modest compared to the capital and operational costs incurred in the Project. An exception to this was the high cost of the seismic work, followed by atmospheric monitoring. Some of the other research areas, such as core characterisation and geomechanics, were relatively low cost but with a high scientific return. In the case of geomechanics, that value was enhanced by using the opportunity offered by the nearby Iona natural gas storage site to obtain additional data. However, because Otway 1 was the first such experiment undertaken in Australia, the tendency was to do as much science as possible, by measuring as many things as possible, because it was all new.

In contrast, Stage 2 science was much more sharply focused. Otway 2A science related almost exclusively to the characterisation of the core obtained through the drilling of CRC-2 and in many ways was a means to an end: to understand heterogeneity within the reservoir. The science undertaken as part of Stage 2B was focused on the issue of residual trapping with very specific experiments planned in great detail (Chapter 17). Planning is still underway for Otway Stage 2C, but that too is sharply focused—on issues relating to seismic imaging of CO_2 in a saline aquifer. Therefore a clear trend emerged at the Otway site, commencing with a very broadly-based scientific programme underpinned by the measurement of a large number of parameters; subsequent science was focused on answering a very specific question. But despite this, because there is still so much that is unknown, the opportunity was taken to obtain new data. For example while not the purpose of Otway 2A or 2B, the opportunity was taken to collect deep subsurface microbiological samples to determine "what was there" rather to answer a specific question. Serendipity still has a place in research.

18.9 Deciding on project monitoring

The point has previously been made that the monitoring programme was developed to meet the needs of regulators and KPIs, including the need to provide assurance on containment. This is expressed in Chapter 9 as "the need to demonstrate 'no leakage' (although this is qualified with a variety of adjectives and adverbs), as well as more routine requirements for health and safety associated with large concentrations of CO_2". However, as also pointed out in Chapter 9, "It is not possible to measure 'no leakage' directly". One of the objectives of monitoring in Otway Stage 1 was also to establish how cost effective the various monitoring methods might be, so that optimal M&V could be devised for large-scale storage projects. Here, a distinction perhaps needs to be drawn between what was outstanding science and what was practical science in terms of monitoring of a storage sites. For example, the work on atmospheric monitoring documented in Chapter 15 was undoubtedly outstanding science. However, it was expensive science and it required the combined efforts of a team of committed experts to maintain the equipment and interpret the results. It is unlikely that such a team of people would be available to most storage projects and therefore there will be a need to automate atmospheric programmes both in terms of data collection and interpretation, if atmospheric monitoring is to be widely applied to future storage projects.

In the case of seismic monitoring, also an expensive component of Otway Stage 1, as expected it was not possible to image the injected CO_2 largely because of the presence of residual methane. But a great deal was learned about sound sources, with explosives being found to be superior to other methods, in the case of the Otway site. Much was also learned about data processing (Chapters 10 and 11). However, perhaps the most important practical observation made as a result of Otway onshore seismic monitoring is that it really was a highly disruptive activity as far as the landowners were concerned, to the extent that in many areas, conventional 3D and 4D seismic will not be an acceptable monitoring option for some CCS projects. Therefore far more effort is required into the options of fixed seismic arrays and downhole seismic monitoring, shown at Otway to be a valuable source of additional information (Chapter 11) and CO2CRC proposes to do this as part of Otway Stage 2C. This is not to say that 4D seismic will not be deployed at large-scale storage sites, because in many ways it does represent the "gold standard" of reservoir monitoring and offshore seismic monitoring is comparatively cheap. However, onshore 4D seismic monitoring is expensive, it is difficult to obtain high quality data in many areas, and attempts to apply it widely and frequently at a site could seriously jeopardise the social licence of a project to operate.

The use of a monitoring well was also expensive, but in the case of Otway 1 there is no doubt that it provided crucially important geochemical data that could not have been obtained any other way (Chapter 12). It was therefore money well spent. The point has been made elsewhere that while money was saved in using an existing well for monitoring purposes, significant operational problems arose which could have been avoided with a new, larger diameter well. Would it have been better to have two or more monitoring wells? Perhaps, in that the decision to have only one such well was a gamble, though in this case the gamble worked and sufficient information was obtained from a single well. Should all storage projects have one or more monitoring wells for sampling the reservoir fluids? Yes, in the case of onshore storage projects undertaken at a significant scale. In the case of offshore projects it could be argued that the very high cost of drilling makes a separate offshore monitoring well impractical. In addition an extra well constitutes a potential pathway for CO_2 to escape and while this can be remediated at modest cost onshore, it could be very expensive and more difficult in the case of offshore remediation.

A comprehensive programme of groundwater monitoring was undertaken as part of Otway 1. A great deal was learned about the groundwaters of the Otway Basin from this work and the sampling and analytical programme was very successful (Chapter 13). But no CO_2 contamination of any groundwaters was detected, nor was any expected, so was it worthwhile? Most if not all storage projects are likely to require some aquifer monitoring as part of the approval process, but the Otway experience suggests that there are limitations to what can be monitored or learned by monitoring existing wells that were drilled for domestic or agricultural purposes. Access to groundwater

through purpose-drilled monitoring wells at critical locations and providing access to key aquifers would have been very useful at Otway and would have meant that fewer sampling campaigns or monitoring sites would have been necessary. This would have had significant cost implications. However, the point was previously made that the monitoring of the geochemistry of reservoir fluids was critically important (Chapter 12) and that drilling one or more wells down to the storage reservoir will be essential for any significant onshore storage project. With careful planning and good downhole engineering, monitoring wells could be constructed that would provide access to not only the reservoir but also to key overlying aquifers.

Soil gas monitoring of the Otway site (Chapter 14) was undertaken on a number of occasions during Otway 1. An automated monitoring programme at a few sampling points was of limited success, largely because of water entering the instrumentation and, for the most part, "traditional" manual sampling was more successful. The monitoring detected some soil gas anomalies, related to natural leakage, but, as expected, detected no anomalies that could be attributed to subsurface leakage from the storage site. The use of tracers (SF_6, R134A, Kr and CD_4) in the injected CO_2, and their absence from the soil gas is evidence of this. Tracers were also useful in establishing the behaviour of fluids in the storage reservoir (Chapter 12) at Otway, and for this reason it has been suggested on occasions that tracers might be added to CO_2 injected in large scale projects However, the handling, sampling and analysis of tracers poses a number of challenges, including the problem of false positives, because of the sensitivity of the measurements. Chapter 12 outlines how these problems can be minimised. But while tracers can be very valuable when used in small quantities in a research environment, their application to large scale commercial projects is dubious. A better option where possible may be to use the natural "tracers" or characteristics of the injected CO_2, such as a unique isotopic composition different to that of background CO_2 (Gilfillan et al. 2008).

Integral to the challenge of undertaking a credible monitoring programme is to have good baseline data at the Otway site against which to measure change; this is discussed in Chapter 9 as well as by Jenkins et al. (2011). A conscious decision was taken to commence background surveys before there was any certainty that an Otway Project would go ahead and this undoubtedly was the right decision. At the same time it could have led to a view amongst local stakeholders that a final decision had already been taken, once scientists started measuring things. Therefore it is important to address this issue as part of the communications strategy. It is also important to make all information arising from the baseline surveys available to landowners who for example might welcome the opportunity to learn more about their water bores.

18.10   Curating project data

The experience of the Otway Project in data handling and curation could be described as "mixed". The processing and analysis of data by the researchers was undertaken to the extent needed to answer particular scientific question, and invariably was done to a high standard. Massive datasets resulted from the seismic and atmospheric programmes for example; other programme areas produced smaller datasets. Meta datasets identify where particular datasets are held and can be sourced. However, for the most part, these datasets are held by the researchers responsible for collecting the data and are not held centrally by CO2CRC or one of its affiliate organisations. Does this matter? Yes it does, to the extent that researchers move on and the data that they might value now may have less value to them in a few years' time and therefore become less accessible. At the same time, given that to date some $70 million has been spent at the Otway site, it is very important that Project data continue to be available for decades to come. Here, national and state geological surveys have a key role to play in the long-term curation of data.

In addition, it was important that not just data but also specimens, particularly core and cuttings but also perhaps fluid samples, are held securely. Fortunately Geoscience Australia has excellent core facilities and, for the most part, the Otway material is held there. Where the Project was remiss, particularly in the early stages, was in not controlling the distribution of core more carefully. As a consequence some valuable samples have not been returned. Projects should ensure that from the start, sampling protocols are in place and are enforced. Better tracking of samples should also be backed up by clear rules on publication

of results. CO2CRC does have an excellent publications tracking system in place, but, even with that in place, cases arose of information being released prematurely.

It is not just the area of science data where there is a need for long-term curation and careful data handling. Where issues of long-term liability may arise, it is important that operational, management and Board decisions are well documented and their basis fully understood. Here, CO2CRC does have a centralised system in place that would enable it to demonstrate that, in the case of the Otway activities, due diligence was observed throughout. The Otway experience showed that careful and well structured data handling is very important for all storage projects, not least because of the possible implications of data or monitoring or engineering or key decisions, to future liability. It is essential that data and decision protocols are in place and enforced from the start of a project, rather than being developed as an afterthought at the conclusion of a project.

18.11   Lessons for the future?

Many of the lessons learned from the Otway Project will be applicable to storage projects around the world in terms of how to do things more efficiently or effectively in the future. Some of the detail of the lessons may only be applicable to comparable small scale projects (up to 100,000 t of CO_2). Some lessons, such as cost, will be very site specific. But some may be of wide applicability, such as the process followed for communicating with stakeholders or aspects of the regulatory regime and the related key performance indicators and the management arrangements. In some instances the major high level benefit arising from the Otway Project will be that it convincingly demonstrated CO_2 can be safely stored in the subsurface. In other words, a major part of the future value of the Otway Project, particularly for large scale projects, will be that it convincingly demonstrated that geological storage "works". However, as the Otway Project has progressed, some other pointers to the future have emerged that may serve to influence the conduct and value of small-scale or demonstration storage projects in particular.

The first of these relates to the number of such projects that might be needed in the future. It would be unwise to try to provide a precise number, but it is clear that there will be benefit in having future projects similar in scale to Otway. In some instances these will be needed to demonstrate storage technology (and its practicality) to particular jurisdictions or communities or to test a regulatory regime. In some cases they will be needed to test particular geological settings and rock types. In some instances they will be needed to test issues (e.g. small-scale leakage or induced seismicity) that no large-scale project would wish to test itself, but which it would very much like to understand.

But the point needs to be made that in planning future storage projects and field experiments, it is necessary to move beyond projects that just demonstrate storage, important though this is. Many projects, and in its very early days the Otway Project was one of these, did not look beyond "demonstrating storage". But far more is now known about what we don't know! Therefore storage projects focused on answering specific questions are now needed. Otway Stage 2B is a very good example of a problem (how to quantify residual trapping) and how to answer it through targeted field experimentation. Stage 2C is another example. However, there are many other questions to be resolved, with mention already made of better understanding the risk of induced seismicity or fault reactivation or the prospect of identifying a small leak. In some cases projects will be needed to help develop better technologies such as much better onshore seismic resolution, and improved use of InSAR or gravity for monitoring CO_2 within the reservoir. Could hydraulic fracturing be used to enhance the storage capacity of low permeability rocks or would it adversely impact on potential seals?

It is important to avoid establishing a new site (at great cost) for each field experiment. By making judicious geological, logistic and other choices it is possible and advisable to concentrate many of the efforts at a limited number of high quality sites with attributes that can serve to answer a range of key questions. The Otway site is one example where multiple question are now being addressed. There is a need for more such sites but also a need for a concentration of effort, expertise and resources to ensure maximum "bang for the buck". This approach will also serve to accelerate the deployment of CCS as a mitigation option for, as Otway and many other sites have demonstrated, it takes several years to develop the necessary infrastructure, obtain approvals and win over the community. Some project consolidation may be in the best interests of not only cost-effectively demonstrating storage but also for answering questions regarding geological storage of CO_2 that are raised by politicians, communities and other scientists and that need to be answered from a firm scientific base, otherwise they might delay large-scale projects at great cost.

The Otway experience also provides an important pointer to conflicts that may arise or are already arising in sedimentary basins as a consequence of increasing use of basin resources, especially the pore space. In the case of the Otway Basin, not only is CO_2 being geologically stored, but there are important aquifers in the basin. The first question posed by farmers when the Otway Project was put forward was invariably: What impact will it have on my groundwater? In addition, the area is one of significant geothermal interest and is of course an area of active exploration for and production of conventional oil and gas. In the future it may well become an area of interest for unconventional gas (shale gas and perhaps coal seam gas). Further, there has been interest from at least one company in subsurface compressed air storage. So, in a relatively small area and within the one sedimentary basin, there are potentially at least half a dozen subsurface resources, each competing for "space" to varying degrees and sometimes competing for exactly the same space.

To make matters even more complicated, each resource is likely to be subject to its own specific regulatory regime, which may be in conflict with some of the regulations for other resources. Along with many of these conflicts is the closely related issue of community concern and securing the licence to operate. This then is the situation not only in the Otway Basin, but in many other basins in Australia and around the world, and it inevitably raises the question of how these conflicts can be resolved and even better, avoided.

So what has this to do with the Otway Project and other comparable sites? Much of the resource conflict arises from a lack of understanding of how basins and their resources "work" and underlying this in many areas is a lack of real data and information. Almost unique in their approach, small-scale storage projects, such as the Otway Project, have been focused on developing a comprehensive scientific understanding of the subsurface and clarifying how one "resource" namely injected CO_2 might impact on other resources and on the subsurface environment in general. Integral to this has been the need to carry out long-term field experiments, to drill deep wells, instrument the wells, obtain fluid samples from the deep subsurface, etc. All this has happened at considerable cost, much of that to the taxpayer, and therefore it is important that the maximum value is obtained form that expenditure.

Usually there is an expectation that, at the end of the storage programme, the site will be decommissioned and abandoned, but for a few carefully selected sites there is perhaps a better way—namely, to turn them in world class subsurface laboratories and experimental sites. Many sites already have millions of dollars worth of infrastructure at the site, offering just the sort of facilities that would provide an exceptional starting point for developing a holistic understanding of deep sedimentary basins—the detailed scientific understanding that is needed to avoid resource (and perhaps community) conflicts in the future. The knowledge obtained will provide a basis for science-based policy and regulation and perhaps also a basis for synergistically developing multiple resources. For example, there has been much speculation about the opportunities that might exist for joint development of CO_2 storage and geothermal resources. A world class subsurface laboratory and experimental site (such as Otway) might offer the opportunity to put such an idea to the test.

It is necessary to no longer think about geological storage of CO_2 in isolation and start to think about it more as just one of a suite of sedimentary basin resources that must be managed in a sensible manner. Some of the existing research storage sites provide precisely the sort of facilities that will be needed to collect the scientific information that will be so essential for the wise management of underground resources, including pore space. It is important therefore that the future use of these sites is carefully considered before they are abandoned. There will be a need for a number of such sites and their subsurface and surface facilities, particularly in parts of the world where sedimentary basins are or will shortly come under increasing resource pressure. There is a need for a network of such observatories around the world, analogous perhaps to the network of astronomical observatories that already exist around the world, although in this case with the "telescope" pointing down!

The Otway site provides an Australian opportunity, and perhaps a building block, for just such a global network.

18.12 Conclusions

A number of important benefits can accrue from projects such as the Otway Project that can be undertaken at a relatively modest cost. These include:

> the opportunity to inform the local community, and the community at large, about CCS and especially storage, through a real-world example

> a low-cost on-the-job learning opportunity for technicians, engineers, scientists, managers

> a chance to decrease technical uncertainty and risk prior to embarking on a large-scale project

> providing impetus to regulators to confront some of the regulatory issues when there is a real project (even if small to medium sized)

> the ability to test equipment (and boundaries) at a modest scale, in a way that could not be contemplated for a large scale project

> providing a real-world working relationship for the partners if the project is being pursued through an industry partnership

> providing tangible evidence that CCS is moving ahead, despite the slow pace of progress on large scale projects

> exposing the CCS "community" to working with a broader community and understanding how to communicate with and listen to that community

> providing something for politicians, bureaucrats and community leaders to visit and understand

> the opportunity to encounter (and overcome) real-world technical and engineering problems

> the ability to test monitoring options under operational conditions and assess the practicality of the various techniques as well as develop new techniques.

As pointed out earlier in this chapter, there is no single all-encompassing "best practice" for small-scale projects that can be followed slavishly. Rather there are lessons to be learned from every project. While each project will have its own specific drivers, there is a high degree of commonality among small-scale projects in terms of the aims and objectives. Similarly there are many common problems that are encountered by projects and it is for this reason it is important that the lessons from projects such as Otway are made readily available. It is hoped this volume will contribute to that learning process as well as to the broader aim of taking CCS forward as a key option for mitigating CO_2 emissions. Projects such as Otway could also serve as a stepping stone to establish subsurface laboratories and test sites that will provide the scientific basis for managing basin resources such as pore space, in a sensible and sustainable manner.

18.13   References

Ashworth, P., Rodriguez, S. and Miller, A. 2010. Case study of the CO2CRC Otway Project. CSIRO report, Pullenvale, Australia, 10.5341/RPT10-2362 EP 103388.

Bowden, A.R. and Rigg, A. 2004. Assessing risk in CO_2 storage projects. Australian Petroleum Production and Exploration Association Journal 42(1), 677–702.

Cook, P.J. 2006. Site characterization for CO_2 geological storage. Berkeley, California, USA. International Symposium on Site Characterization for CO_2 Geological Storage, Lawrence Berkeley National Laboratory (LBNL).

Cook, P.J. 2013. The CO2CRC Otway Project. In: Gluyas, J. and Mathias, S. (Eds.) Geological storage of carbon dioxide (CO_2), Woodhead Publishing, Cambridge, Chapter 11, 251–277.

Cook, P.J., Causebrook, R., Michel, K. and Watson, M. 2013. Developing a small scale CO_2 test injection: Experience to date and best practice. IEAGHG Report.

Dance, T. 2013. Assessment and geological characterisation of the CO2CRC Otway Project CO_2 storage demonstration site: From prefeasibility to injection. Marine and Petroleum Geology 46, 251–269.

Dodds, K., Daley, T., Freifeld, B., Urovsevic, M., Kepic, A. and Sharma, S. 2009. Developing a monitoring and verification plan with reference to the Australian Otway CO_2 pilot project. The Leading Edge 812–818.

Gibson-Poole, C.M., Root, R.S., Lang, S.C., Streit, J.E., Hennig, A.L., Otto, C.J. and Undershultz, J.R. 2005. Conducting comprehensive analyses of potential sites for geological CO_2 storage. In: Rubin, E.S., Keith, D.W. and Gilboy C.F. (Eds.), Greenhouse Gas Control Technologies: Proceedings of the 7th International Conference on Greenhouse Gas Control Technologies (Vols. 1- Peer Reviewed Papers and Overviews, 673–681). Vancouver, Canada: Elsevier

Gifillan, S., Ballentine, C., Holland, G., Blagburn, D., Lollar, B., Stevens, S., Schoell, M. and Cassidy, M. 2008. The noble gas geochemistry of natural CO_2 gas reservoirs from the Colorado Plateau and Rocky Mountain provinces, USA. Geochimica et Cosmochimica Acta 72, 1174–1198.

Gluyas, J. and Mathias, S. 2013. Geological storage of carbon dioxide (CO_2), Woodhead Publishing, Cambridge.

Hitchon, B. (Ed.) 2012. Best practices for validating CO_2 geological storage. Geoscience Publishing.

Hitchon, B. (Ed.) 2013. Pembina Cardium CO_2 monitoring pilot. Geoscience Publishing.

Hovorka, S., Sakurai, S., Kharaka, Y.K., Seay Nance, H., Doughty, C., Benson, S.M. and Daley, T. 2006. Monitoring CO_2 storage in brine formations: lessons learned from the Frio field test one year post injection. UIC Conference of the Groundwater Protection Council, Abstract No.19.

Jenkins, J.R., Cook, P.J., Ennis-King, J., Undershultz, J., Boreham, C., Dance, T. and Urosevic, M. 2011. Safe storage and effective monitoring of CO_2 in depleted gas fields. Proceedings of the National Academy of Sciences of the United States of America 109(2), 35–41.

Kuhn, M. and Munch, U. (Eds.) 2013. CLEAN: Large-scale enhanced gas recovery in the Altmark Natural Gas Field, Science Report 19, Springer Heidelberg.

Liebscher, A., Möller, F., Bannach, A., Köhler, S., Wiebach, J., Schmidt-Hattenberger, C., Weiner, M., Pretschner, C., Ebert, K. and Zemke, J. 2013. Injection operation and operational pressure-temperature monitoring at the CO_2 storage pilot site Ketzin, Germany—Design, results, recommendations. International Journal of Greenhouse Gas Control 15, 163–173.

Litynski, J.T., Rodosta, T., Vikara, D. and Srivastava, R.K. 2013. U.S. DOE's R&D program to develop infrastructure for carbon storage: Overview of the regional carbon sequestration partnerships and other R&D field projects. Energy Procedia.

Michael, K., Golab, A., Shulakova, V., Ennis-King, J., Allinson, G., Sharma, S. and Aiken, T. 2010. Geological storage of CO_2 in saline aquifers – a review of the experience from existing storage operations. International Journal of Greenhouse Gas Control 4, 659–667.

Mito S., Xue, Z. and Sato, T. 2013. Effect of formation water composition on predicting CO_2 behavior: A case study at the Nagaoka post-injection monitoring site. Applied Geochemistry 30, 33–40.

Paterson, L., Boreham, C., Bunch, M., Dance, T., Ennis-King, J., Freifeld, B., Haese, R., Jenkins, C., LaForce, T., Raab, M., Singh, R. and Stalker, L. 2013. Overview of the CO2CRC Otway residual saturation and dissolution test. Energy Procedia 37, 6140–6148.

Stalker, L., Noble, R., Pejcic, B., Leybourne, M., Hortle, A. and Michael, K. 2012. Feasibility of monitoring techniques for subsurface mobilised by CO_2 storage in geological formations. Project for the IEA GHG ref IEA/CON/10/182, CO2CRC Report No: RPT11-2861.

Soroka, M. 2011. A review of existing best practice manuals for carbon dioxide storage and regulation. http://www.globalccsinstitute.com/publications/review-existing-best-practice-manuals-carbon-dioxide-storage-and-regulation.

Wickstrom, L.H., Gupta, N., Ball, D.A. Barnes D.A., Rupp, J.A., Greb, S.F., Sminchak, J.R. and Cumming, L.J. 2009. Geologic storage field demonstrations of the Midwest Regional Carbon Sequestration Partnership Retrieved from http://www.dnr.state.oh.us/portals/10/pdf/Posters/AAPGNatl2009_Wickstrom.pdf.

Würdemann, H., Möller, F., Kühn, M., Heidug, W., Christensen, N.P., Borm, G. and Schilling, F.R. 2010. CO2SINK – From site characterization and risk assessment to monitoring and verification: One year of operational experience with the field laboratory for CO_2 storage at Ketzin, Germany. International Journal of Greenhouse Gas Control 4, 938–951.

INDEX